T0212816

Modern Birkhäuser Classics

Many of the original research and survey monographs in pure and applied mathematics published by Birkhäuser in recent decades have been groundbreaking and have come to be regarded as foundational to the subject. Through the MBC Series, a select number of these modern classics, entirely uncorrected, are being re-released in paperback (and as eBooks) to ensure that these treasures remain accessible to new generations of students, scholars, and researchers.

Klaus Schmidt

Dynamical Systems
of Algebraic Origin

Reprint of the 1995 Edition

 Birkhäuser

Klaus Schmidt
Department of Mathematics
University of Vienna
Nordbergstr. 15
1090 Vienna
Austria

ISBN 978-3-0348-0276-5   e-ISBN 978-3-0348-0277-2
DOI 10.1007/978-3-0348-0277-2
Springer Basel Dordrecht Heidelberg London New York

Library of Congress Control Number: 2011941446

Mathematics Subject Classification (2010): 28-02, 22-02, 22D40, 28D15, 28D20, 37-XX

© Springer Basel AG 1995
Reprint of the 1st edition 1995 by Birkhäuser Verlag, Switzerland
Originally published as volume 128 in the Progress in Mathematics series

Printed on acid-free paper

Springer Basel AG is part of Springer Science+Business Media
(www.birkhauser-science.com)

To Annelise

Contents

Chapter V. Entropy

Chapter VI. Positive entropy

Chapter VII. Zero entropy

Chapter VIII. Mixing

Chapter IX. Rigidity

Introduction

Although the study of dynamical systems is mainly concerned with single transformations and one-parameter flows (i.e. with actions of $\mathbb{Z}$, $\mathbb{N}$, $\mathbb{R}$, or $\mathbb{R}^+$), ergodic theory inherits from statistical mechanics not only its name, but also an obligation to analyze spatially extended systems with multi-dimensional symmetry groups. However, the wealth of concrete and natural examples, which has contributed so much to the appeal and development of classical dynamics, is noticeably absent in this more general theory. A remarkable exception is provided by a class of geometric actions of (discrete subgroups of) semi-simple Lie groups, which have led to the discovery of one of the most striking new phenomena in multi-dimensional ergodic theory: under suitable circumstances orbit equivalence of such actions implies not only measurable conjugacy, but the conjugating map itself has to be extremely well behaved. Some of these rigidity properties are inherited by certain abelian subgroups of these groups, but the very special nature of the actions involved does not allow any general conjectures about actions of multi-dimensional abelian groups.

Beyond commuting group rotations, commuting toral automorphisms and certain other algebraic examples (cf. [39]) it is quite difficult to find non-trivial smooth $\mathbb{Z}^d$-actions on finite-dimensional manifolds. In addition to scarcity, these examples give rise to actions with zero entropy, since smooth $\mathbb{Z}^d$-actions with positive entropy cannot exist on finite-dimensional, connected manifolds. Cellular automata (i.e. shift-commuting homeomorphisms of a subshift of finite type) also generate zero-entropy $\mathbb{Z}^2$-actions, but any attempt to define a reasonably wide class of cellular automata, and to prove non-trivial results about that class, appears to run into logical quicksand (cf. e.g. [31]). The same applies more generally to higher-dimensional subshifts of finite type, where one can easily find examples with positive entropy, but where dynamical questions about even the simplest systems can lead to surprising difficulties (cf. e.g. [25], [14], [15]).

The purpose of this book is to help remedy this lack of examples by introducing a class of continuous $\mathbb{Z}^d$-actions on compact, metric spaces which is diverse enough to exhibit many of the new phenomena encountered in the transition from $\mathbb{Z}$ to $\mathbb{Z}^d$, but which nevertheless lends itself to systematic study: the $\mathbb{Z}^d$-actions by automorphisms of compact, abelian groups. One aspect of these actions, which is a priori not surprising, but is quite striking in its extent and depth, is their connection with commutative algebra and arithmetical algebraic geometry. The algebraic framework resulting from this connection allows the construction of an unlimited supply of examples with specified dynamical properties, and by combining algebraic and dynamical tools one obtains a sufficiently detailed understanding of this class of $\mathbb{Z}^d$-actions to glimpse at least the beginnings of a general theory.

Before describing the contents of this book in any detail I should mention a specific example, which—together with its ramifications—has provided much of the motivation for the work presented here. This example is due to Ledrappier [56], and consists of the shift-action σ of $\mathbb{Z}^2$ on the closed, shift-invariant subgroup

$$X = \big\{x = (x_\mathbf{m}) \in (\mathbb{Z}/2\mathbb{Z})^{\mathbb{Z}^2} : x_{(m_1,m_2)} + x_{(m_1+1,m_2)} + x_{(m_1,m_2+1)} \tag{0.1}$$
$$= 0 \ (\mathrm{mod}\ 2) \ \text{for every}\ \mathbf{m} = (m_1, m_2) \in \mathbb{Z}^2\big\}$$

of the full two-dimensional two-shift. It is clear that σ has zero entropy, since the individual automorphisms $\sigma_\mathbf{m}$, $\mathbf{m} \in \mathbb{Z}^2$, have finite entropy. Ledrappier proved that σ is mixing (cf. Theorem 6.5 in this book), but observed that, for every $x \in X$ and $k \geq 1$, $x_{(0,0)} + x_{(2^k,0)} + x_{(0,2^k)} = 0 \ (\mathrm{mod}\ 2)$. In particular, σ cannot be mixing of order 3. Much of Section 28 will be devoted to this kind of breakdown of higher order mixing for certain mixing $\mathbb{Z}^d$-actions.

There is nothing special about the 'alphabet' $\mathbb{Z}/2\mathbb{Z}$ of the group X in (0.1): one can replace it by any finite, abelian group G. Are the shift-actions $\sigma = \sigma^{(G)}$ on the resulting subgroups of $X = X^{(G)} \subset G^{\mathbb{Z}^2}$ measurably conjugate for different choices of G? By considering the entropies of the individual shifts $\sigma_\mathbf{m}^{(G)}$, $\mathbf{m} \in \mathbb{Z}^2$, one sees immediately that the cardinality $|G|$ of G is a measurable conjugacy invariant of $\sigma^{(G)}$: if G, G' are finite (abelian) groups such that $\sigma^{(G)}$ is measurably conjugate to $\sigma^{(G')}$, then $|G| = |G'|$. Can anything more be said? We refer to the Sections 25–26 for further discussion.

What happens if G is replaced by an uncountable group, like $\mathbb{T} = \mathbb{R}/\mathbb{Z}$? If we define $X^{(\mathbb{T})}$ exactly as in (0.1) by

$$X^{(\mathbb{T})} = \big\{x = (x_\mathbf{m}) \in \mathbb{T}^{\mathbb{Z}^2} : x_{(m_1,m_2)} + x_{(m_1+1,m_2)} + x_{(m_1,m_2+1)} \tag{0.2}$$
$$= 0 \ (\mathrm{mod}\ 1) \ \text{for every}\ \mathbf{m} = (m_1, m_2) \in \mathbb{Z}^2\big\},$$

what can be said about the shift-action $\sigma^{(\mathbb{T})}$ of $\mathbb{Z}^2$ on $X^{(\mathbb{T})}$? Is it mixing of all orders? Does it again have zero entropy? As it turns out, $\sigma^{(\mathbb{T})}$ is not only mixing

(Theorem 6.5), but mixing of all orders (Theorem 27.3), and Proposition 19.7 shows that its entropy is positive and given by

$$h(\sigma^{(\mathrm{T})}) = \frac{3\sqrt{3}}{4\pi} L(2, \chi_3),$$

where

$$L(s, \chi_3) = \sum_{n=1}^{\infty} \frac{\chi_3(n)}{n^s}$$

is Dirichlet's L-function associated with the character

$$\chi_3(n) = \begin{cases} 0 & \text{if } n \equiv 0 \pmod 3, \\ 1 & \text{if } n \equiv 1 \pmod 3, \\ -1 & \text{if } n \equiv 2 \pmod 3. \end{cases}$$

In fact, $\sigma^{(\mathrm{T})}$ is Bernoulli (Theorem 23.1), and hence measurably conjugate to the shift-action σ' of $\mathbb{Z}^2$ on the subgroup

$$X' = \big\{ x = (x_{\mathbf{m}}) \in \mathbb{T}^{\mathbb{Z}^2} : x_{(m_1, m_2)} + x_{(m_1-1, m_2)} + x_{(m_1, m_2+1)} \tag{0.3}$$
$$= 0 \pmod 1 \text{ for every } \mathbf{m} = (m_1, m_2) \in \mathbb{Z}^2 \big\}.$$

However, its zero-dimensional cousin (0.1) does not allow any such change in its defining rule (Proposition 25.7 and Examples 25.8 (1)–(2)), and exhibits quite extraordinary rigidity properties (Chapter 9).

Readers who get bored with abelian groups can carry the problem one step further and consider the shift-action $\sigma^{(\mathrm{SU}(2))}$ of $\mathbb{Z}^2$ on the closed, shift-invariant subset

$$X^{(\mathrm{SU}(2))} = \big\{ x = (x_{\mathbf{m}}) \in \mathrm{SU}(2)^{\mathbb{Z}^2} : x_{(m_1, m_2)} \cdot x_{(m_1+1, m_2)} \cdot x_{(m_1, m_2+1)}$$
$$= 1 \text{ for every } \mathbf{m} = (m_1, m_2) \in \mathbb{Z}^2 \big\},$$

where 1 is the identity element in the group $\mathrm{SU}(2)$. What can be said about the dynamical properties of $\sigma^{(\mathrm{SU}(2))}$? It must have positive entropy, but what is its value? Is $\sigma^{(\mathrm{SU}(2))}$ again Bernoulli? Nothing appears to be known about this problem.

This brief account of Ledrappier's example and the questions raised by it should convey the strong emphasis of this book on explicit examples, which serve not only as illustrations of definitions, results and techniques, but which provide motivation for much of the material presented here.

The contents are organized as follows. Chapter I contains general background on actions of a countable group Γ by automorphisms of a compact (and always metrizable) group X. In Section 1 we review the connections between elementary spectral and dynamical properties of these actions and prove that such an action is ergodic if and only if it is topologically transitive (Theorem

1.1). Sections 2–4 introduce certain finiteness conditions for Γ-actions by auto-morphisms of compact groups. In Proposition 2.2 we establish that the first of these conditions, *expansiveness*, implies another finiteness condition, *conjugacy to a Lie subshift*: every expansive action α of an infinite, discrete group Γ by automorphisms of a compact group X is topologically and algebraically conjugate to the shift-action of Γ on a closed, shift-invariant subgroup $Y \subset G^\Gamma$, where G is a compact Lie group. Furthermore, if X is connected, then G may be chosen to be a finite-dimensional torus (Theorem 2.4). For $\mathbb{Z}$-actions, con-jugacy to a Lie subshift is equivalent to a condition originally introduced in [84] (Proposition 2.15 and Corollary 2.16). The third finiteness condition, in Section 3, is the *descending chain condition* (d.c.c.): the action α of Γ on X satisfies the descending chain condition if every strictly decreasing sequence of closed, α-invariant subgroups of X is finite. The d.c.c. also implies conjugacy to a Lie subshift (Proposition 3.3), but not necessarily expansiveness, and makes life easier in a number of ways (Proposition 3.5 and Theorem 3.6). The most crucial aspect of the d.c.c. is, however, its connection with subshifts of finite type, which is explained in Definition 3.7 and Theorem 3.8. In Section 4 we discuss those countable groups Γ for which every action on a Lie subshift has the d.c.c., and prove that $\mathbb{Z}^d$ has this property for every $d \geq 1$ (Theorem 4.2 and Examples 4.8).

In Chapter II we concentrate on $\mathbb{Z}^d$-actions by automorphisms of com-pact, abelian groups and their connection with commutative algebra. In Sec-tion 5 we show that every $\mathbb{Z}^d$-action α by automorphisms of a compact, abelian group X defines a dual module $\mathfrak{M} = \hat{X}$ over the ring of Laurent polynomi-als $\mathfrak{R}_d = \mathbb{Z}[u_1^{\pm 1}, \ldots, u_d^{\pm 1}]$, and that this module is Noetherian whenever α is expansive or, more generally, whenever α satisfies the d.c.c. (Lemma 5.1 and Proposition 5.4). In Ledrappier's example (0.1) this module turns out to be given by $\mathfrak{M} = \mathfrak{R}_2/(2, 1 + u_1 + u_2)$, where $(2, 1 + u_1 + u_2)$ is the ideal $2\mathfrak{R}_2 + (1 + u_1 + u_2)\mathfrak{R}_2 \subset \mathfrak{R}_2$, and in (0.2) $\mathfrak{M} = \mathfrak{R}_2/(1 + u_1 + u_2)$. Conversely, if $\mathfrak{M}$ is an $\mathfrak{R}_d$-module, then $\mathfrak{M}$ defines a $\mathbb{Z}^d$-action $\alpha = \alpha^{\mathfrak{M}}$ by automorphisms of the compact, abelian group $X = X^{\mathfrak{M}} = \widehat{\mathfrak{M}}$ (Lemma 5.1). The extent of this correspondence between modules and $\mathbb{Z}^d$-actions is discussed in Theorem 5.9 and Corollary 5.10. Section 6 concentrates on $\mathbb{Z}^d$-actions corresponding to Noetherian modules and provides the first entries in a 'dictionary' which trans-lates the dynamical properties of an action α on a compact, abelian group X into algebraic properties of the dual module $\mathfrak{M} = \hat{X}$. Figure 1 illustrates the nature of this dictionary. In the second column of this table we assume that the $\mathfrak{R}_d$-module $\mathfrak{M} = \hat{X}$ defining α is of the form $\mathfrak{R}_d/\mathfrak{p}$, where $\mathfrak{p} \subset \mathfrak{R}_d$ is a prime ideal, and describe the algebraic condition on $\mathfrak{p}$ equivalent to the dynamical condition on $\alpha = \alpha^{\mathfrak{R}_d/\mathfrak{p}}$ appearing in the first column. In the third column we consider a general (countable) $\mathfrak{R}_d$-module $\mathfrak{M}$, and state the algebraic property of $\mathfrak{M}$ corresponding to the dynamical condition on $\alpha = \alpha^{\mathfrak{M}}$ in the first column. As can be seen from the entries in this column, these conditions are in all cases expressed in terms of the prime ideals associated with the module $\mathfrak{M}$, so that many of the dynamical properties of $\alpha^{\mathfrak{M}}$ are, in fact, determined by the dynamics of the $\mathbb{Z}^d$-actions $\alpha^{\mathfrak{R}_d/\mathfrak{p}}$, where $\mathfrak{p}$ varies over the prime ideals associated with $\mathfrak{M}$. The last column gives reference to some of the relevant results in this text.

Property of α	$\alpha = \alpha^{\mathfrak{R}_d/\mathfrak{p}}$: property of $\mathfrak{p}$	$\alpha = \alpha^{\mathfrak{M}}$: property of $\mathfrak{M}$	Reference
α satisfies the descending chain condition	Always true	$\mathfrak{M}$ is Noetherian	Propn. 5.4
α is expansive	$V_{\mathbb{C}}(\mathfrak{p}) \cap \mathbb{S}^d = \varnothing$	$\mathfrak{M}$ is Noetherian and $\alpha^{\mathfrak{R}_d/\mathfrak{p}}$ is expansive for every prime ideal $\mathfrak{p}$ associated with $\mathfrak{M}$	Thm. 6.5
The set of α-periodic points is dense	Always true	$\mathfrak{M}$ is Noetherian	Thm. 5.7
$|\text{Fix}_\Lambda(\alpha)| < \infty$ for a subgroup $\Lambda \subset \mathbb{Z}^d$ of finite index	$V_{\mathbb{C}}(\mathfrak{p}) \cap \{c \in \mathbb{C}^d : c^{\mathbf{n}} = 1$ for every $\mathbf{n} \in \Lambda\} = \varnothing$	$\mathfrak{M}$ is Noetherian and $|\text{Fix}_\Lambda(\alpha^{\mathfrak{R}_d/\mathfrak{p}})| < \infty$ for every prime ideal $\mathfrak{p}$ associated with $\mathfrak{M}$	Thm. 6.5
α is ergodic or topologically transitive	$\{u^{k\mathbf{n}} - 1 : \mathbf{n} \in \mathbb{Z}^d\} \not\subset \mathfrak{p}$ for every $k \geq 1$	$\alpha^{\mathfrak{R}_d/\mathfrak{p}}$ is ergodic for every prime ideal $\mathfrak{p}$ associated with $\mathfrak{M}$	Thm. 6.5, Lemma 6.6
α is strongly mixing	$u^{\mathbf{n}} - 1 \notin \mathfrak{p}$ for every non-zero $\mathbf{n} \in \mathbb{Z}^d$	$\alpha^{\mathfrak{R}_d/\mathfrak{p}}$ is strongly mixing for every prime ideal $\mathfrak{p}$ associated with $\mathfrak{M}$	Thm. 6.5, Lemma 6.6
α is mixing of every order	Either $\mathfrak{p}$ is equal to $p\mathfrak{R}_d$ for some rational prime p, or $\mathfrak{p} \cap \mathbb{Z} = \{0\}$ and $\alpha^{\mathfrak{R}_d/\mathfrak{p}}$ is strongly mixing	For every prime ideal $\mathfrak{p}$ associated with $\mathfrak{M}$, $\alpha^{\mathfrak{R}_d/\mathfrak{p}}$ is mixing of every order	Thm. 27.2
$h(\alpha) > 0$	$\mathfrak{p}$ is principal, and not generated by a generalized cyclotomic polynomial	$h(\alpha^{\mathfrak{R}_d/\mathfrak{p}}) > 0$ for at least one prime ideal $\mathfrak{p}$ associated with $\mathfrak{M}$	Propn. 19.4, Thm. 19.5
$h(\alpha) < \infty$	$\mathfrak{p} \neq \{0\}$	If $\mathfrak{M}$ is Noetherian: $\mathfrak{p} \neq \{0\}$ for every prime ideal $\mathfrak{p}$ associated with $\mathfrak{M}$	Propn. 19.4
α has completely positive entropy	$h(\alpha^{\mathfrak{R}_d/\mathfrak{p}}) > 0$	$h(\alpha^{\mathfrak{R}_d/\mathfrak{p}}) > 0$ for every prime ideal $\mathfrak{p}$ associated with $\mathfrak{M}$	Thm. 20.8
α is Bernoulli	$\alpha^{\mathfrak{R}_d/\mathfrak{p}}$ has completely positive entropy	$\alpha^{\mathfrak{M}}$ has completely positive entropy	Thm. 23.1
α has a unique measure of (finite) maximal entropy	$0 < h(\alpha^{\mathfrak{R}_d/\mathfrak{p}}) < \infty$	$\alpha^{\mathfrak{M}}$ has completely positive entropy and $h(\alpha^{\mathfrak{M}}) < \infty$	Thm. 20.15

FIGURE 1

Apart from the results referred to in Figure 1, Section 6 contains a number of other consequences of this interplay between algebra and dynamics: if an ergodic $\mathbb{Z}^d$-action α on a compact, abelian group X satisfies the d.c.c., then there exists an $\mathbf{n} \in \mathbb{Z}^d$ such that $\alpha_{\mathbf{n}}$ is ergodic (Corollary 6.10); if α is an expansive

$\mathbb{Z}^d$-action by automorphisms of a compact, connected, abelian group X, then α is ergodic (Corollary 6.14); finally, if α is an expansive $\mathbb{Z}^d$-action by automorphisms of a compact, abelian group X, and if $Y \subset X$ is a closed, α-invariant subgroup, then the action induced by α on the quotient group X/Y is again expansive (Corollary 6.15). In Section 7 we consider the dynamical systems arising from prime ideals $\mathfrak{p} \subset \mathfrak{R}_d$ whose variety $V_{\mathbb{C}}(\mathfrak{p})$ is finite, which provide the 'building blocks' for all $\mathbb{Z}^d$-actions by automorphisms of finite-dimensional tori and solenoids, and express the dynamics of $\alpha^{\mathfrak{R}_d/\mathfrak{p}}$ in terms of $V_{\mathbb{C}}(\mathfrak{p})$. This dynamical interpretation of $V_{\mathbb{C}}(\mathfrak{p})$ is in complete analogy to the classical formulae relating ergodicity and expansiveness of the toral automorphism defined by a matrix $A \in \mathrm{GL}(n, \mathbb{Z})$ to the eigenvalues of A. Section 8 analyzes the structure of the dynamical system associated with an arbitrary prime ideal $\mathfrak{p} \subset \mathfrak{R}_d$, and introduces a notion of dimension for such systems which will be related to entropy in Section 24.

In Chapter III, Sections 9–10, we review the classical theory of single automorphisms of compact (and not necessarily abelian) groups, and obtain in particular Lawton's description of all expansive automorphisms of compact, connected (abelian) groups (Theorem 9.7). In Section 10 we present the structure theory of automorphisms developed by Yuzvinskii, Miles and Thomas, which yields in particular that entropy is a complete invariant for topological conjugacy of expansive and ergodic automorphisms of compact, zero-dimensional groups (Remark 10.10). However, the algebraic classification of such automorphisms is much more complicated, even if the underlying group is abelian (Example 10.11). The results of Section 10 are used in Chapter IV, Sections 11–12, where we discuss periodic points of $\mathbb{Z}^d$-actions by automorphisms of compact groups. Theorem 11.2 shows that every $\mathbb{Z}^d$-action α by automorphisms of a compact group X has a dense set of periodic points, provided that α satisfies the d.c.c. and that the connected component X° of the identity in X is abelian. If α satisfies the d.c.c., but X° is non-abelian, the closure of the set of α-periodic points in X may be very small; however, if α is ergodic, and if $\alpha_{\mathbf{n}}$ is ergodic for some $\mathbf{n} \in \mathbb{Z}^d$, then Theorem 12.1 shows that the set of $\alpha_{\mathbf{n}}$-periodic points is dense in X.

Chapter V deals with entropy of $\mathbb{Z}^d$-actions on compact groups. In Section 13 we review the basic concepts of topological and measure theoretic entropy for $\mathbb{Z}^d$-actions, and in Section 14 we prove the higher-dimensional analogue of Yuzvinskii's entropy addition formula, which expresses the entropy $h(\alpha)$ of a $\mathbb{Z}^d$-action α by automorphisms of a compact group X as the sum $h(\alpha^{X/Y}) + h(\alpha^Y)$ for any closed, normal, α-invariant subgroup $Y \subset X$, where $\alpha^{X/Y}$ and α^Y are the $\mathbb{Z}^d$-actions induced by α on X/Y and Y, respectively. Section 15 discusses $\mathbb{Z}^d$-action by automorphisms of compact groups with zero-dimensional centres and shows that every such action α with finite entropy satisfies that $h(\alpha) = \log k$ for some integer $k \geq 1$ (Theorem 15.6). In Section 16 we introduce the tool for calculating entropy for $\mathbb{Z}^d$-actions by automorphisms of compact, abelian groups: the Mahler measure of a Laurent polynomial $f \in \mathfrak{R}_d$. After proving some elementary properties of Mahler measure we use it in the Section 17 to review the classical entropy formulae by Sinai, Rokhlin and Yuzvinskii for automorphisms of compact, abelian groups (Theorem 17.7), and in Section 18 to derive the entropy of an arbitrary $\mathbb{Z}^d$-action α on a compact, abelian group (Theorem 18.1, Corollary 18.5, and Proposition 18.6). By using Yuzvinskii's addition formula one obtains an expression for the entropy of every $\mathbb{Z}^d$-action by automorphisms of a compact group (Corollary 14.3).

The subject of Chapter VI are $\mathbb{Z}^d$-actions on compact, abelian groups with positive entropy. Section 19 begins with a proof of Kronecker's result that a (Laurent) polynomial $f \in \mathfrak{R}_1$ has Mahler measure 1 if and only if it is a product of cyclotomic polynomials (Lemma 19.1). After a brief look at single, non-ergodic group automorphisms in Theorem 19.2 we turn to the generalization of Kronecker's result to higher dimensions by Boyd, Lawton and Smyth, which leads to the characterization of zero and positive entropy appearing in Figure 1 (Proposition 19.4 and Theorem 19.5). In the Propositions 19.7 and 19.10–19.11 we present some explicit entropy formulae in terms of the Mahler measures of certain polynomials calculated by Smyth. These expressions for the entropy of certain $\mathbb{Z}^d$-actions in terms of Dirichlet L-functions are quite mystifying and still await meaningful interpretation. Section 19 ends with a brief discussion of the dynamical implications of Lehmer's conjecture concerning polynomials with minimal Mahler measure. Section 20 contains the characterization of completely positive entropy appearing in Figure 1 (Theorem 20.8), Yuzvinskii's result that every ergodic automorphism of a compact group has completely positive entropy, and Kamiński's proof that every measure preserving $\mathbb{Z}^d$-action with completely positive entropy on a probability space is mixing of all orders (Theorem 20.14). It continues with a generalization of Berg's theorem about the uniqueness of the measure of maximal entropy for ergodic group automorphisms to $\mathbb{Z}^d$-actions: a $\mathbb{Z}^d$-action α by automorphisms of a compact, abelian group X with finite topological entropy has a unique measure of maximal entropy if and only if α has completely positive entropy (Theorem 20.15). In Section 21 we present formulae relating the entropy of a $\mathbb{Z}^d$-action by automorphisms of a compact, abelian group satisfying the d.c.c. to the growth rate of its periodic points. Unlike ergodic automorphisms of tori and solenoids, ergodic $\mathbb{Z}^d$-actions by automorphisms of compact, abelian groups can have uncountably many periodic points with a given period, even if they have finite entropy. In order to arrive at a meaningful 'growth rate' for periodic points one has to count the number of *connected components* in the set of periodic points with a given period, and in Theorem 21.2 we prove that, for a $\mathbb{Z}^d$-action α by automorphisms of a compact, abelian group X with finite entropy and d.c.c., the upper growth rate of the number of connected components in the set of periodic points is indeed equal to the entropy of α. For expansive actions the set of periodic points with given period is always finite, and $h(\alpha)$ is, in fact, equal to the growth rate of the periodic points. The distribution of the periodic points of α is discussed in Section 22: if α is expansive, then all limits of the distributions of periodic points are measures of maximal entropy for α. Finally, in Section 23, we prove the equivalence of Bernoullicity and completely positive entropy (Theorem 23.1) by exhibiting an intriguing connection between the product formula for valuations of global fields and certain asymptotic independence results implying Bernoullicity.

$\mathbb{Z}^d$-actions with zero entropy are investigated in Chapter VII. In Section 24 we study the connection between entropy and dimension for a $\mathbb{Z}^d$-action α on a compact, abelian group X with zero entropy. Section 25 deals with a class of particularly simple examples of zero-entropy actions, the shift-actions of $\mathbb{Z}^2$ on the closed, shift-invariant subgroups of $(\mathbb{Z}/p\mathbb{Z})^{\mathbb{Z}^2}$, where p is a rational prime. If such an action σ of $\mathbb{Z}^2$ on a closed, shift-invariant subgroup $X \subset (\mathbb{Z}/p\mathbb{Z})^{\mathbb{Z}^2}$ is ergodic, then there exists a finite subset $F \subset \mathbb{Z}^2$ such that X consists of all points in $(\mathbb{Z}/p\mathbb{Z})^{\mathbb{Z}^2}$ whose coordinates in every translate of F sum to 0 (mod p) (Proposition 25.5; cf. (0.1)). Since X is determined by the set F it is natural to ask whether F is—up to translation—a measurable conjugacy invariant of the $\mathbb{Z}^2$-action. The main result of Section 25 shows that the convex hull of F is—again up to translation—invariant under measurable conjugacy of such actions (Theorem 25.15), and that this convex hull not only determines the entropies of the individual shifts $\sigma_{\mathbf{n}}$, $\mathbf{n} \in \mathbb{Z}^2$, (Proposition 25.7), but is also related to certain residual sigma-algebras associated with the $\mathbb{Z}^2$-actions, which are studied both in greater detail and in greater generality in Section 26. By developing further ideas of Kamiński on relative entropy we derive two classes of measurable conjugacy invariants for arbitrary measure preserving $\mathbb{Z}^d$-actions on probability spaces, and in particular for d-dimensional subshifts: one related to relative entropy, and the other to certain residual sigma-algebras which measure the amount by which any point in X is determined by its infinitely remote coordinates in a particular direction. These invariants allow us to extend the results of Section 25 to shift-invariant subgroups of $(\mathbb{Z}/p\mathbb{Z})^{\mathbb{Z}^d}$ (Theorem 26.9 and its corollaries).

Chapter VIII returns to the higher order mixing properties of $\mathbb{Z}^d$-actions first studied in connection with completely positive entropy in Section 20, and analyzes a phenomenon discovered by Ledrappier (cf. (0.1)): if α is a mixing $\mathbb{Z}^d$-action by automorphisms of a compact, abelian group X, then it need not be mixing of every order, and the breakdown of higher-order mixing can occur in a particularly striking way (through the existence of non-mixing sets). After proving that all these phenomena are again governed by the prime ideals associated with the $\mathfrak{R}_d$-module $\mathfrak{M} = \hat{X}$ (Theorem 27.2), we exhibit a direct connection between mixing properties and solutions to certain arithmetical equations investigated by—amongst others—Mahler, Masser, van der Poorten and Schlickewei. By making use of this connection we prove that every mixing $\mathbb{Z}^d$-action α by automorphisms of a compact, connected, abelian group X is mixing of all orders (Theorem 27.3). If X is not connected, then α may have non-mixing sets, which are governed by the factorization of the prime ideals associated with $\mathfrak{M} = \hat{X}$ in certain rings of Laurent polynomials in fractional powers of the variables $u_1, \ldots, u_d$ (Theorem 28.7). By studying this factorization we derive further measurable conjugacy invariants which distinguish between some of the $\mathbb{Z}^d$-actions which cannot be told apart with the tools developed in Section 26.

The last chapter, Chapter IX, is devoted to rigidity properties of zero entropy $\mathbb{Z}^d$-actions by automorphisms of compact, abelian groups. As some of the results stated there are still tentative I have restricted myself to stating many of the published results without proofs. In Section 29 we discuss invariant measures of mixing and almost minimal $\mathbb{Z}^d$-actions (i.e. of mixing $\mathbb{Z}^d$-actions α by automorphisms of compact groups X for which every closed, proper, α-invariant subgroup of X is finite). If X is zero-dimensional, then there exist uncountably many distinct, α-invariant, ergodic probability measures μ on X (Remark 29.7); however, if an α-invariant probability measure μ on X is sufficiently mixing, then Theorem 29.4 shows that $\mu = \lambda_X$. If the group X is connected, then every non-atomic, α-invariant, ergodic probability measure $\mu \neq \lambda_X$ on X must—in certain cases—have zero entropy under every $\alpha_{\mathbf{n}}$, $\mathbf{n} \in \mathbb{Z}^d$, but it is not known if there exist non-atomic, ergodic, invariant measures other than λ_X (cf. Theorem 29.9 and Problem 29.10).

Section 30 concerns itself with the first cohomology of $\mathbb{Z}^d$-actions. Since cocycles are used in the construction of skew-products and other 'deformations' of group actions, any statement about scarcity of non-trivial cocycles has an immediate interpretation as a rigidity property (cf. [39]). If α is an expansive and mixing $\mathbb{Z}^d$-action by automorphisms of a compact, abelian group X, then it is fairly easy to see that the first continuous cohomology groups with values in $\mathbb{R}$ and $\mathbb{S}$ are very big and complicated groups; however, a theorem by Livshitz ([65]) shows that the Hölder cohomology of an expansive and ergodic (single) automorphism of compact, abelian group is much nicer than the continuous cohomology, since the Hölder coboundaries form a closed subgroup of the group of Hölder cocycles. For $d > 1$ even more is true: the first Hölder cohomology group of α with values in $\mathbb{R}$ is trivial (Theorem 30.3), and every Hölder cocycle for α with values in $\mathbb{S}$ is Hölder cohomologous to a character-valued cocycle (Theorem 30.4). In particular, every continuous extension of α by a finite, abelian group G is essentially affine (Corollary 30.5).

The final Section 31 consists of a single example: if σ is the shift-action of $\mathbb{Z}^2$ on the group X in (0.1), then every measurable, Haar measure-preserving, shift-commuting isomorphism $\phi \colon X \longmapsto X$ is a.e. equal to $\sigma_{\mathbf{n}}$ for some $\mathbf{n} \in \mathbb{Z}^2$. This form of isomorphism rigidity can be verified for many other examples, but the precise extent of this phenomenon, which is reminiscent of the rigidity properties of horocycle flows in [82], is not yet understood.

Much of the material in this monograph is the result of a stimulating and enjoyable collaboration with Anatole Katok, Bruce Kitchens, Doug Lind, Dan Rudolph and Tom Ward, and large parts of this text are based on joint papers with these authors (in various combinations). Sections 1–4 follow the paper [45] with Bruce Kitchens, Section 8 stems from the article [38] with Anatole Katok, Sections 9–12 again follow [45] to a large extent, and Sections 13–14, 16–18 and 20 are based on the joint work [63] with Doug Lind and Tom Ward. Section 21 is a combination of material in [63] and of the paper [62] with Doug Lind, Section 23 is taken from the article [91] with Dan Rudolph, Sections 25–26 are joint work with Bruce Kitchens ([46], [47]), Section 27 originates from the paper [98] with Tom Ward, and Section 28 again follows [47]. More detailed references to the material presented in each section can be found in the concluding remark(s) of that section.

I would also like to thank François Ledrappier, who attracted my attention to these problems in 1986 by showing me Example (0.1) in [56], and Bill Parry, who took considerable interest in the development of these ideas over the years and suggested that I submit the final manuscript to the Institut d'Estudis Catalans for the Ferran Sunyier i Balaguer Prize. Parts of the manuscript featured in graduate courses at the Universities of Vienna and Warwick, and I am grateful to the participants in these courses, and in particular to Harald Rindler in Vienna, for helping to eradicate some of the misprints and mistakes in the text.

Finally a word about notation. The symbols $\mathbb{R}$ and $\mathbb{C}$ denote the real and complex numbers, $\mathbb{R}^+$ the non-negative reals, $\mathbb{Q}$ the rationals, $\mathbb{Z}$ the integers, $\mathbb{N} = \{0, 1, 2, \dots\}$ the natural numbers, $\mathbb{N}^*$ the non-zero natural numbers, $\mathbb{T} = \mathbb{R}/\mathbb{Z}$ the (one-dimensional) torus, and $\mathbb{S} = \{z \in \mathbb{C} : |z| = 1\} \cong \mathbb{T}$ the multiplicative group of complex numbers of modulus 1. For every $n \geq 1$ we set $\mathbb{Z}[\frac{1}{n}] = \{\frac{k}{n^l} : k \in \mathbb{Z}, l \geq 0\} \subset \mathbb{Q}$ and write SU(n) for the group of all $n \times n$ unitary matrices. If $\mathcal{R}$ is a commutative integral domain with identity 1 we denote by $\mathrm{GL}(n, \mathcal{R})$, $n \geq 1$, the group of all $n \times n$ matrices with entries in $\mathcal{R}$ whose determinant is a unit (i.e. invertible) in $\mathcal{R}$, and $\mathrm{SL}(n, \mathcal{R}) \subset \mathrm{GL}(n, \mathcal{R})$ is the subgroup of matrices with determinant 1. The indicator function of a subset B of a set X is written as

$$1_B(x) = \begin{cases} 1 & \text{if } x \in B, \\ 0 & \text{if } x \in X \smallsetminus B, \end{cases}$$

and the cardinality of B is denoted by $|B|$. A measure space $(X, \mathfrak{S}, \mu)$ is standard if $(X, \mathfrak{S})$ is a standard Borel space and μ a sigma-finite measure on $\mathfrak{S}$. All our measure and probability spaces will be assumed to be standard, even when this is not explicitly stated. If $(X, \mathfrak{S}, \mu)$ is a probability space, then any statement about sets or functions which is true modulo a null-set of μ will be said to hold (mod μ). For every sub-sigma-algebra $\mathfrak{T} \subset \mathfrak{S}$ and $f \in L^1(X, \mathfrak{S}, \mu)$ we denote by $E_\mu(f|\mathfrak{T}) \in L^1(X, \mathfrak{T}, \mu)$ the conditional expectation of f, given $\mathfrak{T}$. The end of each proof is marked with a $\square$, and the end of each example with a $\boxdot$.

CHAPTER I

Group actions by automorphisms of compact groups

1. Ergodicity and mixing

Throughout this book the term *compact group* will denote a compact, metrizable group, and a *compact Lie group* will be a compact (possibly finite) subgroup of some finite-dimensional matrix group over the complex numbers. Any metric δ on a compact group X is assumed to be *invariant*, i.e. $\delta(x, x') = \delta(yx, yx') = \delta(xy, x'y)$ for all $x, x', y \in X$, and to induce the topology of X. The identity element of a group X will usually be written as $\mathbf{1}$, $\mathbf{0}$, $\mathbf{1}_X$, or $\mathbf{0}_X$, depending on whether X is multiplicative or additive, and whether there is danger of confusion. If X is a topological group we write X° for its connected component of the identity, $C(X)$ for its centre, $\mathrm{Aut}(X)$ for the group of continuous automorphisms of X, and $\mathrm{Inn}(X) \subset \mathrm{Aut}(X)$ for the normal subgroup of inner automorphisms of X. The trivial automorphism of X is denoted by $\mathrm{id}_X = \mathbf{1}_{\mathrm{Aut}(X)}$, and we set $\mathrm{Out}(X) = \mathrm{Aut}(X)/\mathrm{Inn}(X)$. If X is compact and δ is a metric on X we define a metric $\boldsymbol{\delta}$ on $\mathrm{Aut}(X)$ by setting

$$\boldsymbol{\delta}(\alpha, \beta) = \max_{x \in X} \left(\delta(\alpha(x), \beta(x)) + \delta(\alpha^{-1}(x), \beta^{-1}(x)) \right)$$

for all $\alpha, \beta \in \mathrm{Aut}(X)$; the topology on $\mathrm{Aut}(X)$ induced by $\boldsymbol{\delta}$ is called the *uniform topology*. If X is a compact Lie group then $\mathrm{Aut}(X)$ is again a Lie group in the uniform topology, $\mathrm{Inn}(X)$ is an open subgroup of X, and $\mathrm{Out}(X)$ is therefore discrete. For an arbitrary compact group X, the group $\mathrm{Out}(X)$ is zero-dimensional in the uniform topology ([32]). In particular, if X is connected, then the connected component of the identity in $\mathrm{Aut}(X)$ is equal to $\mathrm{Inn}(X)$, and if X is abelian, then $\mathrm{Aut}(X) \cong \mathrm{Out}(X)$ is zero-dimensional.

If $\alpha \in \mathrm{Aut}(X)$, and if $V \subset X$ is a closed, α-invariant subgroup, then α^V is the restriction of α to V, and $\alpha^{X/V}$ is the homeomorphism of $X/V = \{xV : x \in X\}$ induced by α. If there is no ambiguity we shall sometimes write α instead of α^V or $\alpha^{X/V}$. The kernel of α is denoted by $\ker(\alpha)$.

Let A be a locally compact, abelian group, and let $\hat{A}$ be the *dual group* or *character group* of A, i.e. the group of all continuous homomorphisms from A into $\mathbb{S} = \{z \in \mathbb{C} : |z| = 1\}$ under point-wise multiplication, furnished with the topology of uniform convergence on compact sets. Then $\hat{A}$ is again a locally compact, abelian group, and $\hat{A}$ is discrete if and only if A is compact. If $a \in A$ and $\chi \in \hat{A}$ we shall write either $\chi(a)$ or $\langle a, \chi \rangle$ to denote the value of the *character* $\chi \in \hat{A}$ at a. The map $\chi \mapsto \langle a, \chi \rangle$ from $\hat{A}$ to $\mathbb{S}$ is obviously an element of the dual group $\hat{\hat{A}}$ of $\hat{A}$; Pontryagin's duality theorem states that the map $a \mapsto \langle a, \cdot \rangle$ is a continuous group isomorphism of A and $\hat{\hat{A}}$, and allows us to identify $\hat{\hat{A}}$ with A (cf. [28]). Every $\alpha \in \mathrm{Aut}(A)$ determines a *dual automorphism* $\hat{\alpha} \in \mathrm{Aut}(\hat{A})$ such that $\langle \alpha(a), \chi \rangle = \langle a, \hat{\alpha}(\chi) \rangle$ for all $a \in A$ and $\chi \in \hat{A}$. The map $\alpha \mapsto \hat{\alpha}$ from $\mathrm{Aut}(A)$ to $\mathrm{Aut}(\hat{A})$ is a homeomorphism, and $\widehat{\alpha\beta} = \hat{\beta}\hat{\alpha}$ for all $\alpha, \beta \in \mathrm{Aut}(A)$. More generally, if A, B are locally compact, abelian groups and $\psi \colon A \longmapsto B$ a continuous group homomorphism, then ψ determines a continuous *dual homomorphism* $\hat{\psi} \colon \hat{B} \longmapsto \hat{A}$ with $\langle \psi(a), \chi \rangle = \langle a, \hat{\psi}(\chi) \rangle$ for every $a \in A$ and $\chi \in \hat{B}$, $\hat{\hat{\psi}} = \psi$, and $\hat{\psi}$ is injective if and only if $\psi(A)$ is dense in B. In particular, if A and B are both either compact or discrete, then $\hat{\psi}$ is injective if and only if ψ is surjective.

We fix a countable group Γ and a compact group X. An *action* of Γ by automorphisms of X is a homomorphism $\alpha \colon \Gamma \longmapsto \mathrm{Aut}(X)$. If $\alpha \colon \gamma \mapsto \alpha_\gamma$ is such an action and $Y \subset X$ a closed, α-invariant subgroup of X (i.e. $\alpha_\gamma(Y) = Y$ for every $\gamma \in \Gamma$), then we denote by $\alpha^Y \colon \gamma \mapsto \alpha_\gamma^Y$ the restriction of α to Y and write $\alpha^{X/Y}$ for the homomorphism $\gamma \mapsto \alpha_\gamma^{X/Y}$ from Γ into the group of homeomorphism of X/Y. If $\Delta \subset \Gamma$ is a subgroup, then $\alpha^{(\Delta)} \colon \gamma \mapsto \alpha_\gamma$, $\gamma \in \Delta$, is the restriction of α to Δ.

The action α is *topologically transitive* if there exists a point $x \in X$ whose orbit $\alpha_\Gamma(x) = \{\alpha_\gamma(x) : \gamma \in \Gamma\}$ is dense in X, and *expansive* if there exists a neighbourhood $N(\mathbf{1}_X)$ of the identity in X such that $\bigcap_{\gamma \in \Gamma} \alpha_\gamma(N(\mathbf{1}_X)) = \{\mathbf{1}_X\}$: in this case $N(\mathbf{1}_X)$ is an *expansive neighbourhood* of $\mathbf{1}_X$. Since the normalized Haar measure λ_X of X is invariant under every α_γ, $\gamma \in \Gamma$, we may regard α as a measure preserving action of Γ on the probability space $(X, \mathfrak{B}_X, \lambda_X)$, where $\mathfrak{B}_X$ is the sigma-algebra of Borel sets in X. We call α *ergodic* if it is ergodic with respect to λ_X, and *(strongly) mixing* if $\lim_{\gamma \to \infty} \lambda_X(B \cap \alpha_\gamma(C)) = \lambda_X(B)\lambda_X(C)$ for all $B, C \in \mathfrak{B}_X$. Finally, if α' is a second Γ-action by automorphisms of a compact group X', then α' is *measurably conjugate* to α if there exists a measure preserving isomorphism $\psi \colon X \longmapsto X'$ of the measure spaces $(X, \mathfrak{B}_X, \lambda_X)$ and $(X', \mathfrak{B}_{X'}, \lambda_{X'})$ such that $\psi \cdot \alpha_\gamma = \alpha_\gamma' \cdot \psi$ λ_X-a.e., for every $\gamma \in \Gamma$, and α' is *topologically* (resp. *algebraically*) *conjugate* to α if there exists a homeomorphism (resp. a continuous group isomorphism) $\phi \colon X \longmapsto X'$ such that $\phi \cdot \alpha_\gamma = \alpha_\gamma' \cdot \phi$ for every $\gamma \in \Gamma$.

If β is a single automorphism of X we say that β is ergodic, mixing, topologically transitive, or expansive, if the $\mathbb{Z}$-action $n \mapsto \beta^n$ is ergodic, mixing, topologically transitive, or expansive.

THEOREM 1.1. *Let α be an action of a countable group Γ by automorphisms of a compact group X. Then α is ergodic if and only if it is topologically transitive.*

The proof of Theorem 1.1 depends on the following lemma.

LEMMA 1.2. *Let α be an action of a countable group Γ by automorphisms of a compact group X. The following conditions are equivalent.*

(1) *α is non-ergodic;*
(2) *There exists a non-trivial, continuous, irreducible, unitary representation τ of X on a (finite-dimensional) complex Hilbert space $\mathcal{K}$ such that the group*

$$\Gamma_\tau = \{\gamma \in \Gamma : \text{the representation } \tau \cdot \alpha_\gamma \text{ is unitarily equivalent to } \tau\}$$

has finite index in Γ;
(3) *There exists a compact Lie group $Y \neq \{1\}$, a continuous, surjective homomorphism $\eta\colon X \longmapsto Y$, and a homomorphism $\beta\colon \Gamma \longmapsto \mathrm{Aut}(Y)$ such that $\eta \cdot \alpha_\gamma = \beta_\gamma \cdot \eta$ for every $\gamma \in \Gamma$, and for which the normal subgroup $\{\gamma \in \Gamma : \beta_\gamma \in \mathrm{Inn}(Y)\}$ has finite index in Γ;*
(4) *There exists a closed, normal, α-invariant subgroup $V \subsetneq X$ such that X/V is a Lie group, and the family of automorphisms $\{\alpha_\gamma^{X/V} : \gamma \in \Gamma\}$ is uniformly equicontinuous;*
(5) *There exists a closed, normal, α-invariant subgroup $V \subsetneq X$ and a metric δ on X/V such that X/V is a Lie group, and δ is invariant under $\alpha^{X/V}$.*

PROOF. Assume that α is non-ergodic. We put $\mathcal{H} = \{f \in L^2(X, \lambda_X) : \int f \, d\lambda_X = 0\}$ and denote by ρ the right regular representation of X on $\mathcal{H}$, defined by $(\rho(x)f)(x') = f(x'x)$ for $f \in \mathcal{H}$, $x, x' \in X$. Since α is non-ergodic we can find a non-zero α-invariant function $f \in \mathcal{H}$ and a ρ-invariant subspace $\mathcal{K} \subset \mathcal{H}$ such that the restriction τ of ρ to $\mathcal{K}$ is irreducible and the projection $f' = P_\mathcal{K} f$ of f onto $\mathcal{K}$ is non-zero. We fix $\gamma \in \Gamma$ for the moment and define a unitary operator W_γ on $\mathcal{H}$ by $W_\gamma h = h \cdot \alpha_\gamma^{-1}$ for every $h \in \mathcal{H}$. Then $W_\gamma f = f$ and hence

$$\|P_\mathcal{K} f\| = \|P_\mathcal{K} W_\gamma^{-1} f\| = \|W_\gamma P_\mathcal{K} W_\gamma^{-1} f\| = \|P_{W_\gamma \mathcal{K}} f\|,$$

where $P_{W_\gamma \mathcal{K}}$ is the projection onto the subspace $W_\gamma \mathcal{K} \subset \mathcal{H}$. Consider the continuous representation τ_γ of X on $\mathcal{K}$ given by $\tau_\gamma(x) = \tau \cdot \alpha_\gamma(x) = W_\gamma^{-1} \tau(x) W_\gamma$ for all $x \in X$. Then

$$\begin{aligned}
\langle \tau_\gamma(x) f', f' \rangle &= \langle \rho \cdot \alpha_\gamma(x) f', f' \rangle = \langle \rho(x) W_\gamma f', W_\gamma f' \rangle \\
&= \langle \rho(x) W_\gamma P_\mathcal{K} f, W_\gamma P_\mathcal{K} f \rangle = \langle \rho(x) W_\gamma P_\mathcal{K} W_\gamma^{-1} f, W_\gamma P_\mathcal{K} W_\gamma^{-1} f \rangle \\
&= \langle \rho(x) P_{W_\gamma \mathcal{K}} f, P_{W_\gamma \mathcal{K}} f \rangle
\end{aligned}$$

for every $x \in X$. Hence the irreducible, unitary representation τ_γ is unitarily equivalent to the restriction of ρ to the invariant subspace $W_\gamma \mathcal{K}$ of $\mathcal{H}$. If τ_γ is unitarily inequivalent to τ then $W_\gamma \mathcal{K}$ must be orthogonal to $\mathcal{K}$; on the other hand we have seen above that $\|P_\mathcal{K} f\| = \|P_{W_\gamma \mathcal{K}} f\|$. By varying γ we see that the group Γ_τ must have finite index in Γ, as claimed in (2).

Now assume that (2) is satisfied. Since every non-trivial, continuous, irreducible, unitary representation τ occurs as a subrepresentation of ρ on $\mathcal{H}$ we may assume that we are in the same situation as in the proof of the implication (1)$\Rightarrow$(2) and use the same notation. Since τ_γ is unitarily equivalent to τ for every $\gamma \in \Gamma_\tau$ there exist unitary operators V_γ, $\gamma \in \Gamma_\tau$, on $\mathcal{K}$ such that $V_\gamma^{-1} \tau(x) V_\gamma = \tau_\gamma(x) = \tau \cdot \alpha_\gamma(x)$ for all $\gamma \in \Gamma_\tau$. Let $F = \Gamma/\Gamma_\tau$ be the (finite) space of left cosets of Γ_τ, $\phi \colon \Gamma \longmapsto F$ the quotient map, and let $c \colon F \longmapsto \Gamma$ be a map with $\phi(c(a)) = a$ for every $a \in F$. We write a typical element v of the Hilbert space $\overline{\mathcal{K}} = \mathcal{K}^F$ as $v = (v_a) = (v_a, a \in F)$ with $v_a \in \mathcal{K}$ for all $a \in F$ and consider the continuous, unitary representation η of X on $\overline{\mathcal{K}}$ given by $(\eta(x)v)_a = \tau_{c(a)^{-1}}(x)v_a = \tau \cdot \alpha_{c(a)}^{-1}(x)v_a$ for all $v = (v_a) \in \overline{\mathcal{K}}$ and $a \in F$. For every $\gamma \in \Gamma$ we define a unitary operator U_γ on $\overline{\mathcal{K}}$ by setting

$$(U_\gamma v)_a = V_{b(\gamma^{-1}, a)} v_{\gamma^{-1} a}$$

for every $v = (v_a) \in \overline{\mathcal{K}}$ and $a \in F$, where

$$b(\gamma, a) = c(\gamma a)^{-1} \gamma c(a) \in \Gamma_\tau.$$

Then

$$(U_\gamma \eta(x) v)_a = V_{b(\gamma^{-1}, a)} \tau(\alpha_{c(\gamma^{-1} a)}^{-1}(x)) v_{\gamma^{-1} a}$$

and

$$\begin{aligned}
(\eta(x) U_\gamma v)_a &= \tau(\alpha_{c(a)}^{-1}(x)) V_{b(\gamma^{-1}, a)} v_{\gamma^{-1} a} \\
&= V_{b(\gamma^{-1}, a)} \tau(\alpha_{b(\gamma^{-1}, a)} \alpha_{c(a)}^{-1}(x)) v_{\gamma^{-1} a} \\
&= V_{b(\gamma^{-1}, a)} \tau(\alpha_{c(\gamma^{-1} a)}^{-1} \alpha_\gamma^{-1}(x)) v_{\gamma^{-1} a}
\end{aligned}$$

for every $x \in X$, $a \in F$, and we conclude that

$$U_\gamma^{-1} \eta(x) U_\gamma = \eta(\alpha_\gamma^{-1}(x)). \tag{1.1}$$

We denote by $Y = \eta(X)$ the image of X under η in the group $\mathcal{U}(\overline{\mathcal{K}})$ of unitary operators on the finite-dimensional Hilbert space $\overline{\mathcal{K}}$. Then Y is a compact Lie group, Γ acts on Y by $\beta_\gamma(\eta(x)) = U_\gamma \eta(x) U_\gamma^{-1} = \eta(\alpha_\gamma(x))$ for every $\gamma \in \Gamma$ and $x \in X$, and the metric δ' on Y induced by the operator norm on $\mathcal{U}(\overline{\mathcal{K}})$ is invariant under the action β of Γ on Y. Since the group $\mathrm{Iso}(Y, \delta')$ of δ'-isometric automorphisms of Y is compact (by the Arzela-Ascoli theorem), and since $\mathrm{Inn}(Y)$ is an open subgroup of $\mathrm{Aut}(Y)$ and hence of $\mathrm{Iso}(Y, \delta')$, $\mathrm{Inn}(Y)$

has finite index in $\mathrm{Iso}(Y, \delta')$, and we conclude that the normal subgroup $\{\gamma \in \Gamma : \beta_\gamma \in \mathrm{Inn}(Y)\}$ has finite index in Γ, as claimed in (3).

In order to prove that (3) implies (5) and hence (4) we set $V = \ker(\eta)$ and fix a metric δ on $X/V \cong Y$. Since δ is invariant under $\mathrm{Inn}(Y)$, and since $\mathrm{Inn}(Y)$ has finite index in $\beta_\Gamma = \{\beta_\gamma : \gamma \in \Gamma\}$, there exists a metric δ' on Y which is invariant under every β_γ, $\gamma \in \Gamma$.

Finally, if (4) is satisfied, we fix a closed neighbourhood $N(\mathbf{1}_{X/V}) \subsetneq X/V$. Since $\alpha^{X/V}$ is uniformly equicontinuous, the set $N' = \bigcap_{\gamma \in \Gamma} \alpha_\gamma(N(\mathbf{1}))$ has non-empty interior, therefore positive Haar measure, and is $\alpha^{X/V}$-invariant. Hence α is non-ergodic, as claimed in (1). $\square$

PROOF OF THEOREM 1.1. If Γ is ergodic it is topologically transitive. Conversely, if Γ is non-ergodic, Lemma 1.2 implies the existence of a closed, normal, α-invariant subgroup $V \subsetneq X$ such that $\alpha^{X/V}$ preserves a metric on X/V. Hence there exists an open, α-invariant subset of X which is not dense, and α is not topologically transitive. $\square$

Our next aim is to show that, if an action α of a countable group Γ by automorphisms of a compact group X is non-ergodic, then X has a maximal closed, normal, α-invariant subgroup X' on which $\alpha^{X'}$ is ergodic. We begin with a version of Lemma 1.2 for normal, α-invariant subgroups.

LEMMA 1.3. *Let α be an action of a countable group Γ by automorphisms of a compact group X, and let $Y \subset X$ be a closed, normal, α-invariant subgroup of X. The following conditions are equivalent.*

(1) *α^Y is non-ergodic;*
(2) *There exists a closed, α-invariant subgroup $W \subsetneq Y$ such that W is normal in X, Y/W is a Lie group, and $\alpha^{Y/W}$ preserves a metric δ on Y/W.*

PROOF. Suppose that α^Y is non-ergodic. According to Lemma 1.2 (2) there exists a non-trivial, continuous, irreducible, unitary representation τ of Y such that Γ_τ has finite index in Γ. As in the proof of the implication (2)$\Rightarrow$(3) in Lemma 1.2 we construct a continuous, unitary representation η of Y on a finite-dimensional, complex Hilbert space $\overline{\mathcal{K}}$ and a map $\gamma \mapsto U_\gamma$ from Γ into the unitary group $\mathcal{U}(\overline{\mathcal{K}})$ of $\overline{\mathcal{K}}$ satisfying (1.1). For every $x \in X$ we write $\eta^{(x)}$ for the representation $y \mapsto \eta^{(x)}(y) = \eta(x^{-1}yx)$ of Y on $\overline{\mathcal{K}}$, and we claim that $\eta^{(x)}$ is unitarily equivalent to η for every x in a subgroup X_η with finite index in X.

In order to verify this claim we denote by $\langle \cdot, \cdot \rangle$ the inner product on $\overline{\mathcal{K}}$ and consider the Hilbert space $\mathcal{H} = L^2(X, \lambda_X, \overline{\mathcal{K}})$ of Borel maps $f \colon X \longmapsto \overline{\mathcal{K}}$ satisfying that $\int_X \langle f(x), f(x) \rangle \, d\lambda_X < \infty$, furnished with the inner product $\langle f, f' \rangle = \int_X \langle f(x), f(x) \rangle \, d\lambda_X(x)$. Let $\overline{\mathcal{H}} \subset \mathcal{H}$ be the subspace consisting of all $f \in \mathcal{H}$ with the property that, for λ_X-a.e. $x \in X$, $f(y^{-1}x) = \eta(y)f(x)$ for λ_Y-a.e. $y \in Y$. In order to see that $\overline{\mathcal{H}} \neq \{0\}$ we denote by $\phi \colon X \longmapsto Z = X/Y$

the quotient map, choose a Borel map $c\colon X/Y \longmapsto X$ with $\phi(c(z)) = z$ for every $z \in Z$ (cf. [78], Section 1.5), and define a non-zero element $f \in \overline{\mathcal{H}}$ by $f(yc(z)) = \eta(y)^{-1}v$ for every $y \in Y$ and $z \in Z$, where v is some unit vector in $\overline{\mathcal{K}}$.

Consider the right regular representation ρ of X on $\overline{\mathcal{H}}$, defined as before by $(\rho(x)f)(x') = f(xx')$ for all $x, x' \in X$, and denote by ρ' the restriction of ρ to Y. Then $\rho' \cong \int_X^\oplus \eta^{(x)} \, d\lambda_X$, acting on $\overline{\mathcal{H}} \cong \int_X^\oplus \overline{\mathcal{K}}_x \, d\lambda_X(x)$, where $\overline{\mathcal{K}}_x = \overline{\mathcal{K}}$ for every $x \in X$, and where the isomorphism of $\overline{\mathcal{H}}$ and $\int_X^\oplus \overline{\mathcal{K}}_x \, d\lambda_X(x)$ is given by $f \mapsto \int_X^\oplus f(x) \, d\lambda_X(x)$ for every $f \in \overline{\mathcal{H}}$. If the dimension of $\overline{\mathcal{K}}$ is equal to k, then we can use Gram-Schmidt orthonormalization to find elements $f_1, \ldots, f_k$ in $\overline{\mathcal{H}}$ such that $f_1(x), \ldots, f_k(x)$ is an orthonormal basis of $\overline{\mathcal{K}}$ for every $x \in X$, and we can use these vectors to calculate the trace $T^{(x)}(y) = \mathrm{tr}(\eta^{(x)}(y))$ for all $x \in X$, $y \in Y$. Let $[\cdot, \cdot]$ be the inner product on $L^2(Y, \lambda_Y)$. Then $[T^{(x)}(\cdot), T^{(x')}(\cdot)]$ is equal to the dimension of the space if linear operators $A\colon \overline{\mathcal{K}} \longmapsto \overline{\mathcal{K}}$ with $A\eta^{(x)}(y) = \eta^{(x')}(y)A$ for all $y \in Y$ (cf. [1]). In particular, $[T^{(x)}(\cdot), T^{(x')}(\cdot)]$ is a non-negative integer, and $\eta^{(x)}$ and $\eta^{(x')}$ are unitarily inequivalent if and only if $1 + [T^{(x)}(\cdot), T^{(x')}(\cdot)] \le [T^{(x)}(\cdot), T^{(x)}(\cdot)] = [T^{(x')}(\cdot), T^{(x')}(\cdot)]$. We fix $\varepsilon > 0$ and use Lusin's theorem (cf. [88]) to find a subset of positive measure $B \subset X$ such that $\lambda_Y(\{y \in Y : \|f_i(xy) - f_i(x'y)\| < \varepsilon\}) > 1 - \varepsilon$ for all $i = 1, \ldots, s$ and $x, x' \in B$, where $\|\cdot\|$ denotes the norm on $\overline{\mathcal{K}}$. If ε is sufficiently small we obtain that $[T^{(x)}(\cdot), T^{(x')}(\cdot)] = [T^{(x)}(\cdot), T^{(x)}(\cdot)]$ for all $x \in B$, i.e. that $\eta^{(x)}$ and $\eta^{(x')}$ are unitarily equivalent for all $x, x' \in B$. Hence the set $X_\eta = \{x \in X : \eta^{(x)} \text{ is unitarily equivalent to } \eta\}$ has positive measure, since it contains $x'^{-1}B$ for every $x' \in B$. As X_η is a closed, normal subgroup of X with positive Haar measure, X/X_η must be finite.

We claim that X_η is α-invariant. Indeed, if $x \in X_\eta$, then there exists a unitary operator V on $\overline{\mathcal{K}}$ such that $\eta(x^{-1}yx) = V^{-1}\eta(y)V$ for every $y \in Y$. According to (1.1) we have, for every $\gamma \in \Gamma$ and $y \in X^\circ$,

$$\begin{aligned}
\eta(\alpha_\gamma^{-1}(x^{-1}yx)) &= U_\gamma^{-1}\eta(x^{-1}yx)U_\gamma = U_\gamma^{-1}V^{-1}\eta(y)VU_\gamma \\
&= U_\gamma^{-1}V^{-1}U_\gamma U_\gamma^{-1}\eta(y)U_\gamma U_\gamma^{-1}VU_\gamma \\
&= U_\gamma^{-1}V^{-1}U_\gamma \eta(\alpha_\gamma^{-1}(y))U_\gamma^{-1}VU_\gamma,
\end{aligned}$$

so that $\eta^{(\alpha_\gamma^{-1}(x))}$ is unitarily equivalent to η.

We write $\psi\colon X \longmapsto X/X_\eta$ for the quotient map and choose a map $c'\colon X/X_\eta \longmapsto X$ with $c'(1_{X/X_\eta}) = 1_X$ and $\psi \cdot c'(v) = v$ for every $v \in X/X_\eta$. Put $\bar{\eta} = \bigoplus_{v \in X/X_\eta} \eta^{(c'(v))}$, and observe that $\bar{\eta}$ is a unitary representation of Y on the finite-dimensional Hilbert space $\mathcal{L} = \overline{\mathcal{K}}^{X/X_\eta}$, and that $W = \ker(\bar{\eta})$ is a closed subgroup of Y which is normal not only in Y, but also in X. Furthermore, since $\bar{\eta}$ is finite-dimensional, $Y/W \cong \bar{\eta}(Y)$ is a Lie group. From (1.1) it is clear that there exists, for every $\gamma \in \Gamma$, a unitary operator $\bar{V}_\gamma$ on $\mathcal{L}$ such that

$\bar{V}_\gamma^{-1}\bar{\eta}(x)\bar{V}_\gamma = \bar{\eta}(\alpha_\gamma^{-1}(x))$, and this implies (2) as in Lemma 1.2. The reverse implication follows from Lemma 1.2. $\square$

THEOREM 1.4. *Let Γ be a countable group, and let α be an action of Γ by automorphisms of a compact group X. Then there exists a countable ordinal ω and a collection $\{V_\xi : \xi < \omega\}$ of closed, normal, α-invariant subgroups of X, indexed by the set of ordinals $\xi < \omega$, with the following properties:*

(1) *$V_0 = X$;*
(2) *If $0 \le \xi < \xi + 1 < \omega$ then $V_{\xi+1} \subsetneqq V_\xi$, $V_\xi/V_{\xi+1}$ is a Lie group, and there exists an $\alpha^{V_\xi/V_{\xi+1}}$-invariant metric on $V_\xi/V_{\xi+1}$;*
(3) *If ξ is a limit ordinal, then $V_\xi = \bigcap_{0 \le \xi' < \xi} V_{\xi'}$;*
(4) *The restriction $\alpha^{X'}$ of α to $X' = \bigcap_{0 \le \xi' < \omega} V_{\xi'}$ is ergodic, and X' is the unique maximal closed, α-invariant subgroup of X on which α is ergodic.*

PROOF. The subgroups $\{V_\xi : \xi < \omega\}$ satisfying (1)–(3) are obtained from Lemma 1.3 through a (necessarily countable) transfinite induction argument. In order to prove (4) we assume that W is a closed, α-invariant subgroup of X with $W \not\subset X'$, and that α^W is ergodic. Then $\lambda_W(W \cap X') = 0$, since α^W is ergodic, and we set

$$\xi_0 = \begin{cases} \min\{\xi : \xi < \omega \text{ and } \lambda_W(W \cap V_\xi) = 0\} & \text{if this set is non-empty,} \\ \omega & \text{otherwise.} \end{cases}$$

If ξ_0 is a limit ordinal there exists a $\xi < \xi_0$ with $0 < \lambda_W(W \cap V_\xi) < 1$, contrary to the ergodicity of α^W. It follows that ξ_0 is not a limit ordinal, and the ergodicity of α^W implies that $W \cap V_{\xi_0-1}$ is an open, α-invariant subgroup of W and therefore equal to W. From the definition of ξ_0 we know that $W \not\subset V_{\xi_0}$; we conclude that W/V_{ξ_0} is a non-trivial, closed, $\alpha^{V_{\xi_0-1}/V_{\xi_0}}$-invariant subgroup of V_{ξ_0-1}/V_{ξ_0}, and (2) implies that $\alpha^{W/V_{\xi_0}}$—and hence α^W—is non-ergodic. This contradiction to our initial assumption implies (4). $\square$

EXAMPLE 1.5. Let α be the automorphism of $X = \mathbb{T}^4$ given by the matrix

$$A = \begin{pmatrix} 0 & 1 & 0 & 0 \\ 0 & 0 & 1 & 0 \\ 0 & 0 & 0 & 1 \\ 1 & 2 & 1 & 0 \end{pmatrix}.$$

The irreducible, unitary representations of X are the characters of X, and condition (2) in Lemma 1.2 is equivalent to the existence of a non-trivial character $\chi \in \hat{X} \cong \mathbb{Z}^4$ such that $\{\chi \cdot A^n = (A^\top)^n \chi : n \in \mathbb{Z}\}$ is finite, where $A^\top$ is the transpose matrix of A. If $\chi = (1, 1, -1, 0)$ then $(A^\top)^3 \chi = \chi$, and the orbit of χ under $A^\top$ consists of the vectors $\chi_1 = \chi = (1, 1, -1, 0)$, $\chi_2 = (0, 1, 1, -1)$, and $\chi_3 = (-1, -2, 0, 1)$. The map $\eta = (\chi_1, \chi_2, \chi_3) \colon X \longmapsto \mathbb{T}^3$ is a homomorphism, and we set $Y = \eta(X) \subset \mathbb{T}^3$ and consider the automorphism β of $Y \subset \mathbb{T}^3$ given by the matrix

$$B = \begin{pmatrix} 0 & 1 & 0 \\ 0 & 0 & 1 \\ 1 & 0 & 0 \end{pmatrix}.$$

Then $\beta \cdot \eta = \eta \cdot \alpha$, and β preserves a metric on Y, since it does so on $\mathbb{T}^3 \supset Y$. The α-invariant group $V = \ker(\eta) \subset \mathbb{T}^4$ is given by $V = \{(t_1, t_2, t_1 + t_2, t_1 + 2t_2) : (t_1, t_2) \in \mathbb{T}^2\} \cong \mathbb{T}^2$, the restriction of α to V is of the form $\begin{pmatrix} 0 & 1 \\ 1 & 1 \end{pmatrix}$ and hence ergodic, and V is the maximal ergodic subgroup for α (cf. Theorem 1.4). $\boxdot$

THEOREM 1.6. *Let α be an action of a countable group Γ by automorphisms of a compact group X.*

(1) *The following conditions are equivalent:*
   (a) *α is mixing;*
   (b) *For every infinite subgroup $\Delta \subset \Gamma$, the restriction of α to Δ is mixing on X;*
   (c) *For every infinite subgroup $\Delta \subset \Gamma$, the restriction of α to Δ is ergodic on X;*
   (d) *The group*

$$\Gamma_\tau = \{\gamma \in \Gamma : \tau \cdot \alpha_\gamma \text{ is unitarily equivalent to } \tau\}$$

   is finite for every non-trivial, continuous, irreducible, unitary representation τ of X.
(2) *If $\Gamma = \mathbb{Z}^d$ the following conditions are equivalent:*
   (a) *α is mixing;*
   (b) *$\alpha_{\mathbf{n}}$ is mixing for every non-zero $\mathbf{n} \in \mathbb{Z}^d$;*
   (c) *$\alpha_{\mathbf{n}}$ is ergodic for every non-zero $\mathbf{n} \in \mathbb{Z}^d$.*

PROOF. Suppose that Γ_τ is finite for every non-trivial, continuous, irreducible, unitary representation τ of X on a complex Hilbert space. As in the proof of Lemma 1.2 we set $\mathcal{H} = \{f \in L^2(X, \lambda_X) : \int f \, d\lambda_X = 0\}$ and denote by ρ the right regular representation of X on $\mathcal{H}$. If τ is the restriction of ρ to a finite-dimensional, ρ-invariant subspace $\mathcal{K}$ of $\mathcal{H}$ we put $\mathcal{K}_\gamma = \{f \cdot \alpha_\gamma : f \in \mathcal{K}\}$ for every $\gamma \in \Gamma$ and conclude as in the proof of Lemma 1.2 that the set $\{\gamma \in \Gamma : \mathcal{K}_\gamma$ is not orthogonal to $\mathcal{K}\}$ is finite. Now assume that f, g are unit vectors in $\mathcal{H}$. For every $\varepsilon > 0$ there exists a finite-dimensional, ρ-invariant subspace $\mathcal{K} \subset \mathcal{H}$ such that the projections $f' = P_{\mathcal{K}} f$, $g' = P_{\mathcal{K}} g$ of f and g onto $\mathcal{K}$ have norms $\geq 1 - \varepsilon$. Then $|\langle f \cdot \alpha_\gamma, g \rangle - \langle f' \cdot \alpha_\gamma, g' \rangle| \leq 2\varepsilon$, and $\langle f' \cdot \alpha_\gamma, g' \rangle = 0$ for all but finitely many $\gamma \in \Gamma$. This proves that α is mixing, i.e. that (1.d)$\Rightarrow$(1.a).

The implication (1.a)$\Rightarrow$(1.b) is trivial, and if we replace Γ by an infinite subgroup $\Delta \subset \Gamma$ in the proof of the implication (1.d)$\Rightarrow$(1.a) we obtain that the restriction of α to Δ is mixing and hence ergodic, i.e. that (1.d)$\Rightarrow$(1.b)$\Rightarrow$(1.c).

Finally, if there exists a non-trivial, continuous, irreducible, unitary representation τ of X on a complex Hilbert space $\mathcal{K}$ such that Γ_τ is infinite, then Lemma 1.2 implies that the restriction of α to the infinite group Γ_τ is non-ergodic. Hence (1.c)$\Rightarrow$(1.d), which completes the proof of (1).

In order to prove (2) we assume that α_n is ergodic for every non-zero $\mathbf{n} \in \mathbb{Z}^d$. According to Lemma 1.2 this implies that, for every non-trivial, continuous, irreducible representation τ of X, and for every non-zero $n \in \mathbb{Z}$, $\tau \cdot \alpha_{\mathbf{n}}$ is unitarily inequivalent to τ. Hence $\Gamma_\tau = \{0\}$ for every non-trivial, continuous, irreducible representation τ of X, and (1) implies that α is mixing. The remaining implications in (2) are obvious from (1). □

CONCLUDING REMARKS 1.7. (1) For further information on automorphisms we refer to [28], §26, and the basic properties of continuous, unitary representations of compact groups can be found in [29]. For the standard results on duality of locally compact, abelian groups see [28] and [87].

(2) Most of the exposition in this section is based on [45]. In [9], Berend proves Theorem 1.1 under the weaker assumptions that Γ is a semi-group of surjective affine transformations of a compact (and not necessarily metrizable) Hausdorff group X. For $\mathbb{Z}$-actions the equivalence of the conditions (1) and (2) in Lemma 1.2 is due to Halmos ([27]) in the abelian case, and to Kaplansky ([35]) in the general case.

(3) If X is abelian, condition (2) in Lemma 1.2 reduces to the assumption that there exists a non-trivial character $\chi \in \hat{X}$ with $\hat{\alpha}_\gamma(\chi) = \chi$ for all γ in a subgroup $\Gamma_\chi \subset \Gamma$ of finite index, where $\hat{\alpha}_\gamma$ is the automorphism of $\hat{X}$ dual to α_γ for every $\gamma \in \Gamma$.

2. Expansiveness and Lie subshifts

Let G be a compact group, S a countable set, and let G^S be the group of all maps $x \colon S \longmapsto G$ with point-wise multiplication as group operation. The group G^S is compact in the product topology. For every $x \in G^S$ we write x_s for the value of x at a point $s \in S$, and we write x as $x = (x_s) = (x_s, s \in S)$. If $F \subset S$ we denote by $\pi_F \colon G^S \longmapsto G^F$ the continuous, surjective group homomorphism obtained by restricting the maps $x \colon S \longmapsto G$ to F.

DEFINITION 2.1. A subgroup $Y \subset G^S$ is *full* if it is closed, and if $\pi_{\{s\}}(Y)$ $= G$ for every $s \in S$.

Let Γ be a countable group, and let G be a compact group. We define the *shift-action* σ of Γ on G^Γ by

$$(\sigma_\gamma(x))_{\gamma'} = x_{\gamma'\gamma} \tag{2.1}$$

for all $x = (x_\gamma) \in G^\Gamma$ and γ, γ' in Γ. A closed, σ-invariant (or *shift-invariant*) subgroup Y of G^Γ is *ergodic*, *mixing*, or *expansive*, if the Γ-action $\sigma = \sigma^Y$ on Y is ergodic, mixing, or expansive. If α is an action of Γ by automorphisms of a compact group X we say that (X, α) is *conjugate to a subshift of* G^Γ if there exists a continuous, injective homomorphism $\phi \colon X \longmapsto G^\Gamma$ such that

$$\phi \cdot \alpha_\gamma = \sigma_\gamma \cdot \phi \tag{2.2}$$

for every $\gamma \in \Gamma$, and (X, α) is *conjugate to a full subshift of* G^Γ if $\phi(X)$ is a full subgroup of G^Γ. If (X, α) is conjugate to a (full) subshift of G^Γ, where G is a compact Lie group, we say that (X, α) is *conjugate to a (full) Lie subshift*.

PROPOSITION 2.2. *Let α be an expansive action of a countable group Γ by automorphisms of a compact group X. Then (X, α) is conjugate to a Lie subshift.*

PROOF. Since α is expansive there exists a neighbourhood N of $\mathbf{1}_X$ with $\bigcap_{\gamma \in \Gamma} \gamma(N) = \mathbf{1}_X$. The compactness of X implies the existence of a finite-dimensional, continuous, unitary representation τ of X and of an $\varepsilon > 0$ such that $N' = \{x \in X : \|\tau(x) - \tau(\mathbf{1}_X)\| < \varepsilon\} \subset N$, where $\|\cdot\|$ denotes the operator norm. We put $G = \tau(X)$, observe that G is a compact Lie group, and define a continuous homomorphism $\boldsymbol{\tau} \colon X \longmapsto G^\Gamma$ by setting

$$(\boldsymbol{\tau}(x))_\gamma = \tau \cdot \alpha_\gamma(x)$$

for every $x \in X, \gamma \in \Gamma$. Our assumption on N' implies that ϕ is injective, and it is clear that ϕ satisfies (2.2), and that $\phi(X)$ is full. □

COROLLARY 2.3. *Let Γ be a countable group, and let α be a Γ-action by automorphisms of a compact, zero-dimensional group X. Then α is expansive if and only if (X, α) is conjugate to a Lie subshift. Furthermore, if α is expansive, and if (X, α) is conjugate to a full subshift of G^Γ for some compact Lie group G, then G is finite. Finally, if G is a finite group and $Y \subset G^\Gamma$ a closed, shift-invariant subgroup, then the shift-action σ^Y of Γ on Y is expansive.*

PROOF. If α is expansive, then Proposition 2.2 implies the existence of a compact Lie group G such that (X, α) is conjugate to a full, shift-invariant subgroup $Y \subset G^\Gamma$. Since $Y \subset G^\Gamma$ is full, the map $\pi_{\{1_\Gamma\}} \colon Y \longmapsto G$ is a continuous, surjective homomorphism from the zero-dimensional group Y to the Lie group G, which implies that G must be finite.

Finally, if G is a finite group and $Y \subset G^\Gamma$ a closed, shift-invariant subgroup, then the set $N = \{y = (y_\gamma) \in Y : y_{1_\Gamma} = \mathbf{1}_g\}$ is an expansive neighbourhood of the identity in Y. This shows that σ^Y is expansive. □

Corollary 2.3 provides a rich source of expansive Γ-actions on compact, totally disconnected groups. In contrast, if X is a compact, *connected* group, then the existence of an expansive Γ-action by automorphisms of X forces X to be abelian.

THEOREM 2.4. *Let α be an expansive action of a countable group Γ by automorphisms of a compact, connected group X. Then X is abelian.*

Before proving Theorem 2.4 we state a corollary.

COROLLARY 2.5. *Let α be an expansive action of a countable group Γ by automorphisms of a compact group X. Then X° is abelian. Furthermore there exists a compact Lie group G, whose connected component of the identity G° is abelian, such that (X, α) is conjugate to a full subshift of G^Γ.*

PROOF OF COROLLARY 2.5. Since X° is a compact, connected group and α^{X° is expansive, Theorem 2.4 implies that X° is abelian. By Proposition 2.2 there exists a compact Lie group G such that (X, α) is conjugate to a full subshift Y of G^Γ. Then Y° is abelian, and $A = \pi_{\{1_\Gamma\}}(Y^\circ)$ is a connected, abelian subgroup of G. As Y° is normal in Y and Y is full in G^Γ we also know that A is a normal subgroup of G. Let $\theta\colon G \longmapsto G/A$ be the quotient map and let $\boldsymbol{\theta}\colon G^\Gamma \longmapsto (G/A)^\Gamma$ be defined by

$$(\boldsymbol{\theta}(x))_\gamma = \theta(x_\gamma), \quad x \in G^\Gamma, \, \gamma \in \Gamma. \tag{2.3}$$

Then $\ker(\boldsymbol{\theta}) \supset Y^\circ$, $Z = Y \cap \ker(\boldsymbol{\theta})$ is abelian, and Y/Z is zero-dimensional. The homomorphism $\pi_{\{1_\Gamma\}}\colon Y \longmapsto G$ induces a surjective homomorphism from the zero-dimensional group Y/Z to the Lie group G/A, and it follows that G/A is zero-dimensional and hence finite. This shows that $G^\circ \subset A$, and the connectedness of A implies that $G^\circ = A$, i.e. that G° is abelian. $\square$

We begin the proof of Theorem 2.4 with an approximation argument which we shall also require for the discussion of periodic points in Chapter 4.

LEMMA 2.6. *Let H be a compact group, $K \subset H$ a normal subgroup, and let $\chi\colon K \longmapsto \mathbb{S}$ be a continuous homomorphism which is central, i.e. which satisfies that $\chi(hkh^{-1}) = \chi(k)$ for every $k \in K$, $h \in H$. Then there exists a homomorphism $\boldsymbol{\chi}\colon H \longmapsto \mathbb{S}$ and an integer $l \geq 1$ such that $\boldsymbol{\chi}(k) = \chi(k)^l$ for every $k \in K$.*

PROOF. Let $\theta\colon H \longmapsto H/K$ be the quotient map, choose a Borel map $c\colon H/K \longmapsto H$ with $\theta \cdot c(x) = x$ for every $x \in H/K$, and put $b(h, x) = c(hx)^{-1}hc(x)$ for every $h \in H$, $x \in H/K$ (cf. [78], Lemma I.5.1). We denote by τ the unitary representation

$$(\tau(h)f)(x) = \chi(b(h, h^{-1}x))f(h^{-1}x), \quad h \in H, \, f \in \mathcal{H}, \, x \in H/K$$

of H on the Hilbert space $\mathcal{H} = L^2(H/K, \lambda_{H/K})$ induced by χ. Since χ is central, $\tau(k) = \chi(k)\mathbf{I}$ for every $k \in K$, where $\mathbf{I}$ is the identity operator on $\mathcal{H}$. Choose an irreducible subrepresentation τ' of τ on a subspace $\mathcal{K} \subset \mathcal{H}$ and put $\boldsymbol{\chi}(h) = \det(\tau'(h))$ for every $h \in H$. Then $\boldsymbol{\chi}\colon H \longmapsto \mathbb{S}$ is a continuous homomorphism, and $\boldsymbol{\chi}(k) = \chi(k)^l$ for every $k \in K$, where l is the dimension of $\mathcal{K}$. $\square$

LEMMA 2.7. *Let $A = \mathbb{T}^n$, $n \geq 2$, $F \subset \mathrm{Aut}(A)$ a finite group, and let $\{0_A\} \neq B \subsetneq A$ be a closed, connected, F-invariant subgroup. Then there exists a closed, connected, F-invariant subgroup $C \subset A$ such that $B + C = A$ and $B \cap C$ is finite.*

PROOF. For every $\beta \in F$ we consider the dual automorphism $\hat{\beta} \in \mathrm{Aut}(\hat{A}) = \mathrm{GL}(n, \mathbb{Z}) \subset \mathrm{GL}(n, \mathbb{R})$, and we set $[\mathbf{u}, \mathbf{v}] = \sum_{\beta \in F} \langle \hat{\beta}(\mathbf{u}), \hat{\beta}(\mathbf{v}) \rangle$, $\mathbf{u}, \mathbf{v} \in \mathbb{R}^n$, where $\langle \cdot, \cdot \rangle$ denotes the Euclidean inner product on $\mathbb{R}^n$. The annihilator $B^\perp = \{\mathbf{n} \in \mathbb{Z}^n = \widehat{\mathbb{T}^n} : \langle b, \mathbf{n} \rangle = 1$ for every $b \in B\} \subset \mathbb{Z}^n \subset \mathbb{R}^n$ of B spans a subspace $S \subset \mathbb{R}^n$ of dimension m with $1 \leq m < n$. The subspace $S^* = \{\mathbf{v} \in \mathbb{R}^n : [\mathbf{u}, \mathbf{v}] = 0$ for all $\mathbf{u} \in S\}$ is obviously invariant under $\hat{F} = \{\hat{\beta} : \beta \in F\}$, and we can find a finite, $\hat{F}$-invariant subset $T \subset S^* \cap \mathbb{Z}^n$ which spans S^* (note that $\hat{F} \subset \mathrm{GL}(n, \mathbb{Z}) \subset \mathrm{GL}(n, \mathbb{R})$). If $\Xi \subset \mathbb{Z}^n$ is the subgroup generated by T then $B^\perp + \Xi$ has finite index in $\mathbb{Z}^n$, $B^\perp \cap \Xi = \{\mathbf{0}\}$, and the connected component C of $\Xi^\perp$ in A is an F-invariant, closed, connected subgroup of A such that $B \cap C$ is finite and $B + C = A$. $\square$

LEMMA 2.8. *Let H be a compact group, and let $A \subset H$ be a closed, normal, abelian Lie group. Then there exists a closed, normal subgroup $H' \subset H$ such that $H' \cap A$ is finite, $H' \cdot A = H$, and H/H' is abelian.*

PROOF. It obviously suffices to prove the lemma with A° replacing A, and we assume for simplicity that A itself is connected. Let $G \subset H$ be the centralizer of A, i.e. $G = \{h \in H : ha = ah$ for every $a \in A\}$. Since $\mathrm{Aut}(A)$ is discrete and X is compact, the quotient group $K = H/G$ is finite, with cardinality n, say. As we are now assuming that A is connected, $A \cong \mathbb{T}^r$ for some $r \geq 0$. If $r = 0$ the lemma is trivial; if $r \geq 1$, we choose characters $\chi_1, \ldots, \chi_r$ in $\hat{A}$ such that the map $a \mapsto \chi(a) = (\chi_1(a), \ldots, \chi_r(a))$ from A into $\mathbb{S}^r$ is bijective. Lemma 2.6 implies the existence of an integer $l \geq 1$ and of continuous homomorphisms $\boldsymbol{\chi}_i \colon G \longrightarrow \mathbb{S}$ such that $\boldsymbol{\chi}_i(a) = \chi_i(a)^l$ for every $a \in A$ and $i = 1, \ldots, r$, and we set $\boldsymbol{\chi} = (\boldsymbol{\chi}_1, \ldots, \boldsymbol{\chi}_r) \colon G \longmapsto \mathbb{S}^r$.

Denote by $\phi \colon H \longmapsto K$ the quotient map, choose a map $c \colon K \longmapsto H$ with $\phi \cdot c(k) = k$ for every $k \in K$, and set $b(h, k) = c(hk)^{-1} hc(k)$, $h \in H$, $k \in K$. Let τ be the unitary representation of H induced by the representation $\boldsymbol{\chi}$ of G on $\mathbb{C}^r$. Then τ acts on the Hilbert space $\mathcal{H} = \ell^2(K)^r$ and can be written as $\tau = \tau_1 \oplus \cdots \oplus \tau_r$, where

$$(\tau_i(h)f)(k) = \boldsymbol{\chi}_i(b(h, h^{-1}k))f(h^{-1}k) \tag{2.4}$$

for every $h \in H$, $f \in \ell^2(K)$, $k \in K$.

For every $k \in K$ we write e_k for the unit vector in $\ell^2(K)$ given by $e_k(k') = 1$ if $k = k'$, and $e_k(k') = 0$ otherwise. Then $\{e_k : k \in K\}$ is an orthonormal basis of $\ell^2(K)$, and we identify each $\tau_i(h)$ with its representation as an $n \times n$-matrix in this basis. This allows us to view τ as a continuous homomorphism $\tau = \tau_1 \oplus \cdots \oplus \tau_r \colon H \longmapsto \mathrm{U}(n)^r \subset \mathrm{U}(nr)$, where $\mathrm{U}(m)$ is the group of unitary

$m \times m$-matrices. The restriction of τ to A has finite kernel $B = \ker(\tau) \cap A$, and $\tau(a) \in D(nr)$ for every $a \in A$, where $D(m) \subset U(m)$ denotes the subgroup of diagonal matrices. From (2.4) it is clear that there exist, for every $h \in H$, unique matrices P_h and $D_h^{(i)}$, $i = 1, \ldots, r$, in $U(n)$ such that P_h is a permutation matrix, $D_h^{(i)}$ is diagonal, and $\tau_i(h) = D_h^{(i)} \cdot P_h$ for $i = 1, \ldots, d$. We set $D_h = D_h^{(1)} \oplus \cdots \oplus D_h^{(r)} \subset D(nr)$, $Q_h = P_h \oplus \cdots \oplus P_h$, and observe that $\tau(h) = D_h \cdot Q_h$, and that the map $h \mapsto Q_h$ is a continuous homomorphism from H into the group of permutation matrices in $U(nr)$. For every $h \in H$ we denote by β_h the automorphism of $D(nr)$ given by $\beta_h(D) = \tau(h)D\tau(h)^{-1} = Q_h D Q_h^{-1}$. Since the group $F = \{\beta_h : h \in H\} \subset \text{Aut}(D(nr))$ is finite and $\mathbb{T}^r \cong \tau(A) = \{\tau(a) : a \in A\} \subset D(nr) \cong \mathbb{T}^{nr}$ is a closed, connected, F-invariant subgroup, Lemma 2.7 implies the existence of a closed, connected, F-invariant subgroup $C \subset D(nr)$ such that $C \cdot \tau(A) = D(nr)$ and $C \cap \tau(A)$ is finite.

Put $Q = \{Q_h : h \in H\}$ and $\Lambda = \tau(H) \cdot D(nr) = Q \cdot D(nr) \subset U(nr)$. Then C and $Q \cdot C$ are normal subgroups of Λ, and we write $\xi : \Lambda \longmapsto \Lambda/Q \cdot C$ for the quotient map. Since $Q \cdot C \cdot \tau(A) = \Lambda$ we can find, for every $h \in H$, an $a \in A$ with $ha \in H' = \ker(\xi \cdot \tau) \subset H$, and we conclude that $H' \cdot A = H$, and that H/H' is therefore abelian. Finally, since $H' \cap A \subset C \cap A$, $H' \cap A$ is finite. $\square$

LEMMA 2.9. *Let H be a compact Lie group. Then there exists an increasing sequence of closed, normal subgroups $H_n \subset H$ such that $\bigcup_{n \geq 1} H_n$ is dense in H, $C(H_n^\circ)$ is finite, H/H_n is abelian, and $H_n \supset \{a \in C(H) : a^n = 1_H\}$ for every $n \geq 1$.*

PROOF. Apply Lemma 2.8 with $A = C(H^\circ)$ to find a subgroup K such that $K \cdot C(H^\circ) = H$ and $C(K^\circ) = K \cap C(H^\circ)$ is finite, and put $H_n = \{h \in H : h^{n!} \in K\}$ for every $n \geq 1$. $\square$

LEMMA 2.10. *Let A be a compact, abelian Lie group, X a closed, shift-invariant subgroup of A^Γ, and let $X^{(m)} = \{x \in X : x^m = 1_X\}$, $m \geq 1$. Then $\bigcup_{m \geq 1} X^{(m)}$ is dense in X.*

PROOF. Since A is isomorphic to a closed subgroup of $\mathbb{T}^n$ for some $n \geq 1$, the dual group $\hat{X}$ is isomorphic to a quotient group of the direct sum of copies of $\mathbb{Z}^n$, indexed by Γ, and we can find a decreasing sequence $(V_m, m \geq 1)$ of subgroups of $\hat{X}$ such that $\hat{X}/V_m$ is finite for every $m \geq 1$ and $\bigcap_{m \geq 1} V_m = \{0\}$. Put $B_m = V_m^\perp \subset X$. Then B_m is finite, and $\bigcup_{m \geq 1} B_m$ is dense in X. For every $m \geq 1$ we can find an $m' \geq 1$ such that $B_m \subset X^{(m')}$, and this proves that $\bigcup_{m \geq 1} X^{(m)}$ is dense in X. $\square$

LEMMA 2.11. *Let H be a compact Lie group, and let $X \subset H^\Gamma$ be a full, shift-invariant subgroup. Then there exists an increasing sequence $(H_n, n \geq 1)$ of closed, normal subgroups of H such that $C(H_n^\circ)$ is finite for every $n \geq 1$, and $\bigcup_{n \geq 1} X_n$ is dense in X, where $X_n = X \cap H_n^\Gamma$.*

PROOF. We choose $(H_n, n \geq 1)$ as in Lemma 2.9, denote by $\psi_n \colon H \longmapsto H/H_n$ the quotient map, and define a shift commuting homomorphism $\boldsymbol{\psi}_n \colon H^\Gamma \longmapsto (H/H_n)^\Gamma$ by $(\boldsymbol{\psi}_n(x))_\gamma = \psi_n(x_\gamma)$, for all $x \in X$ and $\gamma \in \Gamma$. Then $X_n = \ker(\boldsymbol{\psi}_n)$, and ψ_1 embeds X/X_1 injectively in $(H/H_1)^\Gamma$. An application of Lemma 2.10 shows that $\bigcup_{n \geq 1} X_n$ is dense in X. $\square$

LEMMA 2.12. *Let G be a compact, connected Lie group with finite centre, $\theta \colon G \longmapsto G' = G/C(G)$ the quotient map, and define the shift commuting, surjective homomorphism $\boldsymbol{\theta} \colon G^\Gamma \longmapsto G'^\Gamma$ as in (2.3). If $X \subset G^\Gamma$ is a closed, shift-invariant, expansive subgroup, then $\boldsymbol{\theta}(X) \subset G'^\Gamma$ is again expansive.*

PROOF. Let δ be a metric on G and denote by δ' the induced metric on G'. Since X is expansive there exists an $\varepsilon_1 > 0$ such that $x = 1_X = 1_{G^\Gamma}$ whenever $x \in X$ and $\delta(x_\gamma, 1_G) < \varepsilon_1$ for every $\gamma \in \Gamma$. Put $\varepsilon_2 = \min\{\delta(g, 1_G) : 1_G \neq g \in C(G)\}$, let M denote the order of $C(G)$, and choose $\varepsilon_3 > 0$ such that $\{g \in G : g^M = 1_G$ and $\delta(g, 1_G) < \varepsilon_3\} = \{1_G\}$. For every $\varepsilon > 0$ we put

$$X(\varepsilon) = \{x \in X : \delta(x_\gamma, C(G)) < \varepsilon \text{ for every } \gamma \in \Gamma\}.$$

If $\varepsilon < \varepsilon_2/2$ there exists, for every $x \in X(\varepsilon)$, a unique $z(x) \in C(G)^\Gamma$ such that $\delta(x_\gamma, z(x)_\gamma) < \varepsilon$ for every $\gamma \in \Gamma$. If $\varepsilon < \varepsilon_2/4$ then $z(xx') = z(x)z(x')$ for all $x, x' \in X(\varepsilon)$, and if $\varepsilon < \varepsilon_2/2M$ then $z(x^M) = z(x)^M = 1_{G^\Gamma}$ for all $x \in X(\varepsilon)$. We fix $\varepsilon > 0$ with $\varepsilon < \min\{\varepsilon_1/M, \varepsilon_2/2M, \varepsilon_3\}$. For every $x \in X(\varepsilon)$ we have that $z(x^M) = z(x)^M = 1_X$ (since $\varepsilon < \varepsilon_2/2M$), and $\delta(x_\gamma^M, 1_G) < M\varepsilon < \varepsilon_1$ for every $\gamma \in \Gamma$. Our choice of ε_1 implies that $x^M = 1_X$, and since

$$\delta(x_\gamma, C(G)) = \delta(x_\gamma, z(x)_\gamma) = \delta((xz(x)^{-1})_\gamma, 1_G) < \varepsilon < \varepsilon_3$$

and

$$(xz(x)^{-1})_\gamma^M = x_\gamma^M = 1_G$$

for all $\gamma \in \Gamma$, we see that $xz(x)^{-1} = 1_X$ and $x \in C(G)^\Gamma$. This proves that $\boldsymbol{\theta}(X(\varepsilon)) = \{1_{\boldsymbol{\theta}(X)}\}$. We set $B(\varepsilon) = \{g \in G' : \delta'(g, 1) < \varepsilon\}$ and obtain that $\boldsymbol{\theta}(X) \cap B(\varepsilon)^\Gamma = \{1_{\boldsymbol{\theta}(X)}\}$, i.e. that $\boldsymbol{\theta}(X)$ is an expansive subgroup of G'^Γ. $\square$

PROPOSITION 2.13. *Let X be a compact, connected group, and let $Y \subset X$ be a closed, normal subgroup with $Y \supset C(X)$. Then the following is true.*

(1) $C(Y) = C(X)$;
(2) $Y/C(X)$ *is a connected subgroup of* $X/C(X)$;
(3) *If* $Y' = \{x \in X : xy = yx \text{ for all } y \in Y\}$, *then* $Y \cap Y' = C(X)$, *and*

$$X/C(X) = (Y/C(X)) \cdot (Y'/C(X)) \cong (Y/C(X)) \times (Y'/C(X));$$

(4) *If* $F \subset X$ *is a finite, normal subgroup, then* $F \subset C(X)$;
(5) $X/C(X)$ *has trivial centre.*

PROOF. We begin by proving (1). It is clear that $C(Y) \supset C(X)$. In order to prove the reverse inclusion we define, for every $x \in X$, an automorphism γ_x^Y of Y by $\gamma_x^Y(y) = xyx^{-1}$ for all $y \in Y$, and we write $\gamma_x^{C(Y)}$ for the restriction of γ_x^Y to $C(Y)$. The homomorphisms $x \mapsto \gamma_x^Y$ and $x \mapsto \gamma_x^{C(Y)}$ from X into $\mathrm{Aut}(Y)$ and $\mathrm{Aut}(C(Y))$ are both continuous. Since $C(Y)$ is abelian, the group $\mathrm{Aut}(C(Y))$ is zero-dimensional (cf. Section 1), and the connectedness of X implies that $\gamma_x^{C(Y)}$ is trivial for every $x \in X$. Hence $xyx^{-1} = y$ for all $x \in X$ and $y \in C(Y)$, so that $C(Y) \subset C(X)$ and hence $C(Y) = C(X)$.

In order to prove (3) we recall that the group $\mathrm{Out}(Y) = \mathrm{Aut}(Y)/\mathrm{Inn}(Y)$ is zero-dimensional (cf. Section 1). Since X is connected, $\gamma_x^Y \in \mathrm{Inn}(Y)$ for every $x \in X$, and the map $x \mapsto \gamma_x^Y$ induces a continuous, surjective homomorphism $\theta \colon X \longrightarrow Y/C(X) \cong \mathrm{Inn}(Y)$ with kernel $Y' \supset C(X)$. If $\theta' \colon X/C(X) \longrightarrow Y/C(X)$ is the homomorphism induced by θ, then every $x' = xC(X) \in X/C(X)$ can be written as $x' = a' \cdot b'$ with $a' = x'\theta(x)^{-1} \in Y'/C(X)$ and $b' = \theta(x) \in Y/C(X)$; in particular we can find an $a \in Y'$ with $aY = a'Y = xY$. As $x' \in X/C(X)$ is arbitrary this shows that $X/C(X) = (Y/C(X)) \cdot (Y'/C(X))$, and since $Y \cap Y' = C(X)$ we conclude that $X/C(X) \cong (Y/C(X)) \times (Y'/C(X))$, as claimed in (3). The assertion (2) follows from the fact that $Y/C(X) \cong (X/C(X))/(Y'/C(X))$.

Next we turn to (4). If $F \subset X$ a finite, normal subgroup, then we define automorphisms $\gamma_x^F \in \mathrm{Aut}(F)$ as in the proof of (1) (with F replacing Y), and note that $x \mapsto \gamma_x^F$ is a continuous homomorphism from the connected group X into the finite group $\mathrm{Aut}(F)$. Hence γ_x^F to be trivial for every $x \in X$, so that $F \subset C(X)$.

In order to prove (5) we claim that, if G is a compact, connected Lie group, then $G/C(G)$ has trivial centre. According to Theorem XIII.1.3 in [30], $G \cong (A \times B)/D$, where A is a finite-dimensional torus, B is a semi-simple group (which must have finite centre), and D is a finite central subgroup of $A \times B$ such that $A \cap D$ and $B \cap D$ are both trivial. If C' is the centre of the compact, semi-simple Lie group $G/C(G) \cong B/C(B)$, then C' must be finite, so that it is the image of a finite, normal subgroup $F \subset B$ under the quotient map $B \longrightarrow B/C(B)$. As we have seen above, $F \subset C(B) \subset C(G)$, and hence $C' = F/C(G) = \{1\}$. Note that the triviality of the centre of $G/C(G)$ is equivalent to the statement that every $a \in G$ with $a^{-1}gag^{-1} \in C(G)$ for all $g \in G$ lies in $C(G)$.

Now assume that X is an arbitrary compact, connected group, and choose a decreasing sequence $(Y_n, n \geq 1)$ of closed, normal subgroups of X such that $\bigcap_{n \geq 1} Y_n = \{1\}$ and $G_n = X/Y_n$ is a Lie group for every $n \geq 1$. If $X/C(X)$ has non-trivial centre $C(X/C(X))$, choose an element $a \in X \smallsetminus C(X)$ with $aC(X) \in C(X/C(X))$. Then $a^{-1}xax^{-1} \in C(X)$ for every $x \in X$, and the result proved in the preceding paragraph shows that $aY_n \in C(X/Y_n)$ for every $n \geq 1$. Hence $a^{-1}xa \in Y_n$ for every $x \in X$ and $n \geq 1$, which implies that

$a \in C(X)$. This contradiction to our assumption shows that $C(X/C(X))$ is trivial, as claimed in (4). $\square$

COROLLARY 2.14. *Let G be a compact, connected Lie group with trivial centre, and let $X \subset G^\Gamma$ be a full, connected, shift-invariant subgroup. Then there exists a continuous homomorphism $\phi: G \longmapsto X$ such that $\phi(g)_{1_\Gamma} = g$ for every $g \in G$.*

PROOF. Since $X \subset G^\Gamma$ is full, $C(X)$ is trivial, and we set $V = \ker(\pi_{1_\Gamma})$ and observe that V is a closed, normal subgroup of X. Define V' as in Proposition 2.13, note that the restriction π' of $\pi_{\{1_\Gamma\}}: X \longmapsto G$ to V' is an isomorphism of V' and G, and set $\phi = \pi'^{-1}: G \longmapsto V' \subset X$. $\square$

PROOF OF THEOREM 2.4. Proposition 2.2 allows us to assume that X is a full, shift-invariant subgroup of H^Γ, where H is a compact, connected Lie group. Lemma 2.11 yields an increasing sequence $(H_n, n \geq 1)$ of closed, normal subgroups of H such that each H_n has finite centre and the union of the groups $X_n = X \cap H_n^\Gamma$ is dense in X. We fix $n \geq 1$ and put $G = \pi_{1_\Gamma}(X_n^\circ)$. Since X is full and X_n° is normal in X, G is a connected, normal subgroup of H and hence of H_n, the centre of G is finite, and X_n is a full, expansive subgroup of G^Γ. Lemma 2.12 shows that $\boldsymbol{\theta}(X_n) \subset G'^\Gamma$ is expansive, where $\boldsymbol{\theta}: G^\Gamma \longmapsto G'^\Gamma$ is obtained from the quotient map $\theta: G \longmapsto G' = G/C(G)$ via (2.3). If $G' \neq \{1\}$, let $\phi: G' \longmapsto \boldsymbol{\theta}(X_n)$ be the homomorphism constructed in Corollary 2.14. For every $\gamma \in \Gamma$, $\psi_\gamma = \pi_{\{1_\Gamma\}} \cdot \alpha_\gamma \cdot \phi: G' \longmapsto G'$ is a continuous homomorphism from G' into G'. Since G' is a compact, connected Lie group with trivial centre, the group $\text{Aut}(G')$ is compact, and there exists a metric δ on G' which is invariant under every $\beta \in \text{Aut}(G')$. An elementary argument shows that $\delta(\eta(g), \eta(h)) \leq \delta(g, h)$ for every continuous homomorphism $\eta: G' \longmapsto G'$, and in particular $\delta(\psi_\gamma(g), 1_{G'}) \leq \delta(g, 1_{G'})$ for every $\gamma \in \Gamma$, $g \in G'$. This shows that, for every $\varepsilon > 0$, and for every $g \in G'$ with $\delta(g, 1_{G'}) < \varepsilon$, we have that $\delta(\phi(g)_\gamma, 1_{G'}) < \varepsilon$ for all $\gamma \in \Gamma$. In view of the expansiveness of $\boldsymbol{\theta}(X_n)$ this is impossible, and we obtain that $G' = \{1\}$, $\boldsymbol{\theta}(X_n) = \{1\}$, and that X_n is abelian. Since $\bigcup_{n \geq 1} X_n$ is dense in X, X must be abelian, and the theorem is proved. $\square$

Proposition 2.2 shows that every expansive action α of a countable group Γ by automorphisms of a compact group is conjugate to a Lie subshift. The converse is obviously not true: if G is an infinite, compact Lie group, then the shift-action σ of Γ on G^Γ is not expansive. The next proposition shows that conjugacy to a Lie subshift amounts to a *finiteness condition* on the pair (X, α).

PROPOSITION 2.15. *Let α be an action of a countable group Γ by automorphisms of a compact group X. Then (X, α) is conjugate to a Lie subshift if and only if there exist finitely many continuous, irreducible, unitary representations $\tau_1, \dots, \tau_n$ of X such that the family of representations $\{\tau_i \cdot \alpha_\gamma : \gamma \in \Gamma, 1 \leq i \leq n\}$ separates the points of X.*

PROOF. If (X, α) is conjugate to a Lie subshift we assume for simplicity that $X \subset G^\Gamma$, where G is a compact Lie group, and that α is equal to the shift-action σ. Since G possesses finitely many irreducible, unitary representations $\rho_1, \ldots, \rho_n$ which together separate points (or, equivalently, since G has a finite-dimensional, faithful, unitary representation), the representations $\tau_i = \rho_i \cdot \pi_{\{1_\Gamma\}}$ together separate the points of X.

Conversely, if there exist irreducible representations $\tau_1, \ldots, \tau_n$ of X such that $\{\tau_i \cdot \alpha_\gamma : \gamma \in \Gamma, 1 \leq i \leq n\}$ separates the points of X, put $\tau = \tau_1 \oplus \cdots \oplus \tau_n$ and denote by $\mathcal{H}$ the Hilbert space on which τ acts. Then $G = \tau(X)$ is a closed subgroup of the group $\mathcal{U}(\mathcal{H})$ of unitary operators on $\mathcal{H}$ and hence a compact Lie group. The homomorphism $\boldsymbol{\tau} \colon X \longmapsto G^\Gamma$, defined by $(\boldsymbol{\tau}(x))_\gamma = \tau \cdot \alpha_\gamma(x)$, $\gamma \in \Gamma, x \in X$, is injective, embeds X as a full, shift-invariant subgroup of G^Γ, and satisfies that $\boldsymbol{\tau} \cdot \alpha_\gamma = \sigma_\gamma \cdot \boldsymbol{\tau}$ for every $\gamma \in \Gamma$. □

COROLLARY 2.16. *Let α be an action of a countable group Γ by automorphisms of a compact, abelian group X. Then (X, α) is conjugate to a Lie subshift if and only if there exist characters $\chi_1, \ldots, \chi_n$ in $\hat{X}$ such that $\hat{X}$ is generated by $\{\chi_j \cdot \alpha_\gamma : \gamma \in \Gamma, 1 \leq j \leq n\}$.*

PROOF. Proposition 2.15 and the Stone-Weierstrass theorem. □

PROPOSITION 2.17. *Let Γ be a countable group, and let α be a Γ-action by automorphisms of a compact group X. Then there exists a non-increasing sequence $(V_n, n \geq 1)$ of closed, normal, α-invariant subgroups of X such that $\bigcap_{n \geq 1} V_n = \{1_X\}$ and $(X/V_n, \alpha^{X/V_n})$ is conjugate to a Lie subshift for every $n \geq 1$.*

PROOF. Choose a sequence of irreducible, unitary representations $(\rho_n, n \geq 1)$ of X which together separate the points of X. For every $n \geq 1$ we put $\tau_n = \rho_1 \oplus \cdots \oplus \rho_n$, denote by G_n the compact Lie group $\tau_n(X)$ (which is a subgroup of the unitary group of some finite-dimensional Hilbert space), and define a continuous group homomorphism $\boldsymbol{\tau}_n \colon X \longmapsto G_n^\Gamma$ by $(\boldsymbol{\tau}_n(x))_\gamma = \tau_n \cdot \alpha_\gamma(x)$, $\gamma \in \Gamma, x \in X$. Then $\boldsymbol{\tau}_n \cdot \alpha_\gamma = \sigma_\gamma \cdot \boldsymbol{\tau}_n$ for every $\gamma \in \Gamma$, $V_n = \ker(\boldsymbol{\tau}_n)$ is a closed, normal, α-invariant subgroup of X, $V_1 \supset \cdots \supset V_n \supset \ldots$, and $\bigcap_{n \geq 1} V_n = \{1_X\}$. □

EXAMPLES 2.18. (1) Let α be an action of a countable group Γ by automorphisms of a compact Lie group G. Then α is obviously conjugate to the Lie subshift $Y \subset G^\Gamma$ given by $Y = \{y = (y_\gamma) \in G^\Gamma : y_\gamma = \alpha_\gamma(y_{1_\Gamma})$ for all $\gamma \in \Gamma\}$.

(2) Let α be the automorphism of $\mathbb{T}^2$ defined by the matrix $\left(\begin{smallmatrix} 0 & 1 \\ 1 & 1 \end{smallmatrix}\right)$, and let $Y = \{x = (x_n) \in \mathbb{T}^{\mathbb{Z}} : x_n + x_{n+1} - x_{n+2} = 0 \pmod 1$ for all $n \in \mathbb{Z}\}$. Then $Y \subset \mathbb{T}^{\mathbb{Z}}$ is a closed, shift-invariant subgroup, and the coordinate map $\pi_{\{0,1\}} \colon Y \longmapsto \mathbb{T}^2$ is bijective and satisfies that $\pi_{\{0,1\}} \cdot \sigma = \alpha \cdot \pi_{\{0,1\}}$. Note that α is also conjugate to the shift on a closed, shift-invariant subgroup $Y' \subset (\mathbb{T}^2)^{\mathbb{Z}}$, as described in Example (1).

(3) Let $p \geq 2$ be a rational prime, and let $\mathbb{Z}_p$ denote the compact ring of p-adic integers, i.e. the ring of all formal power series $x = \sum_{n \geq 0} x_n p^n$ with $x_n \in \{0, 1, \ldots, p-1\}$ for all $n \geq 0$, furnished with the obvious operations of addition and multiplication, and with the compact topology which makes the bijection $x = \sum_{n \geq 0} x_n p^n \mapsto (x_0, x_1, \ldots)$ of $\mathbb{Z}_p$ and $\mathbb{Z}_{/p}^{\mathbb{N}}$ a homeomorphism (we set $\mathbb{Z}_{/n} = \mathbb{Z}/n\mathbb{Z}$ for every $n \geq 2$). The additive semi-group $\mathbb{N}$ is embedded in X as the set of all power series with only finitely many non-zero terms. Here we consider the case $p = 2$, regard $X = \mathbb{Z}_2$ as an additive group, and define an automorphism α of X as multiplication by 3. The dual group $Y = \hat{X} = \widehat{\mathbb{Z}_2}$ of X is (isomorphic to) the group $\{m2^{-n} : n \geq 1, 0 \leq m < 2^n\}$ under addition modulo 1, and $\hat{\alpha}$ consists of multiplication by 3 (mod 1). For every $n \geq 1$, the subgroup $Y_n = \{m2^{-n} : 0 \leq m < 2^n\} \subset Y$ (with addition modulo 1) is invariant under $\hat{\alpha}$, and this is easily seen to imply that $Y = \hat{X}$ is not finitely generated under $\hat{\alpha}$ in the sense of Corollary 2.16. Hence α is not conjugate to a Lie subshift; since X is zero-dimensional, this also implies that α is not expansive (Corollary 2.3).

For every $n \geq 1$ we set $V_n = Y_n^{\perp}$ and note that V_n is α-invariant, X/V_n is finite, and $V_n \supset V_{n+1}$. As in Example (1) we observe that $(X/V_n, \alpha^{X/V_n})$ is (trivially) conjugate to a Lie subshift, and it is clear that $\bigcap_{n \geq 1} V_n = \{1_X\}$ (cf. Proposition 2.17).

(4) Let Γ be the multiplicative group $\mathbb{Q}^{\times} = \mathbb{Q} \smallsetminus \{0\}$, $X = \hat{\mathbb{Q}}$, and let $\alpha \colon r \mapsto \alpha_r = \hat{\beta}_r$ be the Γ-action on X dual to the action β of Γ on $\hat{X} = \mathbb{Q}$ defined by $\beta_r(s) = rs$ for every $r \in \Gamma, s \in \mathbb{Q}$. The group $\Gamma = \hat{X}$ consists of two β-orbits, $\{0\}$ and $\{\beta_r(1) : r \in \Gamma\} = \Gamma$, so that (X, α) is conjugate to a Lie subshift by Corollary 2.16. In order to make this conjugacy explicit we set

$$Y = \{y = (y_s, s \in \Gamma) \in \mathbb{T}^{\Gamma} : ky_s = ly_{rs} \pmod 1$$
$$\text{for every } r = \tfrac{k}{l}, s \in \Gamma\} \tag{2.5}$$

and denote by σ the shift-action (2.1) of Γ on the shift-invariant subgroup $Y \subset \mathbb{T}^{\Gamma}$. If $\chi \colon X \longmapsto \mathbb{T} \cong \mathbb{S}$ is the surjective homomorphism corresponding to the character $1 \in \mathbb{Q} = \hat{X}$, and if $\boldsymbol{\chi} \colon X \longmapsto \mathbb{T}^{\Gamma}$ is the homomorphism defined by $(\boldsymbol{\chi}(x))_r = \chi \cdot \alpha_r(x)$, $r \in \Gamma$, then $\boldsymbol{\chi}(X) = Y$, and $\boldsymbol{\chi} \cdot \alpha_r = \sigma_r \cdot \boldsymbol{\chi}$ for every $r \in \Gamma$.

More generally, if $\mathbb{K}$ is a countably infinite field, and if $\Gamma = \mathbb{K}^{\times} = \mathbb{K} \smallsetminus \{0\}$, then we obtain an ergodic Γ-action α on $\hat{\mathbb{K}}$ which is dual to the action β on $\mathbb{K}$ by multiplication, and which is conjugate to a Lie subshift. $\boxdot$

CONCLUDING REMARK 2.19. The proof of Theorem 2.4 is a modification of the argument in [50], where Lam obtains the same result under the weaker assumption that Γ is a semi-group of continuous, surjective homomorphisms of X. For $\mathbb{Z}$-actions the finiteness conditions in Proposition 2.15 and Corollary 2.16 were originally introduced by Rokhlin in [85] and have since reappeared in a number of papers (e.g. [103], [104], [55], and [45]).

3. The descending chain condition

Conjugacy to a Lie subshift is implied not only by expansiveness, as we have seen in Proposition 2.2, but also by another finiteness condition on (X, α), the descending chain condition.

DEFINITION 3.1. Let Γ be a countable group, and let α be a Γ-action by automorphisms of a compact group X. The pair (X, α) (or the action α) satisfies the *descending chain condition* (d.c.c.) if there exists, for every non-increasing sequence $X \supset X_1 \supset \cdots \supset X_k \supset \ldots$ of closed, α-invariant subgroups of X, an integer $K \geq 1$ that $X_k = X_K$ for all $k \geq K$.

If α satisfies the d.c.c., then α^V and $\alpha^{X/W}$ satisfy the d.c.c. for all closed, α-invariant subgroups $V \subset X$ and all closed, normal, α-invariant subgroups $W \subset X$.

In Section 4 we shall see that every action of a polycyclic-by-finite group (and, in particular, of $\mathbb{Z}^d$, $d \geq 1$) by automorphisms of a compact group X either satisfies the d.c.c., or is a projective limit of actions satisfying the d.c.c. For the moment we shall content ourselves with a much more basic example.

EXAMPLE 3.2. Let α be an action of a countable group Γ by automorphisms of a compact Lie group X. Then α satisfies the d.c.c., since every decreasing sequence of closed subgroups of X must eventually become constant. In order to have something more specific one can set $X = \mathbb{T}^n$ for some $n \geq 1$ and take Γ to be a subgroup of $\mathrm{GL}(n, \mathbb{Z}) = \mathrm{Aut}(\mathbb{T}^n)$. $\boxdot$

PROPOSITION 3.3. *Let Γ be a countable group, and let α be a Γ-action by automorphisms of a compact group X. If (X, α) satisfies the d.c.c. then it is conjugate to a Lie subshift.*

PROOF. Let $(V_n, n \geq 1)$ be the sequence of closed, normal, α-invariant subgroups of X defined in Proposition 2.17. Since (X, α) satisfies the d.c.c., there has to exist an $N \geq 1$ with $V_N = \{1_X\}$. $\square$

COROLLARY 3.4. *Let Γ be a countable group, and let α be a Γ-action by automorphisms of a compact, zero-dimensional group X satisfying the d.c.c. Then α is expansive.*

PROOF. Proposition 2.2 implies the existence of a compact Lie group G such that (X, α) is conjugate to a full subshift of G^Γ, and Corollary 2.3 does the rest. $\square$

Although the d.c.c. may appear unintuitive, it has a number of very useful dynamical consequences. The first of these concerns the structure of non-ergodic Γ-actions satisfying the d.c.c.

PROPOSITION 3.5. *Let Γ be a countable group, X a compact group, and let α be a Γ-action by automorphisms of X which satisfies the d.c.c. Then there exists a unique maximal, closed, normal, α-invariant subgroup $X' \subset X$ such that X/X' is a Lie group and $\alpha^{X'}$ is ergodic. In particular, if X is zero-dimensional, then X' is open.*

PROOF. The d.c.c. implies that the ordinal ω in Theorem 1.4 is finite, i.e. that there exist closed, normal, α-invariant subgroups $X' = V_n \subset \cdots \subset V_0 = X$ such that $\alpha^{X'}$ is ergodic and, for every $i = 0, \ldots, n-1$, V_i/V_{i+1} is a Lie group with an $\alpha^{V_i/V_{i+1}}$-invariant metric. Hence X/X' is a Lie group. If X is zero-dimensional, then X/X' is a zero-dimensional Lie group and hence finite. □

As a second application of the d.c.c. we consider the restriction of an ergodic action α of a countable group Γ by automorphisms of a compact group X to a closed, α-invariant subgroup $Y \subset X$. While it is clear that $\alpha^{X/Y}$ is ergodic, there is no guarantee that α^Y will be ergodic. The next result describes a situation where the ergodicity of α does imply the ergodicity of α^Y.

THEOREM 3.6. *Let α be an ergodic action of a countable group Γ by automorphisms of a compact group X which satisfies the d.c.c., and let X° be the connected component of the identity in X. Then α^{X° is ergodic.*

PROOF. Suppose that α^{X° is non-ergodic. According to Lemma 1.2 (2) there exists a non-trivial, continuous, irreducible, unitary representation τ of X° such that Γ_τ has finite index in Γ. The closed, normal, α-invariant subgroup $X_\eta \subset X$ constructed in the proof of Lemma 1.3 (with Y replaced by X°) has finite index in X, and the ergodicity of α implies that $X_\eta = X$, i.e. that $\eta^{(x)}$ is unitarily equivalent to η for every $x \in X$. Hence $W = \ker(\eta) \subset X^\circ \subset X$ is an α-invariant, normal subgroup of X, not just of X°, and X°/W is a compact, connected Lie group.

We set $X' = X/W$ and $Y = X^\circ/W$. If Y is non-abelian, let $C(Y)$ be the centre of Y, and set $X'' = X'/C(Y)$ and $Y' = Y/C(Y)$. Then Y' is a compact, connected Lie group with trivial centre which is normal in X'', and $\alpha^{Y'}$ preserves a metric δ on Y', since α^Y preserves the metric on $Y \cong \eta(X^\circ)$ arising from the operator norm. For every $x \in X$ we consider the automorphism β_x of Y' defined by $\beta_x(y) = xyx^{-1}$ for every $y \in Y'$. The map $\beta \colon x \mapsto \beta_x$ is a continuous group homomorphism from X into $\mathrm{Aut}(Y')$, and $\{x \in X : \beta_x \in \mathrm{Inn}(Y')\}$ is a closed, normal, α-invariant subgroup of X'' with finite index in X'' (since Y' has trivial centre, the group of inner automorphisms of Y' has finite index in $\mathrm{Aut}(Y')$). The ergodicity of $\alpha^{X''}$ now implies that $\beta_x \in \mathrm{Inn}(Y')$ for every $x \in X''$, and we denote by $\beta(x) \in Y'$ the unique element with $\beta_x(y) = \beta(x)y\beta(x)^{-1}$ for every $y \in Y'$. It is clear that $\alpha_\gamma^{Y'}(\beta(x)) = \beta(\alpha_\gamma^{X''}(x))$ for all $\gamma \in \Gamma$ and $x \in X''$. In particular, $\beta^{-1}(N)$ is an $\alpha^{X''}$-invariant subset of X'' for every $\alpha^{Y'}$-invariant subset N of Y'. Since Y' has a basis of α-invariant neighbourhoods of the identity we obtain a contradiction to the ergodicity of $\alpha^{X''}$ and hence to the ergodicity of α.

This shows that Y must be abelian, i.e. that $Y = \mathbb{T}^s$ for some $s \geq 1$. As before, we define a homomorphism $\beta\colon x \mapsto \beta_x$ from X' into the discrete group $\mathrm{Aut}(Y)$ by setting $\beta_x(y) = xyx^{-1}$ for all $x \in X'$ and $y \in Y$. The kernel of this homomorphism is an open, α-invariant subgroup of X' and hence—due to the ergodicity of $\alpha^{X'}$—equal to X'. In other words, $Y \subset C(X')$.

Proposition 3.3 allows us to assume that X' is a full, shift-invariant subgroup of H^Γ for some compact Lie group H, and that $\alpha_\gamma^{X'} = \sigma_\gamma$ for all $\gamma \in \Gamma$. Then H° is a homomorphic image of the connected component Y of the identity in X', and $Y \subset (H^\circ)^\Gamma$. We choose an increasing sequence $(F_n, n \geq 1)$ of finite subsets of Γ with $\bigcup_{n \geq 1} F_n = \Gamma$, set $H_n = \pi_{F_n}(X') \subset H^{F_n}$, and define a shift commuting, injective embedding $\psi_n\colon X' \longrightarrow H_n^\Gamma$ by $(\psi_n(x))_\gamma = \pi_{F_n}(\sigma_\gamma(x))$ for all $x \in X'$ and $\gamma \in \Gamma$. For every $n \geq 1$ we have that $\psi_n(Y) \subset (H_n^\circ)^\Gamma$, and $Y_n = \psi_n^{-1}((H_n^\circ)^\Gamma)$ is a decreasing sequence of closed, normal, shift-invariant subgroups of X' with $\bigcap_{n \geq 1} Y_n = Y$. Since $\sigma = \alpha^{X'}$ satisfies the d.c.c., there exists an $m \geq 1$ with $Y_m = Y$, and we assume without loss in generality that $m = 1$ and $Y = X' \cap (H^\circ)^\Gamma$. As $X' \subset H^\Gamma$ is full and Y is central we know that $\pi_{1_\Gamma}(Y) = H^\circ \subset C(H)$.

Lemma 2.8 implies the existence of a closed, normal subgroup $H' \subset H$ such that $A = H/H'$ is abelian, $H' \cap H^\circ$ is finite, and $H' \cdot H^\circ = H$. We write $\theta\colon H \longrightarrow A$ for the quotient map and define a shift commuting homomorphism $\boldsymbol{\theta}\colon X' \longrightarrow A^\Gamma$ by $(\boldsymbol{\theta}(x))_\gamma = \theta(x_\gamma)$ for all $x \in X', \gamma \in \Gamma$. Since $H'' = H' \cap H^\circ$ is finite, the group $Y' = \ker(\boldsymbol{\theta}) \cap Y = \ker(\boldsymbol{\theta}) \cap (H^\circ)^\Gamma$ is a closed, zero-dimensional, and hence finite, subgroup of Y, and $W = \boldsymbol{\theta}(Y) \cong Y/Y'$ is the connected component of the identity in the abelian group $V = \boldsymbol{\theta}(X') \subset A^\Gamma$. Furthermore, the shift-action of Γ preserves a metric on W and is therefore non-ergodic on W.

The shift-action of Γ on $V \subset A^\Gamma$ is algebraically conjugate to $\sigma^{X'/\ker(\boldsymbol{\theta})}$ and thus satisfies the d.c.c. Exactly as before, when we were dealing with X', H, and Y, we may change the group A, if necessary, and assume without loss in generality that $W = V \cap (A^\circ)^\Gamma$. As A/A° is finite, there exists an $m \geq 1$ with $v^m \in V \cap (A^\circ)^\Gamma = W$ for every $v \in V$, and the map $\zeta\colon v \mapsto v^m$ from V to W is a surjective, shift commuting, group homomorphism. Since the shift-action of Γ on W is non-ergodic we see that Γ acts non-ergodically on V. Hence the action $\alpha^{X'}$ cannot be ergodic which, in turn, violates the ergodicity of α. This contradiction shows that α^{X° must be ergodic. $\quad\square$

The appearance of the d.c.c. in Theorem 3.6 is a little unexpected and raises the question whether α^{X° is ergodic for *every* ergodic Γ-action α on X, i.e. independently of whether (X, α) satisfies the d.c.c. We postpone further discussion of this problem to Theorem 4.11 and Example 4.13 and turn instead to the third, and dynamically most interesting, aspect of the d.c.c.: its intrinsic connection with shifts of finite type.

DEFINITION 3.7. Let Γ be a countable group, and let G be a compact group. A subgroup $X \subset G^\Gamma$ is *of finite type* if it is closed and shift-invariant, and if there exists a finite subset $F \subset \Gamma$ such that

$$X = \{x \in G^\Gamma : \pi_F(\sigma_\gamma(x)) \in \pi_F(X) \text{ for every } \gamma \in \Gamma\}. \tag{3.1}$$

In order to interpret condition (3.1) we can think of F as a *window* and of the subgroup $H = \pi_F(X) \subset G^F$ as the set of *allowed words*; in order to decide whether a given point $x \in G^\Gamma$ lies in X we look at x and all its translates through the window F, and $x \in X$ if and only if we always see an allowed word. This definition is consistent with the notion of a classical shift of finite type with finite alphabet, which is the set of all points in a shift space for which every string of coordinates of some fixed length n occurs in a previously specified list of allowed words (usually $n = 2$, but that is irrelevant). For example, the subshift of $\{0, 1\}^{\mathbb{Z}}$ consisting of all sequences with no two adjacent '1'-s is of finite type, whereas the subshift of all sequences in which any two '1'-s have to be separated by an even number of '0'-s is not of finite type; in the first case the window F could be chosen as $\{0, 1\} \subset \mathbb{Z}$ and the set of allowed words as $\{(0, 0), (0, 1), (1, 0)\}$, but in the second case no window F of finite size would work.

THEOREM 3.8. *Let Γ be a countable group, and let G be a compact Lie group. The shift-action σ of Γ on G^Γ satisfies the d.c.c. if and only if every closed, shift-invariant subgroup $X \subset G^\Gamma$ is of finite type.*

PROOF. Suppose that σ satisfies the d.c.c., that $X \subset G^\Gamma$ is a closed, shift-invariant subgroup, and that $(F_n, n \geq 1)$ is an increasing sequence of finite subsets of Γ with $\bigcup_{n \geq 1} F_n = \Gamma$. For every $n \geq 1$ we set $X_n = \{x \in G^\Gamma : \pi_{F_n}(\sigma_\gamma(x)) \in \pi_{F_n}(X) \text{ for every } \gamma \in \Gamma\}$, and we observe that X_n is a closed, shift-invariant subgroup of G^Γ, $X_n \supset X_{n+1} \supset X$ for every $n \geq 1$, and that $\bigcap_{n \geq 1} X_n = X$. The d.c.c. implies that there exists an integer $N \geq 1$ with $X_N = X$, i.e. that X is of finite type.

Conversely, if every closed, shift-invariant subgroup of G^Γ is of finite type, let $(X_n, n \geq 1)$ be a non-increasing sequence of closed, shift-invariant subgroups of G^Γ, and let $X = \bigcap_{n \geq 1} X_n$. Then X is of finite type, and there exists a finite subset $F \subset \Gamma$ satisfying (3.1). We set $H = \pi_F(X)$ and $H_n = \pi_F(X_n)$, $n \geq 1$, and observe that

$$Y_n = \{x \in G^\Gamma : \pi_F(\sigma_\gamma(x)) \in H_n \text{ for every } \gamma \in \Gamma\} \supset X_n \supset X$$

for every $n \geq 1$. Since $(H_n, n \geq 1)$ is a non-increasing sequence of closed subgroups of the compact Lie group G^F, there exists an $N \geq 1$ with $H_N = H$ and hence with $X = Y_N \supset X_N \supset X_n \supset X$ for all $n \geq N$. This proves that σ satisfies the d.c.c. $\square$

We have obtained the following implications for a Γ-action by automorphisms of a compact group X:

$$\text{expansiveness} \Longrightarrow \text{conjugacy to a Lie subshift} \Longleftarrow \text{d.c.c.} \qquad (3.2)$$

As mentioned in Section 2, conjugacy to a Lie subshift need not imply expansiveness. The next example shows that it need not imply the d.c.c., either, and that there is—in general—no connection between expansiveness and the d.c.c.

EXAMPLES 3.9. (1) Consider the direct sum $\Gamma = \sum_{n \geq 2} \mathbb{Z}/n$ of the cyclic groups $\mathbb{Z}/n = \mathbb{Z}/n\mathbb{Z}$ of order n, and let $G = \mathbb{Z}/2$. We write a typical element $\gamma \in \Gamma$ as $\gamma = (\gamma_2, \gamma_3, \dots)$ with $\gamma_n \in \mathbb{Z}/n$ for every $n \geq 2$ and $\gamma_m = 0$ for all but finitely many m, and we set $\Gamma_n = \{\gamma \in \Gamma : \gamma_m = 0 \text{ for all } m \neq n\}$. Let $X = G^{\Gamma}$, and let σ be the shift-action of Γ on X. Then σ is expansive on X.

For every $n \geq 2$ we consider the full, shift-invariant subgroup $X_n = \{x \in X : \sum_{\gamma \in \Gamma_m} \sigma_\gamma(x) = \mathbf{0}_{G^{\Gamma}} \text{ for every } m \leq n\}$. Then $X \supsetneq X_2 \supsetneq \cdots \supsetneq X_n \supsetneq \cdots \supsetneq \bigcap_{n \geq 2} X_n = Y$, say, and $Y \subset X = G^{\Gamma}$ is obviously not of finite type.

(2) Let Γ be a countable group with a strictly increasing sequence $\Gamma_1 \subsetneq \Gamma_2 \subsetneq \cdots \subsetneq \Gamma_n \subsetneq \cdots \subset \Gamma$ of normal subgroups, and let $G \neq \{1\}$ be a compact Lie group. For every $n \geq 1$ we set $X_n = \{x \in G^{\Gamma} : x_{\gamma'\gamma} = x_\gamma \text{ for all } \gamma' \in \Gamma_n, \gamma \in \Gamma\}$. Then $(X_n, n \geq 1)$ is a strictly decreasing sequence of closed, shift-invariant subgroups of G^{Γ}, and the shift-action of Γ on G^{Γ} does not satisfy the d.c.c.

(3) The group $\Gamma = \mathbb{Q}^{\times}$ in Example 2.18 (4) has a sequence of subgroups $\Gamma_1 \subsetneq \Gamma_2 \subsetneq \cdots \subsetneq \Gamma_n \subsetneq \cdots \subset \Gamma$: let $2 = p_1 < 3 = p_2 < \dots$ be the sequence of rational primes, and put, for every $n \geq 1$, $\Gamma_n = \{p_1^{r_1} \cdot \dots \cdot p_n^{r_n} : r_1, \dots, r_n \in \mathbb{Z}\}$. According to Example (2), the shift-action σ of Γ on $\mathbb{T}^{\Gamma}$ does not satisfy the d.c.c. However, if $Y \subset \mathbb{T}^{\Gamma}$ is the closed, shift-invariant subgroup (2.5), then the restriction σ^Y of σ to Y satisfies the d.c.c., since $\hat{Y} = \mathbb{Q}$ has no proper subgroups which are invariant under the Γ-action β on $\hat{Y}$ dual to α. $\square$

CONCLUDING REMARKS 3.10. (1) The descending chain condition and its connection with shifts of finite type first appeared in [45].

(2) The definition of a subgroup $X \subset G^{\Gamma}$ of finite type (Definition 3.7) is natural, but has to be treated with some care: shifts of finite type usually have countable alphabets, and the transition to uncountable alphabets has some unexpected consequences. A subgroup $X \subset G^{\Gamma}$ can be *deterministic* in the sense that every point $x \in X$ is completely determined by its projection onto some fixed finite set of coordinates, and the shift-action of Γ on X may still have positive entropy (cf. Example 2.18 (2)).

4. Groups of Markov type

In this section we exhibit a general class of group actions satisfying the descending chain condition. Motivated by Theorem 3.8 we introduce the following definition.

DEFINITION 4.1. A countable group Γ is *of Markov type* if, for every compact Lie group G, the shift-action of Γ on G^Γ satisfies the d.c.c.

In Example 3.9 (2) we have seen that every countable group Γ of Markov type must satisfy the ascending chain condition on subgroups. In order to describe a class of groups of Markov type we recall that a countable group Γ is *polycyclic-by-finite* if there exist subgroups $\Gamma = \Gamma_d \supset \Gamma_{d-1} \supset \cdots \supset \Gamma_0 = \{1\}$ such that, for every $k = 1, \ldots, d$, Γ_{k-1} is normal in Γ_k, Γ/Γ_{d-1} is finite, and $\Gamma_k/\Gamma_{k-1} \cong \mathbb{Z}$ for $k = 1, \ldots, d-1$. A countable group Γ is polycyclic-by-finite if and only if there exist subgroups $\Gamma = \Gamma_{d'} \supset \Gamma_{d'-1} \supset \cdots \supset \Gamma_0 = \{1\}$ such that, for every $k = 1, \ldots, d'$, Γ_{k-1} is normal in Γ_k and Γ_k/Γ_{k-1} is either finite or isomorphic to $\mathbb{Z}$ (cf. [79]).

THEOREM 4.2. *Every polycyclic-by-finite group Γ is of Markov type.*

We begin the proof of Theorem 4.2 with two lemmas.

LEMMA 4.3. *Let Γ be a countable group, and let $\alpha^{(i)}$ be Γ-actions by automorphisms of compact groups $X^{(i)}$, $i = 1, \ldots, n$. We define the product action $\alpha = \alpha^{(1)} \times \cdots \times \alpha^{(n)}$ of Γ on $X = X_1 \times \cdots \times X_n$ by $\alpha_\gamma(y) = (\alpha_\gamma^{(1)}(y^{(1)}), \ldots, \alpha_\gamma^{(n)}(y^{(n)}))$ for every $\gamma \in \Gamma$ and $y = (y^{(1)}, \ldots, y^{(n)}) \in X$. If $\alpha^{(i)}$ satisfies the d.c.c. for $i = 1, \ldots, n$, then α again satisfies the d.c.c. In other words, the d.c.c. is inherited by finite cartesian products.*

PROOF. We prove the lemma by induction on n and assume that the result has been established for $n < N$, $N \geq 2$. Let $(X^{(i)}, \alpha^{(i)})$, $i = 1, \ldots, N$, be Γ-actions by automorphisms of compact groups satisfying the d.c.c., and let α be the product action on $X = X_1 \times \cdots \times X_N$. Suppose that $(Y_m, m \geq 1)$ is a non-increasing sequence of closed, α-invariant subgroups of X. We denote by $\pi \colon X \longmapsto X_1 \times \cdots \times X_{N-1}$ the projection onto the first $N-1$ coordinates and use the induction hypothesis to obtain an integer $M \geq 1$ with $\pi(Y_m) = \pi(Y_M)$ for all $m \geq M$. For every $m \geq 1$, $Z_m = \ker(\pi) \cap Y_m$ is (isomorphic to) a closed, $\alpha^{(N)}$-invariant subgroup of $X^{(N)}$, and the sequence $(Z_m, m \geq 1)$ is non-increasing. Since $(X^{(N)}, \alpha^{(N)})$ satisfies the d.c.c., there exists an integer $M' \geq 1$ such that $Z_m = Z_{M'}$ for all $m \geq M'$. Hence $Y_m = Y_{\max\{M,M'\}}$ for all $m \geq \max\{M, M'\}$, and this proves the assertion for $n = N$ and hence for all $n \geq 1$. $\square$

LEMMA 4.4. *Let Γ' be a countable group, and let $\Gamma \subset \Gamma'$ be a normal subgroup such that Γ'/Γ is either finite or isomorphic to $\mathbb{Z}$. If Γ is of Markov type, then Γ' is again of Markov type.*

PROOF. Let G be a compact Lie group, and let σ and σ' be the shift-actions of Γ on G^Γ and of Γ' on $G^{\Gamma'}$, respectively. If Γ'/Γ is finite we choose a subset $F = \{\gamma_0 = 1, \gamma_1, \ldots, \gamma_{k-1}\} \subset \Gamma'$ which intersects each coset of Γ in exactly one point. The map $\gamma\gamma_j \mapsto (\gamma, \gamma_j)$ from Γ' to $\Gamma \times F$ is a bijection and induces a group isomorphism $G^{\Gamma'} \longmapsto (G^\Gamma)^k$ which carries the shift-action of Γ on $G^{\Gamma'}$

to the Γ-action $\tau\colon \gamma \mapsto \tau_\gamma = \sigma_\gamma \times \sigma_{\gamma_1^{-1}\gamma\gamma_1} \times \cdots \times \sigma_{\gamma_{k-1}^{-1}\gamma\gamma_{k-1}}$ on $(G^\Gamma)^k$. From Lemma 4.3 we know that the Γ-action τ on $(G^\Gamma)^k$ satisfies the d.c.c., and we conclude that the shift-action of Γ on $G^{\Gamma'}$ satisfies the d.c.c., which obviously implies that σ' satisfies the d.c.c.

If $\Gamma'/\Gamma \cong \mathbb{Z}$ we choose an element $\gamma' \in \Gamma'$ such that $\Gamma' = \bigcup_{n\in\mathbb{Z}} \Gamma\gamma'^n$ and define, for every $k \in \mathbb{Z}$, a continuous, surjective homomorphism $\pi_k\colon G^{\Gamma'} \longmapsto G^\Gamma$ by setting

$$(\pi_k(x))_\gamma = x_{\gamma\gamma'^k} \tag{4.1}$$

for all $x = (x_{\gamma''}, \gamma'' \in \Gamma') \in G^{\Gamma'}$ and $\gamma \in \Gamma$. By setting $x_k = \pi_k(x)$ for every $x \in G^{\Gamma'}$ and $k \in \mathbb{Z}$ we obtain a group isomorphism $\Psi\colon G^{\Gamma'} \longmapsto (G^\Gamma)^{\mathbb{Z}}$ which sends each $x \in G^{\Gamma'}$ to $(\ldots, x_{-1}, x_0, x_1, \ldots) \in (G^\Gamma)^{\mathbb{Z}}$. If β denotes the shift $(\beta(v))_n = v_{n+1}$ on $(G^\Gamma)^{\mathbb{Z}}$, then

$$\beta \cdot \Psi(x) = \Psi \cdot \sigma'_{\gamma'}(x) \tag{4.2}$$

and

$$\Psi \cdot \sigma'_\gamma(x) = (x'_n, \, n \in \mathbb{Z}) \tag{4.3}$$

with

$$x'_n = \sigma_{\gamma'^{-n}\gamma\gamma'^n}(x_n) \tag{4.4}$$

for all $n \in \mathbb{Z}$ and $x \in G^{\Gamma'}$.

Let $V \subset G^{\Gamma'}$ be a σ'-invariant subgroup. For every $k \geq 1$ consider the closed, σ-invariant subgroup

$$X(k) = \pi_k(\{x \in V\colon \pi_j(x) = 1_{G^\Gamma} \text{ for every } j = 0, \ldots, k-1\}) \subset G^\Gamma, \tag{4.5}$$

and set

$$X(0) = \pi_0(V) \subset G^\Gamma. \tag{4.6}$$

Then $X(k) \supset X(k+1)$ for all $k \geq 0$. Since Γ is of Markov type, its shift-action σ on G^Γ satisfies the d.c.c., and there exists a $K \geq 1$ with $X(k) = X(K)$ for all $k \geq K$. We claim that

$$V = \{x \in G^{\Gamma'}\colon (x_m, \ldots, x_{m+K}) \in H \text{ for all } m \in \mathbb{Z}\}, \tag{4.7}$$

where $x_j = \pi_j(x)$ for all $x \in G^{\Gamma'}$ and $j \in \mathbb{Z}$, and $H = \{(x_0, \ldots, x_K)\colon x \in V\}$. In order to prove (4.7) we observe that the Γ-invariant group $X' = X(K) \subset G^\Gamma$ is the follower set of the string $[1_{G^\Gamma}, \ldots, 1_{G^\Gamma}]$ of length K for the shift β on the shift-invariant subgroup $\Psi(V) \subset (G^\Gamma)^{\mathbb{Z}}$. Furthermore,

$$y_K \in y'_K X' \tag{4.8}$$

whenever $(y_0, \ldots, y_K), (y'_0, \ldots, y'_K) \in H$ and $y_i = y'_i$ for $0 \leq i < K$ (in other words, the follower set of $[y_0, \ldots, y_{K-1}]$ is a coset of the follower set X'). From the definition of X' it is also clear that there exists, for every $x \in X'$, $y \in V$, $n \in \mathbb{Z}$, an element $z \in V$ with $z_m = y_m$ for $m < n$ and

$$z_n = y_n x. \tag{4.9}$$

Put

$$Y_H = \{x \in G^{\Gamma'} : (x_m, \ldots, x_{m+K}) \in H \text{ for all } m \in \mathbb{Z}\}.$$

In order to verify that $V = Y_H$ we note that Y_H is a closed, shift-invariant subgroup of $(G^\Gamma)^\mathbb{Z}$, and that $V \subset Y_H$. Conversely, if $z \in Y_H$, there exists an element $y^{(0)} \in V$ with $(z_0, \ldots, z_K) = (y_0^{(0)}, \ldots, y_K^{(0)})$. From (4.8) and (4.9) we know that we can find $y^{(1)} \in V$ with $z_m = y_m^{(0)} = y_m^{(1)}$ for $0 \le m \le K$ and $z_{K+1} = y_{K+1}^{(1)}$. By repeating this argument we obtain a sequence $(y^{(r)}, r \ge 0)$ in V such that $y_m^{(r)} = z_m$ for all $r \ge 0$ and $0 \le m \le K + r$, and since V is compact there exists a $y \in V$ with $y_m = z_m$ for all $m \ge 0$.

We have proved that the set $P(z) = \{y \in V : y_m = z_m \text{ for all } m \ge 0\}$ is a closed, non-empty subset for every $z \in Y_H$. For every $z \in Y_H$, $Q(z) = \bigcap_{n \ge 0} \beta^n(P(\beta^{-n}(z)))$ is again non-empty, and $Q(z) = \{z\}$. Hence $Y_H \subset V$, i.e. $Y_H = V$. This completes the proof of (4.7).

Now assume that $(V(n), n \ge 1)$ is a non-increasing sequence of closed, σ'-invariant subgroups of $G^{\Gamma'}$. We fix n for the moment and define a sequence of closed, σ-invariant subgroups $(X(n, k), k \ge 0)$ of G^Γ by (4.5)–(4.6) with $V(n)$ replacing V. Our assumptions imply that, for all $k, n \ge 1$, $X(n, k) \supset X(n+1, k)$ and $X(n, k) \supset X(n, k+1)$. Since σ satisfies the d.c.c. there exists, for every $n \ge 0$, an integer $K(n) \ge 1$ with $X(n, K(n)) = X(n, k)$ for all $k \ge K(n)$, and we set $X_n = X(n, K(n))$. Since $X_n \supset X_{n+1}$ for all $n \ge 1$, the d.c.c. implies the existence of an integer $N \ge 1$ with $X_n = X_N$ for all $n \ge N$, and we set $X' = X_N$. Put $K = \max_{1 \le n \le N} K(n)$ and note that $X(n, k) = X'$ for all $n \ge N$ and $k \ge K$. For every $n \ge 1$, the group $H(n) = \{(x_0, \ldots, x_K) : x \in V(n)\}$ is invariant under the action $\gamma \mapsto \sigma_\gamma \times \sigma_{\gamma'^{-1}\gamma\gamma'} \times \cdots \times \sigma_{\gamma'^{-K}\gamma\gamma'^K}$ of Γ on $(G^\Gamma)^{K+1}$, and the sequence $(H(n), n \ge 1)$ is non-increasing. By Lemma 4.2 there exists an $N' \ge N$ such that $H(n) = H(N')$ for all $n \ge N'$. The first part of this proof implies that $V(n) = Y_{H(n)}$ for all $n \ge N$, and we conclude that $V(n) = V(N')$ for all $n \ge N'$. This shows that the sequence $(V(n), n \ge 1)$ is eventually constant and that σ' satisfies the d.c.c. $\square$

PROOF OF THEOREM 4.2. Let Γ be a countable, polycyclic-by-finite group, and let $\Gamma = \Gamma_d \supset \Gamma_{d-1} \supset \cdots \supset \Gamma_1 \supset \Gamma_0 = \{1\}$ be a sequence of subgroups of Γ such that Γ_{k-1} is normal in Γ_k and Γ_k/Γ_{k-1} is either finite or isomorphic to $\mathbb{Z}$ for $k = 1, \ldots, d$. The group Γ_0 is of Markov type by Example 3.2, and Lemma 4.4 provides the induction step necessary to show that the groups $\Gamma_1, \ldots, \Gamma_d = \Gamma$ are all of Markov type. $\square$

PROPOSITION 4.5. *Let Γ be a countable group which is of Markov type. Then every subgroup $\Delta \subset \Gamma$ is of Markov type.*

PROOF. If Δ is not of Markov type, there exists a compact Lie group G, a closed, shift-invariant subgroup $X \subset G^{\Delta}$, and an increasing sequence $(F_n, n \geq 1)$ of finite subsets of Δ such that $\bigcup_{n \geq 1} F_n = \Delta$, but

$$X_n = \{x \in G^{\Delta} : \pi_{F_n}(\sigma_{\delta}(x)) \in \pi_{F_n}(X) \text{ for every } \delta \in \Delta\} \supsetneq X$$

for every $n \geq 1$ (Theorem 3.8). We set

$$Y_n = \{x \in G^{\Gamma} : \pi_{F_n}(\sigma_{\gamma}(x)) \in \pi_{F_n}(X) \text{ for every } \gamma \in \Gamma\}$$

and note that $(Y_n, n \geq 1)$ is a non-increasing sequence of closed, shift-invariant subgroup of G^{Γ} with $\pi_{\Delta}(Y_n) = X_n$ for every $n \geq 1$. Since G^{Γ} satisfies the d.c.c. we can find an $N \geq 1$ with $Y_n = Y_N$ for all $n \geq N$ and obtain that $\pi_{\Delta}(Y_n) = X_n = X_N = X$ for all $n \geq N$, contrary to our assumption, and we conclude that Δ must be of Markov type. $\square$

We can interpret Theorem 3.8 as saying that a countable group is of Markov type if and only if it satisfies a certain dynamical finiteness condition, and we shall now prove that this dynamical finiteness condition is, in turn, related to a well known algebraic finiteness condition. Let Γ be a countable group, and let $\mathcal{R}$ be a commutative ring. The *support* of a function $f : \Gamma \longmapsto \mathcal{R}$ is the set $\mathcal{S}(f) = \{\gamma \in \Gamma : f(\gamma) \neq 0\}$. We denote by $\mathcal{R}[\Gamma]$ the *group ring of Γ with coefficients in $\mathcal{R}$*, i.e. the ring of all functions $f : \Gamma \longmapsto \mathcal{R}$ with finite support, furnished with the operations of *addition* $((f_1 + f_2)(\gamma) = f_1(\gamma) + f_2(\gamma))$ and *convolution* $((f_1 * f_2)(\gamma) = \sum_{\gamma' \in \Gamma} f_1(\gamma\gamma'^{-1})f_2(\gamma'))$. The ring $\mathcal{R}[\Gamma]$ is *right Noetherian* if there exists, for every non-decreasing sequence $\mathcal{I}_1 \subset \mathcal{I}_2 \subset \cdots \subset \mathcal{I}_k \subset \ldots$ of right ideals in $\mathcal{R}[\Gamma]$, a $K \geq 1$ with $\mathcal{I}_k = \mathcal{I}_K$ for all $k \geq K$.

PROPOSITION 4.6. *Let Γ be a countable group of Markov type. Then the group ring $\mathbb{Z}[\Gamma]$ is right Noetherian. Conversely, if the group ring $\mathbb{Z}[\Gamma]$ of a countable group Γ is right Noetherian, then G^{Γ} satisfies the d.c.c. for every compact Lie group G which is either finite or abelian.*

PROOF. If Γ is a countable group, the dual group of the abelian, additive group $\mathbb{T}^{\Gamma}$ is equal to $\mathbb{Z}[\Gamma]$; furthermore, if $X \subset \mathbb{T}^{\Gamma}$ is a closed subgroup, then X is invariant under the shift-action of Γ on $\mathbb{T}^{\Gamma}$ if and only if its annihilator $X^{\perp} \subset \mathbb{Z}[\Gamma]$ is a right ideal. In particular, $\mathbb{T}^{\Gamma}$ satisfies the d.c.c. if and only if $\mathbb{Z}[\Gamma]$ is right Noetherian.

Now assume that $\mathbb{Z}[\Gamma]$ is right Noetherian. As we have seen in the first part of this proof, this implies that $\mathbb{T}^{\Gamma}$ satisfies the d.c.c., and Lemma 4.3 shows that $(\mathbb{T}^n)^{\Gamma}$ satisfied the d.c.c. for every $n \geq 1$. Since every compact, abelian Lie group is a closed subgroup of $\mathbb{T}^n$ for some $n \geq 1$, A^{Γ} satisfies the d.c.c. for every compact, abelian Lie group A.

If F is a finite group we set $A = \mathbb{Z}_{/2}^F$, write $B = \hat{A} \cong A$ for the dual group of A, and note that the dual group of A^{Γ} is (isomorphic to) the group $\mathcal{N}$ of all maps $a : \Gamma \longmapsto B$ with finite support (i.e. with $a(\gamma) = 0$ for all but finitely many $\gamma \in \Gamma$) under point-wise addition. The group $\mathcal{N}$ is a $\mathbb{Z}[\Gamma]$-module under the action $f \cdot a = \sum_{\gamma' \in \Gamma} f(\gamma')a(\gamma\gamma'^{-1})$, $\gamma \in \Gamma$, and a subgroup $\mathcal{N}' \subset \mathcal{N}$ is a $\mathbb{Z}[\Gamma]$-submodule of $\mathcal{N}$ if and only if its annihilator $(\mathcal{N}')^{\perp} \subset A^{\Gamma}$ is a shift-invariant (and necessarily closed) subgroup of A^{Γ}. The first part of this proof shows that A^{Γ} satisfies the d.c.c., and we conclude that $\mathcal{N}$ is Noetherian.

For every $g \in F$ we define an automorphism α_g of A by $(\alpha_g(a))_h = a_{g^{-1}h}$, $h \in F$, for all $a \colon F \longmapsto \mathbb{Z}_{/2}$ in A, and we write β_g for the automorphism of $B = \hat{A}$ dual to α_g. For every $x = (x_\gamma) \in F^{\Gamma}$, the map $\bar{\beta}_x \colon \mathcal{N} \longmapsto \mathcal{N}$ given by $(\bar{\beta}_x(a))(\gamma) = \beta_{x_\gamma}(a(\gamma))$ for every $a \in \mathcal{N}$ is an automorphism of $\mathcal{N}$. If $X \subset F^{\Gamma}$ is a closed, shift-invariant subgroup, the set $\mathcal{N}(X) = \{a \in \mathcal{N} \colon \bar{\beta}_x(a) = a$ for all $x \in X\}$ is a submodule of $\mathcal{N}$, and every strictly decreasing sequence $(X_n, n \geq 1)$ of closed, shift-invariant subgroups of F^{Γ} corresponds to a strictly increasing sequence $(\mathcal{N}(X_n), n \geq 1)$ of submodules of $\mathcal{N}$. As we have seen above, $\mathcal{N}$ is Noetherian, and we conclude that F^{Γ} satisfies the d.c.c. $\square$

COROLLARY 4.7. *Let Γ be a countable group such that $\mathbb{Z}[\Gamma]$ is right Noetherian. Then every expansive action α of Γ on a compact group X satisfies the d.c.c.*

PROOF. According to Proposition 2.2 we may assume that α is the shift-action of Γ on a full, shift-invariant subgroup $X \subset G^{\Gamma}$, where G is a compact Lie group. Since X° is abelian by Theorem 2.4, the connected component $G^{\circ} = \pi_{\{1_\Gamma\}}(X^{\circ})$ of G is abelian, and we set $Y = X \cap (G^{\circ})^{\Gamma}$ and observe that the quotient map from G to $F = G/G^{\circ}$ sends X/Y to a closed, shift-invariant subgroup of F^{Γ}. Since F is finite, Proposition 4.6 implies that X/Y satisfies the d.c.c., and a second application of Proposition 4.6 yields that the abelian group $Y \subset (G^{\circ})^{\Gamma}$ satisfies the d.c.c. Since X/Y and Y both satisfy the d.c.c., the same must be true for X. $\square$

EXAMPLES 4.8. (1) Every finitely generated, abelian group Γ is polycyclic, and is therefore of Markov type by Theorem 4.2.

(2) The group

$$\Gamma = \left\{ \begin{pmatrix} 1 & a & c \\ 0 & 1 & b \\ 0 & 0 & 1 \end{pmatrix} : a, b, c \in \mathbb{Z} \right\} \subset \mathrm{SL}(3, \mathbb{Z})$$

is polycyclic and non-abelian, and is of Markov type by Theorem 4.2.

(3) If Γ is a countable group which has a strictly increasing sequence of subgroups $\Gamma_1 \subsetneq \Gamma_1 \subsetneq \cdots \subsetneq \Gamma_n \subsetneq \ldots$ then $\mathbb{Z}[\Gamma]$ is easily seen not to be right Noetherian (cf. [79]) and therefore not of Markov type (Proposition 4.5). In fact, the shift-action of Γ on G^{Γ} does not satisfy the d.c.c. for any compact Lie group $G \neq \{\mathbf{1}\}$ (Example 3.9 (2)).

(4) Let $\Gamma = \left\{ \begin{pmatrix} 2^m & k2^{-n} \\ 0 & 2^{-m} \end{pmatrix} : k, m \in \mathbb{Z}, n \in \mathbb{N} \right\} \subset \mathrm{SL}(2, \mathbb{Q})$. The group Γ is solvable and finitely generated (by the elements $\begin{pmatrix} 2 & 0 \\ 0 & 2^{-1} \end{pmatrix}$ and $\begin{pmatrix} 1 & 1 \\ 0 & 1 \end{pmatrix}$), but Γ has an

infinite, strictly increasing sequence of subgroups $\Gamma_n = \left\{ \left(\begin{smallmatrix} 1 & k2^{-n} \\ 0 & 1 \end{smallmatrix} \right) : k \in \mathbb{Z} \right\}$, $n \geq 1$. Example (3) shows that Γ is not of Markov type.

(5) Let $\Gamma = F(2)$ be the free group on two generators a, b. Then Γ has an infinite, strictly increasing sequence of subgroups and is therefore not of Markov type by Example (3). An example of such a sequence is obtained by taking, for every $n \geq 1$, the group $\Gamma_n \subset \Gamma$ generated by all words of the form $a^{2m}b^m a^{2m+1}$ with $1 \leq m \leq n$. According to Proposition 4.5, no countable group containing $\Gamma = F(2)$ can be of Markov type; in particular, $SL(2, \mathbb{Z})$ is not of Markov type, nor are the groups $SL(n, \mathbb{Z})$, $n \geq 3$, or the free groups $F(n)$ on $n \geq 3$ generators. $\boxdot$

If Γ is a group of Markov type and α a Γ-action by automorphisms of a compact group X, the implications (3.2) take the form

$$\text{expansiveness} \Longrightarrow \text{conjugacy to a Lie subshift} \Longleftrightarrow \text{d.c.c.} \qquad (4.10)$$

The remainder of this section is devoted to deriving further properties of actions of groups of Markov type.

PROPOSITION 4.9. *Let α be an action of a countable group Γ of Markov type by automorphisms of a compact group X. Then there exists a non-increasing sequence $(V_n, n \geq 1)$ of closed, normal, α-invariant subgroups of X such that $\bigcap_{n \geq 1} V_n = \{1_X\}$, and $(X/V_n, \alpha^{X/V_n})$ satisfies the d.c.c. for every $n \geq 1$.*

PROOF. Proposition 2.17 and Definition 4.1. $\square$

PROPOSITION 4.10. *Let α be an action of a group Γ of Markov type by automorphisms of a compact group X. The following conditions are equivalent.*

(1) *(X, α) satisfies the d.c.c.;*
(2) *(X, α) satisfies the d.c.c. on closed, normal, α-invariant subgroups (i.e. for every non-increasing sequence $(X_n, n \geq 1)$ of closed, normal, α-invariant subgroups there exists a $K \geq 1$ such that $X_k = X_K$ whenever $k \geq K$).*

PROOF. It is clear that (1) implies (2). If (2) is satisfied, Proposition 2.17 implies that (X, α) is conjugate to a Lie subshift. $\square$

The next result yields a stronger form of Theorem 3.6 for groups of Markov type.

THEOREM 4.11. *Let Γ be a group of Markov type, and let α be an ergodic Γ action by automorphisms of a compact group X. Then the restriction α^{X° of α to the connected component X° of the identity in X is ergodic.*

PROOF. If α satisfies the d.c.c. the assertion follows from Theorem 3.6. If α does not satisfy the d.c.c., we apply Proposition 4.9 and find a non-increasing sequence $(V_n, n \geq 1)$ of closed, normal, α-invariant subgroups of X such that $\bigcap_{n \geq 1} V_n = \{1_X\}$ and α^{X/V_n} satisfies the d.c.c. for every $n \geq 1$. If α^{X° is non-ergodic, then there exists a closed, normal, α^{X°-invariant subgroup $Y \subset X^\circ$ such that X°/Y is a Lie group carrying an $\alpha^{X^\circ/Y}$-invariant metric (Lemma 1.2). Since every Lie group satisfies the d.c.c. on closed subgroups, and since $(X^\circ \cap V_n)/Y \searrow \{1_{X/Y}\}$ as $n \to \infty$, there exists an $m \geq 1$ with $X^\circ \cap V_m \subset Y$, and we set $X' = X/V_m$, $Y' = Y/V_m \subset X'^\circ$, and obtain a closed, normal, α-invariant subgroup Y' of X'° such that $X'^\circ/Y' \cong X^\circ/Y$ is a compact Lie group with an α-invariant metric. In particular, $\alpha^{X'}$ is ergodic and satisfies the d.c.c., but $\alpha^{X'^\circ}$ is non-ergodic by Lemma 2.1, contrary to the assertion of Theorem 3.6. This contradiction shows that α^{X° must be ergodic. $\square$

COROLLARY 4.12. *Let α be a mixing action of a countable group Γ of Markov type by automorphisms of a compact group X. If $X^\circ \neq \{1_X\}$ then α^{X° is mixing.*

PROOF. If $\Delta \subset \Gamma$ is an infinite subgroup, then Δ is of Markov type by Proposition 4.5. We write β for the restriction of α to Δ and apply Theorem 4.11 to see that β^{X° is ergodic. According to Theorem 1.6 (1) this implies that α^{X° is mixing. $\square$

The following example shows that Theorem 4.11 is incorrect without the condition that Γ is of Markov type, and that the assumption of the d.c.c. in Theorem 3.6 cannot be dropped without further restrictions on the group Γ.

EXAMPLE 4.13. Let $p \geq 2$ be a rational prime, and let $\mathbb{Z}_p$ the group of p-adic integers. The dual group of $\mathbb{Z}_p$ is given by $\widehat{\mathbb{Z}_p} = \{l/p^m : m \geq 0, 0 \leq l < p^m\}$, furnished with addition modulo 1. We set $Y = \mathbb{Z}_p^{\mathbb{Z}}$ and denote by σ the shift on Y given by $(\sigma(y))_n = y_{n+1}$ for every $y = (y_n) = (y_n, n \in \mathbb{Z}) \in Y = \mathbb{Z}_p^{\mathbb{Z}}$. Since the orbit $\{\hat{\sigma}^k(a) : k \in \mathbb{Z}\}$ of every non-zero element $a \in \hat{Y}$ under $\hat{\sigma}$ is infinite, σ is ergodic on Y by Lemma 1.2 and Remark 1.7 (3). We put $X = Y \times \mathbb{S}$, denote by $\alpha \in \mathrm{Aut}(X)$ the map $(y, s) \mapsto (\sigma(y), s)$, $(y, s) \in Y \times \mathbb{S} = X$, and put $\Gamma_1 = \{\alpha^n : n \in \mathbb{Z}\}$.

Every $a \in \hat{Y}$ defines an automorphism $\beta_a \in \mathrm{Aut}(X)$ by

$$\beta_a(y, s) = (y, \langle a, y \rangle s)$$

for every $(y, s) \in Y \times \mathbb{S} = X$, and we set $\Gamma_2 = \{\beta_a : a \in \hat{Y}\} \subset \mathrm{Aut}(X)$ and denote by $\Gamma = \{\alpha^n \beta_a : n \in \mathbb{Z}, a \in \hat{Y}\} \subset \mathrm{Aut}(X)$ the group generated by Γ_1 and Γ_2. By checking that every non-trivial character $\chi \in \hat{X}$ has an infinite orbit under Γ one verifies easily that the action of Γ on X is ergodic (cf. Remark 1.7 (3)). However, for every $\gamma = \alpha^n \beta_a \in \Gamma$, the restriction γ^{X° of γ to $X^\circ = \{(0_Y, s) : s \in \mathbb{S}\}$ is trivial, so that Γ does not act ergodically on X°.

Note that Γ_2 is abelian, $\Gamma/\Gamma_2 = \Gamma_1 \cong \mathbb{Z}$, and that Γ is therefore not far removed from being polycyclic-by-finite and thus of Markov type (cf. Theorem 4.2). $\boxdot$

PROPOSITION 4.14. *Let G be a compact Lie group, and let Γ be a countable group of Markov type.*

(1) *If $Y = Y_0 \supset Y_1 \supset \cdots \supset Y_n$ are closed, shift-invariant subgroups of G^Γ there exists a finite set $F \subset \Gamma$ such that, for every $k = 0, \ldots, n$,*

$$Y_k = \{x \in G^\Gamma : \pi_F(\sigma_\gamma(x)) \in H_k \text{ for every } \gamma \in \Gamma\}$$

with $H_k = \pi_F(Y_k) \subset G^F$;

(2) *If $Y \subset G^\Gamma$ is a closed, shift-invariant subgroup of G^Γ there exists a finite set $F \subset \Gamma$ such that*

$$Y = \{x \in G^\Gamma : \pi_F(\sigma_\gamma(x)) \in H \text{ for every } \gamma \in \Gamma\},$$
$$Y^\circ = \{x \subset G^\Gamma : \pi_F(\sigma_\gamma(x)) \in H^\circ \text{ for every } \gamma \in \Gamma\},$$

and

$$C(Y) = \{x \in G^\Gamma : \pi_F(\sigma_\gamma(x)) \in C(H) \text{ for every } \gamma \in \Gamma\},$$

where $H = \pi_F(Y)$.

PROOF. Choose an increasing sequence $(F_n, n \geq 1)$ of finite subsets of Γ such that $\bigcup_{n \geq 1} F_n = \Gamma$. If $Y \supset Y_1 \supset \cdots \supset Y_n$ are closed, shift-invariant subgroups of G^Γ, put $H_{k,m} = \pi_{F_m}(Y_k)$, $k = 1, \ldots, n$, $m \geq 1$, and note that

$$Y_{k,m} = \{x \in G^\Gamma : \pi_{F_m}(\sigma_\gamma(x)) \in H_{k,m} \text{ for every } \gamma \in \Gamma\} \supset Y_k$$

and $Y_k = \bigcap_{m \geq 1} Y_{k,m}$. The d.c.c. implies that there exists an $M \geq 1$ such that $Y_{k,M} = Y_k$ for $k = 1, \ldots, n$, which proves (1), and (2) is an immediate consequence of (1). $\square$

REMARKS 4.15. (1) Suppose that $\Gamma = \mathbb{Z}^d$, where $d \geq 1$. In the notation of Proposition 4.14 (1), consider the injective homomorphism morphism $\theta \colon Y \longmapsto H^{\mathbb{Z}^d}$ defined by $\theta(x)_{\mathbf{n}} = \pi_F(\sigma_{\mathbf{n}}(x))$ for every $x \in Y$ and $\mathbf{n} \in \mathbb{Z}^d$. Then $W = W_0 = \theta(Y)$ is a full, shift-invariant subgroup of $H^{\mathbb{Z}^d}$, and

$$W_k = \theta(Y_k) = \{w \in H^{\mathbb{Z}^d} : \pi_{\{0,1\}^d}(\sigma_{\mathbf{n}}(w)) \in \pi_{\{0,1\}^d}(W_k) \tag{4.11}$$
$$\text{for every } \mathbf{n} \in \mathbb{Z}^d\}$$

for every $k = 0, \ldots, n$. In other words, by changing the *alphabet* G of the Lie subshift $Y \subset G^{\mathbb{Z}^d}$, we may assume that $F = \{0,1\}^d$. This argument is analogous to the re-coding of a classical subshift of finite type as a one-step Markov shift.

(2) If we apply Remark (1) to the groups Y and $Y^\circ \subset Y$ in Corollary 2.10 (3), then (4.11) shows that $\boldsymbol{\theta}(Y) = \{w \in H^{\mathbb{Z}^d} : \pi_{\{0,1\}^d}(\sigma_{\mathbf{n}}(w)) \in K$ for every $\mathbf{n} \in \mathbb{Z}^d\}$ and $\boldsymbol{\theta}(Y^\circ) = \boldsymbol{\theta}(Y) \cap (H^\circ)^{\mathbb{Z}^d} = \{w \in H^{\mathbb{Z}^d} : \pi_{\{0,1\}^d}(\sigma_{\mathbf{n}}(w)) \in K^\circ$ for every $\mathbf{n} \in \mathbb{Z}^d\}$, where $K = \pi_{\{0,1\}^d}(\boldsymbol{\theta}(Y)) \subset H^{\{0,1\}^d}$.

We end this section with two examples. Since $\mathbb{Z}^d$ is of Markov type for every $d \geq 1$ (Example 4.8 (1)), every closed, shift-invariant subgroup of $G^{\mathbb{Z}^d}$, where G is a compact Lie group, is of finite type. In Remark 3.10 (2) we mentioned a peculiarity of subgroups of finite type with infinite alphabet. However, even if the alphabet G is finite, the subgroups of $G^{\mathbb{Z}^d}$ can have other, unexpected properties.

EXAMPLES 4.16. (1) Let $G = \mathbb{Z}_{/2} \times \mathbb{Z}_{/2}$,

$$F = \{0,1\}^2 = \left\{ \begin{smallmatrix} (0,1) & (1,1) \\ (0,0) & (1,0) \end{smallmatrix} \right\} \subset \mathbb{Z}^2,$$

and let

$$H = \left\{ \left\{ \begin{smallmatrix} (*,k) & (l,*) \\ (*,l) & (k,*) \end{smallmatrix} \right\} : k, l \in \mathbb{Z}_{/2} \right\} \subset G^F,$$

where $*$ indicates that this location can contain an arbitrary element of $\mathbb{Z}_{/2}$. Then H is a full subgroup of G^F, and

$$X = X_{(F,H)} = \{x \in G^{\mathbb{Z}^d} : \pi_F(\sigma_{\mathbf{n}}(x)) \in H \text{ for every } \mathbf{n} \in \mathbb{Z}^d\} \quad (4.12)$$

is a full, shift-invariant subgroup of $G^{\mathbb{Z}^d}$. The shift-action σ of $\mathbb{Z}^2$ on X has the following properties: (i) $\sigma_{(0,2)} = \mathrm{id}_X$; (ii) $\sigma_{(1,0)}$ is ergodic and mixing, and is isomorphic to the full 4-shift. Hence σ is ergodic and expansive, but not mixing. The subset

$$S = \left\{ \left\{ \begin{smallmatrix} (*,h) & (k,0) \\ (*,k) & (h,*) \\ (*,h) & (k,1) \end{smallmatrix} \right\} : h, k \in \mathbb{Z}_{/2} \right\} \subset G^{\{0,1\} \times \{0,1,2\}},$$

satisfies that $S \cap \pi_{\{0,1\} \times \{0,1,2\}}(X) = \varnothing$, while $\pi_D(S) = \pi_D(X)$ for every 2×2 square $D \subset \{0,1\} \times \{0,1,2\}$. In other words, although the size of the defining window F is 2×2, there exist configurations of size 2×3 with the property that every subconfiguration of size 2×2 occurs in a point in X, whereas the configuration itself does not occur in any element of X. This is an indication that the *extension problem* for higher-dimensional subshifts of finite type, which is undecidable in general, is not completely trivial even in this very special setting, where we are dealing with closed, shift-invariant subgroups of $G^{\mathbb{Z}^d}$ for a finite group G. For a discussion of this and related problems involving decidability see [44].

(2) In Example (1) we replace $\mathbb{Z}_{/2}$ by a compact Lie group K with a distinguished closed, normal subgroup $L \subset K$ and set $G = K \times K$ and

$$H = \left\{ \left\{ \begin{smallmatrix} (*,k) & (k'l,*) \\ (*,k') & (kl',*) \end{smallmatrix} \right\} : k, k' \in K, l, l' \in L \right\} \subset G^F.$$

Then the group $X = X_{(F,H)}$ defined by (4.12) is again a full, shift-invariant subgroup of $G^{\mathbb{Z}^2}$. ☐

CONCLUDING REMARKS 4.17. (1) For finitely generated, abelian groups Theorem 4.2 was proved in [45]. Example 4.13 is due to Losert, and Example 4.16 (1) to Coppersmith. Losert has also shown me a proof that, if Γ is a countable, finitely generated group, and if α is an ergodic action of Γ by automorphisms of a compact group X, then $\alpha^{X^{\circ}}$ is ergodic.

(2) It is not known which countable groups Γ are of Markov type. Proposition 4.6 lends plausibility to the conjecture that Γ is of Markov type if and only if $\mathbb{Z}[\Gamma]$ is right Noetherian; however, the latter class of countable groups is just as mysterious as the groups of Markov type. As far as I am aware, all known examples of countable groups Γ for which $\mathbb{Z}[\Gamma]$ is right Noetherian are polycyclic-by-finite; furthermore, the ascending chain conditions on subgroups of Γ (in the sense of Example 3.9 (2)) does not imply that $\mathbb{Z}[\Gamma]$ is right Noetherian (or that Γ is of Markov type).

(3) If G is a Lie group with finite centre, and if $F = \{0,1\}^2 \subset \mathbb{Z}^2$, the choice of full subgroups $H \subset G^F$ is rather limited, and in conjunction with Remark 4.15 this imposes severe restrictions on the possibilities for full, shift-invariant subgroups of $G^{\mathbb{Z}^2}$. In the Sections 5–6 we shall see that the situation is completely different if G is abelian.

CHAPTER II

$\mathbb{Z}^d$-actions on compact abelian groups

5. The dual module

According to Theorem 4.2, $\mathbb{Z}^d$ is of Markov type for every $d \geq 1$, and $\mathbb{Z}^d$-actions by automorphisms of compact groups enjoy the properties described in (4.10), Propositions 4.9–4.10, Remark 4.15, and Theorem 4.11. Just as compact, abelian groups like $\mathbb{T}^n = \mathbb{R}^n/\mathbb{Z}^n$ have automorphisms with very intricate dynamical properties, there is an abundance of examples of interesting $\mathbb{Z}^d$-actions by automorphisms of compact abelian groups. In this section we introduce a general formalism for the investigation of such actions which will also give us a systematic approach to constructing actions with specified properties.

Let $d \geq 1$, and let $\alpha \colon \mathbf{n} \mapsto \alpha_{\mathbf{n}}$ be an action of $\mathbb{Z}^d$ by automorphisms of X. For every $\mathbf{n} = (n_1, \ldots, n_d) \in \mathbb{Z}^d$ we denote by $\hat{\alpha}_{\mathbf{n}}$ the automorphism of $\hat{X}$ dual to $\alpha_{\mathbf{n}}$ and write $\hat{\alpha} \colon \mathbb{Z}^d \longrightarrow \operatorname{Aut}(\hat{X})$ for the resulting $\mathbb{Z}^d$-action dual to α. Under the action $\hat{\alpha}$ the group $\hat{X}$ becomes a $\mathbb{Z}^d$-module, and hence a module over the group ring $\mathbb{Z}[\mathbb{Z}^d]$. In order to make this explicit we denote by

$$\mathfrak{R}_d = \mathbb{Z}[u_1^{\pm 1}, \ldots, u_d^{\pm 1}] \tag{5.1}$$

the ring of Laurent polynomials in the (commuting) variables $u_1, \ldots, u_d$ with coefficients in $\mathbb{Z}$. A typical element $f \in \mathfrak{R}_d$ will be written as

$$f = \sum_{\mathbf{n} \in \mathbb{Z}^d} c_f(\mathbf{n}) u^{\mathbf{n}}, \tag{5.2}$$

where $c_f(\mathbf{n}) \in \mathbb{Z}$ and $u^{\mathbf{n}} = u_1^{n_1} \cdot \ldots \cdot u_d^{n_d}$ for all $\mathbf{n} = (n_1, \ldots, n_d) \in \mathbb{Z}^d$, and where $c_f(\mathbf{n}) \neq 0$ for only finitely many $\mathbf{n} \in \mathbb{Z}^d$. Then $\mathfrak{R}_d \cong \mathbb{Z}[\mathbb{Z}^d]$, $\mathfrak{R}_d$ acts on $\hat{X}$ by

$$(f, a) \mapsto f \cdot a = \sum_{\mathbf{n} \in \mathbb{Z}^d} c_f(\mathbf{n}) \hat{\alpha}_{\mathbf{n}}(a) \tag{5.3}$$

for every $f \in \mathfrak{R}_d$, $a \in \hat{X}$, and $\hat{X}$ is an $\mathfrak{R}_d$-module. Note that

$$\hat{\alpha}_{\mathbf{n}}(a) = \hat{\alpha}_{\mathbf{n}}(a) = u^{\mathbf{n}} \cdot a \tag{5.4}$$

for every $\mathbf{n} \in \mathbb{Z}^d$ and $a \in \hat{X}$. Conversely, if $\mathfrak{M}$ is an $\mathfrak{R}_d$-module (always assumed to be countable), then $\mathbb{Z}^d$ has an obvious action $\hat{\alpha}^{\mathfrak{M}} \colon \mathbf{n} \mapsto \hat{\alpha}_{\mathbf{n}}^{\mathfrak{M}}$ on $\mathfrak{M}$ given by

$$\hat{\alpha}_{\mathbf{n}}^{\mathfrak{M}}(a) = u^{\mathbf{n}} \cdot a \tag{5.5}$$

for every $\mathbf{n} \in \mathbb{Z}^d$ and $a \in \mathfrak{M}$. We write $X = \widehat{\mathfrak{M}}$ for the dual group of $\mathfrak{M}$ and obtain a dual action

$$\alpha^{\mathfrak{M}} \colon \mathbf{n} \mapsto \alpha_{\mathbf{n}}^{\mathfrak{M}} \in \mathrm{Aut}(X) \tag{5.6}$$

of $\mathbb{Z}^d$ on X. For future reference we collect these observations in a lemma.

LEMMA 5.1. *Let $\alpha \colon \mathbf{n} \mapsto \alpha_{\mathbf{n}}$ be a $\mathbb{Z}^d$-action by automorphisms of a compact, abelian group X, and let $\hat{\alpha} \colon \mathbf{n} \mapsto \hat{\alpha}_{\mathbf{n}}$ be the dual action of $\mathbb{Z}^d$ on the dual group $\hat{X}$ of X. If $\mathfrak{R}_d$ is the ring defined in (5.1) then $\hat{X}$ is an $\mathfrak{R}_d$-module under the $\mathfrak{R}_d$-action (5.3). Conversely, if $\mathfrak{M}$ is an $\mathfrak{R}_d$-module, then (5.5) and (5.6) define $\mathbb{Z}^d$-actions $\hat{\alpha}^{\mathfrak{M}} = \hat{\alpha}$ and $\alpha^{\mathfrak{M}} = \alpha$ by automorphisms of $\mathfrak{M}$ and $X^{\mathfrak{M}} = \widehat{\mathfrak{M}}$, respectively.*

EXAMPLES 5.2. Let $d \geq 1$.

(1) Let $\mathfrak{M} = \mathfrak{R}_d$. Since $\mathfrak{R}_d$ is isomorphic to the direct sum $\sum_{\mathbb{Z}^d} \mathbb{Z}$ of copies of $\mathbb{Z}$ indexed by $\mathbb{Z}^d$, the dual group $X = \widehat{\mathfrak{R}_d}$ is isomorphic to the cartesian product $\mathbb{T}^{\mathbb{Z}^d}$ of copies of $\mathbb{T} = \mathbb{R}/\mathbb{Z}$. We write a typical element $x \in \mathbb{T}^{\mathbb{Z}^d}$ as $x = (x_{\mathbf{n}}) = (x_{\mathbf{n}}, \mathbf{n} \in \mathbb{Z}^d)$ with $x_{\mathbf{n}} \in \mathbb{T}$ for every $\mathbf{n} \in \mathbb{Z}^d$ and choose the following identification of $X^{\mathfrak{R}_d} = \widehat{\mathfrak{R}_d}$ and $\mathbb{T}^{\mathbb{Z}^d}$: for every $x = (x_{\mathbf{n}})$ in $\mathbb{T}^{\mathbb{Z}^d}$ and $f \in \mathfrak{R}_d$,

$$\langle x, f \rangle = e^{2\pi i \sum_{\mathbf{n} \in \mathbb{Z}^d} c_f(\mathbf{n}) x_{\mathbf{n}}}, \tag{5.7}$$

where f is given by (5.2). Under this identification the $\mathbb{Z}^d$-action $\alpha^{\mathfrak{R}_d}$ on $X^{\mathfrak{R}_d} = \mathbb{T}^{\mathbb{Z}^d}$ becomes the shift-action

$$\alpha_{\mathbf{n}}^{\mathfrak{R}_d}(x)_{\mathbf{m}} = (\sigma_{\mathbf{n}}(x))_{\mathbf{m}} = x_{\mathbf{m}+\mathbf{n}}, \tag{5.8}$$

with $\mathbf{n} \in \mathbb{Z}^d$ and $x = (x_{\mathbf{m}}) \in X^{\mathfrak{R}_d} = \mathbb{T}^{\mathbb{Z}^d}$.

(2) Let $\mathfrak{a} \subset \mathfrak{R}_d$ be an ideal, and let $\mathfrak{M} = \mathfrak{R}_d/\mathfrak{a}$. Since $\mathfrak{M}$ is a quotient of the additive group $\mathfrak{R}_d$ by a $\hat{\alpha}^{\mathfrak{R}_d}$-invariant subgroup, the dual group $X^{\mathfrak{M}}$ is the $\alpha^{\mathfrak{R}_d}$-invariant subgroup

$$X^{\mathfrak{R}_d/\mathfrak{a}} = \{x \in X^{\mathfrak{R}_d} = \mathbb{T}^{\mathbb{Z}^d} : \langle x, f \rangle = 1 \text{ for every } f \in \mathfrak{a}\}$$

$$= \left\{ x \in \mathbb{T}^{\mathbb{Z}^d} : \sum_{\mathbf{n} \in \mathbb{Z}^d} c_f(\mathbf{n}) x_{\mathbf{m}+\mathbf{n}} = 0 \pmod 1 \right. \tag{5.9}$$
$$\left. \text{for every } f \in \mathfrak{a} \text{ and } \mathbf{m} \in \mathbb{Z}^d \right\},$$

and $\alpha^{\Re_d/\mathfrak{a}}$ is the restriction of $\alpha^{\Re_d}$ to $X^{\mathfrak{M}} \subset \mathbb{T}^{\mathbb{Z}^d}$, i.e.

$$\alpha_{\mathbf{n}}^{\Re_d/\mathfrak{a}} = \sigma_{\mathbf{n}}^{X^{\Re_d/\mathfrak{a}}} \tag{5.10}$$

for every $\mathbf{n} \in \mathbb{Z}^d$.

(3) Let $X \subset \mathbb{T}^{\mathbb{Z}^d} = \widehat{\Re_d}$ be a closed subgroup, and let $X^{\perp} = \{f \in \Re_d : \langle x, f \rangle = 1$ for every $x \in X\}$ be the annihilator of X in $\widehat{\Re_d}$. Then X is shift-invariant if and only if $X^{\perp}$ is an ideal in $\Re_d$: indeed, if $X^{\perp}$ is an ideal, it is obviously invariant under multiplication by the group of units $\{u^{\mathbf{n}} : \mathbf{n} \in \mathbb{Z}^d\} \subset \Re_d$, i.e. $X^{\perp}$ is $\hat{a}^{\Re_d}$-invariant; conversely, if $X^{\perp}$ is $\hat{a}^{\Re_d}$-invariant, then (5.3) shows that $f \cdot a \in X^{\perp}$ for every $f \in \Re_d$ and $a \in X^{\perp}$. In other words, $X^{\perp}$ is an ideal.

(4) Let $\mathfrak{M}$ be a Noetherian $\Re_d$-module, and let $\{a_1, \ldots, a_k\}$ be a set of generators for $\mathfrak{M}$, i.e. $\mathfrak{M} = \Re_d \cdot a_1 + \cdots + \Re_d \cdot a_k$. The surjective homomorphism $(f_1, \ldots, f_k) \mapsto f_1 \cdot a_1 + \cdots + f_k \cdot a_k$ from $\Re_d^k$ to $\mathfrak{M}$ induces a dual injective homomorphism $\phi \colon X^{\mathfrak{M}} \longmapsto X^{\Re_d^k} \cong (\mathbb{T}^k)^{\mathbb{Z}^d} = Y$ such that $\alpha_{\mathbf{n}}^{\mathfrak{M}} \cdot \phi = \sigma_{\mathbf{n}} \cdot \phi$ for every $\mathbf{n} \in \mathbb{Z}^d$, where $\sigma_{\mathbf{n}}$ is the shift on $(\mathbb{T}^k)^{\mathbb{Z}^d}$ defined in (5.8). In particular, ϕ embeds $X^{\mathfrak{M}}$ as a closed, shift-invariant subgroup of $(\mathbb{T}^k)^{\mathbb{Z}^d}$. Conversely, if $X \subset (\mathbb{T}^k)^{\mathbb{Z}^d}$ is a closed, shift-invariant subgroup, then $\hat{X} = \Re_d^k/X^{\perp}$, and $X^{\perp}$ is a submodule of $\Re_d^k$. $\square$

EXAMPLES 5.3. (1) Let α be the automorphism of $\mathbb{T}^2 = \mathbb{R}^2/\mathbb{Z}^2$ determined by the matrix $A = \left(\begin{smallmatrix} 0 & 1 \\ 1 & 1 \end{smallmatrix}\right)$. In Example 2.18 (2) we have seen that α (or, more precisely, the $\mathbb{Z}$-action on $\mathbb{T}^2$ defined by α) is conjugate to $(X^{\Re_1/(f)}, \alpha^{\Re_1/(f)})$, where $(f) \subset \Re_1$ is the principal ideal generated by the characteristic polynomial $f(u_1) = 1 + u_1 - u_1^2$ of A. Indeed, an element $x \in X = \Re_1 = \mathbb{T}^{\mathbb{Z}}$ satisfies that $\langle x, u_1^n f \rangle = 1$ if and only if $x_n + x_{n+1} - x_{n+2} = 0 \pmod 1$, and hence

$$X^{\Re_1/(f)} = \{x \in \mathbb{T}^{\mathbb{Z}} : x_n + x_{n+1} - x_{n+2} = 0 \pmod 1 \text{ for all } n \in \mathbb{Z}\}$$

(cf. (5.7) and (5.9)). The continuous group isomorphism $\phi = \pi_{\{0,1\}} \colon X^{\Re_1/(f)} \longmapsto \mathbb{T}^2$ makes the diagram

$$
\begin{array}{ccc}
X^{\Re_1/(f)} & \xrightarrow{\alpha^{\Re_1/(f)}} & X^{\Re_1/(f)} \\
\phi \downarrow & & \downarrow \phi \\
\mathbb{T}^2 & \xrightarrow{\quad \alpha \quad} & \mathbb{T}^2
\end{array}
\tag{5.11}
$$

commute, and the automorphism $\alpha^{\Re_1/(f)}$ is equal to the shift on $X^{\Re_1/(f)}$.

(2) Example (1) depends on the fact that the matrix A is conjugate (over $\mathbb{Z}$) to the companion matrix of its characteristic polynomial. If α is the automorphism of $\mathbb{T}^2$ defined by $A = \left(\begin{smallmatrix} 3 & 4 \\ 1 & 1 \end{smallmatrix}\right)$, then the characteristic polynomial of A is $f(u_1) = -1 - 4u_1 + u_1^2$, and $AM = MB$, where $B = \left(\begin{smallmatrix} 0 & 1 \\ 1 & 4 \end{smallmatrix}\right)$ and $M = \left(\begin{smallmatrix} 1 & 3 \\ 0 & 1 \end{smallmatrix}\right)$. The

map $\phi: X^{\mathfrak{R}_1/(f)} \longmapsto \mathbb{T}^2$ given by $\phi(x) = (x_0 + 3x_1, x_1)$ for all $x \in X^{\mathfrak{R}_1/(f)} \subset \mathbb{T}^{\mathbb{Z}}$ is a group isomorphism, and the diagram (5.11) commutes.

If $A' = \left(\begin{smallmatrix} 3 & 2 \\ 2 & 1 \end{smallmatrix}\right)$, then the characteristic polynomial of A' is again equal to $f(u_1) = -1 - 4u_1 + u_1^2$, $A'M = MB$ with $M = \left(\begin{smallmatrix} 1 & 3 \\ 0 & 2 \end{smallmatrix}\right)$, but there is no matrix M' with integer entries and determinant 1 such that $A'M' = M'B$. The homomorphism $\phi': X^{\mathfrak{R}_1/(f)} \longmapsto \mathbb{T}^2$ with $\phi'(x) = (x_0 + 3x_1, 2x_1)$ for all $x \in X^{\mathfrak{R}_1/(f)} \subset \mathbb{T}^{\mathbb{Z}}$ is surjective, and we write $\psi' = \hat{\phi}: \mathbb{Z}^2 \longmapsto \mathfrak{R}_1/(f)$ for the dual homomorphism, which is injective, but not bijective. The $\mathfrak{R}_1$-module $\mathfrak{M} = \hat{X}$ arising from the $\mathbb{Z}$-action $n \mapsto (A')^n$ via Lemma 5.1 is (isomorphic to) the submodule $\psi'(\mathbb{Z}^2)$ of $\mathfrak{R}_1/(f)$. We claim that $\mathfrak{M}$ is not isomorphic to $\mathfrak{R}_1/(f)$— in fact, $\mathfrak{M}$ is not even *cyclic*, i.e. not of the form $\mathfrak{M} = \mathfrak{R}_1 \cdot a$ for some $a \in \mathfrak{M}$. Indeed, if $\mathfrak{M}$ were cyclic, there would exist an element $\mathbf{m} = (m_1, m_2) \in \mathbb{Z}^2$ such that $\{(A')^n \mathbf{m} : n \in \mathbb{Z}\}$ generates $\mathbb{Z}^2$, which is equivalent to the condition that

$$\{\mathbf{m}, A'\mathbf{m}\} = \{(m_1, m_2), (3m_1 + 2m_2, 2m_1 + m_2)\}$$

generates $\mathbb{Z}^2$. Hence

$$\det \left(\begin{smallmatrix} m_1 & 3m_1 + 2m_2 \\ m_2 & 2m_1 + m_2 \end{smallmatrix}\right) = 2m_1^2 - 2m_1 m_2 - 2m_2^2 = 1,$$

which is obviously impossible.

(3) Let $f = 2 - u_1 \in \mathfrak{R}_1$, and let (f) be the principal ideal generated by f. According to (5.7) and (5.9),

$$X = X^{\mathfrak{R}_1/(f)} = \{x = (x_n) \in \mathbb{T}^{\mathbb{Z}} : 2x_n = x_{n+1} \pmod 1 \text{ for all } n \in \mathbb{Z}\},$$

and $\alpha^{\mathfrak{R}_1/(f)}$ is equal to the shift-action σ of $\mathbb{Z}$ on X. The zero coordinate projection $\phi = \pi_{\{0\}}: X \longmapsto \mathbb{T}$ is surjective and satisfies that $\phi \cdot \sigma_1 = T \cdot \phi$, where $T: \mathbb{T} \longmapsto \mathbb{T}$ is the surjective homomorphism consisting of multiplication by 2 modulo 1.

(4) Let $f_1 = 2 - u_1$, $f_2 = 3 - u_2$, and let $\mathfrak{a} = (f_1, f_2) = f_1\mathfrak{R}_2 + f_2\mathfrak{R}_2 \subset \mathfrak{R}_2$. Then

$$X = X^{\mathfrak{R}_2/\mathfrak{a}} = \{x = (x_{m,n}) \in \mathbb{T}^{\mathbb{Z}^2} : 2x_{(m,n)} = x_{(m+1,n)} \pmod 1 \text{ and}$$
$$3x_{(m,n)} = x_{(m,n+1)} \pmod 1 \text{ for every } (m,n) \in \mathbb{Z}^2\},$$

and $\alpha^{\mathfrak{R}_2/\mathfrak{a}} = \sigma$ is the shift-action of $\mathbb{Z}^2$ on $X^{\mathfrak{R}_2/\mathfrak{a}}$. The zero coordinate projection $\phi = \pi_{\{(0,0)\}}: X \longmapsto \mathbb{T}$ is again surjective and satisfies that $\phi \cdot \sigma_{\mathbf{n}} = T_{\mathbf{n}} \cdot \phi$ for every $\mathbf{n} \in \mathbb{Z}^2$, where T is the $\mathbb{N}^2$-action on $\mathbb{T}$ defined by $T_{(m,n)}(t) = 2^m 3^n t$ $\pmod 1$ for every $(m,n) \in \mathbb{Z}^2$ and $t \in \mathbb{T}$.

(5) Let

$$X = \{x = (x_{\mathbf{n}}) \in \mathbb{Z}_{/2}^{\mathbb{Z}^2} : x_{(m_1, m_2)} + x_{(m_1+1, m_2)} + x_{(m_1, m_2+1)} = 0 \pmod 2$$
$$\text{for all } \mathbf{m} = (m_1, m_2) \in \mathbb{Z}^2\}.$$

From (5.7) and (5.9) we see that the shift-action σ of $\mathbb{Z}^2$ on the full, shift-invariant subgroup $X \subset \mathbb{Z}_{/2}^{\mathbb{Z}^2}$ is conjugate to $(X^{\mathfrak{R}_2/\mathfrak{a}}, \alpha^{\mathfrak{R}_2/\mathfrak{a}})$, where $\mathfrak{a} = (2, 1 + u_1 + u_2) \subset \mathfrak{R}_2$ is the ideal generated by 2 and $1 + u_1 + u_2$.

(6) Let $d \geq 1$. A Laurent polynomial $f \in \mathfrak{R}_d$ is *primitive* if the highest common factor of its coefficients is equal to 1. Suppose that f is primitive and $m > 1$ an integer, and let (f) and (mf) be the principal ideals in $\mathfrak{R}_d$ generated by f and mf, respectively. The map $h \mapsto mh$ from $\mathfrak{R}_d$ to $\mathfrak{R}_d$ induces an injective homomorphism $\xi \colon \mathfrak{R}_d/(f) \longmapsto \mathfrak{R}_d/(mf)$, the dual homomorphism $\phi \colon X^{\mathfrak{R}_d/(mf)} \longmapsto X^{\mathfrak{R}_d/(f)}$ is surjective, and $\ker(\phi) \cong \mathbb{Z}_{/m}^{\mathbb{Z}^d}$. The group $X^{\mathfrak{R}_d/(f)}$ is connected, and the connected component of the identity in $X^{\mathfrak{R}_d/(mf)}$ is isomorphic to $X^{\mathfrak{R}_d/(f)}$.

More generally, if $\mathfrak{a} \subset \mathfrak{R}_d$ is an arbitrary ideal such that the additive group $\mathfrak{R}_d/\mathfrak{a}$ is torsion-free (or, equivalently, such that $X^{\mathfrak{R}_d/\mathfrak{a}}$ is connected), and if $m \geq 1$ is an integer, then we obtain an exact sequence

$$0 \longrightarrow \mathbb{Z}_{/m}^{\mathbb{Z}^d} \xrightarrow{\psi} X^{\mathfrak{R}_d/m\mathfrak{a}} \xrightarrow{\phi} X^{\mathfrak{R}_d/\mathfrak{a}} \longrightarrow 0,$$

where $\phi \colon X^{\mathfrak{R}_d/m\mathfrak{a}} \longmapsto X^{\mathfrak{R}_d/\mathfrak{a}}$ is the surjection dual to the injective homomorphism $\xi \colon \mathfrak{R}_d/\mathfrak{a} \longmapsto \mathfrak{R}_d/m\mathfrak{a}$ consisting of multiplication by m, and where ψ is the inclusion map. Note that $\psi \cdot \sigma_{\mathbf{n}}(x) = \alpha_{\mathbf{n}}^{\mathfrak{R}_d/m\mathfrak{a}} \cdot \psi(x)$ and $\phi \cdot \alpha_{\mathbf{n}}^{\mathfrak{R}_d/m\mathfrak{a}}(y) = \alpha_{\mathbf{n}}^{\mathfrak{R}_d/\mathfrak{a}} \cdot \phi(y)$ for all $\mathbf{n} \in \mathbb{Z}^d$, $x \in \mathbb{Z}_{/m}^{\mathbb{Z}^d}$, and $y \in X^{\mathfrak{R}_d/m\mathfrak{a}}$, where σ is the shift-action of $\mathbb{Z}^d$ on $\mathbb{Z}_{/m}^{\mathbb{Z}^d}$, and that the map ϕ induces an isomorphism of the connected component of the identity in $X^{\mathfrak{R}_d/m\mathfrak{a}}$ with $X^{\mathfrak{R}_d/\mathfrak{a}}$. $\boxdot$

The next proposition is a straightforward consequence of Theorem 4.2 and Pontryagin duality (cf. also Example 5.2 (4)).

PROPOSITION 5.4. *Let X be a compact, abelian group, α a $\mathbb{Z}^d$-action by automorphisms of X. The following conditions are equivalent.*

(1) *The $\mathfrak{R}_d$-module $\mathfrak{M} = \hat{X}$ obtained via Lemma 5.1 is Noetherian;*

(2) *(X, α) satisfies the d.c.c.;*

(3) *(X, α) is conjugate to a subshift of $(\mathbb{T}^n)^{\mathbb{Z}^d}$ for some $n \geq 1$.*

The Noetherian $\mathfrak{R}_d$-modules form a particularly well-behaved class of $\mathfrak{R}_d$-modules, and it is therefore not surprising that $\mathbb{Z}^d$-actions by automorphisms of compact, abelian groups satisfying the d.c.c. have many exceptional properties. As a first illustration of the rôle played by the descending chain condition, let us consider the set of periodic points for a $\mathbb{Z}^d$-action α on a compact, abelian group X.

DEFINITION 5.5. Let Γ be a countable group and let α be a Γ-action by automorphisms of a compact group X. A point $x \in X$ is *periodic* under α (or *α-periodic*) if its orbit $\alpha_\Gamma(x) = \{\alpha_\gamma(x) : \gamma \in \Gamma\}$ is finite. If $\beta \in \mathrm{Aut}(X)$ then a point $x \in X$ is *periodic* under β if $\beta^n(x) = x$ for some $n \geq 1$.

The following examples show that a $\mathbb{Z}^d$-action by automorphisms of a compact, abelian group need not have any periodic points other than the fixed point $\mathbf{0}_X$, but in Theorem 5.7 we shall see that the set of α-periodic points is dense if (X, α) satisfies the d.c.c.

EXAMPLES 5.6. (1) Let $X = \widehat{\mathbb{Q}}$ be the dual group of the additive group $\mathbb{Q}$, and consider the automorphism α of X dual to multiplication by $\frac{3}{2}$ on $\mathbb{Q}$. If $x \in X$ is a periodic point of α, i.e. if $\alpha^n(x) = x$ for some $n \geq 1$, then $\langle \alpha^n(x) - x, a \rangle = \langle x, (\frac{3^n}{2^n} - 1)a \rangle = 1$ for every $a \in \mathbb{Q}$. However, $(\frac{3^n}{2^n} - 1) \neq 0$, so that $\langle x, a \rangle = 1$ for every $a \in \mathbb{Q}$. This shows that $x = \mathbf{0}_X$.

(2) Let $Y = \mathbb{Z}_{/2}^{\mathbb{Z}}$. For every $n \geq 2$ we define a continuous, shift commuting, surjective homomorphism $\phi_n \colon Y \longmapsto Y$ by setting $(\phi_n(y))_m = \sum_{k=m}^{m+n-1} y_k$ for every $m \in \mathbb{Z}$ and $y = (y_k, k \in \mathbb{Z}) \in Y$. We put $\psi_n = \phi_n$ for every $n \geq 2$ and denote by X the projective limit

$$Y \xleftarrow{\psi_2} Y \xleftarrow{\psi_3} \cdots \xleftarrow{\psi_n} Y \xleftarrow{\psi_{n+1}} \cdots \qquad (5.12)$$

The shift σ on Y commutes with the maps ψ_n and induces an automorphism α of the projective limit X in (5.12). Suppose that α has a periodic point $x \in X$ with period n, say. We can write x as $(x^{(k)}, k \geq 1)$ with $x^{(k)} \in Y$ and $\psi_k(x^{(k)}) = x^{(k-1)}$ for every $k \geq 2$. Since x has period n, $\sigma^n(x^{(k)}) = x^{(k)}$ for every $k \geq 1$. However, $\psi_{nk}(x^{(nk)}) = \phi_{nk}(x^{(nk)}) = x^{(nk-1)} \in \{\mathbf{0}, \mathbf{1}\}$ for every $k \geq 1$, where $\mathbf{0} = (\dots, 0, 0, 0, \dots)$ and $\mathbf{1} = (\dots, 1, 1, 1, \dots)$ are the fixed points of σ in Y. As k can be arbitrarily large we see that $x^{(k)} \in \{\mathbf{0}, \mathbf{1}\}$ for every $k \geq 0$. Finally we observe that, if $k \geq 2$ is even, then $x^{(k-1)} = \psi_k(x^{(k)}) = \mathbf{0}$. This shows that $x^{(k)} = \mathbf{0}$ for every $k \geq 1$, i.e. that $x = \mathbf{0}_X$.

(3) We stay with the notation of Example (2) and set $\psi_n = \phi_2$ for every $n \geq 2$ in (5.12). The projective limit X in (5.12) can be written as $X = \{x = (x_{(m,n)}) \in \mathbb{Z}_{/2}^{\mathbb{Z} \times \mathbb{N}} : x_{(m,n)} = x_{(m,n+1)} + x_{(m+1,n+1)} \pmod 2$ for every $m \in \mathbb{Z}$ and $n \geq 1\}$, and α is the horizontal shift on X defined by $(\alpha(x))_{(m,n)} = x_{(m+1,n)}$ for all $x \in X$ and $(m, n) \in \mathbb{Z} \times \mathbb{N}^*$. The same argument as in Example (2) shows that every point $x \in X$ with period 2^k, $k \geq 0$ is equal to the identity element $\mathbf{0}_X$, but that there exist 2^{k-1} points of period k if $k \geq 1$ is odd (for every sequence $y = (y_m) \in Y$ with $y_{(m+k)} = y_m$ and $\sum_{j=0}^{k-1} x_{m+j} = 0 \pmod 2$ for all $m \in \mathbb{Z}$ there exists a unique point $x \in X$ with $\alpha^k(x) = x$ and $x_{(m,1)} = y_m$ for all $m \in \mathbb{Z}$).

If $\mathfrak{a} \subset \mathfrak{R}_2$ is the ideal $(2, 1 + u_2 + u_1 u_2) = 2\mathfrak{R}_2 + (1 + u_2 + u_1 u_2)\mathfrak{R}_2$, then (5.7) and (5.9) show that $(X^{\mathfrak{R}_2/\mathfrak{a}}, \alpha^{\mathfrak{R}_2/\mathfrak{a}})$ is (conjugate to) the shift-action of $\mathbb{Z}^2$ on

$$X' = \{x = (x_{(m,n)}) \in \mathbb{Z}_{/2}^{\mathbb{Z}^2} : x_{(m,n)} + x_{(m,n+1)} + x_{(m+1,n+1)}$$
$$= 0 \pmod 2 \text{ for every } (m, n) \in \mathbb{Z}^2\},$$

and a comparison of X' with the definition of X in the preceding paragraph reveals that X is equal to the projection of X' onto its coordinates in the upper half plane of $\mathbb{Z}^2$, and that this projection sends the horizontal shift $\sigma_{(1,0)}$ of X' to the automorphism α of X. In particular we see that the shift-action σ of $\mathbb{Z}^2$ on X' has only one point with horizontal period 2^k for every $k \geq 0$ (the identity element). We also refer to Example 5.3 (5): the $\mathbb{Z}^2$-action $\alpha^{\mathfrak{R}_2/\mathfrak{a}}$ appearing there obviously has the same property.

(4) Let $\psi_n = \phi_3$ for every $n \geq$ in (5.12). Then the resulting automorphism α of the projective limit X in (5.12) has only one point with period 3^k, $k \geq 0$, but there exist 2^k points with period k for every k which is not divisible by 3.

(5) Let $(p_n, n \geq 2)$ be a sequence of rational primes in which every prime occurs infinitely often, and let $(q_n, n \geq 2)$ be a sequence of odd primes in which every odd prime occurs infinitely often. If $\psi_n = \phi_{p_n}$ for every $n \geq 2$, then the automorphism α of the projective limit X in (5.12) has no periodic points other than the fixed point $\mathbf{0}_X$. However, if $\psi_n = \phi_{q_n}$, $n \geq 2$, then the resulting automorphism α will have 2^{2^k} periodic points with period 2^k for every $k \geq 0$, but only one point with period $2l + 1$ for every $l \geq 0$ (the fixed point $\mathbf{0}_X$).

None of the automorphisms α in Examples (1)–(5) satisfies the d.c.c. ☐

THEOREM 5.7. *Let X be a compact, abelian group, and let α be a $\mathbb{Z}^d$-action by automorphisms of X. If (X, α) satisfies the d.c.c. then the set of α-periodic points is dense in X.*

PROOF. Let $\mathfrak{M} = \hat{X}$ be the $\mathfrak{R}_d$-module arising from Lemma 5.1. Fix a non-zero element $a \in \mathfrak{M}$ and choose a submodule $\mathfrak{M}_a \subset \mathfrak{M}$ which is maximal with respect to the property that $a \notin \mathfrak{M}_a$. Then the $\mathfrak{R}_d$-module $\mathfrak{M}' = \mathfrak{M}/\mathfrak{M}_a$ has the minimal non-zero submodule $\mathfrak{M}'_1 = (\mathfrak{R}_d \cdot a + \mathfrak{M}_a)/\mathfrak{M}_a$. Consider the ideal $\mathfrak{a} = \{f \in \mathfrak{R}_d : f \cdot \mathfrak{M}'_1 = 0\}$, and let $\mathfrak{b}$ be an ideal with $\mathfrak{a} \subsetneq \mathfrak{b} \subsetneq \mathfrak{R}_d$. The minimality of $\mathfrak{M}'_1$ implies that $\mathfrak{b} \cdot \mathfrak{M}'_1 = \mathfrak{M}'_1$, and Corollary 2.5 in [5] shows that there exists an element $x \in 1 + \mathfrak{b}$ such that $x \cdot \mathfrak{M}'_1 = \{0\}$. This contradicts our definition of $\mathfrak{a}$, and we conclude that the ideal $\mathfrak{a} \subset \mathfrak{R}_d$ is maximal, and that $\mathfrak{k} = \mathfrak{R}_d/\mathfrak{a}$ is a (necessarily finite) field.

For every $m \geq 1$ we write $\mathfrak{a}^m \subset \mathfrak{R}_d$ for the ideal generated by $\{f_1 \cdot \ldots \cdot f_m : f_i \in \mathfrak{a}$ for $i = 1, \ldots, m\}$. If $a' = a + \mathfrak{M}_a \in \mathfrak{a}^m \cdot \mathfrak{M}'$ for every $m \geq 1$, then $a \in \mathfrak{M}'' = \bigcap_{m \geq 1} \mathfrak{a}^m \cdot \mathfrak{M}''$, and $\mathfrak{a} \cdot \mathfrak{M}''/\mathfrak{M}''$. The argument in the preceding paragraph shows that there exists an element $y \in 1 + \mathfrak{a}$ with $y \cdot \mathfrak{M}'' = \{0\}$, and the maximality of $\mathfrak{a}$ implies that $\mathfrak{M}'' = \{0\}$, which is absurd. Hence there exists an integer $m \geq 1$ with $a' \notin \mathfrak{a}^m \cdot \mathfrak{M}'$, and the maximality of $\mathfrak{M}_a$ implies that $\mathfrak{a}^m \cdot \mathfrak{M}' = \{0\}$.

Each of the successive quotients $\mathfrak{a}^r \cdot \mathfrak{M}'/\mathfrak{a}^{r+1} \cdot \mathfrak{M}'$ in the decreasing sequence of $\mathfrak{R}_d$-modules $\mathfrak{M}' \supset \mathfrak{a} \cdot \mathfrak{M}' \supset \cdots \supset \mathfrak{a}^m \cdot \mathfrak{M}' = \{0\}$ is a Noetherian module over $\mathfrak{k}$. Since $\mathfrak{k}$ is finite we conclude that $\mathfrak{M}'$ is finite.

We have found, for every non-zero $a \in \mathfrak{M} = \hat{X}$, a submodule $\mathfrak{M}_a \subset \mathfrak{M}$ such that $a \notin \mathfrak{M}_a$ and $\mathfrak{M}/\mathfrak{M}_a$ is finite. The subgroup $X_a = \mathfrak{M}_a^\perp \subset X$ is finite, α-invariant, and is not annihilated by (the character corresponding to) a. Since every point in X_a must be α-periodic, and since the α-periodic points form a subgroup of X, this shows that the set of α-periodic points is dense in X. $\square$

Before turning to the problem of relating the algebraic properties of a Noetherian $\mathfrak{R}_d$-module $\mathfrak{M}$ to the dynamical properties of $(X^{\mathfrak{M}}, \alpha^{\mathfrak{M}})$ we should discuss the extent to which $\mathfrak{M}$ and $(X^{\mathfrak{M}}, \alpha^{\mathfrak{M}})$ determine each other. Let $d \geq 1$, and let $\mathfrak{M}$ be a Noetherian $\mathfrak{R}_d$-module which is torsion-free when regarded as an additive group or, equivalently, as a $\mathbb{Z}$-module (this is equivalent to the assumption that $X^{\mathfrak{M}} = \widehat{\mathfrak{M}}$ is connected). We define the $\mathbb{Z}^d$-action $\alpha^{\mathfrak{M}}$ on $X^{\mathfrak{M}}$ by (5.5) and (5.6) and consider the action induced by $\alpha^{\mathfrak{M}}$ on the Čech homology group $H_1(X^{\mathfrak{M}}, \mathbb{T})$ (cf. [20]).

LEMMA 5.8. *The group $H_1(X^{\mathfrak{M}}, \mathbb{T})$ is isomorphic to $X^{\mathfrak{M}}$, and the automorphism induced by $\alpha_{\mathbf{n}}^{\mathfrak{M}}$ on $H_1(X^{\mathfrak{M}}, \mathbb{T})$ is equal to $\alpha_{\mathbf{n}}^{\mathfrak{M}}$ for every $\mathbf{n} \in \mathbb{Z}^d$.*

PROOF. In view of Example 5.2 (4) we may assume that $X = X^{\mathfrak{M}}$ is a closed, shift-invariant subgroup of $(\mathbb{T}^k)^{\mathbb{Z}^d}$, and the connectedness of X allows us to assume that X is full. If $F(n) = \{-n, \ldots, n\}^d \subset \mathbb{Z}^d$ then $\pi_{F(n)}(X) \subset (\mathbb{T}^k)^{F(n)}$ is a finite-dimensional torus, and X is equal to the projective limit

$$\pi_{F(1)}(X) \xleftarrow{\ \pi_{F(1)}\ } \pi_{F(2)}(X) \xleftarrow{\ \pi_{F(2)}\ } \pi_{F(3)}(X) \xleftarrow{\ \pi_{F(3)}\ } \cdots . \qquad (5.13)$$

Since $H_1(\pi_{F(k)}(X), \mathbb{T}) \cong \pi_{F(k)}(X)$ ([20]), we see from (5.13) that $H_1(X, \mathbb{T}) \cong X$, and that the automorphism induced by $\alpha_{\mathbf{n}}^{\mathfrak{M}} = \sigma_{\mathbf{n}}$ on $H_1(X, \mathbb{T})$ is equal to $\sigma_{\mathbf{n}}$ for every $\mathbf{n} \in \mathbb{Z}^d$. $\square$

THEOREM 5.9. *Let X and X' be compact, connected, abelian groups, and let α and α' be $\mathbb{Z}^d$-actions by automorphisms of X and X' which satisfy the d.c.c. The following statements are equivalent.*

(1) *The $\mathbb{Z}^d$-actions α and α' are topologically conjugate, i.e. there exists a homeomorphism $\phi \colon X \longmapsto X'$ with $\phi \cdot \alpha_{\mathbf{n}} = \alpha_{\mathbf{n}}' \cdot \phi$ for every $\mathbf{n} \in \mathbb{Z}^d$;*

(2) *The $\mathbb{Z}^d$-actions α and α' are algebraically conjugate, i.e. there exists a continuous group isomorphism $\psi \colon X \longmapsto X'$ such that $\psi \cdot \alpha_{\mathbf{n}} = \alpha_{\mathbf{n}}' \cdot \psi$ for every $\mathbf{n} \in \mathbb{Z}^d$.*

PROOF. The implication $(2) \Rightarrow (1)$ is obvious. If (1) is satisfied we use Lemma 5.1 and Proposition 5.4 to find Noetherian $\mathfrak{R}_d$-modules $\mathfrak{M}$ and $\mathfrak{M}'$ such that (X, α) and (X', α') are conjugate to $(X^{\mathfrak{M}}, \alpha^{\mathfrak{M}})$ and $(X^{\mathfrak{M}'}, \alpha^{\mathfrak{M}'})$, respectively. By Lemma 5.8, $H_1(X^{\mathfrak{M}}, \mathbb{T}) \cong X^{\mathfrak{M}}$, $H_1(X^{\mathfrak{M}'}, \mathbb{T}) \cong X^{\mathfrak{M}'}$, and for every $\mathbf{n} \in \mathbb{Z}^d$ the isomorphisms of $H_1(X^{\mathfrak{M}}, \mathbb{T})$ and $H_1(X^{\mathfrak{M}'}, \mathbb{T})$ defined by $\alpha_{\mathbf{n}}^{\mathfrak{M}}$ and $\alpha_{\mathbf{n}}^{\mathfrak{M}'}$ are equal to $\alpha_{\mathbf{n}}^{\mathfrak{M}}$ and $\alpha_{\mathbf{n}}^{\mathfrak{M}'}$, respectively. The continuous group isomorphism $\psi' \colon H_1(X^{\mathfrak{M}}, \mathbb{T}) \longmapsto H_1(X^{\mathfrak{M}'}, \mathbb{T})$ induced by $\phi \colon X \longmapsto X'$ satisfies that $\psi' \cdot \alpha_{\mathbf{n}}^{\mathfrak{M}} = \alpha_{\mathbf{n}}^{\mathfrak{M}'} \cdot \psi'$ for every $\mathbf{n} \in \mathbb{Z}^d$, and this implies (2). $\square$

COROLLARY 5.10. *Let $d \geq 1$, and let $\mathfrak{M}$ and $\mathfrak{M}'$ be finitely generated $\mathfrak{R}_d$-modules which are torsion-free (as additive groups). The following statements are equivalent.*

(1) *The $\mathbb{Z}^d$-actions $\alpha^{\mathfrak{M}}$ and $\alpha^{\mathfrak{M}'}$ are topologically conjugate;*
(2) *The $\mathbb{Z}^d$-actions $\alpha^{\mathfrak{M}}$ and $\alpha^{\mathfrak{M}'}$ are algebraically conjugate;*
(3) *There exists an $\mathfrak{R}_d$-module isomorphism $\chi \colon \mathfrak{M} \longmapsto \mathfrak{M}'$.*

PROOF. The equivalence of (1) and (2) is stated in Theorem 5.9. If (2) is satisfied, then any group isomorphism $\psi \colon X^{\mathfrak{M}} \longmapsto X^{\mathfrak{M}'}$ with $\psi \cdot \alpha_{\mathbf{n}}^{\mathfrak{M}} = \alpha_{\mathbf{n}}^{\mathfrak{M}'} \cdot \psi$ for all $\mathbf{n} \in \mathbb{Z}^d$ induces a dual isomorphism $\hat{\psi} \colon \mathfrak{M}' \longmapsto \mathfrak{M}$ which is easily seen to be an $\mathfrak{R}_d$-module isomorphism. The implication (3)$\Rightarrow$(2) is obvious. □

CONCLUDING REMARKS 5.11. (1) Most of the material of this section comes from [45], except for Lemma 5.8, Theorem 5.9, and Corollary 5.10, which come from [94]. Example 5.3 (2) is taken from [110], Example 5.3 (4) features in [23] and [89], Example 5.3 (5) comes from [56] (cf. (0.1)), and Example 5.6 (1) appears to be oral tradition attributed to Furstenberg. For $\mathbb{Z}$-actions Theorem 5.7 was first proved in [55], and the general proof presented here is due to Hartley. A more general version of Theorem 5.7 will be proved in Section 10 (Theorem 10.2).

(2) If X and X' are not connected, Theorem 5.9 (or the equivalence of (1) and (2) in Corollary 5.10) is not true in general. The shifts on the groups $\mathbb{Z}_{/4}^{\mathbb{Z}}$ and $(\mathbb{Z}_{/2}^2)^{\mathbb{Z}}$ are topologically, but not algebraically conjugate. However, the equivalence of (2) and (3) in Corollary 5.10 holds for *any* pair of $\mathfrak{R}_d$-modules $\mathfrak{M}$ and $\mathfrak{M}'$, whether they are torsion-free (as additive groups) or not.

6. The dynamical system defined by a Noetherian module

We begin with a little bit of algebra. Let $d \geq 1$, and let $\mathfrak{R}$ be a commutative ring. We denote by $\mathfrak{R}^{\times}$ the set of invertible elements (or units) in $\mathfrak{R}$, write $\mathfrak{R}[u_1, \dots, u_d]$ and $\mathfrak{R}[u_1^{\pm 1}, \dots, u_d^{\pm d}]$ for the rings of polynomials and Laurent polynomials in the commuting variables $u_1, \dots, u_d$ with coefficients in $\mathfrak{R}$, and we define $\mathfrak{R}_d$ by (5.1). For every rational prime p we denote by $\overline{\mathbb{F}}_p$ the algebraic closure of the prime field $\mathbb{F}_p = \mathbb{Z}/p\mathbb{Z} = \mathbb{Z}_{/p}$ and define a homomorphism $f \mapsto f_{/p}$ from $\mathfrak{R}_d$ to

$$\mathfrak{R}_d^{(p)} = \mathbb{F}_p[u_1^{\pm 1}, \dots, u_d^{\pm d}] \tag{6.1}$$

by reducing the coefficients of $f \in \mathfrak{R}_d$ modulo p. An element $f \in \mathfrak{R}_d^{(p)}$ will again be written in the form (5.1) with $c_f(\mathbf{n}) \in \mathbb{F}_p$ for all $\mathbf{n} \in \mathbb{Z}^d$, where $c_f(\mathbf{n}) \neq 0$ for only finitely many $\mathbf{n} \in \mathbb{Z}^d$. For notational consistency we set $\overline{\mathbb{F}}_0$ equal to the algebraic closure $\overline{\mathbb{Q}}$ of $\mathbb{Q}$ and put $\mathfrak{R}_d^{(0)} = \mathfrak{R}_d$ and $f_{/0} = f$ for every $f \in \mathfrak{R}_d$.

Let $\mathfrak{p} \subset \mathfrak{R}_d$ be a prime ideal. We identify $\mathbb{Z}$ with the set of constant polynomials in $\mathfrak{R}_d$, denote by $p(\mathfrak{p})$ the *characteristic* char$(\mathfrak{R}_d/\mathfrak{p})$ of $\mathfrak{R}_d/\mathfrak{p}$, i.e.

the unique non-negative integer such that $\mathfrak{p} \cap \mathbb{Z} = p(\mathfrak{p})\mathbb{Z}$, and define the *variety* of $\mathfrak{p}$ by

$$V(\mathfrak{p}) = \{ c \in (\overline{\mathbb{F}}_{p(\mathfrak{p})}^{\times})^d : f_{/p(\mathfrak{p})}(c) = 0 \text{ for every } f \in \mathfrak{p} \}. \tag{6.2}$$

If $\mathfrak{a} \subset \mathfrak{R}_d$ is an arbitrary ideal we set

$$V_{\mathbb{C}}(\mathfrak{a}) = \{ c \in (\mathbb{C}^{\times})^d : f(c) = 0 \text{ for every } f \in \mathfrak{a} \}. \tag{6.3}$$

Suppose that $\mathfrak{M}$ is an $\mathfrak{R}_d$-module. For every $f \in \mathfrak{R}_d$ we write $f_{\mathfrak{M}} \colon \mathfrak{M} \longmapsto \mathfrak{M}$ for the map $a \mapsto f \cdot a$, $a \in \mathfrak{M}$, and we denote by $\operatorname{ann}(a) = \{ f \in \mathfrak{R}_d : f \cdot a = 0 \}$ the annihilator of an element $a \in \mathfrak{M}$. A prime ideal $\mathfrak{p} \subset \mathfrak{R}_d$ is *associated with* $\mathfrak{M}$ if $\mathfrak{p} = \operatorname{ann}(a)$ for some $a \in \mathfrak{M}$, and the module $\mathfrak{M}$ is *associated with* $\mathfrak{p}$ if $\mathfrak{p}$ is the only prime ideal in $\mathfrak{R}_d$ associated with $\mathfrak{M}$. If $\mathfrak{M}$ is Noetherian then it is associated with $\mathfrak{p}$ if and only if

$$\mathfrak{p} = \{ f \in \mathfrak{R}_d : f_{\mathfrak{M}} \text{ is not injective} \} = \{ f \in \mathfrak{R}_d : f_{\mathfrak{M}} \text{ is nilpotent} \} \tag{6.4}$$

(cf. Corollary VI.4.11 in [51]). If $\mathfrak{M}$ is associated with $\mathfrak{p}$ and $\mathfrak{N} \subset \mathfrak{M}$ is a non-zero submodule, then $\mathfrak{N}$ is again associated with $\mathfrak{p}$. The module $\mathfrak{M}$ is a *torsion module* if the prime ideal $\{0\}$ is not associated with $\mathfrak{M}$. We shall have to be careful to distinguish between $\mathfrak{R}_d$-modules $\mathfrak{M}$ which are not *torsion* and those which are *torsion-free* as additive groups (or $\mathbb{Z}$-modules): $\mathfrak{M}$ is a torsion module if every associated prime ideal is non-zero, $\mathfrak{M}$ is a torsion group if each of its associated primes contains a non-zero constant, and $\mathfrak{M}$ is torsion-free (as an additive group) if none of its associated primes contains a non-zero constant.

A submodule $\mathfrak{W} \subset \mathfrak{M}$ is $\mathfrak{p}$-*primary* (or $\mathfrak{p}$ *belongs to* $\mathfrak{W}$) if $\mathfrak{M}/\mathfrak{W}$ is associated with $\mathfrak{p}$. *From now on we assume that* $\mathfrak{M}$ *is Noetherian.* By Theorem VI.5.3 in [51] there exist primary submodules $\mathfrak{W}_1, \ldots, \mathfrak{W}_m$ of $\mathfrak{M}$ with the following properties:

$$\begin{gathered} \text{the primes } \mathfrak{p}_i \text{ belonging to the submodules } \mathfrak{W}_i \text{ are all distinct;} \\ \mathfrak{W}_1 \cap \cdots \cap \mathfrak{W}_m = \{0\}; \\ \text{for every subset } S \subsetneq \{1, \ldots, m\}, \bigcap_{i \in S} \mathfrak{W}_i \neq \{0\}. \end{gathered} \tag{6.5}$$

A family $\{\mathfrak{W}_1, \ldots, \mathfrak{W}_m\}$ of primary submodules satisfying (6.5) is called a *reduced primary decomposition* of $\mathfrak{M}$, and $\{\mathfrak{p}_1, \ldots, \mathfrak{p}_m\}$ is the *set of associated primes* of $\mathfrak{M}$. According to the Theorems VI.5.2 and VI.5.5 in [51] the set of associated primes of $\mathfrak{M}$ is independent of the specific decomposition (6.5), and

$$\{ f \in \mathfrak{R}_d : f_{\mathfrak{M}} \text{ is not injective} \} = \bigcup_{i=1,\ldots,m} \mathfrak{p}_i. \tag{6.6}$$

PROPOSITION 6.1. *Let* $d \geq 1$, $\mathfrak{q} \subset \mathfrak{R}_d$ *a prime ideal, and let* $\mathfrak{W}$ *be a Noetherian* $\mathfrak{R}_d$-*module associated with* $\mathfrak{q}$. *Then there exist integers* $1 \leq t \leq s$ *and submodules* $\{0\} = \mathfrak{N}_0 \subset \cdots \subset \mathfrak{N}_s = \mathfrak{W}$ *such that, for every* $i = 1, \ldots, s$,

$\mathfrak{N}_i/\mathfrak{N}_{i-1} \cong \mathfrak{R}_d/\mathfrak{q}_i$ for some prime ideal $\mathfrak{q} \subset \mathfrak{q}_i \subset \mathfrak{R}_d$, $\mathfrak{q}_i = \mathfrak{q}$ for $i = 1, \ldots, t$, and $\mathfrak{q}_i \supsetneq \mathfrak{q}$ for $i = t+1, \ldots, s$.

PROOF. Note that, if $\mathfrak{N} \subset \mathfrak{W}$ is a submodule, and if $\mathfrak{p} \subset \mathfrak{R}_d$ is a prime ideal associated with $\mathfrak{W}/\mathfrak{N}$, then $\mathfrak{p} \supset \mathfrak{q}$. Indeed, if $\mathfrak{p} = \operatorname{ann}(a)$ for some $a \in \mathfrak{W}/\mathfrak{N}$, choose $b \in \mathfrak{W}$ such that $a = b + \mathfrak{N}$, and set $\mathfrak{N}' = \mathfrak{p} \cdot b = \{f \cdot b : f \in \mathfrak{p}\} \subset \mathfrak{N}$. If $\mathfrak{N}' \neq \{0\}$ then $\mathfrak{N}'$ is associated with $\mathfrak{q}$, and (6.4) shows that $g^n \in \mathfrak{p}$ for every $g \in \mathfrak{q}$ and every sufficiently large $n \geq 1$. Since $\mathfrak{p}$ is prime we conclude that $\mathfrak{q} \subset \mathfrak{p}$.

Let Ω_1 be the set of submodules $\mathfrak{N} \subset \mathfrak{W}$ with the following property: there exists an integer $r \geq 1$ and submodules $\{0\} = \mathfrak{N}_0 \subset \cdots \subset \mathfrak{N}_r = \mathfrak{N}$ such that $\mathfrak{N}_i/\mathfrak{N}_{i-1} \cong \mathfrak{R}_d/\mathfrak{q}$ for every $i = 1, \ldots, r$. It is clear that $\Omega_1 \neq \varnothing$, since we can find an $a \in \mathfrak{W}$ with $\operatorname{ann}(a) = \mathfrak{q}$ and $\mathfrak{N} = \mathfrak{R}_d a \cong \mathfrak{R}_d/\mathfrak{q}$. Since $\mathfrak{W}$ is Noetherian, Ω_1 contains a maximal element $\mathfrak{W}'$, and we set $\mathfrak{V} = \mathfrak{W}/\mathfrak{W}'$ and consider the set of prime ideals $\{\mathfrak{q}_1, \ldots, \mathfrak{q}_l\}$ associated with the $\mathfrak{R}_d$-module $\mathfrak{V}$. If $\mathfrak{q}_i = \mathfrak{q}$ for some $i \in \{1, \ldots, l\}$, then there exists an element $b \in \mathfrak{W}$ with $b \notin \mathfrak{W}'$ and $\{f \in \mathfrak{R}_d : fb \in \mathfrak{W}'\} = \mathfrak{q}$, and this violates the maximality of $\mathfrak{W}'$.

Let Ω_2 be the set of submodules $\mathfrak{N}$ with $\mathfrak{W}' \subset \mathfrak{N} \subset \mathfrak{W}$, for which there exist submodules $\mathfrak{W}' = \mathfrak{L}_0 \subset \cdots \subset \mathfrak{L}_t = \mathfrak{N}$ such that, for every $i = 1, \ldots, t$, $\mathfrak{L}_i/\mathfrak{L}_{i-1} \cong \mathfrak{R}_d/\mathfrak{q}_i$ for some prime ideal $\mathfrak{q}_i \supsetneq \mathfrak{q}$. Then Ω_2 again has a maximal element $\mathfrak{W}''$. If $\mathfrak{W}'' \neq \mathfrak{W}$ we set $\mathfrak{V}' = \mathfrak{W}/\mathfrak{W}''$, consider the set of prime ideals associated with $\mathfrak{V}'$, all of which are strictly greater than $\mathfrak{q}$ by the argument in the first paragraph of this proof, and obtain a contradiction to the maximality of $\mathfrak{W}''$ exactly as before, where we were dealing with $\mathfrak{W}'$. Hence $\mathfrak{W}'' = \mathfrak{W}$, and the proposition is proved by setting $\mathfrak{N}_0 \subset \cdots \subset \mathfrak{N}_s$ equal to $\{0\} = \mathfrak{N}_0 \subset \cdots \subset \mathfrak{N}_s = \mathfrak{L}_0 \subset \cdots \subset \mathfrak{L}_t = \mathfrak{N}$. $\square$

COROLLARY 6.2. Let $d \geq 1$, $\mathfrak{M}$ a Noetherian $\mathfrak{R}_d$-module with associated primes $\{\mathfrak{p}_1, \ldots, \mathfrak{p}_m\}$ and a corresponding reduced primary decomposition $\{\mathfrak{W}_1, \ldots, \mathfrak{W}_m\}$. Then there exist submodules $\mathfrak{M} = \mathfrak{N}_s \supset \cdots \supset \mathfrak{N}_0 = \{0\}$ such that, for every $i = 1, \ldots, s$, $\mathfrak{N}_i/\mathfrak{N}_{i-1} \cong \mathfrak{R}_d/\mathfrak{q}_i$ for some prime ideal $\mathfrak{q}_i \subset \mathfrak{R}_d$, and $\mathfrak{q}_i \supset \mathfrak{p}_j$ for some $j \in \{1, \ldots, m\}$ (such a sequence $\mathfrak{M} = \mathfrak{N}_s \supset \cdots \supset \mathfrak{N}_0 = \{0\}$ is called a prime filtration of $\mathfrak{M}$).

PROOF. Apply Proposition 6.1 to the successive quotients of the sequence

$$\mathfrak{M} \supset \mathfrak{W}_1 \supset (\mathfrak{W}_1 \cap \mathfrak{W}_2) \supset \cdots \supset (\mathfrak{W}_1 \cap \cdots \cap \mathfrak{W}_m) = \{0\},$$

bearing in mind that

$$(\mathfrak{W}_1 \cap \cdots \cap \mathfrak{W}_i)/(\mathfrak{W}_1 \cap \cdots \cap \mathfrak{W}_{i+1}) \cong (\mathfrak{W}_1 \cap \cdots \cap \mathfrak{W}_i)/\mathfrak{W}_{i+1} \subset \mathfrak{M}/\mathfrak{W}_{i+1}$$

is associated with $\mathfrak{p}_{i+1}$ for every $i = 1, \ldots, m-1$ (if B, C are subgroups of an abelian group A we use the symbol B/C to denote $(B + C)/C$). $\square$

Let $\mathfrak{M}$ be a Noetherian $\mathfrak{R}_d$-module with a prime filtration $\mathfrak{M} = \mathfrak{N}_s \supset \cdots \supset \mathfrak{N}_0 = \{0\}$, and define the $\mathbb{Z}^d$-action $\alpha = \alpha^{\mathfrak{M}}$ on $X = X^{\mathfrak{M}}$ by (5.5) and (5.6). For every $j = 0, \ldots, s$, $Y_j = \mathfrak{N}_j^{\perp}$ is a closed, α-invariant subgroup of X, and the dual group of Y_{j-1}/Y_j is isomorphic to $\mathfrak{R}_d/\mathfrak{q}_j$, where $\mathfrak{q}_j \subset \mathfrak{R}_d$ is a prime ideal containing one of the associated primes of $\mathfrak{M}$. This allows one to build up (X, α) from the successive quotients $(Y_{j-1}/Y_j, \alpha^{Y_{j-1}/Y_j})$, which have the explicit realization (5.9)–(5.10) with $\mathfrak{a} = \mathfrak{q}_j$. However, although the prime ideals $\{\mathfrak{p}_1, \ldots, \mathfrak{p}_m\}$ are canonically associated with $\mathfrak{M}$, the ideals $\mathfrak{q}_j$ appearing in Proposition 6.1 and Corollary 6.2 need no longer be canonical, and may depend on a specific prime filtration of $\mathfrak{M}$. The next corollary can help to overcome this problem.

COROLLARY 6.3. *Let $d \geq 1$, $\mathfrak{M}$ a Noetherian $\mathfrak{R}_d$-module with associated primes $\{\mathfrak{p}_1, \ldots, \mathfrak{p}_m\}$. Then there exists a Noetherian $\mathfrak{R}_d$-module $\mathfrak{N} = \mathfrak{N}^{(1)} \oplus \cdots \oplus \mathfrak{N}^{(m)}$ and an injective $\mathfrak{R}_d$-module homomorphism $\phi \colon \mathfrak{M} \longmapsto \mathfrak{N}$ such that each of the modules $\mathfrak{N}^{(j)}$ has a prime filtration $\mathfrak{N}^{(j)} = \mathfrak{N}_{r_j}^{(j)} \supset \cdots \supset \mathfrak{N}_0^{(j)} = \{0\}$ with $\mathfrak{N}_k^{(j)}/\mathfrak{N}_{k-1}^{(j)} \cong \mathfrak{R}_d/\mathfrak{p}_j$ for $k = 1, \ldots, r_j$.*

If $X = X^{\mathfrak{M}}$ and $Y = X^{\mathfrak{N}} = X^{\mathfrak{N}^{(1)}} \times \cdots \times X^{\mathfrak{N}^{(m)}}$, then the homomorphism $\psi \colon Y \longmapsto X$ dual to ϕ is surjective and satisfies that

$$\psi \cdot \alpha_{\mathbf{n}}^{\mathfrak{N}} = \psi \cdot (\alpha_{\mathbf{n}}^{\mathfrak{N}^{(1)}} \times \cdots \times \alpha_{\mathbf{n}}^{\mathfrak{N}^{(m)}}) = \alpha_{\mathbf{n}}^{\mathfrak{M}} \cdot \psi \qquad (6.7)$$

for every $\mathbf{n} \in \mathbb{Z}^d$.

PROOF. Choose a reduced primary decomposition $\mathfrak{W}_1, \ldots, \mathfrak{W}_m$ of $\mathfrak{M}$ as in (6.5). Then the map $\phi' \colon a \mapsto (a + \mathfrak{W}_1, \ldots, a + \mathfrak{W}_m)$ from $\mathfrak{M}$ into $\mathfrak{K} = \bigoplus_{i=1}^m \mathfrak{M}/\mathfrak{W}_i$ is injective. We fix $j \in \{1, \ldots, m\}$ for the moment and apply Proposition 6.1 to find a prime filtration $\{0\} = \mathfrak{N}_0 \subset \cdots \subset \mathfrak{N}_s = \mathfrak{M}/\mathfrak{W}_j$ such that $\mathfrak{N}_k^{(j)}/\mathfrak{N}_{k-1}^{(j)} \cong \mathfrak{R}_d/\mathfrak{q}_k^{(j)}$ for every $k = 1, \ldots, s_j$, where $\mathfrak{q}_k^{(j)} \subset \mathfrak{R}_d$ is a prime ideal containing $\mathfrak{p}_j$, and where there exists an $r_j \in \{1, \ldots, s_j\}$ such that $\mathfrak{q}_k^{(j)} = \mathfrak{p}_j$ for $k = 1, \ldots, r_j$, and $\mathfrak{q}_k^{(j)} \supsetneq \mathfrak{p}_j$ for $k = r_j + 1, \ldots, s_j$. If $r_j < s_j$ we choose Laurent polynomials $g_k^{(j)} \in \mathfrak{q}_k^{(j)} \smallsetminus \mathfrak{p}_j$ for $k = r_j + 1, \ldots, s_j$, set $g^{(j)} = g_{r_j+1}^{(j)} \cdot \cdots \cdot g_{s_j}^{(j)}$, and note that the map $\psi^{(j)} \colon \mathfrak{M}/\mathfrak{W}_j \longmapsto \mathfrak{N}_{r_j}^{(j)}$ consisting of multiplication by $g^{(j)}$ is injective. Since $\mathfrak{N}_{r_j}^{(j)}$ has the prime filtration $\{0\} = \mathfrak{N}_0^{(j)} \subset \cdots \subset \mathfrak{N}_{r_j}^{(j)}$ whose successive quotients are all isomorphic to $\mathfrak{R}_d/\mathfrak{p}_j$, the module $\mathfrak{N} = \mathfrak{N}_{r_1}^{(1)} \oplus \cdots \oplus \mathfrak{N}_{r_m}^{(m)}$ has the required properties. The last assertion follows from duality. □

EXAMPLE 6.4. In Example 5.3 (2) we considered the automorphism of $\mathbb{T}^2$ given by the matrix $A' = \left(\begin{smallmatrix} 3 & 2 \\ 2 & 1 \end{smallmatrix} \right)$ and obtained that the $\mathbb{Z}$-action on $\mathbb{T}^2$ defined by A' is conjugate to $(X^{\mathfrak{M}}, \alpha^{\mathfrak{M}})$, where $\mathfrak{M}$ is the $\mathfrak{R}_1$-module $\psi'(\mathbb{Z}^2) \subset \mathfrak{R}_1/(f)$ with $f(u_1) = -1 - 4u_1 + u_1^2$ and $\psi'(m_1, m_2) = m_1 + (3m_1 + 2m_2)u_1 \in \mathfrak{R}_1/(f)$ for every $(m_1, m_2) \in \mathbb{Z}^2$. As a submodule of $\mathfrak{R}_1/(f)$, $\mathfrak{M}$ is associated with

(f). Let $a = \psi'(0,1) = 2u_1 \in \Re_1/(f)$, and let $\mathfrak{N} = \Re_1 \cdot a = 2\Re_1/(f)$. Then $\mathfrak{M}/\mathfrak{N} = \Re_1/\mathfrak{a}$, where $\mathfrak{a}$ is the prime ideal $(2, 1+u_1) = 2\Re_1 + \Re_1(1+u_1) \subset \Re_1$, and $\{0\} \subset \mathfrak{N} \subset \mathfrak{M}$ is a prime filtration of $\mathfrak{M}$ with $\mathfrak{M}/\mathfrak{N} \cong \Re_1/\mathfrak{a}$ and $\mathfrak{N}/\{0\} \cong \Re_1/(f)$. $\quad\square$

Our next result shows that certain dynamical properties of the $\mathbb{Z}^d$-action $\alpha^{\mathfrak{M}}$ on $X^{\mathfrak{M}}$ can be expressed purely in terms of the primes associated with $\mathfrak{M}$ and do not require the much more difficult analysis of the primes which may occur in a prime filtration of $\mathfrak{M}$. Recall that an element $g \in \Re_d$ is a *generalized cyclotomic polynomial* if it is of the form $g(u_1, \ldots, u_d) = u^{\mathbf{m}}c(u^{\mathbf{n}})$, where $\mathbf{m}, \mathbf{n} \in \mathbb{Z}^d$, $\mathbf{n} \neq \mathbf{0}$, and c is a cyclotomic polynomial in a single variable.

THEOREM 6.5. *Let $d \geq 1$, let $\mathfrak{M}$ a Noetherian $\Re_d$-module with associated primes $\{\mathfrak{p}_1, \ldots, \mathfrak{p}_m\}$, and let $(X, \alpha) = (X^{\mathfrak{M}}, \alpha^{\mathfrak{M}})$ be defined by (5.5)–(5.6). For every $i = 1, \ldots, m$ we denote by $p(\mathfrak{p}_i) \geq 0$ the characteristic of $\Re_d/\mathfrak{p}_i$.*

(1) *The following conditions are equivalent.*
   (a) *α is ergodic;*
   (b) *$\alpha_{\mathbf{n}}$ is ergodic for some $\mathbf{n} \in \mathbb{Z}^d$;*
   (c) *$\alpha^{\Re_d/\mathfrak{p}_i}$ is ergodic for every $i \in \{1, \ldots, m\}$;*
   (d) *There do not exist integers $i \in \{1, \ldots, m\}$ and $l \geq 1$ with*

$$\{u^{l\mathbf{n}} - 1 : \mathbf{n} \in \mathbb{Z}^d\} \subset \mathfrak{p}_i;$$

   (e) *There do not exist integers $i \in \{1, \ldots, m\}$ and $l \geq 1$ with*

$$V(\mathfrak{p}_i) \subset \{c = (c_1, \ldots, c_d) \in (\overline{\mathbb{F}}^{\times}_{p(\mathfrak{p}_i)})^d : c_1^l = \cdots = c_d^l = 1\}.$$

(2) *The following conditions are equivalent.*
   (a) *α is mixing;*
   (b) *For every $i = 1, \ldots, m$, $\alpha^{\Re_d/\mathfrak{p}_i}$ is mixing;*
   (c) *None of the prime ideals associated with $\mathfrak{M}$ contains a generalized cyclotomic polynomial, i.e. $\{u^{\mathbf{n}} - 1 : \mathbf{n} \in \mathbb{Z}^d\} \cap \mathfrak{p}_i = \{0\}$ for $i = 1, \ldots, m$.*

(3) *Let $\Lambda \subset \mathbb{Z}^d$ be a subgroup with finite index. The following conditions are equivalent.*
   (a) *The set*

$$\mathrm{Fix}_\Lambda(\alpha) = \{x \in X : \alpha_{\mathbf{n}}(x) = x \text{ for every } \mathbf{n} \in \Lambda\}$$

   is finite;
   (b) *For every $i = 1, \ldots, m$, the set $\mathrm{Fix}_\Lambda(\alpha^{\Re_d/\mathfrak{p}_i})$ is finite;*
   (c) *For every $i = 1, \ldots, m$, $V_{\mathbb{C}}(\mathfrak{p}_i) \cap \Omega(\Lambda) = \varnothing$, where*

$$\Omega(\Lambda) = \{c \in \mathbb{C}^d : c^{\mathbf{n}} = 1 \text{ for every } \mathbf{n} \in \Lambda\}$$

   with $c = (c_1, \ldots, c_d)$, $\mathbf{n} = (n_1, \ldots, n_d)$, and $c^{\mathbf{n}} = c_1^{n_1} \cdot \ldots \cdot c_d^{n_d}$.

(4) *The following conditions are equivalent.*
   (a) *α is expansive;*

(b) *For every $i = 1, \ldots, m$, $\alpha^{\mathfrak{R}_d/\mathfrak{p}_i}$ is expansive;*

(c) *For every $i = 1, \ldots, m$, $V_{\mathbb{C}}(\mathfrak{p}_i) \cap \mathbb{S}^d = \varnothing$;*

(d) *For every $i = 1, \ldots, m$ with $p(\mathfrak{p}_i) = 0$, $V(\mathfrak{p}_i) \cap \mathbb{S}^d = \varnothing$.*

We begin the proof of Theorem 6.5 with a general proposition.

PROPOSITION 6.6. *Let $\mathfrak{M}$ a countable $\mathfrak{R}_d$-module.*

(1) *For any $\mathbf{n} \in \mathbb{Z}^d$ the following conditions are equivalent.*

    (a) *$\alpha_{\mathbf{n}}^{\mathfrak{M}}$ is ergodic;*

    (b) *$\alpha_{\mathbf{n}}^{\mathfrak{R}_d/\mathfrak{p}}$ is ergodic for every prime ideal $\mathfrak{p}$ associated with $\mathfrak{M}$;*

    (c) *No prime ideal $\mathfrak{p}$ associated with $\mathfrak{M}$ contains a polynomial of the form $u^{l\mathbf{n}} - 1$ with $l \geq 1$.*

(2) *The following conditions are equivalent.*

    (a) *$\alpha^{\mathfrak{M}}$ is ergodic;*

    (b) *$\alpha^{\mathfrak{R}_d/\mathfrak{p}}$ is ergodic for every prime ideal $\mathfrak{p}$ associated with $\mathfrak{M}$;*

    (c) *No prime ideal $\mathfrak{p}$ associated with $\mathfrak{M}$ contains a set of the form $\{u^{l\mathbf{n}} - 1 : \mathbf{n} \in \mathbb{Z}^d\}$ with $l \geq 1$.*

(3) *The following conditions are equivalent.*

    (a) *$\alpha^{\mathfrak{M}}$ is mixing;*

    (b) *$\alpha_{\mathbf{n}}^{\mathfrak{M}}$ is ergodic for every non-zero element $\mathbf{n} \in \mathbb{Z}^d$;*

    (c) *$\alpha_{\mathbf{n}}^{\mathfrak{M}}$ is mixing for every non-zero element $\mathbf{n} \in \mathbb{Z}^d$;*

    (d) *$\alpha^{\mathfrak{R}_d/\mathfrak{p}}$ is mixing for every prime ideal $\mathfrak{p}$ associated with $\mathfrak{M}$;*

    (e) *None of the prime ideals associated with $\mathfrak{M}$ contains a generalized cyclotomic polynomial.*

PROOF. From Lemma 1.2 and (5.5)–(5.6) it is clear that the $\mathbb{Z}$-action $k \mapsto \alpha_{k\mathbf{n}}^{\mathfrak{M}}$ is non-ergodic if and only if there exists a non-zero element $a \in \mathfrak{M}$ such that $(u^{l\mathbf{n}} - 1)a = 0$ for some $l \geq 1$. Let $\mathfrak{N} = \mathfrak{R}_d \cdot a$, and let $b \in \mathfrak{N}$ be a non-zero element such that $\mathfrak{p} = \text{ann}(b)$ is maximal in the set of annihilators of elements in $\mathfrak{N}$. Then $\mathfrak{p}$ is a prime ideal associated with $\mathfrak{M}$ which contains $u^{l\mathbf{n}} - 1$. This shows that (1.c)$\Rightarrow$(1.a). Conversely, if there exists a prime ideal $\mathfrak{p}$ associated with $\mathfrak{M}$ which contains $u^{l\mathbf{n}} - 1 \in \mathfrak{p}$ for some $l \geq 1$, we choose $a \in \mathfrak{M}$ with $\text{ann}(a) = \mathfrak{p}$, note that $(u^{l\mathbf{n}} - 1)a = 0$, and obtain that (1.a)$\Rightarrow$(1.c).

If we apply the equivalence (1.a)$\Longleftrightarrow$(1.c) to the $\mathfrak{R}_d$-module $\mathfrak{R}_d/\mathfrak{p}$, whose only associated prime is $\mathfrak{p}$, we see that $\alpha_{\mathbf{n}}^{\mathfrak{R}_d/\mathfrak{p}}$ is non-ergodic if and only if $u^{l\mathbf{n}} - 1 \in \mathfrak{p}$ for some $l \geq 1$, which completes the proof of the first part of this lemma.

If $\alpha^{\mathfrak{M}}$ is non-ergodic, then Lemma 1.2 implies that there exists a non-zero element $a \in \mathfrak{M}$ such that the orbit $\{u^{\mathbf{m}} \cdot a : \mathbf{m} \in \mathbb{Z}^d\}$ of the $\mathbb{Z}^d$-action $\hat{\alpha}^{\mathfrak{M}}$ in (5.5) is finite. As in the proof of (1) we set $\mathfrak{N} = \mathfrak{R}_d \cdot a$, choose $0 \neq b \in \mathfrak{N}$ such that $\mathfrak{p} = \text{ann}(b)$ is maximal, and note that $\mathfrak{p}$ is a prime ideal which contains $\{u^{l\mathbf{m}} - 1 : \mathbf{m} \in \mathbb{Z}^d\}$ for some $l \geq 1$. Conversely, if there exists a prime ideal $\mathfrak{p} \subset \mathfrak{R}_d$ associated with $\mathfrak{M}$ which contains $\{u^{l\mathbf{m}} - 1 : \mathbf{m} \in \mathbb{Z}^d\}$ for some $l \geq 1$, and Lemma 1.2 shows that the $\mathbb{Z}^d$-action $\alpha^{\mathfrak{M}}$ cannot be ergodic. This shows that (2.c)$\Longleftrightarrow$(2.a), and the equivalence of (2.b) and (2.c) is obtained by applying the equivalence of (2.a) and (2.c) to the $\mathfrak{R}_d$-module $\mathfrak{R}_d/\mathfrak{p}$.

In order to prove (3) we note that the equivalence (3.a)$\Longleftrightarrow$(3.b)$\Longleftrightarrow$(3.c) follows from Theorem 1.6 (2), and the proof is completed by applying the part (1) of this lemma both to $\alpha^{\mathfrak{M}}$ and to $\alpha^{\mathfrak{R}_d/\mathfrak{p}}$, where $\mathfrak{p}$ ranges over the set of prime ideals associated with $\mathfrak{M}$. $\square$

PROOF OF THEOREM 6.5 (1). The implication (b)$\Rightarrow$(a) is obvious. If (b) does not hold there exists, for every $\mathbf{n} \in \mathbb{Z}^d$, an $l \geq 1$ with $u^{l\mathbf{n}} - 1 \in \bigcup_{1 \leq i \leq m} \mathfrak{p}_i$ (Proposition 6.6). For every $i = 1, \ldots, m$, the set $\Gamma_i = \{\mathbf{n} \in \mathbb{Z}^d : u^{\mathbf{n}} - 1 \in \mathfrak{p}_i\}$ is a subgroup of $\mathbb{Z}^d$. As we have just observed, the set $\Gamma = \bigcup_{i=1}^m \Gamma_i$ contains some multiple of every element of $\mathbb{Z}^d$; if every Γ_i has infinite index in $\mathbb{Z}^d$, then Γ is contained in the intersection with $\mathbb{Z}^d$ of a union of m at most $d - 1$-dimensional subspaces of $\mathbb{R}^d$, which is obviously impossible. Hence Γ_i must have finite index in $\mathbb{Z}^d$ for some $i \in \{1, \ldots, m\}$, and we can find an integer $l \geq 1$ such that $u^{l\mathbf{n}} - 1 \in \mathfrak{p}_i$ for every $\mathbf{n} \in \mathbb{Z}^d$. This proves the implication (d)$\Rightarrow$(b). The implications (a)$\Longleftrightarrow$(c)$\Longleftrightarrow$(d) were proved in Proposition 6.6, and the equivalence of (d) and (e) follows from Hilbert's Nullstellensatz. $\square$

PROOF OF THEOREM 6.5 (2). Use Proposition 6.6. $\square$

LEMMA 6.7. *Let $\mathfrak{a} \subset \mathfrak{R}_d$ be an ideal. Then $\mathfrak{a} \cap \mathbb{Z} \neq \{0\}$ if and only if $V_{\mathbb{C}}(\mathfrak{a}) = \varnothing$.*

PROOF. If $\mathfrak{a} \cap \mathbb{Z} \neq \{0\}$ then $V_{\mathbb{C}}(\mathfrak{a}) = \varnothing$. Conversely, if $V_{\mathbb{C}}(\mathfrak{a}) = \varnothing$, then the Nullstellensatz implies that $\overline{\mathbb{Q}}[u_1^{\pm 1}, \ldots, u_d^{\pm 1}] \cdot \mathfrak{a} = \overline{\mathbb{Q}}[u_1^{\pm 1}, \ldots, u_d^{\pm 1}]$, and there exist polynomials $f_i \in \mathfrak{a}$, $g_i \in \overline{\mathbb{Q}}[u_1^{\pm 1}, \ldots, u_d^{\pm 1}]$, $i = 1, \ldots, n$, with $1 = \sum_{i=1}^n f_i g_i$. The coefficients of the g_i generate a finite extension field $\mathbb{K} \supset \mathbb{Q}$, and $\mathfrak{R}_d^{(\mathbb{K})} = \mathbb{K}[u_1^{\pm 1}, \ldots, u_d^{\pm 1}] = \sum_{j=1}^l v_j \mathfrak{R}_d^{(\mathbb{Q})}$ for suitably chosen elements $\{v_1, \ldots, v_l\} \in \mathfrak{R}_d^{(\mathbb{K})}$, where $\mathfrak{R}_d^{(\mathbb{Q})} = \mathbb{Q}[u_1^{\pm 1}, \ldots, u_d^{\pm 1}]$. Since $\mathfrak{a}^{(\mathbb{Q})} = \mathfrak{R}_d^{(\mathbb{Q})} \cdot \mathfrak{a}$ is an ideal in $\mathfrak{R}_d^{(\mathbb{Q})}$ and $\mathfrak{R}_d^{(\mathbb{K})} \cdot \mathfrak{a}^{(\mathbb{Q})} = \mathfrak{R}_d^{(\mathbb{K})}$, there exist elements $\{h_{j,k} : 1 \leq j, k \leq l\} \subset \mathfrak{a}^{(\mathbb{Q})}$ such that, for every $j = 1, \ldots, l$, $v_j = \sum_{k=1}^l h_{j,k} v_k$. Hence $\det(\delta_{j,k} - h_{j,k}) = 0$, where $\delta_{j,k} = 1$ for $j = k$ and $\delta_{j,k} = 0$ otherwise, and we conclude that $1 \in \mathfrak{a}^{(\mathbb{Q})}$. This proves that $\mathfrak{a} \cap \mathbb{Z} \neq \{0\}$. $\square$

PROOF OF THEOREM 6.5 (3). If $\mathfrak{b}(\Lambda) \subset \mathfrak{R}_d$ is the ideal generated by $\{u^{\mathbf{n}} - 1 : \mathbf{n} \in \Lambda\}$, then

$$V_{\mathbb{C}}(\mathfrak{b}(\Lambda)) = \{c \in \mathbb{C}^d : c^{\mathbf{n}} = 1 \text{ for every } \mathbf{n} \in \Lambda\} = \Omega(\Lambda),$$

$\operatorname{Fix}_\Lambda(\alpha)^{\perp} = \mathfrak{b}(\Lambda) \cdot \mathfrak{M}$, and $\widehat{\operatorname{Fix}_\Lambda(\alpha)} = \mathfrak{M}/\mathfrak{b}(\Lambda) \cdot \mathfrak{M}$ (cf. (5.5)–(5.6)). In particular, $\operatorname{Fix}_\Lambda(\alpha)$ is finite if and only if $\mathfrak{M}/\mathfrak{b}(\Lambda) \cdot \mathfrak{M}$ is finite.

Suppose that $\operatorname{Fix}_\Lambda(\alpha)$ is finite. For every $i = 1, \ldots, m$ we choose $a_i \in \mathfrak{M}$ such that $\mathfrak{p}_i = \operatorname{ann}(a_i)$ and hence $\mathfrak{L}_i = \mathfrak{R}_d \cdot a_i \cong \mathfrak{R}_d/\mathfrak{p}_i$. The Artin-Rees Lemma

(Corollary 10.10 in [5]) implies that

$$\mathfrak{b}(\Lambda)^{(t)} \cdot \mathfrak{M} \cap \mathcal{L}_i = \mathfrak{b}(\Lambda) \cdot (\mathfrak{b}(\Lambda)^{(t-1)} \cdot \mathfrak{M} \cap \mathcal{L}_i) \subset \mathfrak{b}(\Lambda) \cdot \mathcal{L}_i$$

for some $t \geq 1$, where $\mathfrak{b}(\Lambda)^{(t)} \subset \mathfrak{R}_d$ is the ideal generated by $\{f_1 \cdot \ldots \cdot f_t : f_i \in \mathfrak{b}(\Lambda)$ for $i = 1, \ldots, t\}$. By assumption,

$$\widehat{\mathrm{Fix}_\Lambda(\alpha)} = \mathfrak{M}/\mathfrak{b}(\Lambda) \cdot \mathfrak{M}$$

is finite. Since $\mathfrak{b}(\Lambda)$ is finitely generated we can choose $f_1, \ldots f_r$ such that $\mathfrak{b}(\Lambda) = f_1 \mathfrak{R}_d + \cdots + f_r \mathfrak{R}_d$, and we conclude that

$$|\mathfrak{b}(\Lambda) \cdot \mathfrak{M}/\mathfrak{b}(\Lambda)^{(2)} \cdot \mathfrak{M}| \leq \sum_{j=1}^{r} \left| f_j \cdot \mathfrak{M} \middle/ \left(\sum_{j,j'=1}^{r} f_j f_{j'} \cdot \mathfrak{M} \right) \right|$$

$$\leq \sum_{j=1}^{r} \left| f_j \cdot \mathfrak{M} \middle/ \left(\sum_{j'=1}^{r} f_j f_{j'} \cdot \mathfrak{M} \right) \right|$$

$$\leq r \left| \mathfrak{M} \middle/ \left(\sum_{j'=1}^{r} f_{j'} \cdot \mathfrak{M} \right) \right| = r|\mathfrak{M}/\mathfrak{b}(\Lambda) \cdot \mathfrak{M}| < \infty.$$

An induction argument shows that $\mathfrak{b}(\Lambda)^{(k)}\mathfrak{M}/\mathfrak{b}(\Lambda)^{(k+1)} \cdot \mathfrak{M}$ is finite for every $k \geq 1$, and we conclude that $\mathfrak{M}/\mathfrak{b}(\Lambda)^{(k)} \cdot \mathfrak{M}$ is finite for every $k \geq 1$. In particular, the modules $\mathcal{L}_i/\mathfrak{b}(\Lambda)^{(t)} \cdot \mathfrak{M} \cong \mathcal{L}_i/(\mathfrak{b}(\Lambda)^{(t)} \cdot \mathfrak{M} \cap \mathcal{L}_i)$ and $\mathcal{L}_i/\mathfrak{b}(\Lambda) \cdot \mathcal{L}_i \cong \mathfrak{R}_d/(\mathfrak{p}_i + \mathfrak{b}(\Lambda))$ are finite. From Lemma 6.7 we conclude that $V_{\mathbb{C}}(\mathfrak{p}_i + \mathfrak{b}(\Lambda)) = V_{\mathbb{C}}(\mathfrak{p}_i) \cap \Omega(\Lambda) = \varnothing$ for every $i = 1, \ldots, m$, which proves (c).

Conversely, if (c) is satisfied, we choose a prime filtration $\mathfrak{M} = \mathfrak{N}_s \supset \cdots \supset \mathfrak{N}_0 = \{0\}$ of $\mathfrak{M}$ such that, for every $j = 1, \ldots, s$, $\mathfrak{N}_j/\mathfrak{N}_{j-1} \cong \mathfrak{R}_d/\mathfrak{q}_j$ for some prime ideal $\mathfrak{q}_j$ which contains one of the associated primes $\mathfrak{p}_i$ of $\mathfrak{M}$ (cf. Corollary 6.2). Since

$$V_{\mathbb{C}}(\mathfrak{q}_j + \mathfrak{b}(\Lambda)) = V_{\mathbb{C}}(\mathfrak{q}_j) \cap V_{\mathbb{C}}(\mathfrak{b}(\Lambda)) \subset V_{\mathbb{C}}(\mathfrak{p}_i) \cap V_{\mathbb{C}}(\mathfrak{b}(\Lambda)) = \varnothing$$

for every $j = 1, \ldots, s$, the module $\mathfrak{R}_d/(\mathfrak{q}_j + \mathfrak{b}(\Lambda))$ is finite for every j by Lemma 6.7. Hence $\mathfrak{N}_j/(\mathfrak{N}_{j-1} + \mathfrak{b}(\Lambda) \cdot \mathfrak{M})$ is finite for $j = 1, \ldots, s$, since it is (isomorphic to) a quotient of $\mathfrak{R}_d/(\mathfrak{q}_j + \mathfrak{b}(\Lambda))$, and $\mathfrak{M}/\mathfrak{b}(\Lambda) \cdot \mathfrak{M}$ is finite. This implies the finiteness of $\mathrm{Fix}_\Lambda(\alpha)$ and completes the proof of the implication (c)$\Rightarrow$(a). The equivalence of (b) and (c) is obtained by applying what we have just proved to the $\mathbb{Z}^d$-actions $\alpha^{\mathfrak{R}_d/\mathfrak{p}_i}$, $i = 1, \ldots, m$. $\square$

LEMMA 6.8. *Let $\mathfrak{a} \subset \mathfrak{R}_d$ be an ideal with $V_{\mathbb{C}}(\mathfrak{a}) \cap \mathbb{S}^d = \varnothing$. Then $\alpha^{\mathfrak{R}_d/\mathfrak{a}}$ is expansive.*

PROOF. We assume that $X^{\mathfrak{R}_d/\mathfrak{a}} = \widehat{\mathfrak{R}_d/\mathfrak{a}}$ and $\alpha^{\mathfrak{R}_d/\mathfrak{a}}$ are given by (5.9)–(5.10). For every $f \in \mathfrak{R}_d$ of the form (5.2) we set $\|f\| = \sum_{\mathbf{n} \in \mathbb{Z}^d} |c_f(\mathbf{n})|$. Let $\{f_1, \ldots, f_k\}$ be a set of generators for $\mathfrak{a}$, $\varepsilon = (10 \sum_{j=1}^{k} \|f_j\|)^{-1}$, and $N = \{x \in$

$X^{\Re_d/\mathfrak{a}} : \|x_0\| < \varepsilon\}$, where $\|t\| = \min\{|t - n| : n \in \mathbb{Z}\}$ for every $t \in \mathbb{T}$. We claim that N is an expansive neighbourhood of the identity $\mathbf{0}$ in $X^{\Re_d/\mathfrak{a}}$.

If N is not expansive, there exists a point $\mathbf{0} \neq x \in \bigcap_{\mathbf{n} \in \mathbb{Z}^d} \sigma_{\mathbf{n}}(N)$. Let $\mathbf{B} = \ell^\infty(\mathbb{Z}^d)$ be the Banach space of all bounded, complex valued functions $(z_{\mathbf{n}}) = (z_{\mathbf{n}}, \mathbf{n} \in \mathbb{Z}^d)$ on $\mathbb{Z}^d$ in the supremum norm. Since $\|x_{\mathbf{n}}\| < \varepsilon$ for every $\mathbf{n} \in \mathbb{Z}^d$, there exists a unique non-zero point $y \in \mathbf{B}$ with $|y_{\mathbf{n}}| < \varepsilon$ and $y_{\mathbf{n}}$ (mod 1) $= x_{\mathbf{n}}$ for every $\mathbf{n} \in \mathbb{Z}^d$. From (5.7) and (5.9) we know that

$$\langle x, f_j \rangle = e^{2\pi i \sum_{\mathbf{n} \in \mathbb{Z}^d} c_{f_j}(\mathbf{n}) x_{\mathbf{n}}} = 1$$

and hence

$$\sum_{\mathbf{n} \in \mathbb{Z}^d} c_{f_j}(\mathbf{n}) y_{\mathbf{n}} \in \mathbb{Z}$$

for $j = 1, \ldots, k$, and our choice of ε implies that

$$\sum_{\mathbf{n} \in \mathbb{Z}^d} c_{f_j}(\mathbf{n}) y_{\mathbf{n}} = 0 \tag{6.8}$$

for all j. Consider the group of isometries $\{U_{\mathbf{n}} : \mathbf{n} \in \mathbb{Z}^d\}$ of $\mathbf{B}$ defined by $(U_{\mathbf{n}} z)_{\mathbf{m}} = z_{\mathbf{m}+\mathbf{n}}$ for all $\mathbf{m}, \mathbf{n} \in \mathbb{Z}^d$ and $z \in \mathbf{B}$, and put

$$\mathbf{S} = \left\{ z \in \mathbf{B} : \sum_{\mathbf{n} \in \mathbb{Z}^d} c_{f_j}(\mathbf{n}) z_{\mathbf{m}+\mathbf{n}} = 0 \text{ for all } \mathbf{m} \in \mathbb{Z}^d \text{ and } j = 1, \ldots, k \right\}$$

$$= \left\{ z \in \mathbf{B} : \left(\sum_{\mathbf{n} \in \mathbb{Z}^d} c_{f_j}(\mathbf{n}) U_{\mathbf{n}} \right) z = 0 \text{ for } j = 1, \ldots, k \right\}. \tag{6.9}$$

From (6.8) we know that the closed linear subspace $\mathbf{S} \subset \mathbf{B}$ is non-zero. Let $\mathcal{B}(\mathbf{S})$ be the Banach algebra of all bounded, linear operators on $\mathbf{S}$, denote by $V_{\mathbf{n}}$ the restriction of $U_{\mathbf{n}}$ to $\mathbf{S}$, and let $\mathcal{A} \subset \mathcal{B}(\mathbf{S})$ be the Banach subalgebra generated by $\{V_{\mathbf{n}} : \mathbf{n} \in \mathbb{Z}^d\}$. We write $\mathcal{M}(\mathcal{A})$ for the space of maximal ideals of $\mathcal{A}$ in its usual topology. The Gelfand transform $A \mapsto \hat{A}$ from $\mathcal{A}$ to the Banach algebra $\mathcal{C}(\mathcal{M}(\mathcal{A}), \mathbb{C})$ of continuous, complex valued functions on $\mathcal{M}(\mathcal{A})$ is a norm-non-increasing Banach algebra homomorphism (cf. §11 in [75]). For every $\mathbf{n} \in \mathbb{Z}^d$, both $V_{\mathbf{n}}$ and $V_{-\mathbf{n}} = V_{\mathbf{n}}^{-1}$ are isometries of $\mathbf{S}$, and hence $|\widehat{V_{\mathbf{n}}}(\omega)| = 1$ for every $\omega \in \mathcal{M}(\mathcal{A})$. Since $\sum_{\mathbf{n} \in \mathbb{Z}^d} c_{f_j}(\mathbf{n}) V_{\mathbf{n}} = 0$ (cf. (6.9)) we obtain that $\sum_{\mathbf{n} \in \mathbb{Z}^d} c_{f_j}(\mathbf{n}) \widehat{V_{\mathbf{n}}}(\omega) = 0$ for every $j = 1, \ldots, k$ and $\omega \in \mathcal{M}(\mathcal{A})$. Fix $\omega \in \mathcal{M}(\mathcal{A})$ and put $c_i = \widehat{V_{\mathbf{e}^{(i)}}}(\omega)$ for every $i = 1, \ldots, d$, where $\mathbf{e}^{(i)}$ is the i-th unit vector in $\mathbb{Z}^d$. Then $\sum_{\mathbf{n} \in \mathbb{Z}^d} c_{f_j}(\mathbf{n}) c^{\mathbf{n}} = f_j(c) = 0$ for $j = 1, \ldots, k$ with $c = (c_1, \ldots, c_d) \in \mathbb{S}^d$. It follows that $c \in V_{\mathbb{C}}(\mathfrak{a}) \cap \mathbb{S}^d$, contrary to our initial assumption. This proves that $\alpha^{\Re_d/\mathfrak{a}}$ is expansive. $\square$

PROOF OF THEOREM 6.5 (4). We begin by proving the equivalence of (a) and (c). Suppose that (c) is satisfied, but that α is non-expansive. We apply Corollary 6.2 and choose a prime filtration $\mathfrak{M} = \mathfrak{N}_s \supset \cdots \supset \mathfrak{N}_0 = \{0\}$ such that, for every $j = 1, \ldots, s$, $\mathfrak{N}_j/\mathfrak{N}_{j-1} \cong \mathfrak{R}_d/\mathfrak{q}_j$ for some prime ideal $\mathfrak{q}_j \subset \mathfrak{R}_d$ which contains one of the associated primes $\mathfrak{p}_i$. Put $X_j = \mathfrak{N}_j^\perp \subset X$ and observe that $X = X_0 \supset \cdots \supset X_s = \{\mathbf{1}\}$, that X_j is a closed, α-invariant subgroup of X, and that $X_{j-1}/X_j \cong \widehat{\mathfrak{R}_d/\mathfrak{q}_j}$ for $j = 1, \ldots, s$. Then $V_{\mathbb{C}}(\mathfrak{q}_1) \subset \bigcup_{i=1}^m V_{\mathbb{C}}(\mathfrak{p}_i)$, hence $V_{\mathbb{C}}(\mathfrak{q}_1) \cap \mathbb{S}^d = \varnothing$, and Lemma 6.8 shows that $\alpha^{\mathfrak{R}_d/\mathfrak{q}_1}$ is expansive. Since $\alpha^{\mathfrak{R}_d/\mathfrak{q}_1}$ is conjugate to $\alpha^{X/X_1} = \alpha^{X_0/X_1}$ we see that α^{X_0/X_1} is expansive. The non-expansiveness of α implies that α^{X_1} cannot be expansive, and by repeating this argument we eventually obtain that α^{X_s} is non-expansive, which is absurd. This contradiction proves the expansiveness of α.

In order to explain the idea behind the proof of the reverse implication we assume for the moment that $\mathfrak{M}$ is of the form $\mathfrak{R}_d/\mathfrak{a}$ for some ideal $\mathfrak{a} \subset \mathfrak{R}_d$. If $c = (c_1, \ldots, c_d) \in V_{\mathbb{C}}(\mathfrak{a})$ then the evaluation map $f \mapsto f(c)$ defines an $\mathfrak{R}_d$-module homomorphism $\eta_c \colon \mathfrak{R}_d/\mathfrak{a} \longmapsto \mathbb{C}$, where $\mathbb{C}$ is an $\mathfrak{R}_d$-module under the action $(f, z) \mapsto f(c)z$, $f \in \mathfrak{R}_d$, $z \in \mathbb{C}$. If W is the closure of $\eta_c(\mathfrak{R}_d/\mathfrak{a}) \subset \mathbb{C}$, then η_c conjugates the $\mathbb{Z}^d$-action $\hat{\alpha}$ on $\mathfrak{M}$ to the action θ on W, where $\theta_{\mathbf{n}}$ is multiplication by $c^{\mathbf{n}}$ for every $\mathbf{n} \in \mathbb{Z}^d$. If $c \in V_{\mathbb{C}}(\mathfrak{a}) \cap \mathbb{S}^d$ then θ is isometric (with respect to the usual metric on $\mathbb{C}$), and the homomorphism η_c induces an inclusion of $V = \hat{W}$ in $X^{\mathfrak{R}_d/\mathfrak{a}} = \widehat{\mathfrak{R}_d/\mathfrak{a}}$. Since θ is isometric on W, the dual action $\hat{\theta}$ on V is also equicontinuous, and coincides with the restriction of α to V. This shows that α cannot be expansive.

We return to our given module $\mathfrak{M}$ with its associated primes $\mathfrak{p}_1, \ldots, \mathfrak{p}_m$ and a corresponding reduced primary decomposition $\mathfrak{W}_1, \ldots, \mathfrak{W}_m$. If $V_{\mathbb{C}}(\mathfrak{p}_i) \cap \mathbb{S}^d \neq \varnothing$ for some $i \in \{1, \ldots, m\}$ we set $\mathfrak{M}' = \mathfrak{M}/\mathfrak{W}_i$, choose $a_1, \ldots, a_k \in \mathfrak{M}'$ such that $\mathfrak{M}' = \mathfrak{R}_d a_1 + \cdots + \mathfrak{R}_d a_k$, and define a surjective homomorphism $\zeta \colon \mathfrak{R}_d^k \longmapsto \mathfrak{M}'$ by $\zeta(f_1, \ldots, f_k) = f_1 a_1 + \cdots + f_k a_k$.

Choose a point $c = (c_1, \ldots, c_d) \in V_{\mathbb{C}}(\mathfrak{p}_i) \cap \mathbb{S}^d$, denote by $\eta_c \colon \mathfrak{R}_d \longmapsto \mathbb{C}$ the evaluation map at c, and observe that $\mathfrak{a} = \ker(\eta_c) \supset \mathfrak{p}_i$. Let $\mathfrak{L} = \ker(\zeta) + \mathfrak{a}^k \subset \mathfrak{R}_d^k$, and let $\mathfrak{N} = \{(0, \ldots, 0, f) : f \in \mathfrak{R}_d\} \subset \mathfrak{R}_d^k$. From (6.6) (with $\mathfrak{M}$ replaced by $\mathfrak{M}'$) we see that $\mathrm{ann}(a_k) \subset \mathfrak{p}_i$, so that

$$\mathfrak{L} \cap \mathfrak{N} \subset \{(0, \ldots, 0, f) : f \in \mathfrak{p}_i\} \subset \{(0, \ldots, 0, f) : f \in \mathfrak{a}\}.$$

This allows us to define an additive group homomorphism $\xi \colon \mathfrak{L} + \mathfrak{N} \longmapsto \mathbb{C}$ by $\xi(a + b) = \eta_c(f)$ for all $a \in \mathfrak{L}$ and $b = (0, \ldots, 0, f) \in \mathfrak{N}$. Then

$$\xi(a) = 0 \text{ for } a \in \mathfrak{L}, \tag{6.10}$$

and

$$\xi \cdot \hat{\alpha}_{\mathbf{n}}^{\mathfrak{R}_d^k}(a) = c^{\mathbf{n}} \xi(a) \text{ for all } a \in \mathfrak{L} + \mathfrak{N}, \, \mathbf{n} = (n_1, \ldots, n_d) \in \mathbb{Z}^d, \tag{6.11}$$

where $c^{\mathbf{n}} = c_1^{n_1} \cdots c_d^{n_d}$. We claim that ξ can be extended to a homomorphism $\bar{\xi} \colon \mathfrak{R}_d^k \longmapsto \mathbb{C}$ which still satisfies (6.10) and (6.11). Indeed, there exists a maximal extension ξ' of ξ to a submodule $\mathfrak{N}' \subset \mathfrak{R}_d^k$ satisfying (6.11) for every $a \in \mathfrak{N}'$. If $b \in \mathfrak{R}_d^k \setminus \mathfrak{N}'$ and $\xi'(b') = 0$ for every $b' \in \mathfrak{R}_d b \cap \mathfrak{N}'$, then we put $\rho = 0$. If there exists an element $f \in \mathfrak{R}_d$ with $fb \in \mathfrak{R}_d b \cap \mathfrak{N}'$ and $\xi'(fb) \neq 0$, then $f(c) = \eta_c(f) \neq 0$: otherwise $f \in \mathfrak{a}$, $fb \in \mathfrak{a}^k \subset \mathfrak{L}$, and $\xi'(fb) = \xi(fb) = 0$ by (6.10), which is impossible. Hence we can set $\rho = \xi'(fb)/f(c)$. The map $\xi'' \colon \mathfrak{N}'' = \mathfrak{R}_d b + \mathfrak{N}' \longrightarrow \mathbb{C}$, defined by $\xi''(fb + a) = f(c)\rho + \xi'(a)$ for $f \in \mathfrak{R}_d$ and $a \in \mathfrak{N}'$, is a homomorphism which extends ξ' and satisfies (6.11) for all $a \in \mathfrak{N}''$. This contradiction to the maximality of $\mathfrak{N}'$ proves our claim.

We have obtained an extension $\bar{\xi} \colon \mathfrak{R}_d^k \longrightarrow \mathbb{C}$ of ξ satisfying (6.11) for all $a \in \mathfrak{R}_d^k$; this implies that $\ker(\bar{\xi})$ is a submodule of $\mathfrak{R}_d^k$ which contains $\ker(\zeta)$, and that $\bar{\xi}$ induces an $\mathfrak{R}_d$-module homomorphism $\Xi \colon \mathfrak{M}' \cong \mathfrak{R}_d^k/\ker(\zeta) \longrightarrow \mathbb{C}$ with $\Xi(\mathfrak{M}') \supset \eta_c(\mathfrak{R}_d)$ and

$$\Xi \cdot \hat{\alpha}_{\mathbf{n}}^{\mathfrak{M}'} = \theta_{\mathbf{n}} \cdot \Xi \tag{6.12}$$

for every $\mathbf{n} \in \mathbb{Z}^d$, where $\theta_{\mathbf{n}}$ is multiplication by $c^{\mathbf{n}}$. We denote by W the closure of $\Xi(\mathfrak{M}')$ in $\mathbb{C}$ and write $V = \hat{W}$ for the dual group of W. Since Ξ sends $\mathfrak{M}'$ to a dense subgroup of W, there is a dual inclusion $V \subset \widehat{\mathfrak{M}'/\ker(\Xi)} \subset \widehat{\mathfrak{M}'} \subset X$, and (6.12) shows that, for every $v \in V$ and $\mathbf{n} \in \mathbb{Z}^d$,

$$\hat{\theta}_{\mathbf{n}}(v) = \alpha_{\mathbf{n}}(v). \tag{6.13}$$

If the closed subgroup $W \subset \mathbb{C}$ is countable, then the group $\Theta = \{\theta_{\mathbf{n}} : \mathbf{n} \in \mathbb{Z}^d\} \subset \operatorname{Aut}(W)$ is finite, since it consists of isometries of W, and hence $\hat{\Theta} = \{\hat{\theta}_{\mathbf{n}} : \mathbf{n} \in \mathbb{Z}^d\} \subset \operatorname{Aut}(V)$ is finite. From (6.13) it is clear that the restriction of α to the infinite subgroup $V \subset X$ cannot be expansive.

If W is uncountable, but disconnected, we replace W by its infinite, discrete quotient group $W' = W/W^\circ$, and obtain an α-invariant subgroup $V' = \widehat{W/W^\circ} \subset V \subset X$ on which α is not expansive.

If W is connected, it is either equal to $\mathbb{C}$ or isomorphic to $\mathbb{R}$, and the definition of Θ implies that W has a basis of Θ-invariant neighbourhoods of the identity. The dual group V is isomorphic to W, and again possesses a basis of $\hat{\Theta}$-invariant neighbourhoods of the identity. Since the inclusion $V \hookrightarrow X$ is continuous, the $\mathbb{Z}^d$-action $\mathbf{n} \mapsto \hat{\theta}_{\mathbf{n}}$ on $V \subset X$ must also be non-expansive in the subspace topology, i.e. α is not expansive on V.

We have proved that there always exists an infinite, α-invariant, but not necessarily closed, subgroup $V \subset X$ on which α is non-expansive in the induced topology. This shows that α is not expansive and completes the proof that (a)$\Longleftrightarrow$(c).

The equivalence of (b) and (c) is seen by applying the implications (a)$\Longleftrightarrow$(c) already proved to the $\mathbb{Z}^d$-actions $\alpha^{\mathfrak{R}_d/\mathfrak{p}_i}$, $i = 1, \ldots, m$.

It is clear that (c)$\Rightarrow$(d). Conversely, if $V_{\mathbb{C}}(\mathfrak{p}_i) \cap \mathbb{S}^d \neq \varnothing$ for some $i \in \{1, \ldots, m\}$, choose $f_1, \ldots, f_k$ in $\mathfrak{R}_d$ with $\mathfrak{p}_i = f_1\mathfrak{R}_d + \cdots + f_k\mathfrak{R}_d$, and define polynomials $g_j, h_j, j = 1, \ldots, k$, in

$$\mathcal{R}_d = \mathbb{Q}[x_1, \ldots, x_d, y_1, \ldots, y_d]$$

by

$$g_j(a_1,\ldots,a_d,b_1,\ldots,b_d) = \operatorname{Re}(f_j(a_1 + b_1\sqrt{-1},\ldots,a_d + b_d\sqrt{-1}))$$

and

$$h_j(a_1,\ldots,a_d,b_1,\ldots,b_d) = \operatorname{Im}(f_j(a_1 + b_1\sqrt{-1},\ldots,a_d + b_d\sqrt{-1}))$$

for all $j = 1,\ldots,k$ and $(a_1,\ldots,a_d,b_1,\ldots,b_d) \in \mathbb{R}^{2d}$, where $\operatorname{Re}(z)$ and $\operatorname{Im}(z)$ denote the real and imaginary parts of $z \in \mathbb{C}$. For $l = 1,\ldots,d$ we put

$$\chi_l(x_1,\ldots,x_d,y_1,\ldots,y_d) = x_l^2 + y_l^2 - 1 \in \mathcal{R}_d.$$

The ideal $\mathfrak{J} \subset \mathcal{R}_d$ generated by $\{g_1,\ldots,g_k,h_1,\ldots,h_k,\chi_1,\ldots,\chi_k\}$ satisfies that $V_{\mathbb{C}}(\mathfrak{J}) \cap \mathbb{R}^{2d} \neq \varnothing$. Hence $\mathfrak{J}$ does not contain a polynomial of the form $1 + \sum_{j=1}^r \psi_j^2$ with $r \geq 1$ and $\psi_j \in \mathcal{R}_d$, and the real version of Hilbert's Nullstellensatz implies that $V_{\mathbb{C}}(\mathfrak{J}) \cap \mathbb{R}^{2d} \cap \overline{\mathbb{Q}}^{2d} \neq \varnothing$ (proposition 4.1.7 and corollaire 4.1.8 in [11]). In particular we see that (d) cannot be satisfied, and this shows that (d)$\Rightarrow$(c) and completes the proof of Theorem 6.5 (4). $\quad\square$

Before we start listing some useful corollaries of Theorem 6.5 we give an elementary characterization of the connectedness of a group X carrying a $\mathbb{Z}^d$-action by automorphisms in terms of the prime ideals associated with the $\mathfrak{R}_d$-module $\hat{X}$.

PROPOSITION 6.9. *Let α be a $\mathbb{Z}^d$-action by automorphisms of a compact, abelian group X, and let $\mathfrak{M} = \hat{X}$ be the $\mathfrak{R}_d$-module defined by Lemma 5.1. The following conditions are equivalent.*

(1) *X is connected;*
(2) *$V_{\mathbb{C}}(\mathfrak{p}) \neq \varnothing$ for every prime ideal $\mathfrak{p} \subset \mathfrak{R}_d$ associated with $\mathfrak{M}$.*

PROOF. Suppose that X is connected, and let $\mathfrak{p} \subset \mathfrak{R}_d$ be a prime ideal associated with $\mathfrak{M}$. Then there exists an element $a \in \mathfrak{M}$ with $\mathfrak{R}_d \cdot a \cong \mathfrak{R}_d/\mathfrak{p}$, which implies that $X^{\mathfrak{R}_d/\mathfrak{p}}$ is a quotient group of X. In particular, $X^{\mathfrak{R}_d/\mathfrak{p}}$ is connected, so that $\mathfrak{R}_d/\mathfrak{p}$ is a torsion-free, abelian group, and Lemma 6.7 implies that $V_{\mathbb{C}}(\mathfrak{p}) \neq \varnothing$. Conversely, if X is disconnected, then there exists—by duality theory—a non-zero element $a \in \mathfrak{M}$ and a positive integer m with $ma = 0$, and we set $\mathfrak{N} = \mathfrak{R}_d \cdot a$ and observe that $\mathfrak{N}$ (and hence $\mathfrak{M}$) has an associated prime ideal $\mathfrak{p}$ containing a non-zero constant (cf. (6.6)). In particular, $V_{\mathbb{C}}(\mathfrak{p}) = \varnothing$. $\quad\square$

COROLLARY 6.10 (OF THEOREM 6.5). *If α is an ergodic $\mathbb{Z}^d$-action by automorphisms of a compact, abelian group X satisfying the d.c.c., then $\alpha_{\mathbf{n}}$ is ergodic for some $\mathbf{n} \in \mathbb{Z}^d$.*

PROOF. Lemma 5.1, Proposition 5.4, and Theorem 6.5 (1). $\quad\square$

COROLLARY 6.11. *Let $d \geq 2$, and let $(f) \subset \mathfrak{R}_d$ be a principal ideal. Then $\alpha^{\mathfrak{R}_d/(f)}$ is ergodic.*

PROOF. By Theorem 6.5 (1), the non-ergodicity of α implies that $V(\mathfrak{p}_i)$ is finite for at least one of the associated primes of $\mathfrak{M} = \mathfrak{R}_d/(f)$. However, the associate primes of $\mathfrak{M}$ are are all principal (they are given by the prime factors of f in $\mathfrak{R}_d$), and have infinite varieties.   □

COROLLARY 6.12. *Let $d \geq 1$ and $f \in \mathfrak{R}_d$. If f is not divisible by any generalized cyclotomic polynomial then $\alpha^{\mathfrak{R}_d/(f)}$ is mixing.*

PROOF. If $\mathfrak{p}$ is one of the associated primes of $\mathfrak{R}_d/(f)$ then $\mathfrak{p} = (h)$ for a prime factor h of f in $\mathfrak{R}_d$, and $\mathfrak{p}$ contains a polynomial of the form $u^{\mathbf{n}} - 1$ for some (non-zero) $\mathbf{n} \in \mathbb{Z}^d$ if and only if $h = c(u^{\mathbf{n}})$ for some cyclotomic polynomial c (cf. Theorem 6.5 (2)).   □

COROLLARY 6.13. *Let X be a compact, abelian group, and let α be an expansive $\mathbb{Z}^d$-action by automorphisms of X. Then the $\mathfrak{R}_d$-module $\mathfrak{M} = \hat{X}$ is a Noetherian torsion module.*

PROOF. According to (4.10) and Proposition 5.4, $\mathfrak{M}$ is Noetherian, and by Theorem 6.5 (4), $\{0\}$ cannot be an associated prime ideal of $\mathfrak{M}$.   □

COROLLARY 6.14. *Let X be a compact, connected group, and let α be an expansive $\mathbb{Z}^d$-action by automorphisms of X. Then X is abelian and α is ergodic.*

PROOF. Theorem 2.4 shows that X is abelian, and (4.10) and Proposition 5.4 allow us to assume that $(X, \alpha) = (X^{\mathfrak{M}}, \alpha^{\mathfrak{M}})$ for some Noetherian $\mathfrak{R}_d$-module $\mathfrak{M}$. By recalling Proposition 6.9 and comparing the conditions (1.e) and (4.c) in Theorem 6.5 we see that α is ergodic.   □

COROLLARY 6.15. *Let X be a compact group, and let α be an expansive $\mathbb{Z}^d$-action by automorphisms of X. If $Y \subset X$ is a closed, normal, α-invariant subgroup, then α^Y and $\alpha^{X/Y}$ are both expansive.*

PROOF. The expansiveness of α^Y is obvious. In order to see that $\alpha^{X/Y}$ is expansive we note that the connected component of the identity $X^\circ \subset X$ is abelian by Corollary 2.5. The group X/X° is zero-dimensional, and $X/(Y+X^\circ)$ is a quotient of a zero-dimensional group and hence again zero dimensional. Since the $\mathbb{Z}^d$-action $\alpha^{X/(Y+X^\circ)}$ satisfies the d.c.c., Corollary 3.4 implies that $\alpha^{X/(Y+X^\circ)}$ is expansive.

The group $(Y + X^\circ)/Y$ is isomorphic to $X^\circ/(Y \cap X^\circ)$, and this isomorphism carries $\alpha^{(Y+X^\circ)/Y}$ to $\alpha^{X^\circ/(Y\cap X^\circ)}$. We apply Lemma 5.1 to the abelian groups X° and $X^\circ/(Y \cap X^\circ)$, and obtain $\mathfrak{R}_d$-modules $\hat{X}^\circ = \mathfrak{M}$ and $\widehat{X^\circ/(Y \cap X^\circ)} = \mathfrak{N} \subset \mathfrak{M}$ satisfying (5.3)–(5.4). Since α^{X° is expansive, Theorem 6.5 (4) implies that $V_{\mathbb{C}}(\mathfrak{p}) \cap \mathbb{S}^d = \varnothing$ for every prime ideal $\mathfrak{p}$ associated with $\mathfrak{M}$. Every prime ideal associated with $\mathfrak{N}$ is also associated with $\mathfrak{M}$, and Theorem 6.5 (4) implies that $\alpha^{\mathfrak{N}}$ is expansive. This implies the expansiveness of both $\alpha^{X^\circ/(Y\cap X^\circ)}$ and $\alpha^{(Y+X^\circ)/Y}$.

Suppose that $x \in X \smallsetminus Y$. If $x \notin Y + X^\circ$ then the expansiveness of $\alpha^{X/(Y+X^\circ)}$ guarantees the existence of an open neighbourhood $N'(\mathbf{1}_X)$ of the identity in X such that $\alpha_{\mathbf{m}}(x) \notin N'(\mathbf{1}_X) + Y + X^\circ \supset N'(\mathbf{1}_X) + Y$ for some $\mathbf{m} \in \mathbb{Z}^d$. If $x \in Y + X^\circ$ then the expansiveness of $\alpha^{(Y+X^\circ)/Y}$ allows us to choose a neighbourhood $N''(\mathbf{1}_X)$ of the identity in X with $\alpha_{\mathbf{m}}(x) \notin N''(\mathbf{1}_X) + Y$ for some $\mathbf{m} \in \mathbb{Z}^d$. Put $N(\mathbf{1}_X) = N'(\mathbf{1}_X) \cap N''(\mathbf{1}_X)$. Then there exists, for every $x \in X \smallsetminus Y$, an $\mathbf{m} \in \mathbb{Z}^d$ with $\alpha_{\mathbf{m}}(x) \notin N(\mathbf{1}_X) + Y$, which shows that $\alpha^{X/Y}$ is expansive. $\square$

In view of Theorem 6.5 we introduce the following definition, which will help to simplify terminology.

DEFINITION 6.16. Let $d \geq 1$, and let $\mathfrak{p} \subset \mathfrak{R}_d$ be a prime ideal. The ideal $\mathfrak{p}$ will be called *ergodic*, *mixing*, or *expansive* if the $\mathbb{Z}^d$-action $\alpha^{\mathfrak{R}_d/\mathfrak{p}}$ is ergodic, mixing, or expansive.

EXAMPLES 6.17. (1) Let $n \geq 1$, $\alpha = A \in \mathrm{GL}(n, \mathbb{Z}) = \mathrm{Aut}(\mathbb{T}^n)$, and let $\beta = \hat{A} = A^\top \in \mathrm{Aut}(\mathbb{Z}^n)$. The $\mathfrak{R}_1$-module $\mathfrak{M} = \mathbb{Z}^n$ arising from α via Lemma 5.1 is Noetherian, and $\mathrm{ann}(\mathbf{m}) = \{f \in \mathfrak{R}_1 : f(A^\top)\mathbf{m} = 0\}$ for every $\mathbf{m} \in \mathbb{Z}^n$. In particular, the associated primes of $\mathfrak{M}$ are the principal ideals (h), where h runs through the prime factors of the characteristic polynomial $\chi_A = \chi_{A^\top}$ of A (or $A^\top$) in $\mathfrak{R}_1$. In this setting Theorem 6.5 (1) reduces to the following well known facts about toral automorphisms: (i) α is ergodic if and only if no root of χ_A is a root of unity; (ii) α is expansive if and only if no root of χ_A has modulus 1.

(2) The automorphism α in Example 5.6 (1) does not satisfy the d.c.c. (cf. Theorem 5.7), and is therefore non-expansive by (6.10). However, if we replace $\mathbb{Q}$ by $\mathbb{Z}[\frac{1}{6}] = \{k/6^l : k \in \mathbb{Z}, l \geq 0\} \cong \mathfrak{R}_1/(2u_1 - 3) = \mathfrak{M}$, where the isomorphism between $\mathfrak{R}_1/(2u_1 - 3)$ and $\mathbb{Z}[\frac{1}{6}]$ is the evaluation $f \mapsto f(\frac{3}{2})$, then the automorphism β' of $\mathbb{Z}[\frac{1}{6}]$ consisting of multiplication by $\frac{3}{2}$ is conjugate to multiplication by u_1 on $\mathfrak{M}$. Since $\mathfrak{p} = (2u_1 - 3) \subset \mathfrak{R}_1$ is a prime ideal, $\mathfrak{M}$ is associated with $\mathfrak{p}$, $V_{\mathbb{C}}(\mathfrak{p}) = \{\frac{3}{2}\}$, and the automorphism α' on $X = \widehat{\mathbb{Z}[\frac{1}{6}]}$ dual to β' is expansive by Theorem 6.5 (4). An explicit realization of α' can be obtained from Example 5.2 (2) by setting α' equal to the shift σ on $X' = \{(x_k) \in \mathbb{T}^{\mathbb{Z}} : 3x_k = 2x_{k+1}$ for every $k \in \mathbb{Z}\}$.

(3) Let $\mathfrak{p} \subset \mathfrak{R}_1$ be a prime ideal. Since the ring $\mathfrak{R}_1^{(\mathbb{Q})} = \mathbb{Q}[u_1^{\pm 1}]$ of Laurent polynomials with rational coefficients is a principal ideal domain, $\mathfrak{R}_1/\mathfrak{p}$ must be finite if $\mathfrak{p}$ is non-principal. In order to see this, assume that $\mathfrak{p} \subsetneqq \mathfrak{R}_1$ is a non-principal prime ideal, and choose two irreducible elements $g, h \in \mathfrak{p}$ with $g\mathfrak{R}_1 \neq h\mathfrak{R}_1$. We assume without loss in generality that $g\mathfrak{R}_1 \neq m\mathfrak{R}_1$ for any $m \in \mathbb{Z}$. Then $\mathfrak{q} = \{\frac{1}{n}f : n \geq 1, f \in \mathfrak{p}\} \subset \mathfrak{R}_1^{(\mathbb{Q})}$ is an ideal strictly containing the maximal ideal $g\mathfrak{R}_1^{(\mathbb{Q})}$, and therefore equal to $\mathfrak{R}_1^{(\mathbb{Q})}$. We conclude that $\mathfrak{p}$ contains a prime constant p, and hence the ideal $(p, g) = p\mathfrak{R}_1 + g\mathfrak{R}_1$. It follows that $\mathfrak{R}_1/\mathfrak{p}$ is a quotient of the finite ring $\mathfrak{R}_1/(p, g) \cong \mathfrak{R}_1^{(p)}/g_{/p}\mathfrak{R}_1^{(p)}$ (cf. (6.1)). In

particular, if $\mathfrak{p} \subset \mathfrak{R}_1$ is a non-principal prime ideal, then $X^{\mathfrak{R}_1/\mathfrak{p}} = \widehat{\mathfrak{R}_1/\mathfrak{p}}$ is finite, and $\alpha^{\mathfrak{R}_1/\mathfrak{p}}$ is non-ergodic.

If $\mathfrak{p} = (f)$ for some $f \in \mathfrak{R}_1$, the automorphism $\alpha = \alpha^{\mathfrak{R}_1/\mathfrak{p}}$ is non-ergodic if and only if f divides $u_1^n - 1$ for some $n \geq 1$ (Theorem 6.5 (1)) (as f is irreducible this means that $\pm u_1^n f$ is cyclotomic for some $n \in \mathbb{Z}$), and α is expansive if and only if f is non-zero and has no roots of modulus 1 (Theorem 6.5 (4)). Since we can write $X = X^{\mathfrak{R}_1/\mathfrak{p}}$ in the form (5.9) we see that X is (isomorphic to) a finite-dimensional torus if and only if there exists $n \in \mathbb{Z}$ and $s \geq 1$ such that $u_1^n f(u_1) = c_0 + c_1 u_1 + \cdots + c_s u_1^s$ with $|c_0 c_s| = 1$. If $|c_0 c_s| > 1$, then X is a finite-dimensional solenoid, i.e. $\hat{X}$ is isomorphic to a subgroup of $\mathbb{Q}^s$ (Example (2) and Example 5.3 (3)).

(4) Let α be an ergodic automorphism of a compact, abelian group X, and let $\mathfrak{M} = \hat{X}$ be the $\mathfrak{R}_1$-module arising from α via Lemma 5.1. Then every prime ideal $\mathfrak{p} \subset \mathfrak{R}_1$ associated with $\mathfrak{M}$ is principal, and $\mathfrak{p} \neq (f)$ for any cyclotomic polynomial $f \subset \mathfrak{R}_1$ (Proposition 6.6 and Example (3)). $\boxdot$

Further examples of expansive automorphisms of compact, abelian groups will appear in Chapter 3.

EXAMPLES 6.18. In the following illustrations of Theorem 6.5 we consider $\mathfrak{R}_2$-modules of the form $\mathfrak{M} = \mathfrak{R}_2/\mathfrak{a}$, where $\mathfrak{a} \subset \mathfrak{R}_2$ is an ideal, realize $X = X^{\mathfrak{M}} \subset \mathbb{T}^{\mathbb{Z}^2}$ as in Example 5.2 (2), and denote by $\alpha = \alpha^{\mathfrak{M}}$ the shift-action of $\mathbb{Z}^2$ on X.

(1) Let $\mathfrak{a} = (1 + u_1 + u_2)$. Since $\mathfrak{a}$ is prime, $\mathfrak{M}$ is associated with $\mathfrak{a}$. Corollary 6.11 shows that α is ergodic, and Corollary 6.12 implies that α is mixing. Since $((-1 + i\sqrt{-3})/2, (-1 - i\sqrt{-3})/2) \in V_\mathbb{C}(\mathfrak{a}) \cap \mathbb{S}^2$, α is not expansive by Theorem 6.5 (4). Moreover, $V_\mathbb{C}(\mathfrak{a}) \cap \Omega(3\mathbb{Z}^2) \neq \varnothing$, so that $\text{Fix}_{3\mathbb{Z}^2}(\alpha)$ is infinite by Theorem 6.5 (3). note that $\text{Fix}_{3\mathbb{Z}^2}(\alpha)$ consists of all points

$$
\begin{array}{ccccc}
\cdot & \cdot & \cdot & \cdot & \cdot \\
& a & b & c & a \\
\cdot\; a+2b+c & a+b+2c & 2a+b+c & a+2b+c\; \cdot \\
\cdot\; -a-b & -b-c & -a-c & -a-b\; \cdot \\
& a & b & c & a \\
\cdot & \cdot & \cdot & \cdot & \cdot
\end{array}
$$

with $a, b, c \in \mathbb{T}$ and $3a + 3b + 3c = 0 \pmod 1$. In particular, the connected component of the identity $\text{Fix}_{3\mathbb{Z}^2}(\alpha)^\circ \subset \text{Fix}_{3\mathbb{Z}^2}(\alpha)$ is isomorphic to $\mathbb{T}^2$.

(2) Let $\mathfrak{a} = (2 + u_1 + u_2) \subset \mathfrak{R}_2$. The action α is ergodic, mixing, non-expansive, and $(-1, -1) \in V_\mathbb{C}(\mathfrak{a}) \cap \Omega(2\mathbb{Z}^2) \neq \varnothing$. The points in $\text{Fix}_{2\mathbb{Z}^2}(\alpha)$ are of the form

$$
\begin{array}{ccc}
\cdot & \cdot & \cdot \\
a & b & a \\
\cdot\; -2a-b & -a-2b & -2a-b\; \cdot \\
a & b & a \\
\cdot & \cdot & \cdot
\end{array}
$$

with $4a + 4b = 1 \pmod 1$, and $\text{Fix}_{2\mathbb{Z}^2}(\alpha)^\circ$ is isomorphic to $\mathbb{T}$.

(3) Let $\mathfrak{a} = (2 - u_1 - u_2) \subset \mathfrak{R}_2$. Then α is again ergodic, mixing, and non-expansive. Since $(1,1) \in V_{\mathbb{C}}(\mathfrak{a})$, α has uncountably many fixed points, and hence $\mathrm{Fix}_\Lambda(\alpha)$ is uncountable for every subgroup $\Lambda \subset \mathbb{Z}^d$.

(4) If $\mathfrak{a} = (3 + u_1 + u_2) \subset \mathfrak{R}_2$, then α is ergodic, mixing, expansive, and the expansiveness of α implies directly that $\mathrm{Fix}_\Lambda(\alpha)$ is finite for every subgroup $\Lambda \subset \mathbb{Z}^d$ of finite index.

(5) In Example 5.3 (5) we considered the ideal $\mathfrak{a} = (2, 1 + u_1 + u_2) \subset \mathfrak{R}_2$. Then $V_{\mathbb{C}}(\mathfrak{a}) = \varnothing$, and Theorem 6.5 (4) re-establishes the fact that α is expansive. Since the polynomial $1 + u_1 + u_2$ is prime in $\mathfrak{R}_2^{(2)} = \mathbb{Z}_{/2}[u_1^{\pm 1}, u_2^{\pm 2}]$, the ideal $\mathfrak{a}$ is prime, and as in Corollary 6.12 we see that α is mixing (since every prime polynomial in $\mathbb{Z}_{/2}[u]$ divides a polynomial of the form $u^l - 1$ for some $l \geq 1$, (the analogue of) Corollary 6.12 reduces to checking that $1 + u_1 + u_2 \in \mathfrak{R}_2^{(2)}$ is not a polynomial in the single variable $u^{\mathbf{n}}$ for some $\mathbf{0} \neq \mathbf{n} \in \mathbb{Z}^2$).

(6) Let $\mathfrak{a} = (4, 1 + u_1 - u_2 + 2u_2^2 + u_1 u_2) \subset \mathfrak{R}_2$. Since every prime ideal $\mathfrak{p}$ associated with $\mathfrak{M} = \mathfrak{R}_d/\mathfrak{a}$ must contain both the polynomial $1 + u_1 - u_2 + 2u_2^2 + u_1 u_2$ and the constant 2, the prime ideals associated with $\mathfrak{M}$ are given by $\mathfrak{p}_1 = (2, 1 - u_1)$ and $\mathfrak{p}_2 = (2, 1 - u_2)$. In particular, α is ergodic and expansive, but not mixing: the automorphisms $\alpha_{(1,0)}$ and $\alpha_{(0,1)}$ are non-ergodic, whereas $\alpha_{(1,1)}$ is ergodic.

(7) Let $\mathfrak{a} = (6 - 2u_1, 2 - 3u_1 - 5u_2^2)$. The prime ideals associated with $\mathfrak{M} = \mathfrak{R}_2/\mathfrak{a}$ are given by $\mathfrak{p}_1 = (3 - u_1, 7 + 5u_2^2)$, $\mathfrak{p}_2 = (3, 1 + u_2)$, $\mathfrak{p}_3 = (3, 1 - u_2)$, and the $\mathbb{Z}^2$-action α is ergodic and expansive, but non-mixing. In this example $\alpha_{(0,1)}$ is non-ergodic (because of $\mathfrak{p}_3$), but $\alpha_{(1,0)}$ is ergodic.

(8) If $\mathfrak{a} = (1 + u_1 + u_1^2, 1 - u_2)$ then α is non-ergodic, since $\mathfrak{a}$ is prime and contains $\{u^{3\mathbf{n}} - 1 : \mathbf{n} \in \mathbb{Z}^2\}$. $\boxdot$

CONCLUDING REMARKS 6.19. (1) Most of the material in this section is taken from [94]. For Example 6.17 (2) we refer to [71].

(2) If $d \geq 2$, Corollary 6.10 is incorrect without the assumption that (X, α) satisfies the d.c.c.: indeed, let, for every $\mathbf{n} \in \mathbb{Z}^d$, $\mathfrak{N}_{\mathbf{n}} = \mathfrak{R}_d/(u^{\mathbf{n}} - 1)$. Then $\mathfrak{N}_{\mathbf{n}}$ is an $\mathfrak{R}_d$-module, and the $\mathbb{Z}^d$-action $\alpha^{\mathfrak{N}_{\mathbf{n}}}$ is ergodic by Corollary 6.11. We denote by $\mathfrak{M} = \sum_{\mathbf{n} \in \mathbb{Z}^d} \mathfrak{N}_{\mathbf{n}}$ the direct sum of the modules $\mathfrak{N}_{\mathbf{n}}$, $\mathbf{n} \in \mathbb{Z}^d$, and write a typical element $a \in \mathfrak{M}$ as $a = (a_{\mathbf{n}})$ with $a_{\mathbf{n}} \in \mathfrak{N}_{\mathbf{n}}$ for every $\mathbf{n} \in \mathbb{Z}^d$. The $\mathbb{Z}^d$-action $\alpha = \alpha^{\mathfrak{M}}$ arising from the $\mathfrak{R}_d$-module $\mathfrak{M}$ via Lemma 5.1 is ergodic by Lemma 1.2. However, $\alpha_{\mathbf{n}}$ is non-ergodic for every $\mathbf{n} \in \mathbb{Z}^d$: if $\mathbf{n} = \mathbf{0}$, this assertion is obvious, and if $\mathbf{n} \neq \mathbf{0}$, then the non-zero element $a(\mathbf{n}) \in \mathfrak{M}$ defined by

$$a(\mathbf{n})_{\mathbf{m}} = \begin{cases} 1 & \text{for } \mathbf{m} = \mathbf{n} \\ 0 & \text{for } \mathbf{m} \neq \mathbf{n} \end{cases}$$

satisfies that $u^{\mathbf{n}} a(\mathbf{n}) = a(\mathbf{n})$, and hence $\alpha_{\mathbf{n}}$ is non-ergodic by Lemma 1.2 (applied to the $\mathbb{Z}$-action $k \mapsto \alpha_{k\mathbf{n}}$).

(3) Let $\mathfrak{M}$ be a countable $\mathfrak{R}_d$-module, and define $(X^{\mathfrak{M}}, \alpha^{\mathfrak{M}})$ by Lemma 5.1. For every $f = \sum_{\mathbf{n} \in \mathbb{Z}^d} c_f(\mathbf{n}) \in \mathfrak{R}_d$ we define a group homomorphism

$$\alpha_f^{\mathfrak{M}} = \sum_{\mathbf{n} \in \mathbb{Z}^d} c_f(\mathbf{n}) \alpha_{\mathbf{n}}^{\mathfrak{M}} : X^{\mathfrak{M}} \longmapsto X^{\mathfrak{M}} \qquad (6.14)$$

by setting

$$\alpha_f^{\mathfrak{M}}(x) = \sum_{\mathbf{n} \in \mathbb{Z}^d} c_f(\mathbf{n}) \alpha_{\mathbf{n}}^{\mathfrak{M}}(x)$$

for every $x \in X^{\mathfrak{M}}$, and note that $\alpha_f^{\mathfrak{M}}$ commutes with $\alpha^{\mathfrak{M}}$ (i.e. $\alpha_f^{\mathfrak{M}} \cdot \alpha_{\mathbf{n}}^{\mathfrak{M}} = \alpha_{\mathbf{n}}^{\mathfrak{M}} \cdot \alpha_f^{\mathfrak{M}}$ for every $\mathbf{n} \in \mathbb{Z}^d$), and that $\alpha_f^{\mathfrak{M}}$ is dual to the homomorphism

$$f_{\mathfrak{M}} : \mathfrak{M} \longmapsto \mathfrak{M} \qquad (6.15)$$

consisting of multiplication by f. In particular, $\alpha_f^{\mathfrak{M}}$ is surjective if and only if $f_{\mathfrak{M}}$ is injective, i.e. if and only if f does not lie in any prime ideal associated with $\mathfrak{M}$ (cf. (6.4)). If $\mathfrak{M} = \mathfrak{R}_d/\mathfrak{a}$ for some ideal $\mathfrak{a} \subset \mathfrak{R}_d$, then (5.9) shows that

$$X^{\mathfrak{R}_d/\mathfrak{a}} = \{x \in \mathbb{T}^{\mathbb{Z}^d} = X^{\mathfrak{R}_d} : \alpha_f^{\mathfrak{R}_d}(x) = 0_X \text{ for every } f \in \mathfrak{a}\}, \qquad (6.16)$$

and *every* α-commuting homomorphism $\psi : X^{\mathfrak{M}} \longmapsto X^{\mathfrak{M}}$ is of the form $\psi = \alpha_f^{\mathfrak{M}}$ for some $f \in \mathfrak{R}_d$: indeed, if $\hat{\psi} : \mathfrak{R}_d/\mathfrak{a} \longmapsto \mathfrak{R}_d/\mathfrak{a}$ is the homomorphism dual to ψ, then $\hat{\psi}(1) = f + \mathfrak{a}$ for some $f \in \mathfrak{R}_d$, and $\psi = \alpha_f^{\mathfrak{R}_d/\mathfrak{a}}$. For every ideal $\mathfrak{a} \subset \mathfrak{R}_d$ we set $\mathfrak{a}^{\perp} = X^{\mathfrak{R}_d/\mathfrak{a}} = \widehat{\mathfrak{R}_d/\mathfrak{a}} \subset \widehat{\mathfrak{R}_d} = \mathbb{T}^{\mathbb{Z}^d}$, and observe that $\alpha^{\mathfrak{R}_d/\mathfrak{a}}$ is the restriction of the shift-action σ of $\mathbb{Z}^d$ on $\mathbb{T}^{\mathbb{Z}^d}$ to $\mathfrak{a}^{\perp}$. For every $f \in \mathfrak{R}_d$ the sequence

$$0 \longrightarrow (\mathfrak{a} + (f))^{\perp} \longrightarrow \mathfrak{a}^{\perp} \xrightarrow{\alpha_f^{\mathfrak{R}_d}} \mathfrak{b}^{\perp} \longrightarrow 0, \qquad (6.17)$$

is exact, where

$$\mathfrak{b} = \{g \in \mathfrak{R}_d : fg \in \mathfrak{a}\}. \qquad (6.18)$$

In particular, $\alpha_f^{\mathfrak{R}_d/\mathfrak{a}} : \mathfrak{a}^{\perp} \longmapsto \mathfrak{a}^{\perp}$ is surjective if and only if $\mathfrak{a} = \mathfrak{b}$.

(4) Let $p > 1$ be a rational prime, and let α be a $\mathbb{Z}^d$-action by automorphisms of a compact, abelian group X with the property that $px = 0$ for every $x \in X$. If $\mathfrak{M} = \hat{X}$ is the $\mathfrak{R}_d$-module arising from lemma 5.1, then $pa = 0$ for every $a \in \mathfrak{M}$, so that $\mathfrak{M}$ may be viewed as an $\mathfrak{R}_d^{(p)}$-module. Conversely, suppose that $\mathfrak{N}$ is a countable $\mathfrak{R}_d^{(p)}$-module. Exactly as in (5.1)–(5.6) we can define a $\mathbb{Z}^d$-action $\alpha = \alpha^{\mathfrak{N}}$ on the dual group $X = X^{\mathfrak{N}} = \hat{\mathfrak{N}}$ of $\mathfrak{N}$. Since $pa = 0$ for every $a \in \mathfrak{N}$, the group X is totally disconnected, and $x^p = 1_X$ for every $x \in X$. Since $\mathfrak{R}_d^{(p)}$ is a quotient ring of $\mathfrak{R}_d$, $\mathfrak{N}$ is also an $\mathfrak{R}_d$-module, and we write $\mathfrak{N}'$ instead of $\mathfrak{N}$ if we wish to emphasize that $\mathfrak{N}$ is viewed as an $\mathfrak{R}_d$-module. If $\mathfrak{N}$ is Noetherian (either as an $\mathfrak{R}_d$-module or as an $\mathfrak{R}_d^{(p)}$-module—the two conditions

are obviously equivalent), then we can realize $(X^{\mathfrak{N}}, \alpha^{\mathfrak{N}}) = (X^{\mathfrak{N}'}, \alpha^{\mathfrak{N}'})$ as the shift-action σ of $\mathbb{Z}^d$ on a closed, shift-invariant subgroup $X \subset (\mathbb{T}^k)^{\mathbb{Z}^d}$ for some $k \geq 1$ (Example 5.2 (3)–(4)). Since $px = \mathbf{0}_X$ for every $x \in X$, we know that $x_{\mathbf{n}} \in (F_p)^k$ for every $\mathbf{n} \in \mathbb{Z}^d$, where $F_p = \{\frac{k}{p} : k = 0, \ldots, p-1\} \subset \mathbb{T}$, and the obvious identification of F_p with the prime field $\mathbb{F}_p$ allows us to regard X (and hence $X^{\mathfrak{N}}$) as a closed, shift-invariant subgroup of $(\mathbb{F}_p^k)^{\mathbb{Z}^d}$, and $\alpha^{\mathfrak{N}}$ as the shift-action on X.

In particular, if $\mathfrak{a} \subset \mathfrak{R}_d^{(p)}$ is an ideal, and if $\mathfrak{N} = \mathfrak{R}_d^{(p)}/\mathfrak{a}$, then we may regard $\alpha^{\mathfrak{N}} = \alpha^{\mathfrak{R}_d^{(p)}/\mathfrak{a}}$ as the shift-action of $\mathbb{Z}^d$ on the subgroup

$$
X^{\mathfrak{R}_d^{(p)}/\mathfrak{a}} = \left\{ x = (x_{\mathbf{m}}) \in \mathbb{F}_p^{\mathbb{Z}^d} : \sum_{\mathbf{n} \in \mathbb{Z}^d} c_f(\mathbf{n}) x_{\mathbf{m}+\mathbf{n}} = \mathbf{0}_{\mathbb{F}_p} \atop \text{for all } f \in \mathfrak{a}, \, \mathbf{m} \in \mathbb{Z}^d \right\} \tag{6.19}
$$

of $\mathbb{F}_p^{\mathbb{Z}^d}$. Conversely, if $X \subset \mathbb{F}_p^{\mathbb{Z}^d}$ is a closed, shift-invariant subgroup, then

$$
X^{\perp} = \mathfrak{a} \subset \mathfrak{R}_d^{(p)} \cong \widehat{\mathbb{F}_p^{\mathbb{Z}^d}} \tag{6.20}
$$

is an ideal, $X \cong X^{\mathfrak{R}_d^{(p)}/\mathfrak{a}}$, and the isomorphism between X and $X^{\mathfrak{R}_d^{(p)}/\mathfrak{a}}$ carries the shift-action σ of $\mathbb{Z}^d$ on X to $\alpha^{\mathfrak{R}_d^{(p)}/\mathfrak{a}}$.

Every prime ideal $\mathfrak{p} \subset \mathfrak{R}_d^{(p)}$ associated with an $\mathfrak{R}_d^{(p)}$-module $\mathfrak{N}$ defines a prime ideal $\mathfrak{p}' = \{f \in \mathfrak{R}_d : f_{/p} \in \mathfrak{p}\} \subset \mathfrak{R}_d$, and $\mathfrak{p}'$ varies over the set of prime ideals in $\mathfrak{R}_d$ associated with $\mathfrak{N}'$ as $\mathfrak{p}$ varies over the prime ideals in $\mathfrak{R}_d^{(p)}$ associated with $\mathfrak{N}$. As we have seen in Example 6.18 (5), the dynamical properties of $\alpha^{\mathfrak{N}'}$ expressed in terms of the associated primes $\mathfrak{p}' \subset \mathfrak{R}_d$ of $\mathfrak{N}'$ have an analogous expression in terms of the prime ideals $\mathfrak{p} \subset \mathfrak{R}_d^{(p)}$ associated with $\mathfrak{N}$. In particular, $\alpha = \alpha^{\mathfrak{N}} = \alpha^{\mathfrak{N}'}$ is non-ergodic if and only if $V(\mathfrak{p})$ is finite for some prime ideal $\mathfrak{p} \subset \mathfrak{R}_d^{(p)}$ associated with $\mathfrak{N}$, and α is mixing if and only if no prime ideal $\mathfrak{p} \subset \mathfrak{R}_d^{(p)}$ associated with $\mathfrak{N}$ contains a polynomial in a single variable $u^{\mathbf{n}}, \mathbf{0} \neq \mathbf{n} \in \mathbb{Z}^d$. Furthermore, if $\mathfrak{N}$ is Noetherian, then $\mathrm{Fix}_\Lambda(\alpha)$ is finite for every subgroup $\Lambda \subset \mathbb{Z}^d$ of finite index, and α is expansive.

The algebraic advantage in viewing an $\mathfrak{R}_d$-module $\mathfrak{M}$ with $pa = 0$ for all $a \in \mathfrak{M}$ as an $\mathfrak{R}_d^{(p)}$-module is that $\mathfrak{R}_d^{(p)}$ is a ring of polynomials with coefficients in the field $\mathbb{F}_p$, which simplifies the ideal structure of $\mathfrak{R}_d^{(p)}$ when compared with that of $\mathfrak{R}_d$. As far as the dynamics are concerned there is, of course, no difference between viewing $\mathfrak{M}$ as a module over either of the rings $\mathfrak{R}_d$ or $\mathfrak{R}_d^{(p)}$.

7. The dynamical system defined by a point

The results in Section 6 show that many questions about $\mathbb{Z}^d$-actions by automorphisms of compact, abelian groups can be reduced to questions about $\mathbb{Z}^d$-actions of the form $\alpha^{\mathfrak{R}_d/\mathfrak{p}}$, where $\mathfrak{p} \subset \mathfrak{R}_d$ is a prime ideal. In this section we consider prime ideals of the form $\mathfrak{p} = \mathfrak{j}_c = \{f \in \mathfrak{R}_d : f(c) = 0\}$ with $c =$

$(c_1, \ldots, c_d) \in (\overline{\mathbb{Q}}^\times)^d$. The groups $X^{\mathfrak{R}_d/\mathfrak{j}_c}$ arising from these ideals via Lemma 5.1 turn out to be connected and finite-dimensional (i.e. finite-dimensional tori or solenoids); conversely, if $\mathfrak{p} \subset \mathfrak{R}_d$ is a prime ideal such that $X^{\mathfrak{R}_d/\mathfrak{p}}$ is connected and finite-dimensional, then $\mathfrak{p} = \mathfrak{j}_c$ for some $c \in (\overline{\mathbb{Q}}^\times)^d$ (Corollary 7.4).

Let $\mathbb{K}$ be an algebraic number field, i.e. a finite extension of $\mathbb{Q}$. A *valuation* of $\mathbb{K}$ is a homomorphism $\phi \colon \mathbb{K} \longmapsto \mathbb{R}^+$ with the property that $\phi(a) = 0$ if and only if $a = 0$, $\phi(ab) = \phi(a)\phi(b)$, and $\phi(a+b) \leq c \cdot \max\{\phi(a), \phi(b)\}$ for all $a, b \in \mathbb{K}$ and some $c \in \mathbb{R}$ with $c \geq 1$. The valuation ϕ is *non-trivial* if $\phi(\mathbb{K}) \supsetneq \{0, 1\}$, *non-archimedean* if ϕ is non-trivial and we can set $c = 1$, and *archimedean* otherwise. Two valuations ϕ, ψ of $\mathbb{K}$ are *equivalent* if there exists an $s > 0$ with $\phi(a) = \psi(a)^s$ for all $a \in \mathbb{K}$. An equivalence class v of non-trivial valuations of $\mathbb{K}$ is called a *place* of $\mathbb{K}$, and v is *finite* if v contains a non-archimedean valuation, and *infinite* otherwise. If v is finite, all valuations $\phi \in v$ are non-archimedean.

Let v be a place of $\mathbb{K}$, and let $\phi \in v$ be a valuation. A sequence $(a_n, \ n \geq 1)$ is *Cauchy* with respect to ϕ if there exists, for every $\varepsilon > 0$, an integer $N \geq 1$ such that $\phi(a_m - a_n) < \varepsilon$ whenever $m, n \geq N$. It is clear that this definition does not depend on the valuation $\phi \in v$, so that we may call (a_n) a Cauchy sequence for v. Two Cauchy sequences (a_n) and (b_n) for v are *equivalent* if $\lim_{n \to \infty} \phi(a_n - b_n) = 0$, and this notion of equivalence again only depends on v and not on ϕ. With respect to the obvious operations the set of equivalence classes of Cauchy sequences for v is a field, denoted by $\mathbb{K}_v$, which contains $\mathbb{K}$ as a dense subfield (every $a \in \mathbb{K}$ is identified with the equivalence class of the constant Cauchy sequence $(a, a, a, \ldots)$ in $\mathbb{K}_v$). The field $\mathbb{K}_v$ is the *completion of $\mathbb{K}$ in the v-adic topology*.

Ostrowski's Theorem (Theorem 2.2.1 in [16]) states that every non-trivial valuation ϕ of $\mathbb{Q}$ is either equivalent to the absolute value (i.e. there exists a $t > 0$ with $\phi(a)^t = |a|$ for every $a \in \mathbb{Q}$), or to the p-adic valuation for some rational prime $p \geq 2$ (i.e. there exists a $t > 0$ such that $\phi(\frac{m}{n})^t = p^{(n'-m')} = |\frac{m}{n}|_p$ for all $\frac{m}{n} \in \mathbb{Q}$, where $m = p^{m'}m''$, $n = p^{n'}n''$, and neither m'' nor n'' are divisible by p). It is easy to see that the valuations $|\cdot|_\infty, \ |\cdot|_p, \ |\cdot|_q$ are mutually inequivalent whenever p, q are distinct rational primes, i.e. that the places of $\mathbb{Q}$ are indexed by the set $\Pi \cup \{\infty\}$, where $\Pi \subset \mathbb{N}$ denotes the set of rational primes. The completion $\mathbb{Q}_\infty$ of $\mathbb{Q}$ is equal to $\mathbb{R}$, and for every rational prime p the completion $\mathbb{Q}_p$ of $\mathbb{Q}$ is the field of p-adic rationals.

For every valuation ϕ of $\mathbb{K}$, the restriction of ϕ to $\mathbb{Q} \subset \mathbb{K}$ is a valuation of $\mathbb{Q}$ and is equivalent either to $|\cdot|_\infty$ or to $|\cdot|_p$ for some rational prime p. In the first case the place $v \ni \phi$ is infinite (or *lies above* ∞), and in the second case v *lies above* p (or p *lies below* v). We denote by w the place of $\mathbb{Q}$ below v and observe that $\mathbb{K}_v$ is a finite-dimensional vector space over the locally compact, metrizable field $\mathbb{Q}_w$ and hence locally compact and metrizable in its own right. Choose a Haar measure λ_v on $\mathbb{K}_v$ (with respect to addition), fix a compact set $C \subset \mathbb{K}_v$ with non-empty interior, and write $\text{mod}_{\mathbb{K}_v}(a) = \lambda_v(aC)/\lambda_v(C)$ for the *module* of an element $a \in \mathbb{K}_v$. The map $\text{mod}_{\mathbb{K}_v} \colon \mathbb{K} \longmapsto \mathbb{R}^+$ is continuous, independent of the choice of λ_v, and its restriction to $\mathbb{K}$ is a valuation in v which is denoted by $|\cdot|_v$.

Above every place v of $\mathbb{Q}$ there are at least one and at most finitely many places of $\mathbb{K}$. Indeed, if $\mathbb{K} = \mathbb{Q}(a_1,\ldots,a_n)$ with $\{a_1,\ldots,a_n\} \subset \overline{\mathbb{Q}}$, and if f is the minimal polynomial of a_1 over $\mathbb{Q}$, then f is irreducible over $\mathbb{Q}$, but f may be reducible over $\mathbb{Q}_v$; we write $f = f_1 \cdot \ldots \cdot f_k$ for the decomposition of f into irreducible factors over $\mathbb{Q}_v$ and consider the field $\mathbb{Q}_v[x]/(f_i)$, where (f_i) denotes the principal ideal in the ring $\mathbb{Q}_v[x]$ generated by f_i. We define an injective field homomorphism $\zeta \colon \mathbb{K}^{(1)} = \mathbb{Q}_v(a_1) \longmapsto \mathbb{Q}_v[x]/(f_i)$ by setting $\zeta(a_1) = x$ and $\zeta(b) = b$ for every $b \in \mathbb{Q}_v$ and put $\phi_i(a) = \mathrm{mod}_{\mathbb{Q}_v[x]/(f_i)}(\zeta(a))$ for every $a \in \mathbb{K}^{(1)}$. Then ϕ_i is a valuation of $\mathbb{K}^{(1)}$ whose place w_i lies above v. The places $w_1,\ldots,w_k$ are all distinct, and they are the only places of $\mathbb{K}^{(1)}$ above v (Theorem III.1 in [109]). In exactly the same way we find finitely many places of $\mathbb{K}^{(2)} = \mathbb{K}^{(1)}(a_2) = \mathbb{Q}(a_1,a_2)$ above each place of $\mathbb{K}^{(1)}$, and after n steps we obtain that there are at least one and at most finitely many places of $\mathbb{K}$ above each place of $\mathbb{Q}$. A place v of $\mathbb{K}$ is infinite if and only it lies above ∞; in this case v is either *real* (if $\mathbb{K}_v = \mathbb{R}$) or *complex* (if $\mathbb{K}_v = \mathbb{C}$).

We write $P^{\mathbb{K}}$, $P_{\mathrm{f}}^{\mathbb{K}}$, and $P_{\infty}^{\mathbb{K}}$, for the sets of places, finite places, and infinite places of $\mathbb{K}$. For every $v \in P^{\mathbb{K}}$, $\mathcal{R}_v = \{r \in \mathbb{K}_v : |r|_v \le 1\}$ is a compact subset of $\mathbb{K}_v$. If $v \in P_{\mathrm{f}}^{\mathbb{K}}$, then $\mathcal{R}_v$ is, in addition, open, and is the unique maximal compact subring of $\mathbb{K}_v$; furthermore there exists a *prime element* $\pi_v \in \mathcal{R}_v$ such that $\pi_v \mathcal{R}_v$ is the unique maximal ideal of $\mathcal{R}_v$. For every $v \in P_{\mathrm{f}}^{\mathbb{K}}$ we set $\mathfrak{o}_v = \mathbb{K} \cap \mathcal{R}_v$, and we note that $\mathfrak{o}_{\mathbb{K}} = \bigcap_{v \in P_{\mathrm{f}}^{\mathbb{K}}} \mathfrak{o}_v$ is the ring of integral elements in $\mathbb{K}$ (Theorem V.1 in [109]). The set

$$
\mathbb{K}_{\mathbb{A}} = \Big\{ \omega = (\omega_v, \, v \in P^{\mathbb{K}}) \in \prod_{v \in P^{\mathbb{K}}} \mathbb{K}_v :
$$
$$
|\omega_v|_v \le 1 \text{ for all but finitely many } v \in P^{\mathbb{K}} \Big\}, \tag{7.1}
$$

furnished with that topology in which the subgroup

$$
\big\{ \omega = (\omega_v, \, v \in P^{\mathbb{K}}) \in \mathbb{K}_{\mathbb{A}} : |\omega_v|_v \le 1 \text{ for every } v \in P_{\mathrm{f}}^{\mathbb{K}} \big\}
$$
$$
\cong \prod_{v \in P_{\infty}^{\mathbb{K}}} \mathbb{K}_v \times \prod_{v \in P_{\mathrm{f}}^{\mathbb{K}}} \mathcal{R}_v
$$

carries the product topology and is open in $\mathbb{K}_{\mathbb{A}}$, is the locally compact *adele ring* of $\mathbb{K}$. The diagonal embedding $i \colon \xi \mapsto (\xi, \xi, \ldots)$ of $\mathbb{K}$ in $\mathbb{A}_{\mathbb{K}}$ maps $\mathbb{K}$ to a discrete, co-compact subring of $\mathbb{K}_{\mathbb{A}}$ (cf. [16], [109]).

We fix a non-trivial character $\chi \in i(\mathbb{K})^{\perp} \subset \widehat{\mathbb{K}_{\mathbb{A}}}$ and define, for every $a \in \mathbb{K}$, a character $\chi_a \in i(\mathbb{K})^{\perp} \subset \widehat{\mathbb{K}_{\mathbb{A}}}$ by setting

$$
\chi_a(\omega) = \chi(i(a)\omega)
$$

for every $\omega \in \mathbb{K}_\mathbb{A}$. By [16] or [109], the map $a \mapsto \chi_a$ is an isomorphism of the discrete, additive group $\mathbb{K}$ onto $i(\mathbb{K})^\perp \subset \widehat{\mathbb{K}}_\mathbb{A}$. The resulting identification

$$\widehat{\mathbb{K}} \cong \mathbb{K}_\mathbb{A}/i(\mathbb{K}) \tag{7.2}$$

depends, of course, on the chosen character χ. In order to make the isomorphism (7.2) a little more canonical we consider, for every $w \in P^\mathbb{K}$, the subgroup

$$\Omega(\{w\})' = \{\omega = (\omega_v) \in \mathbb{K}_\mathbb{A} : \omega_v = 0 \text{ for every } v \neq w\} \cong \mathbb{K}_w$$

of $\mathbb{K}_\mathbb{A}$ and denote by $\chi^{(w)} \in \widehat{\mathbb{K}_w}$ the character induced by the restriction of χ to $\Omega(\{w\})'$. After replacing χ by a suitable χ_a, $a \in \mathbb{K}$, if necessary, we may assume that the induced characters $\chi^{(w)} \in \widehat{\mathbb{K}_w}$, $w \in P^\mathbb{K}_\mathsf{f}$, satisfy that

$$\begin{aligned} \mathcal{R}_w \subset \ker(\chi^{(w)}) &= \{\omega \in \mathbb{A}_w : \chi^{(w)}(\omega) = 1\}, \\ \pi_w^{-1}\mathcal{R}_w &\not\subset \ker(\chi^{(w)}) \end{aligned} \tag{7.3}$$

for every $w \in P^\mathbb{K}_\mathsf{f}$, where $\pi_w \in \mathcal{R}_w$ is the prime element appearing in the preceding paragraph (cf. [109]). With this choice of χ we have that

$$\chi \in \left(i(\mathbb{K}) + \Omega(P^\mathbb{K}_\mathsf{f})'\right)^\perp,$$

where

$$\Omega(P^\mathbb{K}_\mathsf{f})' = \{\omega = (\omega_v) \in \mathbb{K}_\mathbb{A} : \omega_v = 0 \text{ for every } v \in P^\mathbb{K}_\infty = P^\mathbb{K} \smallsetminus P^\mathbb{K}_\mathsf{f}\}.$$

Now consider a finite subset $F \subset P^\mathbb{K}$ which contains $P^\mathbb{K}_\infty$, denote by

$$i_F : \mathbb{K} \longmapsto \prod_{v \in F} \mathbb{K}_v \tag{7.4}$$

the diagonal embedding $r \mapsto (r, \ldots, r)$, $r \in \mathbb{K}$, put

$$R_F = \{a \in \mathbb{K} : |a|_v \leq 1 \text{ for every } v \notin F\}, \tag{7.5}$$

and observe that $i_F(R_F)$ is a discrete, additive subgroup of $\prod_{v \in F} \mathbb{K}_v$. If

$$\begin{aligned} \Omega = \Omega(F) &= \{\omega = (\omega_v) \in \mathbb{K}_\mathbb{A} : |\omega_v|_v \leq 1 \text{ for every } v \in P^\mathbb{K} \smallsetminus F\}, \\ \Omega' = \Omega(P^\mathbb{K} \smallsetminus F)' &= \{\omega = (\omega_v) \in \mathbb{K}_\mathbb{A} : \omega_v = 0 \text{ for every } v \in F\}, \\ \Omega'' &= \Omega \cap \Omega', \end{aligned}$$

then $i(\mathbb{K}) + \Omega'' = i(\mathbb{K}) + \Omega'$, and (7.3) implies that $\chi \in (i(\mathbb{K}) + \Omega'')^\perp = (i(\mathbb{K}) + \Omega')^\perp$ and

$$R_F = \{a \in \mathbb{K} : \chi_a \in (i(\mathbb{K}) + \Omega')^\perp\}.$$

Hence

$$\widehat{R_F} = \mathbb{K}_\mathbb{A}/(i(\mathbb{K}) + \Omega') \cong \left(\prod_{v \in F} \mathbb{K}_v\right) \bigg/ i_F(R_F). \tag{7.6}$$

Let $d \geq 1$, $c = (c_1, \ldots, c_d) \in (\overline{\mathbb{Q}}^{\times})^d$, and $\mathfrak{j}_c = \{f \in \mathfrak{R}_d : f(c) = 0\}$. We wish to investigate the dynamical system $(X, \alpha) = (X^{\mathfrak{R}_d/\mathfrak{j}_c}, \alpha^{\mathfrak{R}_d/\mathfrak{j}_c})$ determined by c. Denote by $\mathbb{K} = \mathbb{Q}(c)$ the algebraic number field generated by $\{c_1, \ldots, c_d\}$ and put

$$F(c) = \{v \in P_{\mathfrak{f}}^{\mathbb{K}} : |c_i|_v \neq 1 \text{ for some } i \in \{1, \ldots, d\}\}, \tag{7.7}$$

which is finite by Theorem III.3 in [109], and

$$R_c = R_{P(c)}, \tag{7.8}$$

where $P(c) = P_{\infty}^{\mathbb{K}} \cup F(c)$. Then R_c is an $\mathfrak{R}_d$-module under the action $(f, a) \mapsto f(c)a$, and we define the $\mathbb{Z}^d$-action

$$\alpha^{(c)} = \alpha^{R_c} \tag{7.9}$$

on the compact group

$$Y^{(c)} = \widehat{R_c} = \left(\prod_{v \in P(c)} \mathbb{K}_v \right) \bigg/ i_F(R_c) \tag{7.10}$$

by (5.5)–(5.6), where we use (7.6) to identify $\widehat{R_c}$ and $\left(\prod_{v \in P(c)} \mathbb{K}_v \right) / i_F(R_c)$.

THEOREM 7.1. *There exists a continuous, surjective, finite-to-one homomorphism* $\phi: Y^{(c)} \longmapsto X^{\mathfrak{R}_d/\mathfrak{j}_c}$ *such that the diagram*

$$
\begin{array}{ccc}
Y^{(c)} & \xrightarrow{\;\alpha_{\mathbf{m}}^{(c)}\;} & Y^{(c)} \\
\phi \downarrow & & \downarrow \phi \\
X^{\mathfrak{R}_d/\mathfrak{j}_c} & \xrightarrow[\;\alpha_{\mathbf{m}}^{\mathfrak{R}_d/\mathfrak{j}_c}\;]{} & X^{\mathfrak{R}_d/\mathfrak{j}_c}
\end{array}
\tag{7.11}
$$

commutes for every $\mathbf{m} \in \mathbb{Z}^d$.

PROOF. The evaluation map $\eta_c \colon f \mapsto f(c)$ induces an isomorphism η of the $\mathfrak{R}_d$-module $\mathfrak{R}_d/\mathfrak{j}_c$ with the submodule $\eta_c(\mathfrak{R}_d) \subset R_c \subset \mathbb{K}$; in particular

$$\eta(\hat{\alpha}_{\mathbf{m}}^{\mathfrak{R}_d/\mathfrak{j}_c}(a)) = \hat{\alpha}_{\mathbf{m}}^{\eta_c(\mathfrak{R}_d)}(\eta(a)) = \hat{\alpha}_{\mathbf{m}}^{R_c}(\eta(a)) \tag{7.12}$$

for every $a \in \mathfrak{R}_d/\mathfrak{j}_c$ and $\mathbf{m} \in \mathbb{Z}^d$.

We claim that $R_c/\eta_c(\mathfrak{R}_d)$ is finite. Indeed, since $\mathbb{K} = \mathbb{Q}(c)$ is algebraic, every $a \in \mathbb{K}$ can be written as $a = b/m$ with $b \in \mathbb{Z}[c] = \mathbb{Z}[c_1, \ldots, c_d]$ and $m \geq 1$. In particular, since the ring of integers $\mathfrak{o}(c) = \mathfrak{o}_{\mathbb{K}} \subset \mathbb{K}$ is a finitely generated $\mathbb{Z}$-module, there exist positive integers m_0, M_0 with $m_0\mathfrak{o}(c) \subset \mathbb{Z}[c] \subset \eta_c(\mathfrak{R}_d)$ and $|\mathfrak{I}_c/\eta_c(\mathfrak{R}_d)| \leq |\mathfrak{o}(c)/m_0\mathfrak{o}(c)| = M_0 < \infty$.

According to the definition of $F(c)$ there exists, for every $v \in F(c)$, an element $a_v \in \eta_c(\mathfrak{R}_d)$ such that $|a_v|_v > 1$ and $|a_v|_w = 1$ for all $w \in P_{\mathfrak{f}}^{\mathbb{K}} \setminus F(c)$. Then $|a_v^n \mathfrak{o}(c)/\eta_c(\mathfrak{R}_d)| \leq M_0$ and $|(\sum_{v \in F(c)} a_v^n \mathfrak{o}(c))/\eta_c(\mathfrak{R}_d)| \leq M_0^{|F(c)|}$ for all

$n > 0$. As $n \to \infty$, $\sum_{v \in F(c)} a_v^n \mathfrak{o}(c)$ increases to R_c, and we conclude that $|R_c / \eta_c(\mathfrak{R}_d)| \le M_0^{|F(c)|} < \infty$.

The inclusion map $\mathfrak{R}_d / \mathfrak{j}_c \cong \eta_c(\mathfrak{R}_d) \hookrightarrow R_c$ induces a dual, surjective, finite-to-one homomorphism $\phi \colon Y^{(c)} \longmapsto X = \widehat{\mathfrak{R}_d / \mathfrak{j}_c}$, and the diagram (7.11) commutes by (7.12). $\square$

This comparison between R_c and $\eta_c(\mathfrak{R}_d)$ shows that the $\mathbb{Z}^d$-actions $\alpha^{(c)}$ and $\alpha^{\mathfrak{R}_d / \mathfrak{j}_c}$ are closely related. The group R_c can be determined much more easily than $\eta_c(\mathfrak{R}_d)$ and has other advantages, e.g. for the computation of entropy in Section 7; on the other hand R_c may not be a cyclic $\mathfrak{R}_d$-module, in contrast to $\eta_c(\mathfrak{R}_d) \cong \mathfrak{R}_d / \mathfrak{j}_c$. Since R_c is torsion-free (as an additive group), $Y^{(c)}$ and $X^{\mathfrak{R}_d / \mathfrak{j}_c}$ are both connected.

PROPOSITION 7.2. *Let* $d \ge 1$, $c = (c_1, \dots, c_d) \in (\overline{\mathbb{Q}}^\times)^d$, *and let* $(X^{\mathfrak{R}_d / \mathfrak{j}_c}, \alpha^{\mathfrak{R}_d / \mathfrak{j}_c})$ *and* $(Y^{(c)}, \alpha^{(c)})$ *be defined as in Theorem 7.1.*

(1) *For every* $\mathbf{m} \in \mathbb{Z}^d$, *the following conditions are equivalent.*
   (a) $\alpha_{\mathbf{m}}^{(c)}$ *is ergodic;*
   (b) $\alpha_{\mathbf{m}}^{\mathfrak{R}_d / \mathfrak{j}_c}$ *is ergodic;*
   (c) $c^{\mathbf{m}}$ *is not a root of unity.*
(2) *The following conditions are equivalent.*
   (a) $\alpha^{(c)}$ *is ergodic;*
   (b) $\alpha^{\mathfrak{R}_d / \mathfrak{j}_c}$ *is ergodic;*
   (c) *At least one coordinate of c is not a root of unity.*
(3) *The following conditions are equivalent.*
   (a) $\alpha^{(c)}$ *is mixing;*
   (b) $\alpha^{\mathfrak{R}_d / \mathfrak{j}_c}$ *is mixing;*
   (c) $c^{\mathbf{m}} \ne 1$ *for all non-zero* $\mathbf{m} \in \mathbb{Z}^d$.
(4) *If* $\alpha^{(c)}$ *is ergodic then the groups* $\mathrm{Fix}_\Lambda(\alpha^{(c)})$ *and* $\mathrm{Fix}_\Lambda(\alpha^{\mathfrak{R}_d / \mathfrak{j}_c})$ *are finite for every subgroup* $\Lambda \subset \mathbb{Z}^d$ *with finite index.*
(5) *The following conditions are equivalent.*
   (a) $\alpha^{(c)}$ *is expansive;*
   (b) $\alpha^{\mathfrak{R}_d / \mathfrak{j}_c}$ *is expansive;*
   (c) *The orbit of c under the diagonal action of the Galois group* $\mathrm{Gal}[\overline{\mathbb{Q}} : \mathbb{Q}]$ *on* $(\overline{\mathbb{Q}}^\times)^d$ *does not intersect* $\mathbb{S}^d$.

PROOF. The $\mathfrak{R}_d$-modules R_c and $\mathfrak{R}_d / \mathfrak{j}_c$ are both associated with the prime ideal $\mathfrak{j}_c$, $V_\mathbb{C}(\mathfrak{j}_c) = \mathrm{Gal}[\overline{\mathbb{Q}} : \mathbb{Q}](c)$, and all assertions follow from Theorem 6.5. $\square$

PROPOSITION 7.3. *Let* $N(c)$ *be the cardinality of the orbit* $\mathrm{Gal}[\overline{\mathbb{Q}} : \mathbb{Q}](c)$ *of c under the Galois group. Then* $Y^{(c)} \cong \mathbb{T}^{N(c)}$ *if and only if* c_i *is an algebraic unit for every* $i = 1, \dots, d$ *(i.e.* c_i *and* c_i^{-1} *are integral in* $\mathbb{Q}(c)$ *for* $i = 1, \dots, d$). *If at least one of the coordinates of c is not a unit, then* $Y^{(c)}$ *is a projective limit of copies of* $\mathbb{T}^{N(c)}$.

PROOF. We use the notation established in (7.1)–(7.8). The number $N(c)$ is equal to the degree $[\mathbb{Q}(c) : \mathbb{Q}]$. If $N_{\mathbb{R}}(c)$ and $N_{\mathbb{C}}(c)$ are the numbers of real and complex (infinite) places of $\mathbb{Q}(c)$ then $N(c) = N_{\mathbb{R}}(c) + 2N_{\mathbb{C}}(c)$, and the connected component of the identity in $\prod_{v \in P(c)} \mathbb{K}_v$ is isomorphic to $\mathbb{R}^{N(c)}$. The condition that every coordinate of c be a unit is equivalent to the assumption that $F(c) = \varnothing$; in this case $Y^{(c)}$ is isomorphic to the quotient of $\mathbb{R}^{N(c)}$ by the discrete, co-compact subgroup $i_{P(c)}(R_c)$, i.e. $Y^{(c)} \cong \mathbb{T}^{N(c)}$. If $F(c) \neq \varnothing$ then $Y^{(c)}$ is isomorphic to the quotient of $\mathbb{R}^{N(c)} \times \prod_{v \in F(c)} \mathbb{K}_v$ by $i_{P(c)}(R_c)$. In order to prove the assertion about the projective limit we choose, for every $v \in F(c)$, a prime element $p_v \in \mathbb{K}_v$ (i.e. an element with $p_v \mathcal{R}_v = \{a \in \mathbb{K}_v : |a|_v < 1\}$), and set $\Delta_n = i_{P(c)}(R_c) + \prod_{v \in F(c)} p_v^n \bar{\mathcal{R}}_v$ for every $n \geq 1$. Then $\bigcap_{n \geq 1} \Delta_n = i_{P(c)}(R_c)$, and $Y^{(c)}$ is the projective limit of the groups $Y_n = Y^{(c)}/\Delta_n \cong \mathbb{T}^{N(c)}$, $n \geq 1$, where the last isomorphism is established by meditation. $\square$

If X is a compact, connected, abelian group with dual group $\hat{X}$, then $\hat{X}$ is torsion-free, and the map $a \mapsto 1 \otimes a$ from $\hat{X}$ into the tensor product $\mathbb{Q} \otimes_{\mathbb{Z}} \hat{X}$ is therefore injective. We denote by $\dim X$ the dimension of the vector space $\mathbb{Q} \otimes_{\mathbb{Z}} \hat{X}$ over $\mathbb{Q}$ and note that this definition of $\dim X$ is consistent with the usual topological dimension of X: in particular, $0 < \dim Y^{(c)} = N(c) < \infty$ in Proposition 7.3. With this terminology we obtain the following corollary of Theorem 7.1 and Proposition 7.3.

COROLLARY 7.4. *Let $\mathfrak{p} \subset \mathfrak{R}_d$ be a prime ideal, and let $(X^{\mathfrak{R}_d/\mathfrak{p}}, \alpha^{\mathfrak{R}_d/\mathfrak{p}})$ be defined as in Lemma 5.1. The following conditions are equivalent.*

(1) *$X^{\mathfrak{R}_d/\mathfrak{p}}$ is a connected, finite-dimensional, abelian group;*

(2) *$\mathfrak{p} = \mathfrak{j}_c$ for some $c \in (\overline{\mathbb{Q}}^{\times})^d$.*

Furthermore, if α is an ergodic $\mathbb{Z}^d$-action by automorphisms of a compact, connected, finite-dimensional abelian group X, then the $\mathfrak{R}_d$-module $\mathfrak{M} = \hat{X}$ has only finitely many associated prime ideals, each of which is of the form $\mathfrak{p} = \mathfrak{j}_c$ for some $c \in (\overline{\mathbb{Q}}^{\times})^d$.

PROOF. The implication (2)$\Rightarrow$(1) is clear from Theorem 7.1, Proposition 7.3, and the definition of $\dim X$. Conversely, if $\mathfrak{p} \subset \mathfrak{R}_d$ is a prime ideal such that $X^{\mathfrak{R}_d/\mathfrak{p}} = \widehat{\mathfrak{R}_d/\mathfrak{p}}$ is connected, then $\mathfrak{p}$ does not contain any non-zero constants, and the map $a \mapsto 1 \otimes a$ from $\mathfrak{R}_d/\mathfrak{p}$ into the tensor product $\mathbb{Q} \otimes_{\mathbb{Z}} (\mathfrak{R}_d/\mathfrak{p})$ is injective. This allows us to regard $\mathfrak{R}_d/\mathfrak{p}$ as a subring of $\mathbb{Q} \otimes_{\mathbb{Z}} (\mathfrak{R}_d/\mathfrak{p})$. The variety $V(\mathfrak{p})$ is non-empty by Proposition 6.9, and is finite if and only if each of the elements $u_i + \mathfrak{p} \in \mathbb{Q} \otimes_{\mathbb{Z}} (\mathfrak{R}_d/\mathfrak{p})$, $i = 1, \ldots, d$, is algebraic over the subring $\mathbb{Q} \subset \mathbb{Q} \otimes_{\mathbb{Z}} (\mathfrak{R}_d/\mathfrak{p})$. In particular, if $V(\mathfrak{p})$ is finite, then $\mathfrak{p} = \mathfrak{j}_c$ for every $c \in V(\mathfrak{p})$, which implies (2). If $V(\mathfrak{p})$ is infinite, then at least one of the elements $u_j + \mathfrak{p}$ is transcendental over $\mathbb{Q} \subset \mathbb{Q} \otimes_{\mathbb{Z}} (\mathfrak{R}_d/\mathfrak{p})$, and the powers $u_j^k + \mathfrak{p}$, $k \in \mathbb{Z}$, are rationally independent. This is easily seen to imply that $\dim X^{\mathfrak{R}_d/\mathfrak{p}} = \infty$.

In order to prove the last assertion we assume that $\mathfrak{p} \subset \mathfrak{R}_d$ is a prime ideal associated with $\mathfrak{M}$. Then $X^{\mathfrak{R}_d/\mathfrak{p}}$ is (isomorphic to) a quotient group of

X, hence connected and finite-dimensional, and Proposition 6.9 and the first part of this corollary together imply that $\mathfrak{p} = \mathfrak{j}_c$ for some $c \in (\overline{\mathbb{Q}}^\times)^d$. If $\mathfrak{M}$ has infinitely many distinct associated prime ideals $\{\mathfrak{j}_{c^{(1)}}, \mathfrak{j}_{c^{(2)}}, \dots\}$, then we can find, for every $i \geq 1$, an element $a_i \in \mathfrak{M}$ with $\mathfrak{R}_d \cdot a_i \cong \mathfrak{R}_d / \mathfrak{j}_{c^{(i)}}$. If $b \in (\sum_{i=1}^{j-1} \mathfrak{R}_d \cdot a_i) \cap \mathfrak{R}_d \cdot a_j \neq \{0\}$ for some $j > 1$, then the submodule $\mathfrak{R}_d \cdot b \subset \mathfrak{M}$ has an associated prime ideal $\mathfrak{j}$ which strictly contains $\mathfrak{j}_{c^{(j)}}$; in particular, $\mathfrak{j}$ must contain a non-zero constant, in violation of the fact that every prime ideal $\mathfrak{p}$ associated with $\mathfrak{R}_d \cdot b$ (and hence with $\mathfrak{M}$) must satisfy that $V_{\mathbb{C}}(\mathfrak{p}) \neq \varnothing$. It follows that $\mathfrak{M}$ has a submodule isomorphic to $\mathfrak{R}_d / \mathfrak{j}_{c^{(1)}} \oplus \mathfrak{R}_d / \mathfrak{j}_{c^{(2)}} \oplus \cdots$, and hence that $\dim X = \infty$. This contradiction proves that there are only finitely many distinct prime ideals associated with $\mathfrak{M}$. $\square$

EXAMPLE 7.5. If α is a $\mathbb{Z}^d$-action by automorphisms of a compact, connected, finite-dimensional, abelian group, then the $\mathfrak{R}_d$-module $\mathfrak{M} = \hat{X}$ need not be Noetherian (cf. Corollary 7.4): if α is the automorphism of $X = \hat{\mathbb{Q}}$ in Example 5.6 (1) consisting of multiplication by $\frac{3}{2}$, then $\dim(X) = 1$, but $\mathfrak{M} = \hat{X} = \mathbb{Q}$ is not Noetherian (cf. Example 6.17 (2). $\boxdot$

The following Examples 7.6 show that the $\mathbb{Z}^d$-actions $\alpha^{(c)}$ and $\alpha^{\mathfrak{R}/\mathfrak{j}_c}$ may be, but need not be, topologically conjugate.

EXAMPLES 7.6. (1) If $c = 2$ then $F(c) = \{2\}$, $R_c = \mathbb{Z}[\frac{1}{2}]$, and we claim that the automorphism $\alpha_1^{(c)}$ on $Y^{(c)} = \widehat{R_c} = (\mathbb{R} \times \mathbb{Q}_2)/i_{F(c)}(\mathbb{Z}[\frac{1}{2}])$, which is multiplication by 2, is conjugate to the shift $\alpha_1^{\mathfrak{R}_1/(2-u_1)}$ on the group $X^{\mathfrak{R}_1/(2-u_1)}$ described in Example 5.3 (3). In order to verify this we note that there exists, for every $(s,t) \in \mathbb{R} \times \mathbb{Q}_2$, a unique element $r \in \mathbb{Z}[\frac{1}{2}]$ with $r + s \in [0,1)$ and $r + t \in \mathbb{Z}_2$. This allows us to identify $Y^{(c)} = \widehat{\mathbb{Z}[\frac{1}{2}]}$ with $(\mathbb{R} \times \mathbb{Z}_2)/i_{F(c)}(\mathbb{Z})$. An element $a = \frac{k}{2^l} \in \mathbb{Z}[\frac{1}{2}]$ defines a character on $Y^{(c)} = (\mathbb{R} \times \mathbb{Z}_2)/i_{F(c)}(\mathbb{Z})$ by $\langle a, (s,t) + i_{F(c)}(\mathbb{Z})\rangle = e^{2\pi i(\mathrm{Int}(as) + \mathrm{Frac}(at))}$ for every $s \in \mathbb{R}$ and $t \in \mathbb{Z}_2$, where $\mathrm{Int}(as)$ is the integral part of $as \in \mathbb{R}$ and $\mathrm{Frac}(at) \in [0,1)$ is the (well-defined) fractional part of $at \in \mathbb{Q}_2$. Consider the homomorphism $\phi \colon Y^{(c)} \longmapsto \mathbb{T}^{\mathbb{Z}}$ defined by $e^{2\pi i(\phi(y))_m} = \langle 2^m, y\rangle$ for every $y \in Y^{(c)}$ and $m \in \mathbb{Z}$. Then ϕ is injective, $\phi(Y^{(c)}) \subset X^{\mathfrak{R}_1/(2-u_1)}$, and it is not difficult to see that $\phi \colon Y^{(c)} \longmapsto X^{\mathfrak{R}_1/(2-u_1)}$ is a continuous group isomorphism which makes the diagram (7.11) commute. In particular, if we write a typical element $y \in Y^{(c)}$ as $y = (s,t) + i_{F(c)}(\mathbb{Z})$ with $s \in \mathbb{R}$ and $t \in \mathbb{Z}_2$, then

$$(\phi((0,t) + i_{F(c)}(\mathbb{Z})))_m = 0 \quad \text{and} \quad (\phi((s,0) + i_{F(c)}(\mathbb{Z})))_m = 2^m s \pmod 1$$

for every $m \geq 0$.

Proposition 7.2 shows that the automorphism $\alpha^{(c)} = \alpha^{\mathfrak{R}_1/\mathfrak{j}_c}$ is expansive and hence ergodic.

(2) If $c = 3/2$ then $F(c) = \{2,3\}$, $R_c = \mathbb{Z}[\frac{1}{6}]$, and we see as in Example (1) that multiplication by $\frac{3}{2}$ on

$$Y^{(c)} = \widehat{R_c} = (\mathbb{R} \times \mathbb{Q}_2 \times \mathbb{Q}_3)/i_{F(c)}(\mathbb{Z}[\tfrac{1}{6}]) \cong (\mathbb{R} \times \mathbb{Z}_2 \times \mathbb{Z}_3)/i_{F(c)}(\mathbb{Z})$$

is conjugate to the shift $\alpha^{\mathfrak{R}_1/\mathfrak{J}_c}$ on $X^{\mathfrak{R}_1/\mathfrak{J}_c}$ in Example 6.18 (2). The $\mathbb{Z}$-action $\alpha^{(c)} = X^{\mathfrak{R}_1/\mathfrak{J}_c}$ is expansive and ergodic by Proposition 7.2.

(3) Let $c = 2 + \sqrt{5}$. Then $\eta_c(\mathfrak{R}_1) = \{k + l\sqrt{5} : k, l \in \mathbb{Z}\} \cong \mathbb{Z}^2$, $F(c) = \varnothing$, and R_c is equal to the set $\mathfrak{o}(c) = \mathfrak{o}_{\mathbb{Q}(c)}$ of integral elements in $\mathbb{Q}(c)$. Since $\mathfrak{o}_{\mathbb{Q}(c)} = \{k\frac{1+\sqrt{5}}{2} + l\frac{1-\sqrt{5}}{2} : k, l \in \mathbb{Z}\}$ (cf. Lemma 10.3.3 in [16]), $R_c \neq \eta_c(\mathfrak{R}_1)$. By Proposition 7.2, the $\mathbb{Z}$-actions $\alpha^{(c)}$ and $\alpha^{\mathfrak{R}_1/\mathfrak{J}_c}$ are both expansive (and hence ergodic), but we claim that they are not topologically conjugate. According to Corollary 5.10 this amounts to showing that R_c and $\mathfrak{R}_1/\mathfrak{J}_c$ are not isomorphic as $\mathfrak{R}_1$-modules, and we establish this by showing that R_c is not cyclic. In terms of the $\mathbb{Z}$-basis $\{\frac{1+\sqrt{5}}{2}, \frac{1-\sqrt{5}}{2}\}$ for R_c, multiplication by c is represented by the matrix $A = \left(\begin{smallmatrix} 5 & -2 \\ 2 & -1 \end{smallmatrix}\right)$. If the module R_c is cyclic, then there exists a vector $\mathbf{m} = (m_1, m_2) \in \mathbb{Z}^2$ such that $\{\mathbf{m}, A\mathbf{m}\} = \{(m_1, m_2), (5m_1 - 2m_2, 2m_1 - m_2)$ generates $\mathbb{Z}^2$, and as in Example 5.3 (2) we see that this impossible.

In this example $X^{\mathfrak{R}_1/\mathfrak{J}_c} \cong Y^{(c)} \cong \mathbb{T}^2$. The matrix $A' = \left(\begin{smallmatrix} 2 & 5 \\ 1 & 2 \end{smallmatrix}\right)$ represents multiplication by c in terms of the $\mathbb{Z}$-basis $\{1, \sqrt{5}\}$ of $\eta_c(\mathfrak{R}_1)$, and the matrices A and A' define non-conjugate automorphisms of $\mathbb{T}^2$ with identical characteristic polynomials (cf. Example 5.3 (2)).

(4) Let $c = \frac{1+\sqrt{5}}{2}$. Then $\eta_c(\mathfrak{R}_1) = \mathfrak{o}(\mathbb{Q}(c)) = R_c$, and the $\mathbb{Z}$-actions $\alpha^{(c)}$ and $\alpha^{\mathfrak{R}_1/\mathfrak{J}_c}$ are algebraically conjugate. However, a little care is needed in identifying $\widehat{R_c}$ with $Y^{(c)}$ in (7.10). The set $P(c) = P_\infty^{\mathbb{Q}(c)}$ consists of the two real places determined by the embeddings $\sqrt{5} \mapsto \sqrt{5}$ and $\sqrt{5} \mapsto -\sqrt{5}$ of $\mathbb{Q}(c) = \mathbb{Q}(\sqrt{5})$ in $\mathbb{R}$, so that $Y^{(c)} = \mathbb{R}^2/i_{P(c)}(R_c)$ with $i_{P(c)}(R_c) = \{(k + l\frac{1+\sqrt{5}}{2}, k+l\frac{1-\sqrt{5}}{2}) : (k, l) \in \mathbb{Z}^2\} \subset \mathbb{R}^2$. Under the usual identification of $\widehat{\mathbb{R}^2}$ with $\mathbb{R}^2$ given by $\langle (t_1, t_2), (s_1, s_2) \rangle = e^{2\pi i(s_1 t_1 + s_2 t_2)}$ for every $(s_1, s_2), (t_1, t_2) \in \mathbb{R}^2$, the annihilator $i_{P(c)}(R_c)^\perp \subset \widehat{\mathbb{R}^2} = \mathbb{R}^2$ is of the form $i_{P(c)}(R_c)^\perp = \frac{1}{\sqrt{5}} \cdot i_{P(c)}(R_c)$, and

$$\widehat{Y^{(c)}} = i_{P(c)}(R_c)^\perp = \frac{1}{\sqrt{5}} \cdot i_{P(c)}(R_c) = i_{P(c)}\left(\frac{1}{\sqrt{5}} \cdot R_c\right) \cong \frac{1}{\sqrt{5}} \cdot R_c \cong R_c.$$

(5) Let $\omega = (-1 + \sqrt{-3})/2$ and $c = 1 + 3\omega \in \overline{\mathbb{Q}}$. Then $\mathbb{K} = \mathbb{Q}(\omega)$ and $F(c) = \{7\}$. We claim that $R_c \neq \eta_c(\mathfrak{R}_1)$. Indeed, since the minimal polynomial $f(u) = u^2 + u + 1$ of ω is irreducible over the field $\mathbb{Q}_3$ of triadic rationals, there exists a unique place v of $\mathbb{K}$ above 3, and $\mathbb{K}_v = \mathbb{Q}_3(\omega)$. Let $\mathcal{R}_v = \{a \in \mathbb{K}_v : |a|_v \leq 1\}$ and $\mathfrak{o}_v = \mathbb{K} \cap \mathcal{R}_v$. As $|3|_v = 1/9$, every $a \in S = \mathbb{Z} + 3\mathfrak{o}_v \subset \mathfrak{o}_v$ with $|a|_v < 1$ satisfies that $|a|_v \leq 3^{-2}$. In particular, $\zeta = 1 - \omega \in \mathfrak{o}_v \smallsetminus S$, since $\zeta^2 = (1-\omega)^2 = -3\omega$ and hence $|\zeta|_v = 1/3$ (cf. p.139 in [16]). Since $\eta_c(\mathfrak{R}_1) \subset S$ and $\zeta \in \mathfrak{o}(c) \subset R_c$ we conclude that $\zeta \in R_c \smallsetminus \eta_c(\mathfrak{R}_1) \neq \varnothing$.

In order to verify that $\eta_c(\mathfrak{R}_1) \cong \mathfrak{R}_1/\mathfrak{j}_c$ and R_c are non-isomorphic we take an arbitrary, non-zero element $a \in R_c$ and note that

$$\{|b|_v : b \in \eta_c(\mathfrak{R}_1) \cdot a\} = \{|f(c)|_v|a|_v : f \in \mathfrak{R}_1\} \subset \{|a|_v|b|_v : b \in S\}$$
$$\subsetneq \{3^{-n} : n \geq 0\} = \{|b|_v : b \in R_c\}.$$

Hence R_c is not cyclic, in contrast to $\mathfrak{R}_1/\mathfrak{j}_c$. Corollary 5.10 shows that the $\mathbb{Z}^d$-actions $\alpha^{(c)}$ and $\alpha^{\mathfrak{R}_1/\mathfrak{j}_c}$ are not topologically conjugate. In this example the isomorphic groups $Y^{(c)}$ and $X^{\mathfrak{R}_1/\mathfrak{j}_c}$ are projective limits of two-dimensional tori, and the automorphisms $\alpha^{(c)}$ and $\alpha^{\mathfrak{R}_1/\mathfrak{j}_c}$ are expansive (and ergodic) by Proposition 7.2. $\quad\boxdot$

EXAMPLES 7.7. (1) Let $c = (2,3) \subset (\overline{\mathbb{Q}}^{\times})^2$. Then $\mathfrak{j}_c = (u_1 - 2, u_2 - 3) \subset \mathfrak{R}_2$, $F(c) = \{2,3\}$, $R_c = \mathbb{Z}[\frac{1}{6}]$, and as in Example 7.6 (1) one sees that the $\mathbb{Z}^2$-action $\alpha^{(c)}$ on $Y^{(c)}$ is conjugate to shift-action $\alpha^{\mathfrak{R}_2/\mathfrak{j}_c}$ on the group $X^{\mathfrak{R}_2/\mathfrak{j}_c}$ appearing in in Example 5.3 (4). Note that $\alpha^{\mathfrak{R}_2/\mathfrak{j}_c}$ is expansive and mixing; in fact, $\alpha_\mathbf{n}^{\mathfrak{R}_2/\mathfrak{j}_c}$ is expansive for every non-zero $\mathbf{n} \in \mathbb{Z}^2$ (Proposition 7.2). The group $Y^{(c)} = (\mathbb{R} \times \mathbb{Q}_2 \times \mathbb{Q}_3)/i_{F(c)}(\mathbb{Z}[\frac{1}{6}]) \cong (\mathbb{R} \times \mathbb{Z}_2 \times \mathbb{Z}_3)/i_{F(c)}(\mathbb{Z})$ is the same as in Example 7.6 (2), but $X^{\mathfrak{R}_2/\mathfrak{j}_c}$ is now a closed, shift-invariant subgroup of $\mathbb{T}^{\mathbb{Z}^2}$. In order to describe an explicit isomorphism $\phi: Y^{(c)} \longmapsto X^{\mathfrak{R}_2/\mathfrak{j}_c}$ we proceed as in Example 7.6 (1): identify $Y^{(c)}$ with $(\mathbb{R} \times \mathbb{Z}_2 \times \mathbb{Z}_3)/i_{F(c)}(\mathbb{Z})$, and write the character of $Y^{(c)}$ defined by an element $a = \frac{j}{2^k 3^l} \in \mathbb{Z}[\frac{1}{6}]$ as $\langle a, (r,s,t) + i_{F(c)}(\mathbb{Z})\rangle = e^{2\pi i(\text{Int}(ar) + \text{Frac}(as) + \text{Frac}(at))}$ for every $r \in \mathbb{R}$, $s \in \mathbb{Z}_2$ and $t \in \mathbb{Z}_3$. If $\phi: Y^{(c)} \longmapsto \mathbb{T}^{\mathbb{Z}^2}$ is the map given by $e^{2\pi i(\phi(y))_{(n_1, n_2)}} = \langle 2^{n_1} 3^{n_2}, y\rangle$ for every $y \in Y$ and $(n_1, n_2) \in \mathbb{Z}^2$, then ϕ is injective, $\phi(Y^{(c)}) = X^{\mathfrak{R}_2/\mathfrak{j}_c}$, and ϕ makes the diagram (7.11) commute.

(2) Let $\mathbb{K} \supset \mathbb{Q}$ be an algebraic number field. We denote by $\mathfrak{o}_{\mathbb{K}} \subset \mathbb{K}$ the ring of integers and write $\mathcal{U}_{\mathbb{K}} \subset \mathfrak{o}_{\mathbb{K}}$ for the group of units (i.e. $\mathcal{U}_{\mathbb{K}} = \{a \in \mathfrak{o}_{\mathbb{K}} : a^{-1} \in \mathfrak{o}_{\mathbb{K}}\}$). By Theorem 10.8.1 in [16], $\mathcal{U}_{\mathbb{K}}$ is isomorphic to the cartesian product $F \times \mathbb{Z}^{r+s-1}$, where F is a finite, cyclic group consisting of all roots of unity in $\mathbb{K}$ and r and s are the numbers of real and complex places of $\mathbb{K}$. We set $d = r + s - 1$, choose generators $c_1, \ldots, c_d \in \mathcal{U}_{\mathbb{K}}$ such that every $a \in \mathcal{U}_{\mathbb{K}}$ can be written as $a = u c_1^{k_1} \cdot \ldots \cdot c_d^{k_d}$ with $u \in F$ and $k_1, \ldots, k_d \in \mathbb{Z}$, and set $c = (c_1, \ldots, c_d)$. Then $X^{\mathfrak{R}_d/\mathfrak{j}_c} \cong Y^{(c)} \cong \mathbb{T}^{r+2s}$, and the $\mathbb{Z}^d$-actions $\alpha^{\mathfrak{R}_d/\mathfrak{j}_c}$ and $\alpha^{(c)}$ are mixing by Proposition 7.2.

(3) Let $d \geq 1$, and let $\mathfrak{a} \subset \mathfrak{R}_d$ be an ideal with $V(\mathfrak{a}) \neq \varnothing$ (or, equivalently, with $V_{\mathbb{C}}(\mathfrak{a}) \neq \varnothing$). For every $c \in V(\mathfrak{a})$ the evaluation map $\eta_c: f \mapsto f(c)$ from $\mathfrak{R}_d/\mathfrak{a}$ to $\mathbb{Q}(c)$ induces a dual, injective embedding of $X^{\mathfrak{R}_d/\mathfrak{j}_c}$ in $X^{\mathfrak{R}_d/\mathfrak{a}}$, so that we may regard $X^{\mathfrak{R}_d/\mathfrak{j}_c}$ as a subgroup of $X^{\mathfrak{R}_d/\mathfrak{a}}$; in this picture $\alpha^{\mathfrak{R}_d/\mathfrak{j}_c}$ is the restriction of $\alpha^{\mathfrak{R}_d/\mathfrak{a}}$ to $X^{\mathfrak{R}_d/\mathfrak{j}_c}$. In fact, if $\mathfrak{a}$ is *radical*, i.e. if $\mathfrak{a} = \sqrt{\mathfrak{a}} = \{f \in \mathfrak{R}_d : f^k \in \mathfrak{a}$ for some $k \geq 1\}$, then $\mathfrak{a} = \{f \in \mathfrak{R}_d : f(c) = 0$ for every $c \in V(\mathfrak{a})\}$, and the group generated by $X^{\mathfrak{R}_d/\mathfrak{j}_c}$, $c \in V_{\mathbb{C}}(\mathfrak{a})$, is dense in $X^{\mathfrak{R}_d/\mathfrak{a}}$. In general, $\alpha^{\mathfrak{R}_d/\mathfrak{a}}$ is expansive if and only if $\alpha^{\mathfrak{R}_d/\mathfrak{j}_c}$ is expansive for every $c \in V(\mathfrak{a})$, but

$\alpha^{\Re_d/\mathfrak{a}}$ may be mixing in spite of $\alpha^{\Re_d/\mathfrak{j}_c}$ being non-ergodic for some $c \in V(\mathfrak{a})$: take, for example, $d = 2$, $\mathfrak{a} = (1+u_1+u_2) \subset \Re_2$, and $c = ((-1+i\sqrt{-3})/2, (-1-i\sqrt{-3})/2) \in V(\mathfrak{a})$ (Theorem 6.5, Proposition 7.2, and Example 6.18 (1)). $\boxdot$

CONCLUDING REMARK 7.8. Theorem 7.1, Proposition 7.2, and Example 7.6 (5) are taken from [94], and Example 7.6 (4) was pointed out to me by Jenkner. The possible difference between $\alpha^{(c)}$ and $\alpha^{\Re_d/\mathfrak{j}_c}$ for $c \in (\overline{\mathbb{Q}}^\times)^d$ allows the construction of analogues to Williams' Example 5.3 (2) for $\mathbb{Z}^d$-actions.

8. The dynamical system defined by a prime ideal

In this section we continue our investigation of the structure of the $\mathbb{Z}^d$-actions $\alpha^{\Re_d/\mathfrak{p}}$, where $\mathfrak{p} \subset \Re_d$ is a prime ideal. For prime ideals of the form $\mathfrak{j}_c$, $c \in (\overline{\mathbb{Q}}^\times)^d$, the work was done in Section 7, and for $\mathfrak{p} = \{0\}$ we already know that $\alpha^{\Re_d/\mathfrak{p}}$ is the shift-action of $\mathbb{Z}^d$ on $X^{\Re_d/\mathfrak{p}} = \mathbb{T}^{\mathbb{Z}^d}$. Another case which can be dealt with easily are the non-ergodic prime ideals (Definition 6.16).

PROPOSITION 8.1. *Let $\mathfrak{p} \subset \Re_d$ be a prime ideal. Then $\mathfrak{p}$ is non-ergodic if and only if $\mathfrak{p}$ is either maximal, or of the form $\mathfrak{j}_c$ for a point $c = (c_1, \ldots, c_d) \in \overline{\mathbb{Q}}^d$ with $c_1^l = \cdots = c_d^l = 1$ for some $l \geq 1$. Furthermore, if $\alpha^{\Re_d/\mathfrak{p}}$ is non-ergodic, then $X^{\Re_d/\mathfrak{p}}$ is either finite or a finite-dimensional torus, and there exists an integer $L \geq 1$ such that $\alpha_{L\mathbf{n}}^{\Re_d/\mathfrak{p}} = id_{X^{\Re_d/\mathfrak{p}}}$ for every $\mathbf{n} \in \mathbb{Z}^d$.*

PROOF. This is just a re-wording of Theorem 6.5 (1). An ideal $\mathfrak{p} \subset \Re_d$ is maximal if and only if $\Re_d/\mathfrak{p}$ is a finite field; in particular, the characteristic $p(\mathfrak{p})$ is positive for any maximal ideal $\mathfrak{p}$.

Let $\mathfrak{p} \subset \Re_d$ be a prime ideal such that $\alpha = \alpha^{\Re_d/\mathfrak{p}}$ is non-ergodic. If $p = p(\mathfrak{p}) > 0$, then Theorem 6.5 (1.e) implies that $V(\mathfrak{p}) \subset (\overline{\mathbb{F}}_{p(\mathfrak{p})}^\times)^d$ is finite and that $\mathfrak{p}$ is therefore maximal. In particular, $\Re_d/\mathfrak{p} \cong \mathbb{F}_{p^l}$ for some $l \geq 1$, where $\mathbb{F}_{p^l}$ is the finite field with p^l elements, and $\alpha_{(p^l-1)\mathbf{n}}$ is the identity map on $X^{\Re_d/\mathfrak{p}} \cong \mathbb{F}_{p^l}$ for every $\mathbf{n} \in \mathbb{Z}^d$. Conversely, if $\mathfrak{p}$ is maximal, then $|X^{\Re_d/\mathfrak{p}}| = |\Re_d/\mathfrak{p}|$ is finite, and α is therefore non-ergodic.

If $p(\mathfrak{p}) = 0$, then Theorem 6.5 (1.e) guarantees the existence of an integer $l \geq 1$ with $c_1^l = \cdots = c_d^l = 1$ for every $c = (c_1, \ldots, c_l) \in V(\mathfrak{p}) = V_{\mathbb{C}}(\mathfrak{p})$, so that $V(\mathfrak{p})$ is finite, and the primality of $\mathfrak{p}$ allows us to conclude that $\mathfrak{p} = \mathfrak{j}_c$ for some $c = (c_1, \ldots, c_d) \in \overline{\mathbb{Q}}^d$ with $c_1^l = \cdots = c_d^l = 1$. From the definition of $\alpha^{(c)}$ in (7.9)–(7.10), Theorem 7.1, and Proposition 7.3, it is clear that $X^{\Re_d/\mathfrak{p}}$ is a finite-dimensional torus, and that $\alpha_{l\mathbf{n}}$ is the identity map on $X^{\Re_d/\mathfrak{p}}$ for every $\mathbf{n} \in \mathbb{Z}^d$. Conversely, if $\mathfrak{p} = \mathfrak{j}_c$ for some $c = (c_1, \ldots, c_d) \in \overline{\mathbb{Q}}^d$ with $c_1^l = \cdots = c_d^l = 1$, then Theorem 6.5 (1.e) shows that α is non-ergodic. $\square$

Next we consider ergodic prime ideals $\mathfrak{p} \subset \Re_d$ with $p(\mathfrak{p}) > 0$. We call a subgroup $\Gamma \subset \mathbb{Z}^d$ primitive if $\mathbb{Z}^d/\Gamma$ is torsion-free; a non-zero element $\mathbf{n} \in \mathbb{Z}^d$ is primitive if the subgroup $\{k\mathbf{n} : k \in \mathbb{Z}\} \subset \mathbb{Z}^d$ is primitive. The following proposition shows that there exists, for every ergodic prime ideal $\mathfrak{p} \subset \Re_d$ with

$p(\mathfrak{p}) > 0$, a maximal primitive subgroup $\Gamma \subset \mathbb{Z}^d$ and a finite, abelian group G such that the restriction α^Γ of $\alpha^{\mathfrak{R}_d/\mathfrak{p}}$ to Γ is topologically and algebraically conjugate to the shift-action of Γ on G^Γ.

PROPOSITION 8.2. *Let* $\mathfrak{p} \subset \mathfrak{R}_d$ *be an ergodic prime ideal with* $p = p(\mathfrak{p}) > 0$, *and assume that* $\alpha = \alpha^{\mathfrak{R}_d/\mathfrak{p}}$ *is the shift-action of* $\mathbb{Z}^d$ *on the closed, shift-invariant subgroup* $X = X^{\mathfrak{R}_d/\mathfrak{p}} \subset \mathbb{F}_p^{\mathbb{Z}^d}$ *defined by* (6.19). *Then there exists an integer* $r = r(\mathfrak{p}) \in \{1, \ldots, d\}$, *a primitive subgroup* $\Gamma = \Gamma(\mathfrak{p}) \subset \mathbb{Z}^d$, *and a finite set* $Q = Q(\mathfrak{p}) \subset \mathbb{Z}^d$ *with the following properties.*

(1) $\Gamma \cong \mathbb{Z}^r$;
(2) $\mathbf{0} \in Q$, *and* $Q \cap (Q + \mathbf{m}) = \varnothing$ *whenever* $\mathbf{0} \neq \mathbf{m} \in \Gamma$;
(3) *If* $\bar{\Gamma} = \Gamma + Q = \{\mathbf{m} + \mathbf{n} : \mathbf{m} \in \Gamma, \mathbf{n} \in Q\}$, *then the coordinate projection* $\pi_{\bar{\Gamma}} : X \longmapsto \mathbb{F}_p^{\bar{\Gamma}}$, *which restricts any point* $x \in X \subset \mathbb{F}_p^{\mathbb{Z}^d}$ *to its coordinates in* $\bar{\Gamma}$, *is a continuous group isomorphism; in particular, the* Γ-*action* $\alpha^\Gamma : \mathbf{n} \mapsto \alpha_{\mathbf{n}}$, $\mathbf{n} \in \Gamma$, *is (isomorphic to) the shift-action of* Γ *on* $(\mathbb{F}_p^Q)^\Gamma$.

PROOF. This is Noether's normalization lemma in disguise. Consider the prime ideal $\mathfrak{p}' = \{f_{/p} : f \in \mathfrak{p}\} \subset \mathfrak{R}_d^{(p)}$ defined in Remark 6.19 (4), and write $\mathbf{e}^{(i)}$ for the i-th unit vector in $\mathbb{Z}^d$. We claim that there exists a matrix $A \in \mathrm{GL}(d, \mathbb{Z})$ and an integer r, $1 \leq r \leq d$, such that the elements $v_i = u^{A\mathbf{e}^{(i)}} + \mathfrak{p}'$ are algebraically independent in the ring $\mathfrak{R} = \mathfrak{R}_d^{(p)}/\mathfrak{p}'$ for $i = 1, \ldots, r$, and $v_j = u^{A\mathbf{e}^{(j)}} + \mathfrak{p}'$ is an algebraic unit over the subring $\mathbb{F}_p[v_1^{\pm 1}, \ldots, v_{j-1}^{\pm 1}] \subset \mathfrak{R}$ for $j = r + 1, \ldots, d$. Indeed, if $u_1' = u_1 + \mathfrak{p}', \ldots, u_d' = u_d + \mathfrak{p}'$ are algebraically independent elements of $\mathfrak{R}$, then $\mathfrak{p}' = \{0\}$, and the assertion holds with $r = d$, and with A equal to the $d \times d$ identity matrix. Assume therefore (after renumbering the variables, if necessary) that there exists an irreducible Laurent polynomial $f \in \mathfrak{p}'$ of the form $f = g_0 + g_1 u_d + \cdots + g_l u_d^l$, where $g_i \in \mathbb{F}_p[u_1^{\pm 1}, \ldots, u_{d-1}^{\pm 1}]$ and $g_0 g_l \neq 0$. If the supports of g_0 and g_l are both singletons, then u_d and u_d^{-1} are both integral over the subring $\mathbb{F}_p[u_1'^{\pm 1}, \ldots, u_{d-1}'^{\pm 1}] \subset \mathfrak{R}$. If the support of either g_0 or g_l is not a singleton one can find integers $k_1, \ldots, k_d$ such that substitution of the variables $w_i = u_i u_d^{k_i}$, $i = 1, \ldots, d-1$, in f leads to a Laurent polynomial $g(w_1, \ldots, w_{d-1}, u_d) = u_d^{k_d} f(u_1, \ldots, u_d)$ of the form $g = g_0' + g_1' u_d + \cdots + g_{l'}' u_d^{l'}$, where $g_i' \in \mathbb{F}_p[w_1^{\pm 1}, \ldots, w_{d-1}^{\pm 1}]$, and where the supports of g_0' and $g_{l'}'$ are both singletons. We set

$$B = \begin{pmatrix} 1 & 0 & \ldots & 0 & k_1 \\ 0 & 1 & \ldots & 0 & k_2 \\ \vdots & & \ddots & & \vdots \\ 0 & 0 & \ldots & 1 & k_{d-1} \\ 0 & 0 & \ldots & 0 & 1 \end{pmatrix},$$

$w_i' = w_i + \mathfrak{p}' = u^{B\mathbf{e}^{(i)}} + \mathfrak{p}'$, $i = 1, \ldots, d-1$, and note that w_d' and $w_d'^{-1}$ are integral over $\mathbb{F}_p[w_1'^{\pm 1}, \ldots, w_{d-1}'^{\pm 1}] \subset \mathfrak{R}$. If the elements $w_1', \ldots, w_{d-1}'$ are algebraically independent in $\mathfrak{R}$, then our claim is proved; if not, then we can apply the same argument to $w_1, \ldots, w_{d-1}$ instead of $u_1, \ldots, u_d$, and iteration of this procedure leads to a matrix $A \in \mathrm{GL}(d, \mathbb{Z})$ and an integer $r \geq 0$ such that the elements $v'_j = u^{A\mathbf{e}^{(j)}} + \mathbf{p}' \in \mathcal{R}$ satisfy that $v'_1, \ldots, v'_r$ are algebraically independent, and v'_j and v'^{-1}_j are integral over $\mathcal{R}^{(j-1)} = \mathbb{F}_p[v'^{\pm 1}_1, \ldots, v'^{\pm 1}_{j-1}] \subset \mathcal{R}$ for $j > r$, where $\mathcal{R}^{(0)} = \mathbb{F}_p$ if $r = 0$ (in which case $\mathcal{R}$ must be finite). From Theorem 3.2 it is clear that the ergodicity of α implies that $r \geq 1$, and this completes the proof of our claim.

For the remainder of this proof we assume for simplicity that A is the $d \times d$ identity matrix, so that $v_i = u_i$ for $i = 1, \ldots, d$ (this is—in effect—equivalent to replacing α by the $\mathbb{Z}^d$-action $\alpha' \colon \mathbf{n} \mapsto \alpha'_{\mathbf{n}} = \alpha_{A\mathbf{n}}$). The argument in the preceding paragraph gives us, for each $j = r + 1, \ldots, d$, an irreducible polynomial $f_j(x) = \sum_{k=0}^{l_j} g_k^{(j)} x^k$ with coefficients in the ring $\mathbb{F}_p[u_1^{\pm 1}, \ldots, u_{j-1}^{\pm 1}] \subset \mathcal{R}_d$ such that $h_j(u_j) = h_j(u_1, \ldots, u_{j-1}, u_j) \in \mathbf{p}'$ and the supports of $g_0^{(j)}$ and $g_{l_j}^{(j)}$ are singletons. Let $\Gamma \subset \mathbb{Z}^d$ be the group generated by $\{\mathbf{e}^{(1)}, \ldots, \mathbf{e}^{(r)}\}$, $Q = \{0\} \times \cdots \times \{0\} \times \{0, \ldots, l_{r+1} - 1\} \times \{0, \ldots, l_d - 1\} \subset \mathbb{Z}^d$, and let $\bar{\Gamma} = \Gamma + Q = \{\mathbf{m} + \mathbf{n} : \mathbf{m} \in \Gamma, \mathbf{n} \in Q\}$. We write $\pi_{\bar{\Gamma}} \colon X \longmapsto \mathbb{F}_p^{\bar{\Gamma}}$ for the coordinate projection which restricts every $x \in X$ to its coordinates in $\bar{\Gamma}$ and note that $\pi_{\bar{\Gamma}} \colon X \longmapsto \mathbb{F}_p^{\bar{\Gamma}}$ is a continuous group isomorphism. In other words, the restriction of α to the group $\Gamma \cong \mathbb{Z}^r$ is conjugate to the shift-action of Γ on $(\mathbb{F}_p^Q)^{\Gamma}$. $\square$

If the prime ideal $\mathbf{p} \subset \mathcal{R}_d$ satisfies that $p(\mathbf{p}) = 0$, then the analysis of the action $\alpha^{\mathcal{R}_d/\mathbf{p}}$ becomes somewhat more complicated. We denote by $\kappa \colon \hat{\mathbb{Q}} \longmapsto \mathbb{T}$ the surjective group homomorphism dual to the inclusion $\hat{\kappa} \colon \mathbb{Z} \longmapsto \mathbb{Q}$. If $\mathbf{p} \subset \mathcal{R}_d$ is a prime ideal with $p(\mathbf{p}) = 0$ we regard $X^{\mathcal{R}_d/\mathbf{p}}$ as the subgroup (5.9) of $\mathbb{T}^{\mathbb{Z}^d}$, and define a closed, shift-invariant subgroup $\bar{X}^{\mathcal{R}_d/\mathbf{p}} \subset \hat{\mathbb{Q}}^{\mathbb{Z}^d}$ by

$$\bar{X}^{\mathcal{R}_d/\mathbf{p}} = \left\{ x = (x_{\mathbf{n}}) \in \hat{\mathbb{Q}}^{\mathbb{Z}^d} : \sum_{\mathbf{n} \in \mathbb{Z}^d} c_f(\mathbf{n}) x_{\mathbf{m}+\mathbf{n}} = 0_{\hat{\mathbb{Q}}^{\mathbb{Z}^d}} \atop \text{for every } f \in \mathbf{p} \right\}. \tag{8.1}$$

The restriction of the shift-action σ of $\mathbb{Z}^d$ on $\hat{\mathbb{Q}}^{\mathbb{Z}^d}$ to $\bar{X}^{\mathcal{R}_d/\mathbf{p}}$ will be denoted by $\bar{\alpha}^{\mathcal{R}_d/\mathbf{p}}$ (cf. (2.1)). Define a continuous, surjective homomorphism $\kappa \colon \hat{\mathbb{Q}}^{\mathbb{Z}^d} \longmapsto \mathbb{T}^{\mathbb{Z}^d}$ by $(\kappa(x))_{\mathbf{n}} = \kappa(x_{\mathbf{n}})$ for every $x = (x_{\mathbf{m}}) \in \hat{\mathbb{Q}}^{\mathbb{Z}^d}$ and $\mathbf{n} \in \mathbb{Z}^d$, and write

$$\kappa^{\mathcal{R}_d/\mathbf{p}} \colon \bar{X}^{\mathcal{R}_d/\mathbf{p}} \longmapsto X^{\mathcal{R}_d/\mathbf{p}} \tag{8.2}$$

for the restriction of κ to $\bar{X}^{\mathcal{R}_d/\mathbf{p}}$. The map $\kappa^{\mathcal{R}_d/\mathbf{p}}$ is surjective, and the diagram

$$
\begin{array}{ccc}
\bar{X}^{\mathcal{R}_d/\mathbf{p}} & \xrightarrow{\bar{\alpha}_{\mathbf{n}}^{\mathcal{R}_d/\mathbf{p}}} & \bar{X}^{\mathcal{R}_d/\mathbf{p}} \\
\kappa \downarrow & & \downarrow \kappa \\
X^{\mathcal{R}_d/\mathbf{p}} & \xrightarrow[\alpha_{\mathbf{n}}^{\mathcal{R}_d/\mathbf{p}}]{} & X^{\mathcal{R}_d/\mathbf{p}}
\end{array}
\tag{8.3}
$$

commutes for every $\mathbf{n} \in \mathbb{Z}^d$.

In order to explain this construction in terms of the dual modules we consider the ring $\mathfrak{R}_d^{(\mathbb{Q})} = \mathbb{Q}[u_1^{\pm 1}, \ldots, u_d^{\pm 1}] = \mathbb{Q} \otimes_{\mathbb{Z}} \mathfrak{R}_d$, regard $\mathfrak{R}_d$ as the subring of $\mathfrak{R}_d^{(\mathbb{Q})}$ consisting of all polynomials with integral coefficients, and denote by $\mathfrak{p}^{(\mathbb{Q})} = \mathbb{Q} \otimes_{\mathbb{Z}} \mathfrak{p} \subset \mathfrak{R}_d^{(\mathbb{Q})}$ the prime ideal in $\mathfrak{R}_d^{(\mathbb{Q})}$ corresponding to $\mathfrak{p}$. Since $p(\mathfrak{p}) = 0$, every $\mathfrak{R}_d$-module $\mathfrak{N}$ associated with $\mathfrak{p}$ is embedded injectively in the $\mathfrak{R}_d^{(\mathbb{Q})}$-module $\mathfrak{N}^{(\mathbb{Q})} = \mathbb{Q} \otimes_{\mathbb{Z}} \mathfrak{N}$ by

$$\hat{i}^{\mathfrak{N}} : a \mapsto 1 \otimes_{\mathbb{Z}} a, \, a \in \mathfrak{N}, \tag{8.4}$$

and $\mathfrak{N}^{(\mathbb{Q})}$ is associated with $\mathfrak{p}^{(\mathbb{Q})}$. Since $\mathfrak{R}_d \subset \mathfrak{R}_d^{(\mathbb{Q})}$, $\mathfrak{N}^{(\mathbb{Q})}$ is an $\mathfrak{R}_d$-module, and we can define the $\mathbb{Z}^d$-action $\alpha^{\mathfrak{N}^{(\mathbb{Q})}}$ on $X^{\mathfrak{N}^{(\mathbb{Q})}}$ as in Lemma 5.1. Note that the set of prime ideals associated with the $\mathfrak{R}_d$-module $\mathfrak{N}^{(\mathbb{Q})}$ is the same as that of $\mathfrak{N}$; in particular, $\alpha^{\mathfrak{N}^{(\mathbb{Q})}}$ is ergodic if and only if $\alpha^{\mathfrak{N}}$ is ergodic and, for every $\mathbf{n} \in \mathbb{Z}^d$, $\alpha_{\mathbf{n}}^{\mathfrak{N}^{(\mathbb{Q})}}$ is ergodic if and only if $\alpha_{\mathbf{n}}^{\mathfrak{N}}$ is ergodic. The homomorphism

$$i^{\mathfrak{N}} : X^{\mathfrak{N}^{(\mathbb{Q})}} \longmapsto X^{\mathfrak{N}} \tag{8.5}$$

dual to

$$\hat{i} : \mathfrak{N} \longmapsto \mathfrak{N}^{(\mathbb{Q})} \tag{8.6}$$

is surjective, and the diagram

$$
\begin{array}{ccc}
X^{\mathfrak{N}^{(\mathbb{Q})}} & \xrightarrow{\alpha_{\mathbf{n}}^{\mathfrak{N}^{(\mathbb{Q})}}} & \bar{X}^{\mathfrak{N}^{(\mathbb{Q})}} \\
\kappa \downarrow & & \downarrow \kappa \\
X^{\mathfrak{N}} & \xrightarrow{\alpha_{\mathbf{n}}^{\mathfrak{N}}} & X^{\mathfrak{N}}
\end{array}
\tag{8.7}
$$

commutes for every $\mathbf{n} \in \mathbb{Z}^d$. For $\mathfrak{N} = \mathfrak{R}_d / \mathfrak{p}$ we obtain that

$$X^{(\mathfrak{R}_d/\mathfrak{p})^{(\mathbb{Q})}} = \bar{X}^{\mathfrak{R}_d/\mathfrak{p}},$$
$$\alpha^{(\mathfrak{R}_d/\mathfrak{p})^{(\mathbb{Q})}} = \bar{\alpha}^{\mathfrak{R}_d/\mathfrak{p}}, \tag{8.8}$$
$$i^{\mathfrak{R}_d/\mathfrak{p}} = \kappa^{\mathfrak{R}_d/\mathfrak{p}}.$$

PROPOSITION 8.3. *Let $\mathfrak{p} \subset \mathfrak{R}_d$ be a prime ideal with $p(\mathfrak{p}) = 0$ which is not of the form $\mathfrak{p} = \mathfrak{j}_c$ for any $c \in \overline{\mathbb{Q}}^d$. Then the $\mathbb{Z}^d$-action $\alpha = \alpha^{\mathfrak{R}_d/\mathfrak{p}}$ on $X = X^{\mathfrak{R}_d/\mathfrak{p}}$ is ergodic, and there exists an integer $r = r(\mathfrak{p}) \in \{1, \ldots, d\}$, a primitive subgroup $\Gamma = \Gamma(\mathfrak{p}) \subset \mathbb{Z}^d$, and a finite set $Q = Q(\mathfrak{p}) \subset \mathbb{Z}^d$ with the following properties.*

(1) $\Gamma \cong \mathbb{Z}^r$;
(2) $\mathbf{0} \in Q$, and $Q \cap (Q + \mathbf{m}) = \varnothing$ whenever $\mathbf{0} \neq \mathbf{m} \in \Gamma$;

(3) *If* $\bar{\Gamma} = \Gamma + Q = \{\mathbf{m} + \mathbf{n} : \mathbf{m} \in \Gamma, \mathbf{n} \in Q\}$, *then the coordinate projection* $\pi_{\bar{\Gamma}} : \bar{X}^{\mathfrak{R}_d/\mathfrak{p}} \longmapsto \widehat{\mathbb{Q}}^{\bar{\Gamma}}$, *which restricts any point* $x \in \bar{X}^{\mathfrak{R}_d/\mathfrak{p}} \subset \widehat{\mathbb{Q}}^{\mathbb{Z}^d}$ *to its coordinates in* $\bar{\Gamma}$, *is a continuous group isomorphism; in particular, the* Γ-*action* $\mathbf{n} \mapsto \bar{\alpha}_{\mathbf{n}}^{\mathfrak{R}_d/\mathfrak{p}}$, $\mathbf{n} \in \Gamma$, *is (isomorphic to) the shift-action of* Γ *on* $(\widehat{\mathbb{Q}^Q})^{\Gamma}$.

PROOF. The proof is completely analogous to that of Proposition 8.2. We find a matrix $A \in \mathrm{GL}(d, \mathbb{Z})$ and an integer $r \in \{1, \ldots, d\}$ with the following properties: if $v_j = u^{A\mathbf{e}^{(j)}}$ and $v_j' = v_j + \mathfrak{p}$ for $j = 1, \ldots, d$, then $v_1', \ldots, v_r'$ are algebraically independent elements of $\mathfrak{R} = \mathfrak{R}_d/\mathfrak{p}$, and there exists, for each $j = r+1, \ldots, d$, an irreducible polynomial $f_j(x) = \sum_{k=0}^{l_j} g_k^{(j)}(x^k)$ with coefficients in the ring $\mathbb{Z}[v_1^{\pm 1}, \ldots, v_{j-1}^{\pm 1}] \subset \mathfrak{R}_d$ such that $f_j(v_1, \ldots, v_{j-1}, v_j) \in \mathfrak{q}$ and the supports of $g_0^{(j)}$ and $g_{l_j}^{(j)}$ are singletons.

We assume again that A is the $d \times d$ identity matrix, so that $v_j = u_j$ for $j = 1, \ldots, d$ and $\Gamma \cong \mathbb{Z}^r$ is generated by $\mathbf{e}^{(1)}, \ldots, \mathbf{e}^{(r)}$, set $Q = \{0\} \times \cdots \times \{0\} \times \{0, \ldots, l_{r+1} - 1\} \times \cdots \times \{0, \ldots, l_d - 1\} \subset \mathbb{Z}^d$, and complete the proof in the same way as that of Proposition 3.4, using (8.1) instead of (6.19). The ergodicity of $\bar{\alpha}^{\mathfrak{R}_d/\mathfrak{p}}$ is obvious from the conditions (1)–(3), and from (8.3) we conclude the ergodicity of $\alpha^{\mathfrak{R}_d/\mathfrak{p}}$. $\square$

REMARKS 8.4. (1) We can extend the definition of $r(\mathfrak{p})$ in Proposition 8.2 and 8.3 to ergodic prime ideals of the form $\mathfrak{p} = \mathfrak{j}_c, c \in (\overline{\mathbb{Q}}^{\times})^d$, by setting $r(\mathfrak{j}_c) = 0$. Then the integer $r(\mathfrak{p})$ is a well-defined property of the prime ideal $\mathfrak{p}$, and is in particular independent of the choice of the primitive subgroup $\Gamma \subset \mathbb{Z}^d$ in Proposition 8.2 or 8.3 (it is easy to see that there is considerable freedom in the choice of Γ): if r', Γ', Q' are a positive integer, a primitive subgroup of $\mathbb{Z}^d$, and a finite subset of $\mathbb{Z}^d$, satisfying the conditions (1)–(3) in either of the Propositions 8.2 or 8.3, then $r' = r(\mathfrak{p})$. This follows from Noether's normalization theorem; a dynamical proof using entropy will be given in Section 24.

(2) If $\mathfrak{p} \subset \mathfrak{R}_d$ is an ergodic prime ideal with $p(\mathfrak{p}) > 0$, then the subgroup $\Gamma \subset \mathbb{Z}^d$ in Proposition 8.2 is a maximal subgroup of $\mathbb{Z}^d$ for which the restriction α^{Γ} of $\alpha^{\mathfrak{R}_d/\mathfrak{p}}$ to Γ is expansive. In particular, $r(\mathfrak{p})$ is the smallest integer for which there exists a subgroup $\Gamma \cong \mathbb{Z}^r$ in $\mathbb{Z}^d$ such that α^{Γ} is expansive.

(3) Even if the $\mathbb{Z}^d$-action $\alpha^{\mathfrak{R}_d/\mathfrak{p}}$ in Proposition 8.3 is expansive, the action $\alpha^{(\mathfrak{R}_d/\mathfrak{p})^{(Q)}}$ is non-expansive. By proving a more intricate version of Proposition 8.3 one can analyze the structure of the group $X^{\mathfrak{R}_d/\mathfrak{p}}$ directly, without passing to $X^{(\mathfrak{R}_d/\mathfrak{p})^{(Q)}}$: if $X^{\mathfrak{R}_d/\mathfrak{p}}$ is written as a shift-invariant subgroup of $\mathbb{T}^{\mathbb{Z}^d}$ (cf. (5.9)), and if $r = r(\mathfrak{p}), \Gamma, Q$ are given as in Proposition 8.3, then the projection $\pi_{\bar{\Gamma}} : X^{\mathfrak{R}_d/\mathfrak{p}} \longmapsto \mathbb{T}^{\bar{\Gamma}}$ is still surjective, but need no longer be injective; the kernel of $\pi_{\bar{\Gamma}}$ is of the form Y^{Γ} for some compact, zero-dimensional group Y (cf. Example 8.5 (2)).

EXAMPLES 8.5. (1) Let $\mathfrak{p} = (2, 1 + u_1 + u_2) \subset \mathfrak{R}_2$ (cf. Example 5.3 (5)). Then $p(\mathfrak{p}) = 2$, $r(\mathfrak{p}) = 1$, and we may set $\Gamma = \{(k, k) : k \in \mathbb{Z}\} \cong \mathbb{Z}$ and $Q = \{(0, 0), (1, 0)\} \subset \mathbb{Z}^2$ in Proposition 8.2. If $X = X^{\mathfrak{R}_2/\mathfrak{p}}$ is written in the form (6.19) as

$$X = \left\{ x = (x_{\mathbf{m}}) \subset \mathbb{F}_2^{\mathbb{Z}^d} : x_{(m_1, m_2)} + x_{(m_1+1, m_2)} + x_{(m_1, m_2+1)} = 0_{\mathbb{F}_2} \right.$$
$$\left. \text{for all } (m_1, m_2) \in \mathbb{Z}^2 \right\},$$

then the projection $\pi_{\bar{\Gamma}} \colon X \longmapsto \mathbb{F}_2^{\bar{\Gamma}}$ sends the shift $\alpha_{(1,1)}^{\mathfrak{R}_2/\mathfrak{p}} = \alpha_{(1,1)}$ on X to the shift on $\mathbb{F}_2^{\bar{\Gamma}} \cong (\mathbb{Z}_{/2} \times \mathbb{Z}_{/2})^{\mathbb{Z}}$. Note that, although $\alpha_{(1,1)}$ acts expansively on X, other elements of $\mathbb{Z}^2$ may not be expansive; for example, $\alpha_{(1,0)}$ is non-expansive.

(2) Let $\mathfrak{p} = (3 + u_1 + 2u_2) \subset \mathfrak{R}_2$. Then $p(\mathfrak{p}) = 0$, $r(\mathfrak{p}) = 1$, and Γ and Q may be chosen as in Example (1). Note that $X^{\mathfrak{R}_2/\mathfrak{p}} = X = \{x = (x_{\mathbf{m}}) \subset \mathbb{T}^{\mathbb{Z}^d} : x_{(m_1, m_2)} + x_{(m_1+1, m_2)} + x_{(m_1, m_2+1)} = 0_{\mathbb{T}}$ for all $(m_1, m_2) \in \mathbb{Z}^2\}$; the coordinate projection $\pi_{\bar{\Gamma}} \colon X \longmapsto \mathbb{T}^{\bar{\Gamma}}$ in Proposition 8.3 is not injective; for every $x \in X$, the coordinates $x_{(m_1, m_2)}$ with $m_1 \geq m_2$ are completely determined by $\pi_{\bar{\Gamma}}(x)$, but each of the coordinates $x_{(k, k+1)}$, $k \in \mathbb{Z}$, has two possible values. Similarly, if we know the coordinates $x_{(m_1, m_2)}$, $m_1 \geq m_2 - r$ of a point $x = (x_{\mathbf{m}}) \in X$ for any $r \geq 0$, then there are exactly two (independent) choices for each of the coordinates $x_{(k, k+r+1)}$, $k \in \mathbb{Z}$. This shows that the kernel of the surjective homomorphism $\pi_{\bar{\Gamma}} \colon X \longmapsto \mathbb{T}^{\bar{\Gamma}} \cong (\mathbb{T}^2)^{\mathbb{Z}}$ is isomorphic to $\mathbb{Z}_2^{\Gamma}$, where $Y = \mathbb{Z}_2$ denotes the group of dyadic integers.

If $\mathfrak{p}$ is replaced by the prime ideal $\mathfrak{p}' = (1 + 3u_1 + 2u_2) \subset \mathfrak{R}_2$, then Γ and Q remain unchanged, but the kernel of $\pi_{\bar{\Gamma}}$ becomes isomorphic to $(\mathbb{Z}_2 \times \mathbb{Z}_3)^{\Gamma}$, where $\mathbb{Z}_3$ is the group of tri-adic integers. Finally, if $\mathfrak{p}'' = (1 + u_1 + u_2) \subset \mathfrak{R}_2$, and if Γ and Q are as in Example (1), then $\pi_{\bar{\Gamma}} \colon X^{\mathfrak{R}_2/\mathfrak{p}''} \longmapsto (\mathbb{T}^Q)^{\mathbb{Z}}$ is a group isomorphism. $\square$

CONCLUDING REMARK 8.6. The material in this section (with the exception of Proposition 8.1) is taken from [38].

CHAPTER III

Expansive automorphisms of compact groups

9. Expansive automorphisms of compact connected groups

In the Sections 7–8 we investigated the structure of $\mathbb{Z}^d$-actions of the form $\alpha^{\mathfrak{R}_d/\mathfrak{p}}$, where $\mathfrak{p} \subset \mathfrak{R}_d$ is a prime ideal. Although we can find, for every $\mathbb{Z}^d$-action α by automorphisms of a compact, abelian group X, a sequence of closed, α-invariant subgroups $X = Y_0 \supset Y_1 \supset \cdots$ such that $\alpha^{Y_j/Y_{j+1}}$ is of the form $\alpha^{\mathfrak{R}_d/\mathfrak{q}_j}$ for every $j \geq 0$, where $(\mathfrak{q}_j)$ is a sequence of prime ideals in $\mathfrak{R}_d$ (Corollary 6.2), the reconstruction of α from these quotient-actions is a problem of formidable difficulty. Only when $d = 1$ can one 'almost' re-build the action α from the quotient actions $\alpha^{\mathfrak{R}_d/\mathfrak{q}_i}$ (Corollary 9.4), due to the fact that $\mathbb{Q} \otimes_{\mathbb{Z}} \mathfrak{R}_1 = \mathbb{Q}[u_1^{\pm 1}]$ is a principal ideal domain. The main tool in this reconstruction is the following Lemma 9.1.

LEMMA 9.1. *Let $\mathfrak{M}$ be a Noetherian torsion $\mathfrak{R}_1$-module, which is torsion-free as an additive group. Then there exist primitive polynomials $f_1, \ldots, f_r$ in $\mathfrak{R}_1$ such that f_j divides f_{j+1} for all $j = 1, \ldots, r-1$, and an injective $\mathfrak{R}_1$-module homomorphism $\theta \colon \mathfrak{M} \longmapsto \mathfrak{R}_1/(f_1) \oplus \cdots \oplus \mathfrak{R}_1/(f_r) = \mathfrak{N}$ such that $\mathfrak{N}/\theta(\mathfrak{M})$ is finite.*

PROOF. Since $\mathfrak{M}$ is torsion-free, none of the prime ideals associated with $\mathfrak{M}$ contains a non-zero constant, and the embedding $a \mapsto 1 \otimes a$ of $\mathfrak{M}$ in $\mathfrak{M}^{(\mathbb{Q})} = \mathbb{Q} \otimes_{\mathbb{Z}} \mathfrak{M}$ is injective. This allows us to identify $\mathfrak{M}$ with the subset $\{1 \otimes a : a \in \mathfrak{M}\} \subset \mathfrak{M}^{(\mathbb{Q})}$ and to assume for simplicity that $\mathfrak{M} \subset \mathfrak{M}^{(\mathbb{Q})}$. We write $\mathfrak{R}_1^{(\mathbb{Q})} = \mathbb{Q}[u^{\pm 1}]$ for the ring of Laurent polynomials in the variable u with rational coefficients, regard $\mathfrak{R}_1$ as a subring of $\mathfrak{R}_1^{(\mathbb{Q})}$, and observe that $\mathfrak{M}^{(\mathbb{Q})}$ is a module over $\mathfrak{R}_1^{(\mathbb{Q})}$ and hence over $\mathfrak{R}_1$. By Theorem XV.2.6 in [51] there exist polynomials $f_1, \ldots, f_r$ in $\mathfrak{R}_1^{(\mathbb{Q})}$ such that f_j divides f_{j+1} for $j = 1, \ldots, r-1$, and an $\mathfrak{R}_1^{(\mathbb{Q})}$-module isomorphism $\theta' \colon \mathfrak{M}^{(\mathbb{Q})} \longmapsto \mathfrak{R}_1^{(\mathbb{Q})}/f_1\mathfrak{R}_1^{(\mathbb{Q})} \oplus \cdots \oplus \mathfrak{R}_1^{(\mathbb{Q})}/f_r\mathfrak{R}_1^{(\mathbb{Q})} = \mathfrak{N}^{(\mathbb{Q})}$. We assume without loss in generality that each f_i lies in $\mathfrak{R}_1$, and that

$f_i \in \mathfrak{R}_1$ is primitive. Since $\theta'(\mathfrak{M}) \subset \mathfrak{N}^{(\mathbb{Q})}$ is a Noetherian submodule over $\mathfrak{R}_1$, there exists an integer $k \geq 1$ with

$$\theta'(\mathfrak{M}) \subset \frac{1}{k}\mathfrak{N}, \tag{9.1}$$

where $\mathfrak{N} = \mathfrak{R}_1/f_1\mathfrak{R}_1 \oplus \cdots \oplus \mathfrak{R}_1/f_r\mathfrak{R}_1 \subset \mathfrak{N}^{(\mathbb{Q})}$. Similarly we conclude the existence of an integer $l \geq 1$ such that

$$\mathfrak{N} \subset \frac{1}{l}\theta'(\mathfrak{M}). \tag{9.2}$$

We claim that $\frac{1}{k}\mathfrak{N}/l\mathfrak{N}$ is finite. Indeed, if $f = f_1 \dots f_r$, then every element $a \in \frac{1}{k}\mathfrak{N}/l\mathfrak{N}$ is annihilated by the ideal $(kl, f) = kl\mathfrak{R}_1 + f\mathfrak{R}_1$, so that every prime ideal associated with $\frac{1}{k}\mathfrak{N}/l\mathfrak{N}$ must be non-principal. Since $\mathfrak{R}_1/\mathfrak{p}$ is finite for every non-principal prime $\mathfrak{p} \subset \mathfrak{R}_1$ (Example 6.17 (3)), Proposition 6.1 implies that $\frac{1}{k}\mathfrak{N}/l\mathfrak{N}$ is finite. Hence $\frac{1}{k}\mathfrak{N}/\theta'(\mathfrak{M})$ is finite by (9.2), and the proof is completed by setting $\theta(a) = k\theta'(a)$ for every $a \in \mathfrak{M}$ and by noting that $\theta(\mathfrak{M}) \subset \mathfrak{N}$, by (9.1). $\square$

THEOREM 9.2. *Let α be an expansive automorphism of a compact, connected group X. Then there exist primitive polynomials $f_1, \dots, f_r$ in $\mathfrak{R}_1$ such that f_j divides f_{j+1} for $j = 1, \dots, r-1$, and continuous, surjective, finite-to-one group homomorphisms $\eta: Y = X^{\mathfrak{R}_1/(f_1)} \times \cdots \times X^{\mathfrak{R}_1/(f_r)} \longmapsto X$ and $\eta': X \longmapsto Y$, such that $\alpha \cdot \eta = \eta \cdot \alpha'$ and $\alpha' \cdot \eta' = \eta' \cdot \alpha$, where α' is the automorphism $\alpha^{\mathfrak{R}_1/f_1\mathfrak{R}_1} \times \cdots \times \alpha^{\mathfrak{R}_1/f_r\mathfrak{R}_1}$ of Y.*

PROOF. The group X is abelian by Theorem 2.4, and we write $\mathfrak{M} = \hat{X}$ for the $\mathfrak{R}_1$-module arising from Lemma 5.1. Then $\mathfrak{M}$ is Noetherian by (4.10) and Proposition 5.4, and the connectedness of X implies that $\mathfrak{M}$ is torsion-free as an additive group. We apply Lemma 9.1 to find the polynomials $f_1, \dots, f_r$ and set $Y = X^{\mathfrak{N}}$, $\eta = \hat{\theta}$, and $\eta' = \theta'^{-1} \cdot \psi$, where $\psi: \mathfrak{N} \longmapsto \theta'(\mathfrak{M})$ is the map $\psi(a) = la$ (cf. (9.2)). In the proof of Lemma 9.1 we have seen that $l\mathfrak{N} \subset \theta'(\mathfrak{M}) \subset \frac{1}{k}\mathfrak{N}$, and that $\frac{1}{k}\mathfrak{N}/l\mathfrak{N}$ is finite. Hence $\mathfrak{N}/\theta(\mathfrak{M}) \cong \frac{1}{k}\mathfrak{N}/\theta'(\mathfrak{M})$ and $\theta'(\mathfrak{M})/l\mathfrak{N}$ are finite, and duality shows that $\ker(\eta)$ and $\ker(\eta')$ are finite. $\square$

REMARK 9.3. Although the polynomials $f_1, \dots, f_r$ in Lemma 9.1 and Theorem 9.2 are obviously not unique, the ideals $(f_j) = f_j\mathfrak{R}_1$, $j = 1, \dots, r$, are unique by Theorem XV.2.6 in [51]. Hence the automorphisms $\alpha^{\mathfrak{R}_1/(f_i)}$, $i = 1, \dots, r$, are determined uniquely up to topological conjugacy (Theorem 5.9).

COROLLARY 9.4. *Let α be an automorphism of a compact, connected, abelian group X. The following conditions are equivalent.*

(1) *α is expansive;*
(2) *There exist primitive polynomials $f_1, \dots, f_r$ in $\mathfrak{R}_1$ such that f_j divides f_{j+1} for $j = 1, \dots, r-1$ and f_r has no roots of modulus 1, and a finite, α-invariant subgroup $F \subset X$, such that $\alpha^{X/F}$ is algebraically conjugate to $\alpha^{\mathfrak{R}_1/(f_1)} \times \cdots \times \alpha^{\mathfrak{R}_1/(f_r)}$;*

(3) *There exist primitive polynomials $f_1, \ldots, f_r$ in $\mathfrak{R}_1$ such that f_j divides f_{j+1} for $j = 1, \ldots, r-1$ and f_r has no roots of modulus 1, and a finite, $(\alpha^{\mathfrak{R}_1/(f_1)} \times \cdots \times \alpha^{\mathfrak{R}_1/(f_r)})$-invariant subgroup $F' \subset X' = X^{\mathfrak{R}_1/(f_1)} \times \cdots \times X^{\mathfrak{R}_1/(f_r)}$, such that the automorphism induced by $\alpha' = \alpha^{\mathfrak{R}_1/(f_1)} \times \cdots \times \alpha^{\mathfrak{R}_1/(f_r)}$ on X'/F' is algebraically conjugate to α.*

PROOF. If α is expansive, $\alpha^{X/F}$ is expansive for every closed, α-invariant subgroup $F \subset X$ (by Corollary 6.15), and hence the automorphism $\alpha^{\mathfrak{R}_1/(f_1)} \times \cdots \times \alpha^{\mathfrak{R}_1/(f_r)}$ in (2) is expansive. According to Theorem 6.5 (4) this means that f_r has no roots of modulus 1. If (2) is satisfied, then $\alpha' = \alpha^{\mathfrak{R}_1/(f_1)} \times \cdots \times \alpha^{\mathfrak{R}_1/(f_r)}$ is expansive by Theorem 6.5 (4), hence $\alpha^{X/F}$ is expansive for some finite, α-invariant subgroup $F \subset X$, and this obviously implies that α is expansive. The equivalence (1)$\Longleftrightarrow$(3) is proved similarly. $\square$

Automorphisms of the form $\alpha^{\mathfrak{R}_1/(f)}$ with $f \subset \mathfrak{R}_1$ have already been discussed in Examples 5.2 and 6.17, and Corollary 9.4 shows how to obtain all expansive automorphisms of compact, connected, abelian groups from automorphisms of the form $\alpha^{\mathfrak{R}_1/(f)}$. In [52] there is a different realization of all expansive automorphisms of compact, connected, abelian groups in terms of rational matrices. In order to explain Lawton's description we consider the following generalization of Examples 5.6 (1) and 6.17 (2), where the underlying matrix was of the form $\left(\frac{3}{2}\right)$.

EXAMPLES 9.5. (1) Let $A^\top \in \mathrm{GL}(n, \mathbb{Q}) = \mathrm{Aut}(\mathbb{Q}^n)$, and let $X = \widehat{\mathbb{Q}^n}$. Since $\mathbb{Q}^n$ is torsion-free, the group X is connected. The automorphism α of X dual to $\beta = A^\top$ on $\mathbb{Q}^n$ is given by the transpose matrix A of $A^\top$. It is clear that α does not satisfy the d.c.c., since there does not exist a finite subset $S = \{v_1, \ldots, v_m\} \subset \mathbb{Q}^n$ such that $\mathbb{Q}^n$ is generated by $\bigcup_{k \in \mathbb{Z}} (A^k)^\top (S)$ (Corollary 2.16). In particular, α is non-expansive by (4.10).

We write $\mathfrak{M}^* = \mathbb{Q}^n$ for the $\mathfrak{R}_1$-module arising from α via Lemma 5.1. In order to determine the prime ideals associated with $\mathfrak{M}^*$ we consider, for every $a \in \mathbb{Q}^n$, the annihilator $\mathrm{ann}(a) = \{f \in \mathfrak{R}_1 : f(A) \cdot a = 0\}$. Since χ_A has coefficients in $\mathbb{Q}$, we multiply χ_A by the smallest integer k such that $k\chi_A \in \mathfrak{R}_1$, and choose a prime decomposition of $k\chi_A = h_1 \ldots h_m$ in $\mathfrak{R}_1$. Then the set of prime ideals associated with $\mathfrak{M}^*$ is equal to $\{(h_1), \ldots, (h_m)\}$. From Proposition 6.6 it is clear that α is non-ergodic if and only if one of the polynomials h_i divides $u^l - 1$ for some $l \geq 1$, i.e. if and only if A has an eigenvalue which is a root of unity.

(2) In Example (1), let $\mathfrak{N} \subset \mathfrak{M}^*$ be an $\mathfrak{R}_1$-submodule. Then the set of prime ideals associated with $\mathfrak{N}$ is contained in $\{(h_1), \ldots, (h_m)\}$. The group $X^{\mathfrak{N}} = \widehat{\mathfrak{N}}$ is the quotient of X by a closed, α-invariant subgroup, and hence the automorphism $\alpha^{\mathfrak{N}}$ on $X^{\mathfrak{N}}$ is ergodic whenever α is ergodic. We claim that $\alpha^{\mathfrak{N}}$ is expansive if and only if $\mathfrak{N}$ is a Noetherian $\mathfrak{R}_1$-module, and A has no eigenvalue of modulus 1. Indeed, if $\alpha^{\mathfrak{N}}$ is expansive, then (4.10) and Proposition

5.4 together imply that $\mathfrak{N}$ is Noetherian, and Theorem 6.5 (4) shows that none of the polynomials h_i can have a root of modulus 1. Conversely, if $\mathfrak{N}$ is Noetherian, and if χ_A has no root of modulus 1, then $\alpha^{\mathfrak{N}}$ is expansive by Theorem 6.5 (4). The condition for ergodicity is unchanged from Example (1): $\alpha^{\mathfrak{N}}$ is non-ergodic if and only if A has an eigenvalue which is a root of unity.

(3) In Example (2), let $\mathfrak{M}^A = \mathfrak{N} = \mathbb{Z}^n[A^\top, (A^{-1})^\top]$ be the subgroup of $\mathfrak{M}^* = \mathbb{Q}^n$ generated by

$$\bigcup_{m \in \mathbb{Z}} (A^m)^\top \mathbb{Z}^n.$$

Since $\mathfrak{M}^A$ is invariant under left multiplication by $A^\top$, $\mathfrak{M}^A$ is an $\mathfrak{R}_1$-submodule of $\mathfrak{M}^*$, and we write β^A for the automorphism of (the additive group) $\mathfrak{M}^A$ defined by $A^\top$, and denote by α^A the automorphism of $X^{\mathfrak{M}^A} = \widehat{\mathfrak{M}^A}$ dual to β^A. In order to realize α^A explicitly we consider the subgroup $\Xi \subset Z = \sum_{\mathbb{Z}} \mathbb{Z}^n$ generated by all elements of the form $\boldsymbol{\xi} = (\xi_k) \in \sum_{\mathbb{Z}} \mathbb{Z}^n$ such that $\xi_k = 0$ for $k \notin \{l, l+1\}$ and $B^\top \xi_{l+1} = -m \xi_l$ for some integers $l \in \mathbb{Z}$ and $m \geq 1$ for which $B = mA$ has integer entries. The group $X^A = \Xi^\perp \subset (\mathbb{T}^n)^{\mathbb{Z}}$ is closed, shift-invariant, and is given by

$$X^A = \{x = (x_k) \in (\mathbb{T}^n)^{\mathbb{Z}} : l x_{k+1} = B x_k \text{ for all } k \in \mathbb{Z} \text{ and all}$$
$$m \in \mathbb{Z} \text{ for which } B = mA \text{ has integer entries}\}. \tag{9.3}$$

We claim that the shift σ on X^A is conjugate to α^A. Indeed, the homomorphism $\psi \colon Z \longmapsto \mathfrak{M}^A$, given by $\psi(\boldsymbol{\xi}) = \sum_{k \in \mathbb{Z}} (A^k)^\top \xi_k$ for every $\boldsymbol{\xi} \in Z$, is well-defined, and $\ker(\psi) = \Xi$. Hence ψ induces a dual isomorphism $\eta \colon X^A = \Xi^\perp = \widehat{Z/\Xi} \longmapsto \widehat{\mathfrak{M}^A}$ with $\eta \cdot \sigma = \alpha^A \cdot \eta$. This also proves that α^A satisfies the d.c.c., and that $\mathfrak{M}^A \subset \mathfrak{M}^*$ is Noetherian (cf. Proposition 5.4). From Example (2) we see that α^A is expansive if and only if A has no eigenvalue of modulus 1, and ergodic if and only if no eigenvalue of A is a root of unity. $\quad\boxdot$

REMARKS 9.6. (1) If we apply Lemma 9.1 to the $\mathfrak{R}_1$-module $\mathfrak{M}^A$ in Example 9.5 (3), we obtain the primitive polynomials $f_1, \ldots, f_r$ in $\mathfrak{R}_1$ described in the statement of Lemma 9.1. Choose $m_i \in \mathbb{Z}$ so that $g_i = m_i^{-1} f_i \in \mathbb{R}$ is monic for $i = 1, \ldots, m$. Then there exist integers l, l' such that $u^l g_r(u)$ is the minimal polynomial of A, and $u^{l'} g_1(u) \ldots g_r(u) = \chi_A(u)$ is the characteristic polynomial of A (cf. [51]). The presence of the monomials u^l and $u^{l'}$ is due to the fact that we are dealing with Laurent polynomials rather than polynomials.

(2) The relation between $\mathfrak{M} = \mathfrak{M}^A$ and $\mathfrak{R}_1/(f_1) \oplus \cdots \oplus \mathfrak{R}_1/(f_r)$ in Lemma 9.1 and Remark (1) is completely analogous to that between the matrices A and B in Example 5.3 (2).

THEOREM 9.7. *An automorphism α of a compact, connected group X is expansive if and only if it is algebraically conjugate to an automorphism of the* *form α^A for some matrix $A \in \mathrm{GL}(n, \mathbb{Q})$, $n \geq 1$, without eigenvalues of modulus 1* (cf. Example 9.5 (3)).

Although Theorem 9.7 can be derived from Theorem 9.2, we give a different proof more closely related to the material in Section 4. The following definition and lemmas are more general than is necessary for the proof of Theorem 9.7, but we shall need them for the discussion of automorphisms of general compact groups in the Sections 10–12.

DEFINITION 9.8. Let G be a compact group. A full, shift-invariant subgroup $Y \subset G^{\mathbb{Z}}$ is called a *Markov subgroup* if there exists a (necessarily full) subgroup $H \subset G \times G$ such that

$$Y = Y_H = \{y = (y_n) \in G^{\mathbb{Z}} : (y_n, y_{n+1}) \in H \text{ for every } n \in \mathbb{Z}\}. \quad (9.4)$$

LEMMA 9.9. *Let X be a compact group, and let α be a continuous automorphism of X which satisfies the d.c.c. Then we can find a compact Lie group G and a full subgroup $H \subset G \times G$ with the following properties.*

(1) *If $Y = Y_H$ is the Markov subgroup defined in (9.4) and σ is the shift on Y, then there exists a continuous isomorphism $\phi \colon X \longmapsto Y$ with $\phi \cdot \alpha = \sigma \cdot \phi$;*
(2) *H° is a full subgroup of $G^\circ \times G^\circ$, $H^\circ = H \cap (G^\circ \times G^\circ)$, and $Y^\circ = \{y = (y_n) \in Y : (y_n, y_{n+1}) \in H^\circ \text{ for every } n \in \mathbb{Z}\}$. In particular, Y is connected if and only if H is connected.*

PROOF. According to (4.10) we may assume that α is the shift on a full, shift-invariant subgroup $X \subset G^{\mathbb{Z}}$, where G is a compact Lie group. Proposition 4.14 and Remark 4.15 (1) allow us to assume furthermore that $X = Y_H$, where $H \subset G \times G$ is a full subgroup and Y_H is the Markov subgroup defined in (9.4), and that the connected components X°, G°, and H°, of the identity in X, G, and H, satisfy that $X^\circ = X \cap (G^\circ)^{\mathbb{Z}} = \{x \in G^{\mathbb{Z}} : (x_n, x_{n+1}) \in H^\circ \text{ for every } n \in \mathbb{Z}\}$. Since Y/Y° is zero-dimensional and G and H are Lie groups, $\pi_{\{0\}}(Y^\circ)$ and $\pi_{\{0,1\}}(Y^\circ)$ are connected, open subgroups of G and H, respectively. Hence $\pi_{\{0\}}(Y^\circ) = G^\circ$, and $\pi_{\{0,1\}}(Y^\circ) = H^\circ$ is a full subgroup of $G^\circ \times G^\circ$. If $H^\circ \neq H \cap (G^\circ \times G^\circ)$ we choose a point $(g, g') \in H \cap (G^\circ \times G^\circ)$ which does not lie in H°, and use the fullness of $H^\circ \subset G^\circ \times G^\circ$ to find a point $x = (x_n) \in G^{\mathbb{Z}}$ with $x_0 = g$, $x_1 = g'$, and $(x_n, x_{n+1}) \in H^\circ \subset H \cap (G^\circ \times G^\circ)$ whenever $0 \neq n \in \mathbb{Z}$. Then $x \in X \cap (G^\circ)^{\mathbb{Z}} = X^\circ$, but $\pi_{\{0,1\}}(x) \notin \pi_{\{0,1\}}(X^\circ)$, which is absurd. It follows that $H^\circ = H \cap (G^\circ \times G^\circ)$. The last statement is obvious. $\square$

LEMMA 9.10. *Let G be a compact Lie group, $H \subset G \times G$ a full subgroup, and let*

$$F_H^+ = \{g \in G : (1_G, g) \in H\}, \quad F_H^- = \{g \in G : (g, 1_G) \in H\}. \quad (9.5)$$

Then F_H^+ and F_H^- are closed, normal subgroups of G. If the group $X = Y_H \subset G^{\mathbb{Z}}$ in (9.4) is expansive, then F_H^+ and F_H^- are finite.

PROOF. It is clear that the groups $F_H^{\pm}$ are closed, and their normality is a consequence of the fullness of H. Now assume that Y_H is expansive. From (9.4) and the proof of Lemma 9.9 is is clear that $G^{\circ} = \pi_{\{0\}}(Y_H^{\circ})$, $H^{\circ} = \pi_{\{0,1\}}(Y_H^{\circ})$. In particular, H° is a full subgroup of $G^{\circ} \times G^{\circ}$. Since Y_H° is abelian by Theorem 2.4, H° and G° are abelian, and the finiteness of $F_H^{\pm}$ will follow from the finiteness of $F_{H^{\circ}}^{\pm}$. In order to establish this lemma it will thus suffice to assume from now on that G is connected and abelian, i.e. that $G = \mathbb{T}^n$ for some $n \geq 1$, and that $H \subset G \times G$ is a full, connected subgroup such that $X = Y_H$ is expansive.

The dual group of $Z = (\mathbb{T}^n)^{\mathbb{Z}}$ is (isomorphic to) the direct sum $\hat{Z} = \sum_{\mathbb{Z}} \mathbb{Z}^n$ of copies of $\mathbb{Z}^n$, indexed by $\mathbb{Z}$, and the automorphism $\hat{\sigma}$ of $\hat{Z}$ dual to the shift σ on Z is again the shift. The annihilator $X^{\perp} \subset \hat{Z}$ is shift-invariant, $\hat{X} = \hat{Z}/X^{\perp}$, and $\hat{\alpha}$ is the automorphism of $\hat{X}$ induced by the shift $\hat{\sigma}$. As described in Lemma 5.1 and Proposition 5.4, $\hat{X} = \mathfrak{M}$ is a Noetherian $\mathfrak{R}_1$-module under the action $(f, a) \mapsto f * a$, where $f * a = f(\hat{\alpha})(a) = \sum_{k \in \mathbb{Z}} c_f(k) \hat{\alpha}^k(a)$ for every $f \in \mathfrak{R}_1$ and $a \in \mathfrak{M}$ (cf. (5.2)). For every $\mathbf{k} \in \mathbb{Z}^n$ we write $\mathbf{t} \mapsto \chi_{\mathbf{k}}(\mathbf{t}) = \langle \mathbf{t}, \mathbf{k} \rangle$ for the character of $\mathbb{T}^n$ corresponding to $\mathbf{k}$. If the group F_H^+ is infinite, there exists a $\mathbf{k} \in \mathbb{Z}^n$ such that $\chi_{c\mathbf{k}}$ does not annihilate F_H^+ for any $0 \neq c \in \mathbb{Z}$. We define an element $a \in \hat{X} = \mathfrak{M}$ by $x \mapsto \langle x, a \rangle = \chi_{\mathbf{k}}(\pi_{\{0\}}(x))$, and claim that the annihilator $\mathrm{ann}(a) = \{f \in \mathfrak{R}_1 : f * a = 0\}$ is equal to zero. Indeed, if $0 \neq f \in \mathfrak{R}_1$ and $f * a = 0$, then $f = c_s u_1^s + c_{s+1} u_1^{s+1} + \cdots + c_t u_1^t$ with $-\infty < s \leq t < \infty$, $c_i \in \mathbb{Z}$, $c_s c_t \neq 0$, and $\langle x, f * a \rangle = \langle \sum_{j=s}^{t} c_j \sigma^j(x), a \rangle = \chi_{\mathbf{k}}(\sum_{j=s}^{t} c_j x_j) = 1$ for all $x = (x_m) \in X \subset (\mathbb{T}^n)^{\mathbb{Z}}$. According to the definition of $X = Y_H$ there exists, for every $\mathbf{t} \in F_H^+$, an element $x(\mathbf{t}) = (x(\mathbf{t})_m, m \in \mathbb{Z}) \in X$ with $x(\mathbf{t})_j = \mathbf{0}_{\mathbb{T}^n}$ for $j < t$, and $x(\mathbf{t})_t = \mathbf{t}$. Then $\langle x(\mathbf{t}), f * a \rangle = \chi_{\mathbf{k}}(c_t \mathbf{t}) = \chi_{c_t \mathbf{k}}(\mathbf{t}) = 1$ for all $\mathbf{t} \in F_H^+$, contrary to our choice of $\mathbf{k}$. This shows that $\mathrm{ann}(a) = \{0\}$, i.e. that $\{0\}$ is one of the prime ideals associated with the Noetherian $\mathfrak{R}_1$-module $\mathfrak{M}$. By Theorem 6.5 (4) this is impossible in view of the expansiveness of α, so that F_H^+ must be finite. Similarly one can show that F_H^- must be finite. $\square$

PROOF OF THEOREM 9.7. Suppose that α is expansive. Then X is abelian by Theorem 2.4, and Lemma 9.9 and (4.10) allow us to assume without loss in generality that α is equal to the shift on a Markov subgroup $X = Y_H \subset (\mathbb{T}^n)^{\mathbb{Z}}$, where $H \subset \mathbb{T}^n \times \mathbb{T}^n$ is a full, connected subgroup. The groups $F_H^{\pm}$ are finite by Lemma 9.10, so that H is an n-dimensional subtorus of $\mathbb{T}^n \times \mathbb{T}^n$ whose projection onto each copy of $\mathbb{T}^n$ is surjective. An elementary argument shows that there exists a unique matrix $A \in \mathrm{GL}(n, \mathbb{Q})$ such that H is the image of the subspace $\{(v, Av) : v \in \mathbb{R}^n\} \subset \mathbb{R}^n \times \mathbb{R}^n$ under the natural quotient map from $\mathbb{R}^n \times \mathbb{R}^n$ onto $\mathbb{T}^n \times \mathbb{T}^n$. A brief glance at Example 9.5 (3) reveals that $X = Y_H = X^A$, and that α is algebraically conjugate to α^A. From Example 9.5 (3) we know that A cannot have any eigenvalues of modulus 1. This proves that (1)$\Rightarrow$(2), and the reverse implication is contained in Example 9.5 (3). $\square$

CONCLUDING REMARK 9.11. The exposition in this section follows [45]. Theorem 9.7 is due to [52].

10. The structure of expansive automorphisms

We fix a compact group G and a full subgroup $H \subset G \times G$, and define the Markov subgroup $Y_H \subset G^{\mathbb{Z}}$ by (9.4). For every $g \in G$ and $k \in \mathbb{Z}$, put

$$F_H(g, k) = \{y_k : y = (y_n) \in Y_H \text{ and } y_0 = g\} \subset G \qquad (10.1)$$

and

$$F_H(k) = F_H(1_G, k). \qquad (10.2)$$

Then $F_H(\pm 1) = F_H^{\pm}$ (cf. (9.5)), and $F_H(k)$ is a closed, normal subgroup of G with

$$F_H(k) \subset F_H(k+1) \text{ and } F_H(-k) \subset F_H(-k-1) \qquad (10.3)$$

for every $k \geq 0$, since $(1_G, 1_G) \in H$. For $k \geq 1$, the sets $F_H(g, k)$ and $F_H(g, -k)$ are the usual k-th follower and predecessor sets of symbolic dynamics. From the definition of $F_H(g, k)$ it is clear that the map $\theta'_n : G \longmapsto G/F_H(k)$, obtained by setting $\theta'_n(g) = F_H(g, k)/F_H(k)$, is a well-defined, continuous group homomorphism with kernel $F_H(-k)$, and that θ'_k induces a continuous isomorphism

$$\theta_k : G/F_H(-k) \longmapsto G/F_H(k). \qquad (10.4)$$

LEMMA 10.1. Put $\Lambda = F_H(-1) \cap F_H(1)$, $G' = G/F_H(1)$, $H' = H/(F_H(1) \times F_H(1)) \subset G' \times G'$, and denote by $\eta : G \longmapsto G'$ the quotient map. The shift commuting homomorphism $\boldsymbol{\eta} : Y_H \longmapsto G'^{\mathbb{Z}}$ given by $\boldsymbol{\eta}(y)_n = \eta(y_n)$, $n \in \mathbb{Z}$, $y \in Y_H$, has the following properties.

(1) $\boldsymbol{\eta}(Y_H) = Y_{H'}$, where $Y_{H'} \subset G'^{\mathbb{Z}}$ is defined as in (9.4);
(2) $Y_H \cap F_H(1)^{\mathbb{Z}} = \ker(\boldsymbol{\eta}) = \Lambda^{\mathbb{Z}}$;
(3) If $F_{H'}(k)$, $k \in \mathbb{Z}$, is defined as in (10.1)–(10.2) with $Y_{H'} \subset G'^{\mathbb{Z}}$ replacing Y_H, then $F_{H'}(k) = F_H(k+1)/F_H(1)$ for every $k \geq 1$;
(4) There exists a Haar measure preserving Borel isomorphism $\psi : Y_H \longmapsto \Lambda^{\mathbb{Z}} \times Y_{H'}$ which carries the shift on Y_H to the cartesian product of the shifts on $\Lambda^{\mathbb{Z}}$ and $Y_{H'}$;
(5) If G/Λ is finite, the map ψ in (4) can be chosen to be a homeomorphism;
(6) If Y_H is expansive, then Λ is finite and $Y_{H'}$ is again expansive.

PROOF. From (10.1)–(10.2) it is clear that $h \in F_H(1)$ if and only if $(1_G, h) \in H$, and $h \in F_H(-1)$ if and only if $(h, 1_G) \in H$. In particular, if $\eta' : H \longmapsto H'$ is the quotient map, and if $(g, h) \in H$ and $\eta'(g, h) = (\eta(g), \eta(h)) = (u, v) \in H'$, then $(g, hk) \in H$ and $\eta'(g, hk) = (u, v)$ for every $k \in F_H(1)$. It follows that there exists, for every $(u, v) \in H'$, and for every $h \in G$ with $\eta(h) = v$, an element $g \in G$ with $\eta(g) = u$ and $(g, h) \in H$. This implies (1).

Now assume that $(g, h) \in \ker(\eta') \subset H$. Then $g, h \in F_H(1)$, $(1_G, h) \in H$ (according to the definition of $F_H(1)$), $(g, 1_G) \in H$ (since H is a group), and $g \in F_H(-1) \cap F_H(1) = \Lambda$. Since $\Lambda^{\mathbb{Z}} \subset Y_H$ this shows that $y = (y_n) \in \ker(\boldsymbol{\eta})$ if and only if $y \in \Lambda^{\mathbb{Z}}$, as claimed in (2).

In order to prove (3) we proceed by induction and assume that $F_{H'}(k) = F_H(k+1)/F_H(k)$ for $0 \le k < K$, where $K \ge 1$. If $h \in F_H(K+1)$, choose $1_G = h_0, h_1, \ldots, h_{k-1}$ in G such that $(h_i, h_{i+1}) \in H$ for $i = 0, \ldots, K$, where $h_{K+1} = h$. By setting $h'_i = \eta(h_i)$ we see that $h'_0 = h'_1 = 1_{G'}$ and $(h'_i, h_{i+1'}) \in H'$ for all $i = 0, \ldots, K$, so that $h' \in F_{H'}(K)$. Conversely, if $h' \in F_{H'}(K)$, then we choose $h'_K \in F_{H'}(K-1)$ with $(h'_K, h') \in H'$, and the induction hypothesis allows us to find elements $1_G = h_0, h_1, \ldots, h_K$ in G such that $(h_i, h_{i+1}) \in H$ and $\eta(h_i) = h'_i$ for $i = 0, \ldots, K-1$, and $\eta(h_K) = h'_K$. There also exists an element $(\bar{h}_K, h) \in H$ with $\eta(\bar{h}_K) = h'_K = \eta(h'_K)$ and $\eta(h) = h'$. Then $\bar{h}_K$ differs from h_K by an element of $F_H(1)$, so that $(h_{K-1}, \bar{h}_K) \in H$, and by considering the sequence $1_G = h_0, \ldots, h_{K-1}, \bar{h}_K, h$ we see that $h \in F_H(K+1)$. This shows that $F_{H'}(K) = F_H(K+1)/F_H(1)$ and completes the induction step for (3).

We denote by $\tau \colon G \longmapsto G'' = G/\Lambda$ the quotient map, choose a Borel map $\omega \colon G'' \longmapsto G$ with $\omega(1_{G''}) = 1_G$ and $\tau(\omega(u)) = u$ for every $u \in G''$, and define $\boldsymbol{\omega} \colon G''^{\mathbb{Z}} \longmapsto G^{\mathbb{Z}}$ by $\boldsymbol{\omega}(w)_n = \omega(w_n)$ for every $w = (w_n) \in G''^{\mathbb{Z}}$. From (2) we see that the homomorphism $\boldsymbol{\tau} \colon Y_H \longmapsto G''^{\mathbb{Z}}$, given by $\boldsymbol{\tau}(y)_n = \tau(y_n)$ for every $y = (y_n) \subset Y_H$, satisfies that $\ker(\boldsymbol{\tau}) = \ker(\boldsymbol{\eta})$, and that $y\boldsymbol{\omega}(\boldsymbol{\tau}(y))^{-1} \in \Lambda^{\mathbb{Z}}$ for every $y \in Y_H$. Since $\Lambda^{\mathbb{Z}} \subset Y_H$ we conclude that $\boldsymbol{\omega}(u) \in Y_H$ for every $u \in \boldsymbol{\tau}(Y_H)$, and that the map $\boldsymbol{\theta} \colon Y_H \longmapsto \Lambda^{\mathbb{Z}} \times Y_{H'}$, given by $\boldsymbol{\theta}(y) = (y\boldsymbol{\omega}(\boldsymbol{\tau}(y))^{-1}, \boldsymbol{\eta}(y))$, is a Haar measure preserving Borel isomorphism which sends the shift on Y_H to the cartesian product of the shifts on $\Lambda^{\mathbb{Z}}$ and $Y_{H'}$. If G'' is finite, the maps ω and $\boldsymbol{\omega}$ are both continuous, and we have proved (4) and (5).

If Y_H is expansive, then $\Lambda^{\mathbb{Z}} \subset Y_H$ is again expansive, and Λ is finite. In this case the quotient map $\tau \colon G \longmapsto G/\Lambda$ is a homeomorphism of a neighbourhood N of the identity in G onto a neighbourhood N' of the identity in G/Λ, and by decreasing N, if necessary, we may assume that $Y_H \cap N^{\mathbb{Z}} = \{1_{G^{\mathbb{Z}}}\}$ (the existence of such a neighbourhood $N \subset G$ is equivalent to the expansiveness of Y_H). If $Y' = \boldsymbol{\tau}(Y_H) \subset (G/\Lambda)^{\mathbb{Z}}$ is not expansive, there exists a point $u = (u_n) \in Y' \cap N'^{\mathbb{Z}}$ with $u \ne 1_{(G/\Lambda)^{\mathbb{Z}}}$. If $u = \boldsymbol{\tau}(y)$ for some $y \in Y_H$ we can choose a point $z \in \Lambda^{\mathbb{Z}} \subset Y_H$ with $u_n z_n \in N$ for all $n \in \mathbb{Z}$. Since $1_{G^{\mathbb{Z}}} \ne uz \in Y_H \cap N^{\mathbb{Z}}$ we have arrived at a contradiction. Hence Y' is expansive. The continuous homomorphism $\boldsymbol{\eta} \colon Y_H \longmapsto G'^{\mathbb{Z}}$ induces a continuous, shift commuting isomorphism $\boldsymbol{\eta}' \colon Y' = Y_H/\Lambda^{\mathbb{Z}} \longmapsto Y_{H'}$, so that $Y_{H'}$ is again expansive. This proves (6). $\quad\square$

PROPOSITION 10.2. *Let G be a compact group, $H \subset G \times G$ a full subgroup, $Y = Y_H \subset G^{\mathbb{Z}}$ the Markov subgroup (9.4), and let σ be the shift $\sigma(y)_n = y_{n+1}$ on Y. We define $F_H(k)$, $k \in \mathbb{Z}$, by (10.1)–(10.2) and set, for every $k \ge 0$, $Y_k = Y \cap F_H(k)^{\mathbb{Z}}$, $G_k = G/F_H(k)$, denote by $\eta^{(k)} \colon G \longmapsto G_k$ the quotient map, and define a shift commuting map $\boldsymbol{\eta}^{(k)} \colon Y \longmapsto G_k^{\mathbb{Z}}$ with $\ker(\boldsymbol{\eta}^{(k)}) =$*

Y_k by $\eta^{(k)}(y)_n = \eta^{(k)}(y_n)$, $n \in \mathbb{Z}$. The maps $\boldsymbol{\eta}^{(k)}$, $k \geq 0$, have the following properties.

(1) $\boldsymbol{\eta}^{(k)}(Y) = Y_{H_k} \subset G_k^{\mathbb{Z}}$, where $H_k = H/(F_H(k) \times F_H(k))$, and where Y_{H_k} is defined as in (9.4);

(2) $Y_{k+1}/Y_k \cong \boldsymbol{\eta}^{(k)}(Y_{k+1}) = \Lambda_{k+1}^{\mathbb{Z}}$ for some closed, normal subgroup $\Lambda_{k+1} \subset F_H(k+1)/F_H(k)$.

(3) There exists a Haar measure preserving Borel isomorphism $\phi \colon Y \longmapsto \Lambda_1^{\mathbb{Z}} \times \cdots \times \Lambda_k^{\mathbb{Z}} \times Y_{k+1}$ which carries σ to $\sigma^{(1)} \times \cdots \times \sigma^{(k)} \times \sigma'$, where $\sigma^{(i)}$ denotes the shift on $\Lambda_i^{\mathbb{Z}}$ and σ' is the automorphism of Y_{k+1} induced by σ. The map ϕ can be chosen to be a homeomorphism if G is finite.

If G is a compact Lie group such that $C(G^\circ)$ is finite, the following stronger assertions are true.

(4) There exists an integer $K \geq 1$ with $F_H(K) = F_H(k)$ for all $k \geq K$, and an automorphism $\beta \in \text{Aut}(G_K)$ such that $H_K = \{(u, \beta(u)) : u \in G_K\}$,

$$Y/Y_K \cong \boldsymbol{\eta}^{(K)}(Y) = Y_{H_K} = \{v \in G_K^{\mathbb{Z}} : v_{n+1} = \beta(v_n) \text{ for all } n \in \mathbb{Z}\}$$
$$\cong G_K,$$

and the isomorphism $\boldsymbol{\eta}^{(K)}(Y) \cong G_K$ sends the shift on $\boldsymbol{\eta}^{(K)}(Y)$ to β;

(5) There exists a Haar measure preserving Borel isomorphism $\phi \colon Y \longmapsto \Lambda_1^{\mathbb{Z}} \times \cdots \times \Lambda_K^{\mathbb{Z}} \times G_K$ which carries σ to $\sigma^{(1)} \times \cdots \times \sigma^{(K)} \times \beta$, where $\sigma^{(i)}$ denotes the shift on $\Lambda_i^{\mathbb{Z}}$, and ϕ can be chosen to be a homeomorphism if G is finite;

(6) The restriction of σ to Y_k is ergodic for $k = 1, \ldots, K$, and σ is ergodic if and only if $Y_K = Y$;

(7) If Y_H is expansive, then G is finite.

PROOF. In order to prove (1) we apply Lemma 10.1 to see that $\boldsymbol{\eta}^{(1)}(Y) = Y_{H_1}$ and note that the groups $F_{H_1}(k)$, defined by (10.2) with Y_{H_1} and G_1 replacing Y_H and G, satisfy that $F_{H_1}(k) = F_H(k+1)/F_H(1)$ for all $k \geq 1$. Repeated application of Lemma 10.1 shows that $\boldsymbol{\eta}^{(k)}(Y) = Y_{H_k} \subset G_k^{\mathbb{Z}}$ and $F_{H_k}(m) = F_H(m+k)/F_H(k)$ for $k, m \geq 1$. All the assertion in (1) are now obvious.

For every $k \geq 0$, put $\Lambda_{k+1} = F_{H_k}(-1) \cap F_{H_k}(1) \subset F_H(k+1)/F_H(k) \subset G_k$, where $H_0 = H$ and $G_0 = G$. From Lemma 10.1 we know that $\boldsymbol{\eta}^{(k)}(Y_{k+1}) = \boldsymbol{\eta}^{(k)}(Y) \cap F_{H_k}(1)^{\mathbb{Z}} = \Lambda_{k+1}^{\mathbb{Z}}$, as claimed in (2), and the assertions in (3) follow from repeated applications of Lemma 10.1 (4)–(5).

If $C(G^\circ)$ is finite then G has only finitely many closed, normal subgroups, and the sequence $F_H(k)$, $k \geq 1$, must eventually become constant. Hence there exists a $K \geq 1$ with $F_H(k) = F_H(K)$ for all $k \geq K$. Since the group G_K° is a quotient of a compact, connected Lie group with finite centre, $C(G_K^\circ)$ is again finite, and there exists an $L \geq 1$ with $F_{H_K}(-l) = F_{H_K}(-L)$ for all $l \geq L$. For every $m \geq 1$ we denote by $\theta_m : G_K/F_{H_K}(-m) \longmapsto G_K/F_{H_K}(m)$ the isomorphism in (10.4) and put $\beta = \theta_{L+1}\theta_L^{-1} : G_K \longmapsto G_K$. From the definition of β it is clear that $H_K = \{(u, \beta(u)) : u \in G_K\}$, and the other statements in (4) are immediate consequences of this.

The assertion (5) is obvious from (3), and (6) follows from (5) and from the fact that G_K has no ergodic automorphisms (since $\mathrm{Aut}(G_K)$ is compact, there exists a metric on G_K which is invariant under every $\alpha \in \mathrm{Aut}(G_K)$).

Finally, if Y is expansive, then Lemma 10.1 (6) implies that Y_{H_k} is expansive and Λ_k is finite for every $k = 1, \ldots, K$, and that β is an expansive automorphism of G_K. Since there exists a metric on G_K which is invariant under every $\alpha \in \mathrm{Aut}(G_K)$ we conclude that G_K must be finite. From (2) we conclude that Y_H is zero-dimensional, and Corollary 2.3 implies that G is finite. $\square$

COROLLARY 10.3. *Let α be an automorphism of a compact group X such that $C(X^\circ)$ is zero-dimensional, and assume that α satisfies the d.c.c. Then there exist compact groups $\Lambda_1, \ldots, \Lambda_K, G_K$, an automorphism β of G_K, and a Haar measure preserving Borel isomorphism $\phi \colon X \longmapsto \Lambda_1^{\mathbb{Z}} \times \cdots \times \Lambda_K^{\mathbb{Z}} \times G_K$ which carries α to $\sigma^{(1)} \times \cdots \times \sigma^{(K)} \times \beta$, where $\sigma^{(i)}$ denotes the shift on $\Lambda_i^{\mathbb{Z}}$. The automorphism α is ergodic if and only if $G_K = \{\mathbf{1}\}$.*

If X itself is zero-dimensional, then α is expansive, the groups $\Lambda_1, \ldots, \Lambda_K$, G_K, are finite, and the Borel isomorphism ϕ can be chosen to be a homeomorphism.

PROOF. Lemma 9.9 allows us to assume that $X = Y_H \subset G^{\mathbb{Z}}$ for some compact, connected Lie group G and some full subgroup $H \subset G \times G$. Corollary 2.3 implies that $\alpha^{C(X^\circ)}$ is expansive, and that $C(G^\circ) = \pi_{\{0\}}(C(X^\circ))$ is finite. Our assertion follows from Proposition 10.2. $\square$

COROLLARY 10.4. *Let α be an automorphism of a compact group X which satisfies the d.c.c. If $Z \subset X$ is a closed, normal, α-invariant subgroup such that $C(Z^\circ)$ is zero-dimensional and α^Z is ergodic, then there exists a Haar measure preserving Borel isomorphism $\phi \colon X \longmapsto Z \times X/Z$ with $\phi \cdot \alpha = (\alpha^Z \times \alpha^{X/Z}) \cdot \phi$. If X is zero-dimensional, ϕ can be chosen to be a homeomorphism.*

PROOF. According to (4.10) and Remark 4.15 (1) we may assume that $X = Y_H \subset G^{\mathbb{Z}}$, where G is a compact Lie group and $H \subset G \times G$ a full subgroup, and that $Z = Y_L \subset K^{\mathbb{Z}}$, where $K \subset G$ is a closed, normal subgroup, $L = H \cap K \times K$ is a full subgroup, and Y_H and Y_L are given by (9.4). Since $Z \subset K^{\mathbb{Z}}$ is full, the group $C(K^\circ)$ is finite by Corollary 2.3. The assertions follow exactly as in the proof of Proposition 10.2 by applying Lemma 10.1 repeatedly to Y_H, with $F_H(k)$ replaced by $F_L(k) \subset G$ for every $k \in \mathbb{Z}$. $\square$

EXAMPLES 10.5. (1) Let K be a compact group, $G = K \times K$, and let $H = \{((k_1, k_2), (k_3, k_4)) \in G \times G \cong K^4 : k_2 = k_3\}$. Then $H \subset G \times G$ is a full subgroup. In the notation of Proposition 10.2 we have that $F_H(1) = \{\mathbf{1}_K\} \times K$, $F_H(2) = G$, $F_H(-1) = K \times \{\mathbf{1}_K\}$, and $F_H(-2) = G$. The group $\Lambda_1 = F_H(1) \cap F_H(-1)$ is equal to $\{\mathbf{1}_G\}$, $G_1 = K \times \{\mathbf{1}_K\} \cong K$, and the homomorphism $\eta^{(1)} \colon Y_H \longmapsto G_1^{\mathbb{Z}}$ has trivial kernel. Furthermore, $H_1 = G_1 \times G_1$, $\Lambda_2 = G_1$, and $Y_H \cong \eta^{(1)}(Y_H) = Y_{H_1} = G_1^{\mathbb{Z}} \cong K^{\mathbb{Z}}$. Note that Y_H is the two-block representation of the Bernoulli shift $X = K^{\mathbb{Z}}$, and that Proposition 10.2 has led us back from Y_H to X.

(2) If G has infinite centre, the groups Λ_k in Proposition 10.2 may be trivial for every $k \geq 1$ even if Y_H is ergodic. Consider the shift $\sigma = \alpha^{\mathfrak{R}_1/(f)}$ on $X = X^{\mathfrak{R}_1/(f)} = \{(x_n) : 2x_n = x_{n+1} \pmod 1 \text{ for all } n \in \mathbb{Z}\}$, where $f = 2 - u_1 \in \mathfrak{R}_1$. Then $X = Y_H$ with $H = \{(2t, t) : t \in \mathbb{T}\} \subset \mathbb{T} \times \mathbb{T}$ is full, $F_H(k) = \{0\}$, and $F_H(-k) = \{l 2^{-k} : 0 \leq l < 2^k\}$ for all $k \geq 1$. In particular, $G_k = G$, and $\eta^{(k)}$ is bijective for every $k \geq 1$. The ergodicity of Y_H was established in Theorem 6.5.   ⊡

THEOREM 10.6. *Let α be an ergodic automorphism of a compact group X satisfying the d.c.c. Then there exist an integer $n \geq 1$, a matrix $A \in \mathrm{GL}(n, \mathbb{Q})$, none of whose eigenvalues is a root of unity, compact Lie groups $\Lambda_1, \ldots, \Lambda_K$, $K \geq 1$, and a Haar measure preserving Borel isomorphism $\phi \colon X \longmapsto \Lambda_1^{\mathbb{Z}} \times \cdots \times \Lambda_K^{\mathbb{Z}} \times X^A$, such that $\phi \cdot \alpha = \sigma^{(1)} \times \cdots \times \sigma^{(K)} \times \alpha^A$, where $\sigma^{(i)}$ and α^A are the shifts on $\Lambda_i^{\mathbb{Z}}$ and X^A, respectively, and where X^A is defined in (9.3).*

The automorphism α is expansive if and only if A has no eigenvalue of modulus 1 and every Λ_i is finite.

PROOF. Lemma 9.9 allows us to assume that X is a Markov subgroup of $G^{\mathbb{Z}}$, where G is a compact Lie group, that α is equal to the shift on X, and that $X^{\circ} = X \cap (G^{\circ})^{\mathbb{Z}}$. We define $F_H(k)$, $k \in \mathbb{Z}$, as in (10.2), set $F^+ = \bigcup_{k \geq 1} F_H(k)$, and note that F^+ is normal in G. From the definition of F^+ it is clear that, for every $x \in X$ with $x_n \in F^+$ for some $n \in \mathbb{Z}$, $x_{m+n} \in F^+$ for all $m \geq 0$. If $\bar{F}^+$ is the closure of F^+ and $B = \{x \in X : x_0 \in \bar{F}^+\}$ then the continuity of α implies that $\alpha^{-m}(B) \subset B$ for all $m \geq 1$, i.e. that $F_H(g, k) \subset \bar{F}^+$ for all $g \in \bar{F}^+$ and $k \geq 1$ (cf. (10.1)). Let $\eta \colon G \longmapsto G' = G/\bar{F}$ be the quotient map, and define $\eta \colon G^{\mathbb{Z}} \longmapsto G'^{\mathbb{Z}}$ by $(\eta(x))_n = \eta(x_n)$ for all $x = (x_n) \in \mathbb{Z}$ and $n \in \mathbb{Z}$. The restriction to $\eta(X)$ of the coordinate projection $\pi_{\{0\}} \colon G'^{\mathbb{Z}} \longmapsto G'$ induces a commutative diagram

$$
\begin{array}{ccc}
\eta(X) & \xrightarrow{\ \sigma\ } & \eta(X) \\
{\scriptstyle \pi_{\{0\}}} \downarrow & & \downarrow {\scriptstyle \pi_{\{0\}}} \\
G' & \xrightarrow[\ \tau\ ]{} & G'
\end{array}
$$

where $\tau \colon G' \longmapsto G'$ is a continuous, surjective group homomorphism, and the ergodicity of α forces τ to be ergodic. If G' is not connected, then $\tau(G'^{\circ}) = G'^{\circ}$ and, since G'/G'° is finite, $\tau^{-1}(G'^{\circ}) = G'^{\circ}$. Hence G' must be connected or, equivalently, $G^{\circ} \cdot F^+ = G$.

If G' is non-abelian, then $G'/C(G')$ is a compact, connected Lie group with trivial centre (cf. Proposition 2.13), $\tau(C(G')) \subset C(G')$, and τ induces a homomorphism $\tau': G'/C(G') \longmapsto G'/C(G')$. Since τ' is surjective, $\ker(\tau')$ must be a finite, and therefore central, subgroup of $G'/C(G')$, and we conclude that $\tau' \subset \operatorname{Aut}(G'/C(G'))$. However, $G'/C(G')$ has a metric which is invariant under $\operatorname{Aut}(G'/C(G'))$, and the ergodicity of τ' leads to a contradiction unless G' is abelian.

We have proved that $G' = G/\bar{F}^+ \cong \mathbb{T}^m$ for some $m \geq 0$, and by looking at dimensions of Lie algebras we see that there exist integers $K \geq 1$ and $n \geq 0$ such that $G_k = G/F_H(k) \cong \mathbb{T}^n$ for all $k \geq K$. In the notation of Proposition 10.2 we obtain that $\boldsymbol{\eta}^{(K)}(X) = Y_{H_K} \subset (\mathbb{T}^n)^{\mathbb{Z}}$, and that $F_{H_K}(\pm 1)$ is finite, since $G_K = G/F_H(K) \cong G/F_H(K+1) = G_{K+1}$, $G/F_H(-k) \cong G/F_H(k)$ for all $k \geq 1$, and $F_{H_K}(1) \cong F_H(K+1)/F_H(K)$. In other words, $H_K \subset \mathbb{T}^n \times \mathbb{T}^n$ must be isomorphic to $\mathbb{T}^n$, which forces $\boldsymbol{\eta}^{(K)}$ to be of the form (9.3) for some matrix $A \in \operatorname{GL}(n, \mathbb{Q})$. The remaining assertions all follow from Proposition 10.2. $\square$

If the automorphism α in Theorem 10.6 is non-ergodic, then Proposition 3.5 implies the existence of a maximal closed, normal, α-invariant subgroup $X' \subset X$ such that $\alpha^{X'}$ is ergodic and X/X' is a compact Lie group. If α is expansive, a stronger assertion can be made.

THEOREM 10.7. *Let α be an expansive automorphism of a compact group X. Then we can find an open, normal, α-invariant subgroup $X' \subset X$ such that $\alpha^{X'}$ is ergodic, and a Haar measure preserving homeomorphism $\psi: X \longmapsto X' \times X/X'$ with $\psi \cdot \alpha = (\alpha^{X'} \times \alpha^{X/X'}) \cdot \psi$.*

By combining the Theorems 10.6 and 10.7 we obtain the following corollary.

COROLLARY 10.8. *Let α be an expansive automorphism of a compact group X. Then there exist an integer $n \geq 1$, a matrix $A \in \operatorname{GL}(n, \mathbb{Q})$, none of whose eigenvalues has modulus 1, finite groups $\Lambda_0, \Lambda_1, \ldots, \Lambda_K$, $K \geq 1$, and a Haar measure preserving Borel isomorphism $\phi: X \longmapsto \Lambda_0 \times \Lambda_1^{\mathbb{Z}} \times \cdots \times \Lambda_K^{\mathbb{Z}} \times X^A$ such that $\phi \cdot \alpha = \beta \times \sigma^{(1)} \times \cdots \times \sigma^{(K)} \times \alpha^A$, where β is an automorphism of Λ_0, and where $\sigma^{(i)}$ and (X^A, α^A) are defined as in Theorem 10.6.*

For the proof of Theorem 10.7 we need a lemma.

LEMMA 10.9. *Let α be an ergodic automorphism of a compact group X. Then there exists, for every $x \in X$ and $s \geq 1$, an element $y \in X$ with $x = \alpha^s(y)y^{-1}$.*

PROOF. We begin by proving the assertion under the assumption that X is abelian. Since α is ergodic, the automorphism $\hat{\alpha}$ of $\hat{X}$ dual to α satisfies that $\hat{\alpha}^s(\chi) \neq \chi$ whenever $s \geq 1$ and $\mathbf{1}_{\hat{X}} \neq \chi \in \hat{X}$ (Lemma 1.2 and Remark 1.7 (2)).

Hence the homomorphism $\chi \mapsto \hat{\alpha}^s(\chi)\chi^{-1}$ from $\hat{X}$ to $\hat{X}$ is injective, and the dual homomorphism $x \mapsto \alpha^s(x)x^{-1}$ from X to X is surjective, as claimed.

Next we assume that α is the shift on $X = \Lambda^{\mathbb{Z}}$, where Λ is a compact group. If $x = (x_n) \in X = \Lambda^{\mathbb{Z}}$ and $s \geq 1$ are fixed, there exists a unique point $y = (y_n) \in X$ with $x_i = y_i$ for $i = 0, \ldots, s-1$ and $x_n = y_{n+s}y_n^{-1}$ for all $n \in \mathbb{Z}$, so that the lemma also holds in this case.

Now assume that we have proved the lemma in general, but under the additional assumption that α satisfies the d.c.c. According to Proposition 4.5 there exists a non-increasing sequence $(V_m, m \geq 1)$ of closed, normal, α-invariant subgroups such that $\bigcap_{m \geq 1} V_m = \{1_X\}$ and α^{X/V_m} satisfies the d.c.c. for every $m \geq 1$. We fix $x \in X$ and $s \geq 1$ and observe that the set $B_m = \{y \in X : \alpha^s(y)y^{-1} \in xV_m\}$ is closed and non-empty, and that $B_m \supset B_{m+1}$ for all $m \geq 1$. The compactness of X implies that $B = \bigcap_{m \geq 1} B_m \neq \varnothing$, and $x = \alpha^s(y)y^{-1}$ for every $y \in B$.

In order to prove our assertion for an automorphism α of a compact group X satisfying the d.c.c. we assume that X is a Markov subgroup of $G^{\mathbb{Z}}$, where G is a compact Lie group. We apply Proposition 10.2 and the proof of Theorem 10.6 and obtain an integer $K \geq 1$ such that the group G_K (in the notation of Proposition 10.2) is isomorphic to $\mathbb{T}^n$ for some $n \geq 0$, and $\eta^{(K)}(X) = Y_{H_K} \cong X^A$ for some $A \in \mathrm{GL}(n, \mathbb{Q})$ which has no root of unity as an eigenvalue. Furthermore, the closed, normal, α-invariant subgroups $Y_k = X \cap F_H(k)^{\mathbb{Z}}$ satisfy that $\{1_X\} = Y_0 \subset Y_1 \subset \cdots \subset Y_K \subset X$, $X/Y_K \cong \eta^{(K)}(X)$, and $Y_k/Y_{k-1} \cong \Lambda_k^{\mathbb{Z}}$ for $k = 1, \ldots, K$.

We fix $x \in X$ and $s \geq 1$, apply the first part of this proof, where the assertion was proved for abelian groups, to $X/Y_K \subset (\mathbb{T}^n)^{\mathbb{Z}}$, and obtain a $y^{(K+1)} \in X$ with $\alpha^s(y^{(K+1)})(y^{(K+1)})^{-1} \in xY_K$ or, equivalently, with $x^{(K)} = \alpha^s((y^{(K+1)})^{-1})xy^{(K+1)} \in Y_K$. Since $Y_K/Y_{K-1} \cong \Lambda_K^{\mathbb{Z}}$, and since $\alpha^{Y_K/Y_{K-1}}$ corresponds to the shift on $\Lambda_K^{\mathbb{Z}}$, the second part of this proof implies the existence of a point $y^{(K)} \in Y_K$ such that $\alpha^s(y^{(K)})(y^{(K)})^{-1} \in x^{(K)}Y_{K-1}$. By repeating this argument we construct points $y^{(i)} \in Y_i$, $i = 1, \ldots, K$, with $\alpha^s((y^{(1)})^{-1}) \cdot \ldots \cdot \alpha^s((y^{(K+1)})^{-1})xy^{(K+1)} \cdot \ldots \cdot y^{(1)} \in Y_0 = \{1\}$, which completes the proof of the lemma. $\square$

PROOF OF THEOREM 10.7. We choose a finite subset $E \in X$ such that $\{xX' : x \in E\}$ intersects each orbit of $\alpha^{X/X'}$ in exactly one point. Fix $x \in E$ for the moment and denote by $s(x)$ the smallest positive integer with $\alpha^{s(x)}(x) \in xX'$. Lemma 10.9 guarantees the existence of a $y(x) \in X'$ with $\alpha^{s(x)}(x)x^{-1} = \alpha^{s(x)}(y(x))y(x)^{-1}$, and we put $x' = y(x)^{-1}x$ and note that $\alpha^{s(x)}(x') = x'$. The set $E' = \{\alpha^k(x') : x \in E, 0 \leq k < s(x)\}$ intersects each coset of X' in exactly one point, and we write $z(v)$ for the unique element in $E' \cap vX'$ for every $v \in X$. Then $\alpha(z(v)) = z(\alpha(v))$ for every $v \in X$, and the map $v \mapsto \psi(v) = (vz(v)^{-1}, z(v)X')$ from X to $X' \times X/X'$ is a homeomorphism with the required properties. $\square$

REMARK 10.10. Corollary 10.3 gives a complete topological classification of ergodic and expansive automorphisms of compact, zero-dimensional groups (cf. [43]): two ergodic and expansive automorphisms α and α' of compact, zero-dimensional groups X and X' are topologically conjugate if and only if they have the same entropy (cf. Chapter V). Indeed, if $h(\alpha) = h(\alpha')$, then Corollary 10.3 shows that $h(\alpha) = h(\alpha') = \log n$ for some $n \geq 2$, and that α and α' are topologically conjugate to the shift σ on $(\mathbb{Z}/n\mathbb{Z})^{\mathbb{Z}}$. Whereas the algebraic and topological classification of expansive automorphisms of compact, connected groups coincides by Theorem 2.4 and Theorem 5.9, the algebraic classification of expansive automorphisms of compact, zero-dimensional groups is much more complicated than the topological one, even if the groups are abelian. According to Corollary 5.10 and Example 6.17 (3), the latter problem is equivalent to the classification—up to module-isomorphism—of all Noetherian $\mathfrak{R}_1$-modules whose associated prime ideals are all of the form $\mathfrak{p} = p\mathfrak{R}_1$ for some rational prime p. The following examples may give some idea of the complexity of such a classification.

EXAMPLES 10.11. Let p be a rational prime.

(1) Let X be a compact group with $px = 0$ for every $x \in X$, and let α be an ergodic automorphism of X. Then $\mathfrak{M} = \hat{X}$ is a module over the principal ideal domain $\mathfrak{R}_1^{(p)}$ (cf. Remark 6.19 (4)). By Example 6.17 (3) and the ergodicity of α, no non-zero ideal $\mathfrak{p} \subset \mathfrak{R}_1^{(p)}$ can be associated with $\mathfrak{M}$, so that $\mathfrak{M}$ is a torsion-free $\mathfrak{R}_1^{(p)}$-module. Theorem 2.2 in [51] implies that $\mathfrak{M}$ is *free*, i.e. isomorphic to $(\mathfrak{R}_1^{(p)})^k$ for some $k \geq 1$, and duality shows that α is algebraically conjugate to the shift on $(\mathbb{F}_p^k)^{\mathbb{Z}}$. The integer k is obviously an algebraic conjugacy invariant; moreover, since α has entropy $h(\alpha) = k \log p$ (cf. Chapter 5), k is also preserved by measurable conjugacies of such automorphisms.

(2) If $k \geq 2$ then there exist infinitely many algebraically non-isomorphic $\mathfrak{R}_1$-modules leading to algebraically non-conjugate ergodic group automorphisms with equal entropy $k \log p$. In the following proof of this claim we denote by $M_n(\mathfrak{R}_1)$ the ring of $n \times n$ matrices with coefficients in $\mathfrak{R}_1$.

If $\mathfrak{M}_1 = \mathfrak{R}_1/p^k\mathfrak{R}_1$, then $\alpha^{\mathfrak{M}_1}$ is (algebraically conjugate to) the shift-action of $\mathbb{Z}$ on $\mathbb{Z}_{/p^k}^{\mathbb{Z}}$, and $h(\alpha^{\mathfrak{M}_2}) = k \log p$ (cf. (5.5)–(5.6)). Note that $\alpha^{\mathfrak{M}_1}$ is topologically, but not algebraically, conjugate to the $\mathbb{Z}$-action in Example (1).

Now choose a non-zero element $\mathbf{f} = (f_1, f_2) \in \mathfrak{R}_1^2 = \mathfrak{R}_1 \times \mathfrak{R}_1$, consider the $\mathfrak{R}_1$-module $\mathfrak{M}_{\mathbf{f}} = \mathfrak{R}_1^2/(p^k\mathfrak{R}_1^2 + \mathfrak{R}_1 \cdot \mathbf{f})$, and set $\mathfrak{a} = (p^k, f_1, f_2) = p^k\mathfrak{R}_1 + f_1\mathfrak{R}_1 + f_2\mathfrak{R}_1 \subset \mathfrak{R}_1$ and $\mathfrak{b} = (p, f_1, f_2) = p\mathfrak{R}_1 + f_1\mathfrak{R}_1 + f_2\mathfrak{R}_1 \subset \mathfrak{R}_1$. We claim that the $\mathfrak{R}_1$-module $\mathfrak{M}_{\mathbf{f}}$ is cyclic (and hence isomorphic to $\mathfrak{M}_1$) if and only if $\mathfrak{a} = \mathfrak{b} = \mathfrak{R}_1$.

Indeed, $\mathfrak{M}_{\mathbf{f}}$ is cyclic if and only if there exists an element $\mathbf{g} = (g_1, g_2) \in \mathfrak{R}_1^2$ with $\mathfrak{R}_1 \cdot \mathbf{f} + \mathfrak{R}_1 \cdot \mathbf{g} + p^k\mathfrak{R}_1^2 = \mathfrak{R}_1^2$, i.e. if and only if the matrix $A = \begin{pmatrix} f_1 & g_1 \\ f_2 & g_2 \end{pmatrix} \in M_2(\mathfrak{R}_1)$ defines a surjective $\mathfrak{R}_1$-linear map $\phi_A : \mathfrak{N} \longmapsto \mathfrak{N}$, where $\mathfrak{N} = \mathfrak{R}_1^2/p^k\mathfrak{R}_1^2$.

Suppose that ϕ_A is surjective. Since the kernel of ϕ_A is an $\mathfrak{R}_1$-submodule of $\mathfrak{N}$, an elementary rank- (or, by duality, entropy-) argument shows that ϕ_A is invertible. The adjoint matrix $\mathrm{adj}(A) = \left(\begin{smallmatrix} g_2 & -g_1 \\ -f_2 & f_1 \end{smallmatrix} \right) \in M_2(\mathfrak{R}_1)$ of A satisfies that $A \cdot \mathrm{adj}(A) = \mathrm{adj}(A) \cdot A = \det A \cdot \left(\begin{smallmatrix} 1 & 0 \\ 0 & 1 \end{smallmatrix} \right)$, so that $\det A \cdot \phi_A^{-1}$ is equal to the $\mathfrak{R}_1$-linear map $\phi_{\mathrm{adj}(A)} \colon \mathfrak{N} \longmapsto \mathfrak{N}$ defined by adj A. Hence $\phi_A(\mathfrak{N}) \subset \det A \cdot \mathfrak{N}$, which implies that multiplication by $\det A$ is invertible on $\mathfrak{R}_1/p^k\mathfrak{R}_1$, $\det A \cdot \mathfrak{R}_1 + p^k\mathfrak{R}_1 = \mathfrak{R}_1$, and $\mathfrak{a} = \mathfrak{b} = \mathfrak{R}_1$.

Conversely, if $\mathfrak{b} = \mathfrak{R}_1$, then there exist elements g_1, g_2 in $\mathfrak{R}_1$ such that $f_1 g_2 - g_1 f_2 = 1 - ph$ for some $h \in \mathfrak{R}_1$. Hence there exist elements g_1', g_2', h' in $\mathfrak{R}_1$ with $f_1 g_2' - g_1' f_2 = 1 - p^k h'$, which proves that $\mathfrak{a} = \mathfrak{R}_1$ and that $\mathfrak{M}_{\mathbf{f}}$ has the cyclic element $\bar{\mathbf{g}} = \mathbf{g} + \mathfrak{R}_1 \cdot \mathbf{f} + p^k\mathfrak{R}_1^2 \in \mathfrak{M}_{\mathbf{f}}$.

Having verified that $\mathfrak{M}_{\mathbf{f}}$ is cyclic if and only if $\mathfrak{a} = \mathfrak{b} = \mathfrak{R}_1$ we turn to the case where $\mathfrak{M}_{\mathbf{f}}$ is not cyclic. If $\mathbf{f} \in p\mathfrak{R}_1^2$, then $\mathfrak{M}_{\mathbf{f}}$ is obviously not cyclic, $\mathfrak{M}_{\mathbf{f}}/\mathfrak{M}'$ is infinite for every cyclic submodule $\mathfrak{M}' \subset \mathfrak{M}_{\mathbf{f}}$, and $h(\alpha^{\mathfrak{M}_{\mathbf{f}}}) > h(\alpha^{\mathfrak{R}_1/p^k\mathfrak{R}_1}) = k \log p$. Now suppose that $\mathfrak{M}_{\mathbf{f}}$ is not cyclic, but that $\mathbf{f} \in \mathfrak{R}_1^2 \smallsetminus p\mathfrak{R}_1^2$. Then there exists an element $\mathbf{g} = (g_1, g_2) \in \mathfrak{R}_1^2$ such that $\det A \in \mathfrak{R}_1 \smallsetminus p\mathfrak{R}_1$, where $A = \left(\begin{smallmatrix} f_1 & g_1 \\ f_2 & g_2 \end{smallmatrix} \right) \in M_2(\mathfrak{R}_1)$. If $\mathfrak{M}' \subset \mathfrak{M}_{\mathbf{f}}$ is the cyclic submodule of $\mathfrak{M}_{\mathbf{f}}$ generated by $\bar{\mathbf{g}} = \mathbf{g} + \mathfrak{R}_1 \cdot \mathbf{f} + p^k\mathfrak{R}_1^2 \in \mathfrak{M}_{\mathbf{f}}$, then the argument in the preceding paragraph shows that $\mathfrak{K} = \mathfrak{M}_{\mathbf{f}}/\mathfrak{M}' \cong \mathfrak{N}/\phi_A(\mathfrak{N})$, where $\phi_A \colon \mathfrak{N} \longmapsto \mathfrak{N}$ is the $\mathfrak{R}_1$-linear map defined by A. By using the adjoint matrix $\mathrm{adj}(A) = \left(\begin{smallmatrix} g_2 & -f_1 \\ f_2 & g_1 \end{smallmatrix} \right) \in M_2(\mathfrak{R}_1)$ of A as above we see that every prime ideal $\mathfrak{p}$ in $\mathfrak{R}_1$ associated with the $\mathfrak{R}_1$-module $\mathfrak{K}$ must contain the ideal $\mathfrak{q} = p\mathfrak{R}_1 + \det A \cdot \mathfrak{R}_1$, so that $\mathfrak{p}$ induces a (principal) prime ideal $\bar{\mathfrak{p}} \subset \mathfrak{R}_1^{(p)}$ generated by one of the irreducible factors of $(\det A)_{/p}$; moreover, every irreducible factor of $(\det A)_{/p}$ occurs in this manner. In particular $\mathfrak{R}_1/\mathfrak{p}$ is finite for every prime ideal $\mathfrak{p} \subset \mathfrak{R}_1$ associated with the Noetherian $\mathfrak{R}_1$-module $\mathfrak{K}$, so that $\mathfrak{K}$ is finite and $h(\alpha^{\mathfrak{M}_{\mathbf{f}}}) = h(\alpha^{\mathfrak{M}'}) = k \log p$. Note, however, that $\alpha^{\mathfrak{M}_{\mathbf{f}}}$ may not be ergodic: if $\mathbf{f} = (1 - u_1, 1 - u_1)$, then the polynomial $1 - u_1$ lies in a prime ideal associated with $\mathfrak{M}_{\mathbf{f}}$, and $\alpha^{\mathfrak{M}_{\mathbf{f}}}$ is non-ergodic by Theorem 6.5.

It is now obvious how to construct infinitely many non-isomorphic modules $\mathfrak{M}_{\mathbf{f}}$ such that $\alpha^{\mathfrak{M}_{\mathbf{f}}}$ is ergodic and $h(\alpha^{\mathfrak{M}_{\mathbf{f}}}) = k \log p$: for any fixed, non-zero, irreducible element $1 \neq h \in \mathfrak{R}_1^{(p)}$ we choose an $f \in \mathfrak{R}_1$ with $f_{/p} = h$ and set $\mathbf{f} = (p, f) \in \mathfrak{R}_1^2$. It is clear that $p\mathfrak{R}_1$ is the only prime ideal associated with $\mathfrak{M}_{\mathbf{f}}$, so that $\alpha^{\mathfrak{M}_{\mathbf{f}}}$ is ergodic by Theorem 6.5 and Proposition 6.6. Put $\mathbf{g} = (0, 1) \in \mathfrak{R}_1^2$ and consider the cyclic submodule $\mathfrak{M}'$ of $\mathfrak{M}_{\mathbf{f}}$ generated by $\bar{\mathbf{g}} = \mathbf{g} + \mathfrak{R}_1 \cdot \mathbf{f} + p^k\mathfrak{R}_1^2 \in \mathfrak{M}_{\mathbf{f}}$. If $A = \left(\begin{smallmatrix} f_1 & g_1 \\ f_2 & g_2 \end{smallmatrix} \right)$, then the discussion in the preceding paragraph shows that

$$|\mathfrak{M}_{\mathbf{f}}/\mathfrak{M}'| = |\mathfrak{N}/A\mathfrak{N}| \le |\mathfrak{N}/\mathrm{adj}(A)A\mathfrak{N}| = |\mathfrak{R}_1/(\det A\mathfrak{R}_1 + p^k\mathfrak{R}_1)|^2 < \infty,$$

and that the only prime ideal $\mathfrak{p} \subset \mathfrak{R}_1$ associated with $\mathfrak{M}_{\mathbf{f}}/\mathfrak{M}'$ satisfies that $\bar{\mathfrak{p}} = h\mathfrak{R}_1^{(p)}$. In particular, $h(\alpha^{\mathfrak{M}_{\mathbf{f}}}) = h(\alpha^{\mathfrak{R}_1/p^k\mathfrak{R}_1}) = k \log p$. Furthermore, if $\mathfrak{M}''$

is any other cyclic submodule of $\mathfrak{M}_{\mathfrak{f}}$ such that $\mathfrak{M}_{\mathfrak{f}}/\mathfrak{M}''$ is finite, then $\mathfrak{p}$ is also associated with $\mathfrak{M}_{\mathfrak{f}}/\mathfrak{M}''$.

For different choices of h (and hence of $\bar{\mathfrak{p}}$) we obviously obtain non-isomorphic modules $\mathfrak{M}_{\mathfrak{f}}$, which proves our assertion. ⊡

CONCLUDING REMARK 10.12. Lemma 10.1 and Proposition 10.2 are taken from [111], [71] and [45], Corollary 10.3 for compact, zero-dimensional groups is due to [111] and [43], Example 10.5 (1) is taken from [43], and Theorem 10.6 can be found in [71] (cf. also [111]). Example 10.11 (2) is based on [22].

CHAPTER IV

Periodic points

11. Periodic points of $\mathbb{Z}^d$-actions

As we have seen in Theorem 5.7, every $\mathbb{Z}^d$-action by automorphisms of a compact, abelian group satisfying the d.c.c. has a dense set of periodic points (Definition 5.5), and Example 5.6 (1) shows that the d.c.c. cannot be dropped in general. In this section we investigate the density of the set of periodic points for a $\mathbb{Z}^d$-action α by automorphisms of compact group X satisfying the d.c.c. We begin with two examples which show that—if X is non-abelian—the d.c.c. does not necessarily imply that the set of α-periodic points is dense.

EXAMPLES 11.1. (1) Let $X = \mathrm{SU}(2)$, and let $h \in X$ be an element of infinite order (i.e. $h^n \neq \mathbf{1}_X$ for all $n \geq 1$). The inner automorphism $\alpha(x) = hxh^{-1}$, $x \in X$, has no periodic points other than fixed points, and the set of fixed points of α is the closure of $\{h^n : n \in \mathbb{Z}\}$, which is a maximal torus in X. It is clear that α satisfies the d.c.c., but that α is not ergodic.

(2) Let $X = \mathrm{SU}(2)^{\mathbb{Z}}$, and let $h \in \mathrm{SU}(2)$ be an element of infinite order. We define $\beta \in \mathrm{Aut}(X)$ by $(\beta(x))_n = hx_nh^{-1}$ for every $x = (x_n) \in X$ and $n \in \mathbb{Z}$, and consider the shift σ on X. Since β and σ commute, and since σ is ergodic, β and σ together define an ergodic $\mathbb{Z}^2$-action α with $\alpha_{(1,0)} = \beta$ and $\alpha_{(0,1)} = \sigma$, which satisfies the d.c.c. by (4.10). Example (1) shows that the set of periodic points for this $\mathbb{Z}^2$-action cannot be dense. ⊡

We shall prove the following result.

THEOREM 11.2. *Let α be a $\mathbb{Z}^d$-action by automorphisms of a compact group X such that X° is abelian. If α satisfies the d.c.c., then the set of α-periodic points is dense in X.*

In conjunction with Theorem 2.4 and Corollary 4.7, Theorem 11.2 implies the following corollary.

COROLLARY 11.3. *Let α be an expansive $\mathbb{Z}^d$-action by automorphisms of a compact group X. Then the set of α-periodic points is dense in X.*

For the proof of Theorem 11.2 we need several lemmas. If X is a compact group and $\alpha \in \mathrm{Aut}(X)$ we set

$$\mathrm{Fix}(\alpha) = \{x \in X : \alpha(x) = x\},$$
$$\mathrm{Per}(\alpha) = \bigcup_{n \geq 1} \mathrm{Fix}(\alpha^n). \tag{11.1}$$

LEMMA 11.4. *Let X be a compact group, α an automorphism of X, and V a closed, normal, α-invariant subgroup of X. Suppose that the following conditions are satisfied.*

(1) *For every $v \in V$ and $s \geq 1$ there exists a point $w \in V$ with $v = \alpha^s(w)w^{-1}$;*
(2) *$\mathrm{Per}(\alpha^V)$ is dense in V;*
(3) *$\mathrm{Per}(\alpha^{X/V})$ is dense in X/V.*

Then $\mathrm{Per}(\alpha)$ is dense in X.

PROOF. Let δ be a metric on X, and let δ' be the metric on X/V induced by δ. We fix $x \in X$ and $\varepsilon > 0$, and choose a point $u \in \mathrm{Per}(\alpha^{X/V})$ such that $\delta'(x, u) < \varepsilon/2$. If $u = yV$ for some $y \in X$ then $\alpha^m(y)y^{-1} = v \in V$ for some $m \geq 1$, and assumption (1) implies the existence of a $w \in V$ with $\alpha^m(w)w^{-1} = v$. Then $\alpha^m(w^{-1}y) = w^{-1}y$, and $u = w^{-1}yV$. By assumption (3), there exists an α-periodic point $z \in V$ with $\delta(w^{-1}yz, x) < \varepsilon$, and $w^{-1}yz \in \mathrm{Per}(\alpha)$. This shows that $\mathrm{Per}(\alpha)$ is dense in X. $\square$

LEMMA 11.5. *Let G be a compact group, $H \subset G \times G$ a full subgroup, $Y = Y_H$ the Markov subgroup defined by (9.4), and let $\alpha = \sigma^{Y_H}$ be the shift on Y_H. Then $\mathrm{Per}(\alpha^{Y_k})$ is dense in Y_k for every $k \geq 1$, where $Y_k \subset Y$ is defined in Proposition 10.2. If G is a compact Lie group such that $C(G^\circ)$ is finite, and if α is ergodic, then $\mathrm{Per}(\alpha)$ is dense in Y_H.*

PROOF. We use the notation of Proposition 10.2. According to Proposition 10.2 (6), the sequence $\{\mathbf{1}_Y\} = Y_0 \subset Y_1 \subset \cdots \subset Y_K = Y$ of closed, normal α-invariant subgroups of Y satisfies that $Y_k/Y_{k-1} \cong \Lambda_k^{\mathbb{Z}}$ for every $k = 1, \ldots, K$. Since $\alpha^{Y_k/Y_{k-1}}$ is the shift on Y_k/Y_{k-1}, $\mathrm{Per}(\alpha^{Y_k/Y_{k-1}})$ is dense in Y_k/Y_{k-1} for every k, and Lemma 10.9 and the ergodicity of α^{Y_k} imply that there exists, for every $k = 1, \ldots, K$, every $y \in Y_k$, and every $s \geq 1$, an element $z \in Y_k$ with $\alpha^s(z)z^{-1} = y$. Repeated application of Lemma 11.4 shows that $\mathrm{Per}(\alpha^{Y_k})$ is dense in Y_k for every $k \geq 1$. If G is a compact Lie group such that $C(G^\circ)$ is finite, and if α is ergodic, then $Y_H = Y_K$ by Proposition 10.2 (6), so that $\mathrm{Per}(\alpha)$ is dense in $Y = Y_K$. $\square$

LEMMA 11.6. *Let G be a compact group such that $C(G^\circ)$ is finite, let $H \subset G \times G$ be a full subgroup, and let $Y = Y_H$ be the Markov subgroup defined by (9.4). The following conditions are equivalent.*

   (1) *The set of shift-periodic points is dense in Y_H;*
   (2) *The automorphism β of G_K in Proposition 10.2 (4) has finite order.*

PROOF. We use the notation of Proposition 10.2. The restriction of the shift σ to Y_K is ergodic by Proposition 10.2 (6), and Lemma 10.9 implies that the shift σ satisfies condition (1) in Lemma 11.4 (with σ and Y_H replacing α and V). Lemma 11.5 shows that α^{Y_K} satisfies condition (2) of Lemma 11.4, and condition (3) is trivially satisfied if β has finite order. This proves that $(2) \Rightarrow (1)$. If β has infinite order, the finiteness of $\mathrm{Inn}(G_K)$ in $\mathrm{Aut}(G_K)$ implies that there exists an $m \geq 1$ such that $\beta^m(g) = hgh^{-1}$ for some element $h \in G_K$ of infinite order. We denote by A the connected component of the identity in the closure of $\{h^n : n \in \mathbb{Z}\}$. Every β-periodic $g \in G_K$ must commute with A, and hence the closure P of $\mathrm{Per}(\beta)$ is not equal to G_K, since $A \cap C(G_K) = \{1_G\}$. Since $yY_K \in \mathrm{Per}(\beta) \subset P$ for every α-periodic point $y \in Y$, $\mathrm{Per}(\alpha)$ cannot be dense. This proves that $(1) \Rightarrow (2)$. $\square$

LEMMA 11.7. *Let α be an expansive automorphism of a compact, zero dimensional group X, and let δ be a metric on X. Then $\mathrm{Per}(\alpha)$ is dense in X. Furthermore, if α is ergodic, there exists, for every $\varepsilon > 0$, an integer $N(\varepsilon) \geq 1$ such that $\mathrm{Fix}(\alpha^n)$ is ε-dense in X for every $n \geq N(\varepsilon)$ (a set $B \subset X$ is ε-dense if $\delta(x, B) < \varepsilon$ for every $x \in X$).*

PROOF. By Lemma 9.9 and Corollary 2.3 we may assume that $X = Y_H \subset G^{\mathbb{Z}}$, where G is a finite group, $H \subset G \times G$ a full subgroup, and Y_H the Markov subgroup defined in (9.4), and that α is the shift on $X = Y_H$. We use the notation of Proposition 10.2 and note that $X/Y_K \cong G/F_H(K) = G_K$ is finite, and that α^{Y_K} is ergodic. Since the automorphism $\beta = \alpha^{X/Y_K}$ is obviously of finite order, Lemma 11.6 shows that $\mathrm{Per}(\alpha)$ is dense in X. If α is ergodic, $X = Y_K$, and $F_H(K) = G$. Hence we can find, for every $s \geq 1$ and $x = (x_n) \in X$, a point $y \in \mathrm{Fix}(\alpha^{2s+2K})$ such that $x_n = y_n$ for $|n| \leq s$, and $y_{s+K} = y_{-s-K} = 1_G$. By choosing t sufficiently large we can ensure that $\mathrm{Fix}(\alpha^m)$ is ε-dense in X whenever $m \geq 2t + 2K = N(\varepsilon)$. $\square$

LEMMA 11.8. *Let α be an automorphism of a compact, zero dimensional group, $V \subset X$ be a closed, normal, α-invariant subgroup, and let $\eta \colon X \longmapsto X/V$ be the quotient map. If α^V is ergodic, then $\eta(\mathrm{Fix}(\alpha^m)) = \mathrm{Fix}((\alpha^{X/V})^m)$ for every $m \geq 1$.*

PROOF. Let $u \in \mathrm{Fix}(\alpha^m)$, and choose $x \in X$ with $u = \eta(x)$. Then $v = \alpha^m(x)x^{-1} \in V$, and Lemma 10.9 guarantees the existence of a point $w \in V$ with $v = \alpha^m(w)w^{-1}$. Then $w^{-1}v \in \mathrm{Fix}(\alpha^m)$, and $\eta(w^{-1}v) = u$. $\square$

LEMMA 11.9. *Let α be an expansive $\mathbb{Z}^d$-action by automorphisms of a compact, zero dimensional group X, $V \subset X$ a closed, normal, α-invariant subgroup, and let $\mathbf{n} \in \mathbb{Z}^d$ be an element such that $\alpha_{\mathbf{n}}^V$ is non-ergodic on V. Then there exists a closed, normal, α-invariant subgroup $W \subsetneq V$ such that W is normal in X, $\alpha_{\mathbf{n}}^W$ is ergodic on W, and $\alpha_{\mathbf{n}}^{V/W}$ has finite order.*

PROOF. Lemma 1.3 implies the existence of a closed, α-invariant subgroup $Y \subset V$, which is normal in X, such that $\alpha_{\mathbf{n}}^Y$ is ergodic on Y, and $F = V/Y$ is finite (Lemma 1.3 states that V/Y is a Lie group; however, since V is zero dimensional, V/Y must be finite). We write $\eta \colon X \longmapsto X/Y$ for the quotient map, denote by $\eta' \colon V \longmapsto F$ the restriction of η to V, define a continuous homomorphism $\boldsymbol{\eta} \colon X \longmapsto (X/Y)^{\mathbb{Z}^d}$ by setting $(\boldsymbol{\eta}(x))_{\mathbf{m}} = \eta(\alpha_{\mathbf{m}}(x))$, $x \in X, \mathbf{m} \in \mathbb{Z}^d$, and write $\boldsymbol{\eta}' \colon V \longmapsto F^{\mathbb{Z}^d}$ for the restriction of $\boldsymbol{\eta}$ to V. Then $\boldsymbol{\eta} \cdot \alpha_{\mathbf{m}} = \sigma_{\mathbf{m}} \cdot \boldsymbol{\eta}$ for every $\mathbf{m} \in \mathbb{Z}^d$, where $\sigma_{\mathbf{m}}$ is the shift on $(X/Y)^{\mathbb{Z}^d}$, and

$$\alpha_{\mathbf{n}}^F \cdot \eta(\alpha_{\mathbf{m}}(v)) = \eta(\alpha_{\mathbf{n}+\mathbf{m}}(v)) = (\sigma_{\mathbf{n}} \cdot \boldsymbol{\eta}'(v))_{\mathbf{m}} \tag{11.2}$$

for every $v \in V$ and $\mathbf{m} \in \mathbb{Z}^d$. Equation (11.2) implies that $\sigma_{\mathbf{n}}^{\boldsymbol{\eta}'(V)}$ has finite order. The proof is completed by setting $W = \ker(\boldsymbol{\eta}) \cap V$ and by noting that $W \subset V$ is a closed, normal, α-invariant subgroup of X, and that $\alpha_{\mathbf{n}}^{V/W}$ is conjugate to $\sigma_{\mathbf{n}}^{\boldsymbol{\eta}'(V)}$ and hence of finite order. $\square$

For any compact group X, let

$$r(x) = \begin{cases} \min\{r \geq 1 : x^r = \mathbf{1}_X \text{ for all } x \in X\} & \text{if this set is non-empty,} \\ \infty & \text{otherwise.} \end{cases}$$

If X is a compact, zero dimensional group carrying an expansive $\mathbb{Z}^d$-action α, then $r(X) < \infty$ by Corollary 2.3.

LEMMA 11.10. *Let X be a compact, zero dimensional group with $r(X) < \infty$, and let α be an automorphism of X with the following property: there exists an integer $Q \geq 1$ and closed, normal, α-invariant subgroups $X \supset V = V(0) \supset V(1) \supset \cdots \supset V(n) \supset \cdots$ such that $\bigcap_{n \geq 1} V(n) = \{\mathbf{1}_X\}$ and, for every $n \geq 0$, $\alpha^{V(k)/V(k+1)}$ is either ergodic, or of order Q. Let $p \geq 2$ be an integer with $\gcd\{p, Qr(X)\} = 1$, $\eta \colon X \longmapsto X/V$ the quotient map, and let $\beta = \alpha^{X/V}$. Then*

$$\eta(\mathrm{Fix}(\alpha^p)) \supset \{\beta(u)u^{-1} : u \in \mathrm{Fix}(\beta^p)\}. \tag{11.3}$$

PROOF. Let $\beta_l = \alpha^{X/V(l)}$, and let $\eta_l \colon X \longmapsto X/V(l)$ and $\eta_{k,l} \colon X/V(k) \longmapsto X/V(l)$ be the quotient maps, where $0 \leq l < k$. We fix $u \in \mathrm{Fix}(\beta^p) \subset X/V$, choose $x(0) \in X$ such that $\eta(x(0)) = \eta_0(x(0)) = u$, and construct inductively points $x(k) \in X$, $k \geq 1$, such that

$$\eta_l(x(k)) = \eta_l(x(l)) \tag{11.4}$$

and

$$\beta_k^{Qr(X)^k}(u(k))u(k)^{-1} \in \text{Fix}(\beta_k^p) \tag{11.5}$$

whenever $0 \le k \le l$, where $u(k) = \eta_k(x(k))$. Suppose that we have found $x(0), \ldots, x(j)$, $j \ge 0$, with the required properties.

If $\alpha^{V(j)/V(j+1)}$ is ergodic we apply Lemma 11.8 and choose a point $w \in \text{Fix}(\beta_{j+1}^p) \subset X/V(j+1)$ such that $\eta_{j+1,j}(w) = \beta_j^{Qr(X)^j}(u(j))u(j)^{-1}$. Then

$$w = \beta_{j+1}^{Qr(X)^j}(\eta_{j+1}(x(j)))v\eta_{j+1}(x(j))^{-1}$$

for some $v \in V(j)/V(j+1)$, and Lemma 10.9 allows us to find a $v' \in V(j)/V(j+1)$ with $v = \beta_{j+1}^{Qr(X)^j}(v')v'^{-1} = (\alpha^{V(j)/V(j+1)})^{Qr(X)^j}(v')v'^{-1}$. We choose $x(j+1) \in X$ such that $u(j+1) = \eta_{j+1}(x(j+1)) = \eta_{j+1}(x(j))v'$ and note that $\eta_k(x(j+1)) = \eta_k(x(j))$ for $0 \le k \le j$,

$$\beta_{j+1}^{Qr(X)^j}(u(j+1))u(j+1)^{-1} = w \in \text{Fix}(\beta_{j+1}^p),$$

and hence

$$\beta_{j+1}^{Qr(X)^{j+1}}(u(j+1))u(j+1)^{-1} \in \text{Fix}(\beta_{j+1}^p).$$

If $\alpha^{V(j)/V(j+1)}$ has order Q we put $u' = \eta_{j+1}(x(j))$ and observe that

$$\beta_{j+1}^p(\beta_{j+1}^{Qr(X)^j}(u')u'^{-1}) = \beta_{j+1}^{Qr(X)^j}(u')vu'^{-1}$$

for some $v \in V(j)/V(j+1)$. Hence

$$\beta_{j+1}^{Qr(X)^j}(u'^{-1}\beta_{j+1}^p(u')) = vu'^{-1}\beta_{j+1}^p(u'),$$

and

$$\beta_{j+1}^{Qir(X)^j}(u'^{-1}\beta_{j+1}(u')) = v^i u'^{-1}\beta_{j+1}^p(u')$$

for every $i \ge 1$, since $\beta_{j+1}^{Qr(X)^j}(v) = \beta_{j+1}^Q(v) = (\alpha^{V(j)/V(j+1)})^Q(v)$. By setting $i = r(X)$ we see that

$$\beta_{j+1}^{Qr(X)^{j+1}}(u'^{-1}\beta_{j+1}^p(u')) = u'^{-1}\beta_{j+1}^p(u')$$

or, equivalently, that

$$\beta_{j+1}^p(\beta_{j+1}^{Qr(X)^{j+1}}(u')u'^{-1}) = \beta_{j+1}^{Qr(X)^{j+1}}(u')u'^{-1}.$$

In other words, the point $x(j+1) = x(j)$ satisfies (11.4)–(11.5).

This induction process yields, for every $k \ge 1$, a point $x(k) \in X$ satisfying (11.4)–(11.5). We fix k for the moment and note that there exist integers a, b with $aQr(X)^k + bp = 1$, since $\gcd\{p, Qr(X)\} = 1$. Then

$$\beta_k^{aQr(X)^k}(u(k))u(k)^{-1} = \eta_k(\alpha^{aQr(X)^k}(x(k))x(k)^{-1}) \in \text{Fix}(\beta_k^p),$$

and

$$\eta(\alpha^{aQr(X)^k}(x(k))x(k)^{-1}) = \beta^{aQr(X)^k}(u)u^{-1} = \beta^{aQr(X)^k+bp}(u)u^{-1} = \beta(u)u^{-1},$$

since $\eta(x(k)) = \eta_0(x(k)) = \eta_0(x(0)) = u$ and $\beta^p(u) = u$. We have proved that the closed set

$$S(k) = \{x \in X : \eta_k(x) \in \text{Fix}(\beta_k^p) \text{ and } \eta(x) = \beta(u)u^{-1}\}$$

is non-empty for every $k \geq 1$. As X is compact and $S(1) \supset S(2) \supset \cdots \supset S(k) \supset \cdots$, $S = \bigcap_{k \geq 1} S(k) \neq \varnothing$, and $S \subset \text{Fix}(\alpha^p) \cap \eta^{-1}(\beta(u)u^{-1})$. This proves (11.3), since $u \in \text{Fix}(\beta^p)$ was arbitrary. $\square$

LEMMA 11.11. *Let α be an expansive $\mathbb{Z}^d$-action by automorphisms of a compact, zero dimensional group X. Then the set of α-periodic points is dense in X.*

PROOF. (A) INITIAL REDUCTION. According to Corollary 2.3 we may assume that α is the shift-action of $\mathbb{Z}^d$ on a closed, shift-invariant subgroup of $G^{\mathbb{Z}^d}$, where G is a finite group. If $d = 1$, and if δ is a metric on $X \subset G^{\mathbb{Z}}$, Lemma 11.7 shows that the set of α-periodic points is dense in X, and that there exists, for every $\varepsilon > 0$, an $N(\varepsilon) \geq 1$ such that $\text{Fix}(\alpha_n)$ is ε-dense in X for all $n \geq N(\varepsilon)$.

We shall use induction on d and assume that, for some $d \geq 1$, every finite group G, and every closed, shift-invariant subgroup $X \subset G^{\mathbb{Z}^d}$, the following conditions are satisfied.

(1) The set of σ-periodic points is dense in X, where σ is the shift-action of $\mathbb{Z}^d$ on X;
(2) If δ is a metric on X, $\varepsilon > 0$, and if $\sigma_{\mathbf{e}^{(d)}}$ is ergodic on X, where $\mathbf{e}^{(d)} = (0,\ldots,0,1) \in \mathbb{Z}^d$, then there exists an integer $L \geq 1$ such that $\text{Fix}(\sigma_{k\mathbf{e}^{(d)}})$ is ε-dense for all sufficiently large $k \geq 1$ with $\gcd\{k,L\} = 1$.

Suppose that we can prove the following for every finite group G, every closed, shift-invariant subgroup $X \subset G^{\mathbb{Z}^{d+1}}$, every metric δ on X, and every $\varepsilon > 0$.

(3) $\text{Per}(\sigma_{\mathbf{e}^{(d+1)}})$ is dense in X, where $\mathbf{e}^{(d+1)} = (0,\ldots,0,1) \in \mathbb{Z}^{d+1}$;
(4) If $\sigma_{\mathbf{e}^{(d+1)}}$ is ergodic there exists an integer $L \geq 1$ such that $\text{Fix}(\sigma_{k\mathbf{e}^{(d+1)}})$ is ε-dense in X for all sufficiently large $k \geq 1$ with $\gcd\{k,L\} = 1$.

We set $\Gamma = \{\mathbf{m} = (m_1,\ldots,m_{d+1}) \in \mathbb{Z}^{d+1} : m_{d+1} = 0\} \cong \mathbb{Z}^d$ and write $\beta \colon \mathbf{n} \mapsto \sigma_{\mathbf{n}}, \mathbf{n} \in \Gamma$ for the restriction of the shift-action σ of $\mathbb{Z}^{d+1}$ to Γ. For every $k \geq 1$, the set $Y(k) = \text{Fix}(\sigma_{k\mathbf{m}})$ is a closed, shift-invariant subgroup of X, and $\beta^{Y(k)}$ satisfies the d.c.c. Our induction hypothesis (1) implies that the set of β-periodic points is dense in $Y(k)$ for every $k \geq 1$, and we conclude that the $\mathbb{Z}^{d+1}$-action σ on X satisfies (1). Hence (1) and (2) are satisfied with $d+1$ replacing d, and the lemma is proved for every $d \geq 1$.

(B) THE PROOF OF (3) AND (4) IN THE ERGODIC CASE. Assume that the automorphism $\sigma_{\mathbf{e}(d+1)}$ is ergodic. Remark 4.15 allows us to assume that

$$X = \{x \in G^{\mathbb{Z}^{d+1}} : \pi_{\{0,1\}^{d+1}}(\sigma_{\mathbf{n}}(x)) \in \pi_{\{0,1\}^{d+1}}(X)\}, \tag{11.6}$$

and we set

$$\Delta = \{\mathbf{n} = (n_1, \ldots, n_{d+1}) \in \mathbb{Z}^{d+1} : n_1 = 0\} \cong \mathbb{Z}^d. \tag{11.7}$$

For every $k \geq 0$ we put $\Delta(k) = \{\mathbf{n} = (n_1, \ldots, n_{d+1}) \in \mathbb{Z}^{d+1} : |n_1| \leq k\}$, $Y(k) = \pi_{\Delta(k)}(X) \subset G^{\Delta(k)}$, and $W(k) = \ker(\pi_{\Delta(k)}) \subset X$. For any subsets $Z' \subset Z \subset \mathbb{Z}^{d+1}$ we shall not distinguish notationally between the projection $\pi_{Z'}: X \longmapsto G^{Z'}$ and the projection $\pi_{Z'}: \pi_Z(X) \longmapsto G^{Z'}$ induced by $\pi_{Z'}$, and we write $\alpha: \mathbf{m} \mapsto \alpha_{\mathbf{m}} = \sigma_{\mathbf{m}}, \mathbf{m} \in \Delta$, for the restriction of the shift-action σ of $\mathbb{Z}^{d+1}$ on X to Δ. From Lemma 4.3 it is clear that $\alpha^{Y(k)}$ satisfies the d.c.c. for every $k \geq 0$. According to (11.6), the Δ-actions $\alpha^{W(k)/W(k+1)}$ are all algebraically conjugate, and Lemma 11.9, applied to the action $\alpha^{W(0)/W(1)}$ of $\Delta \cong \mathbb{Z}^d$ on $W(0)/W(1)$, and to $\mathbf{n} = \mathbf{e}^{(d+1)} \in \Delta$, allows us to find a closed, normal, α-invariant subgroup $V(1) \subset X$ with $W(1) \subset V(1) \subset V(0) = W(0)$, such that $\alpha_{\mathbf{e}(d+1)}$ is ergodic on $V(1)/W(1)$ and has finite order Q on $V(0)/V(1)$. By applying this argument repeatedly to the inclusions

$$W(0) \supset W(1) \supset \cdots \supset W(k) \supset \cdots$$

we obtain a sequence

$$W(0) = V(0) \supset V(1) \supset W(1) = V(2) \supset \cdots \supset W(k-1)$$
$$= V(2k-2) \supset V(2k-1) \supset W(k) = V(2k) \supset \cdots$$

of closed, normal, α-invariant subgroups of X such that

$$\alpha_{\mathbf{e}(d+1)}^{V(k)/V(k+1)}$$

is conjugate to

$$\alpha_{\mathbf{e}(d+1)}^{V(k+2)/V(k+3)}$$

for every $k \geq 0$. We can thus apply Lemma 11.10 and conclude that

$$\pi_{\Delta(k)}(\{x \in X : \alpha_{p\mathbf{e}(d+1)}(x) = x\})$$
$$\supset \{\alpha_{\mathbf{e}(d+1)}^{Y(k)}(u)u^{-1} : u \in Y(k) \text{ and } \alpha_{p\mathbf{e}(d+1)}^{Y(k)}(u) = u\} \tag{11.8}$$

for all $k \geq 0$ and all $p \geq 2$ with $\gcd\{p, Qr(X)\} = 1$ (note that $r(X) \leq r(G) < \infty$).

Let $\varepsilon > 0$, and let δ be a metric on X. There exists an integer $K \geq 1$ such that $\delta(x, x') < \varepsilon/2$ for all $x, x' \in X$ with $\pi_{\Delta(K)}(x) = \pi_{\Delta(K)}(x')$. Since $\alpha_{\mathbf{e}(d+1)}$ is ergodic on $Y(K)$, the map $\partial_{\mathbf{e}(d+1)}: u \mapsto \alpha_{\mathbf{e}(d+1)}(u)u^{-1}, u \in Y(K)$, is continuous and surjective on $Y(K)$ (the latter by Lemma 10.9). Hence there exists an $\varepsilon' > 0$ such that $\partial_{\mathbf{e}(d+1)}(B)$ is $\varepsilon/2$-dense in X for every ε'-dense set

$B \subset Y(K)$. Since the Δ-action $\alpha^{Y(K)}$ satisfies the d.c.c., and since $\Delta \cong \mathbb{Z}^d$, our induction hypothesis (2) implies that there exists an integer $L \geq 1$ such that $\{u \in Y(K) : \alpha_{n\mathbf{e}(d+1)}(u) = u\}$ is ε'-dense in $Y(K)$ for all sufficiently large $n \geq 1$ with $\gcd\{n, L\} = 1$. Then $\{\alpha_{\mathbf{e}(d+1)}(u)u^{-1} : u \in Y(K)$ and $\alpha_{n\mathbf{e}(d+1)}(u) = u\}$ is $\varepsilon/2$-dense in $Y(K)$, and our choice of K, together with (11.8), implies that $\mathrm{Fix}(\alpha_{n\mathbf{e}(d+1)}) = \mathrm{Fix}(\sigma_{n\mathbf{e}(d+1)})$ is ε-dense in X for all sufficiently large $n \geq 1$ with $\gcd\{n, LQr(X)\} = 1$. We have proved (4), hence (3), and consequently (1) and (2) if $\sigma_{\mathbf{e}(d+1)}$ is ergodic.

(C) THE PROOF OF (3) IN THE NON-ERGODIC CASE. If $\sigma_{\mathbf{e}(d+1)}$ is non-ergodic on X we apply Lemma 11.9 and find a closed, normal, σ-invariant subgroup $W \subset X$ such that $\sigma_{\mathbf{e}(d+1)}$ is ergodic on W and has finite order on X/W. From part (b) of this proof we know that the set of $\sigma_{\mathbf{e}(d+1)}$-periodic points is dense in W, and the Lemmas 10.9 and 11.4 show that the set of $\sigma_{\mathbf{e}(d+1)}$-periodic points is dense in X. The proof of the lemma is complete. $\square$

LEMMA 11.12. *Let α be a $\mathbb{Z}^d$-action by automorphisms of a compact group X satisfying the d.c.c., and assume that X° is abelian. Then there exists in increasing sequence $(X_n, n \geq 1)$ of closed, normal, zero dimensional, α-invariant subgroups of X such that $\bigcup_{n \geq 1} X_n$ is dense in X.*

PROOF. According to (4.10) we may assume that X is a full, shift-invariant subgroup of $G^{\mathbb{Z}^d}$, where G is a compact Lie group whose connected component G° of the identity is abelian, and that α is the shift-action of $\mathbb{Z}^d$ on X. Lemma 2.11 allows us to find an increasing sequence $(G_n, n \geq 1)$ of closed, normal subgroups of G such that $C(G_n^\circ)$ is finite for every $n \geq 1$, and $\bigcup_{n \geq 1} X_n$ is dense in X, where $X_n = X \cap G_n^{\mathbb{Z}^d}$. Since we are assuming that $G^\circ = C(G^\circ)$ we have that $C(G_n^\circ) = G_n^\circ$, so that G_n is finite for every $n \geq 1$. In particular, X_n is zero dimensional, and the d.c.c. implies that σ^{X_n} is expansive for every n (Corollary 3.4). $\square$

PROOF OF THEOREM 11.2. Lemmas 11.11 and 11.12. $\square$

CONCLUDING REMARKS 11.13. (1) All the material in this section is taken from [45]. Lemma 11.4 is due to [111].

(2) If α is an automorphism of a compact, zero dimensional group X (or, more generally, of a group X for which X° is abelian), then $\mathrm{Per}(\alpha)$ will generally not be dense unless α satisfies the d.c.c. However, Theorem 11.2 implies that $\mathrm{Per}(\alpha)$ is dense whenever there exists a $\mathbb{Z}^d$-action β on X satisfying the d.c.c. such that α commutes with every $\beta_{\mathbf{n}}$, $\mathbf{n} \in \mathbb{Z}^d$. The rôle of this action β is to provide a certain amount of homogeneity in X. For example, if X is zero dimensional, then the existence of such a β will guarantee that $r(X) < \infty$, where $r(X)$ was defined at the beginning of this section. In the Examples 5.6 (1)–(2), this homogeneity is obviously missing; this is particularly apparent in 5.6 (2), where the choice of the maps ψ_n in (5.12) is quite crucial. If $\psi_n = \phi_n$ for

every $n \geq 2$, then $\mathbf{1}_Y$ is the only shift-periodic point in Y. However, if one fixes $m \geq 2$ and sets $\psi_n = \phi_m$ for every $n \geq 2$ in (5.12), then the set of shift-periodic points is dense in the resulting group Y; one way of seeing this is by considering the group $X = \{x = (x_{k,l}) \in \mathbb{Z}_{/2}^{\mathbb{Z}^2} : x_{k,l} = \sum_{r=k}^{k+m-1} x_{r,l+1} \text{ for all } k, l \in \mathbb{Z}\}$, and by noting that $\pi_{\mathbb{Z} \times \mathbb{N}^*}(X) = Y$ and $\pi_{\mathbb{Z} \times \mathbb{N}^*} \cdot \sigma_{(1,0)} = \sigma \cdot \pi_{\mathbb{Z} \times \mathbb{N}^*}$, where $\sigma_{(1,0)}$ is the horizontal shift on X and σ is the shift on Y. Theorem 11.2 shows that the shift-action of $\mathbb{Z}^2$ on X has a dense set of periodic points, and hence the same must be true for σ on Y.

12. Periodic points of ergodic group automorphisms

If α is an automorphism of a compact group X whose connected component of the identity is abelian, then Theorem 11.2 and Remark 11.13 (2) show that the set of α-periodic points is dense whenever α can be embedded in a $\mathbb{Z}^d$-action β by automorphisms of X such that β satisfies the d.c.c. In this case even the set of β-periodic points is dense in X, which is a much stronger assertion than the density of the set $\mathrm{Per}(\alpha)$ of α-periodic points. The condition on X° in Theorem 11.2 cannot be dropped: the Examples 11.1 (1)–(2) provide evidence for this. However, although the $\mathbb{Z}^2$-action α in Example 11.1 (2) does not have a dense set of periodic points, the ergodic automorphism $\alpha_{(0,1)}$ obviously has a dense set of periodic points. In this section we prove that this situation is typical.

THEOREM 12.1. *Let α be an ergodic $\mathbb{Z}^d$-action by automorphisms of a compact group X which satisfies the d.c.c. Then $\mathrm{Per}(\alpha_{\mathbf{n}})$ is dense in X for every $\mathbf{n} \in \mathbb{Z}^d$ such that $\alpha_{\mathbf{n}}$ is ergodic.*

COROLLARY 12.2. *Let α be an ergodic automorphism of a compact group X. If there exists a $d \geq 1$ and a $\mathbb{Z}^d$-action β by automorphisms of X such that β satisfies the d.c.c. and $\alpha\beta_{\mathbf{n}} = \beta_{\mathbf{n}}\alpha$ for all $\mathbf{n} \in \mathbb{Z}^d$, then $\mathrm{Per}(\alpha)$ is dense in X.*

Corollary 12.2 is obvious from Theorem 12.1. The proof of Theorem 12.1 requires three lemmas.

LEMMA 12.3. *For every primitive element $\mathbf{v} = (v_1, \ldots, v_d) \in \mathbb{Z}^d$ there exists a matrix $A = (a_{ij}) \in \mathrm{SL}(d, \mathbb{Z})$ such that $a_{id} = v_i$ for $i = 1, \ldots, d$.*

PROOF. For $d = 2$, the existence of the matrix A is a trivial consequence of Euclid's algorithm: if $(v_1, v_2) \in \mathbb{Z}^2$ is non-zero and primitive, there exist $a, b \in \mathbb{Z}$ with $av_1 - bv_2 = -1$, and the matrix $A = \begin{pmatrix} b & v_1 \\ a & v_2 \end{pmatrix}$ has the required properties. If $d > 2$ we employ the argument just described and choose a matrix $A^{(1)} \in \mathrm{SL}(2, \mathbb{Z})$ such that $A^{(1)} \begin{pmatrix} v_1 \\ v_2 \end{pmatrix} = \begin{pmatrix} 0 \\ \gcd(v_{d-1}, v_d) \end{pmatrix}$. Next we choose $A^{(2)} \in \mathrm{SL}(2, \mathbb{Z})$ such that $A^{(2)} \begin{pmatrix} \gcd\{v_1, v_1\} \\ v_3 \end{pmatrix} = \begin{pmatrix} 0 \\ \gcd(v_1, v_1, v_3) \end{pmatrix}$. After $d - 1$ steps we have that $B^{(d-1)} \cdots B^{(1)}\mathbf{v} = \mathbf{e}^{(d)}$, where each $B^{(k)} \in \mathrm{SL}(d, \mathbb{Z})$ is obtained by replacing the entries $\begin{bmatrix} c_{k,k} & c_{k,k+1} \\ c_{k+1,k} & c_{k+1,k+1} \end{bmatrix}$ in the $d \times d$ identity matrix $I = (c_{ij})$ with the entries $\begin{bmatrix} a_{11}^{(k)} & a_{12}^{(k)} \\ a_{21}^{(k)} & a_{22}^{(k)} \end{bmatrix}$ of $A^{(k)}$, and where $\mathbf{e}^{(d)}$ is the last unit vector in $\mathbb{Z}^d$. By setting $A = B^{-1}$ we have proved the lemma. $\square$

LEMMA 12.4. *Let α be a $\mathbb{Z}^d$-action by automorphisms of a compact, connected group X with trivial centre, and assume that α satisfies the d.c.c. Then α satisfies the* ascending chain condition *on closed, normal, α-invariant subgroups of X, i.e. there exists, for every non-decreasing sequence $(V_n, n \geq 1)$ of closed, normal, α-invariant subgroups of X, an integer $N \geq 1$ with $V_n = V_N$ for every $n \geq N$.*

PROOF. Let $V \subset X$ be a closed, normal, α-invariant subgroup, and define $V' \subset X$ as in Proposition 2.13. Then V' is obviously α-invariant, $V \cap V' = \{1_X\}$, and $V \cdot V' = X$. Since $V'' \supset V$, $V'' \cap V' = V \cap V' = \{1_X\}$, and $V \cdot V' = V'' \cdot V' = X$, V'' must be equal to V.

If follows that, if $V \subsetneq W \subset X$ are closed, normal, α-invariant subgroups, then $W' \subsetneq V' \subset X$. If $(V_n, n \geq 1)$ is a strictly increasing sequence of closed, normal, α-invariant subgroups of X, then $(V_n', n \geq 1)$ is a strictly decreasing sequence, in violation of the d.c.c. $\square$

LEMMA 12.5. *Let G be a compact, connected Lie group, $d \geq 1$, and let $X \subset G^{\mathbb{Z}^d}$ be a closed, connected, shift-invariant subgroup. If σ is the shift-action of $\mathbb{Z}^d$ on X, then $\mathrm{Per}(\sigma_\mathbf{n})$ is dense in X for every primitive $\mathbf{n} \in \mathbb{Z}^d$ such that $\sigma_\mathbf{n}$ is ergodic.*

PROOF. According to Lemma 12.3 there exists a matrix $A \in \mathrm{SL}(d, \mathbb{Z})$ such that $A\mathbf{n} = \mathbf{e}^{(d)} = (0, \ldots, 0, 1) \in \mathbb{Z}^d$. The automorphism ξ of $G^{\mathbb{Z}^d}$ given by $(\xi(x))_\mathbf{m} = x_{A^{-1}\mathbf{m}}$ for every $x = (x_\mathbf{m}) \in G^{\mathbb{Z}^d}$ and $\mathbf{m} \in \mathbb{Z}^d$ satisfies that $\xi \cdot \sigma_{\mathbf{e}^{(d)}} \cdot \xi = \sigma_\mathbf{n}$, $\sigma_{\mathbf{e}^{(d)}}$ is ergodic on $\xi(X)$ if and only if $\sigma_\mathbf{n}$ is ergodic on X, and the set of $\sigma_\mathbf{n}$-periodic points is dense in X if and only if the set of $\sigma_{\mathbf{e}^{(d)}}$-periodic points is dense in $\xi(X)$. This allows us to assume without loss in generality that $\mathbf{n} = \mathbf{e}^{(d)}$.

We assume without loss in generality that X is full in $G^{\mathbb{Z}^d}$, write $\theta \colon G \longmapsto G' = G/C(G)$ for the quotient map, and define a shift commuting homomorphism $\boldsymbol{\theta} \colon X \longmapsto G'^{\mathbb{Z}^d}$ by $(\boldsymbol{\theta}(x))_\mathbf{n} = \theta(x_\mathbf{n})$ for every $x \in X$ and $\mathbf{n} \in \mathbb{Z}^d$. Remark 4.15 allows us to change G, if necessary, and to assume that

$$X = \{x \in G^{\mathbb{Z}^d} : \pi_{\{0,1\}^d}(\sigma_\mathbf{n}(x)) \in \pi_{\{0,1\}^d}(X) \text{ for every } \mathbf{n} \in \mathbb{Z}^d\}, \quad (12.1)$$

and

$$X' = \boldsymbol{\theta}(X) = \{x \in G'^{\mathbb{Z}^d} : \pi_{\{0,1\}^d}(\sigma_\mathbf{n}(x)) \in \pi_{\{0,1\}^d}(X') \\ \text{for every } \mathbf{n} \in \mathbb{Z}^d\}. \quad (12.2)$$

Let $\Delta = \{\mathbf{m} = (m_1, \ldots, m_d) \in \mathbb{Z}^d : m_d = 0\}$, and denote by α and α' the shift-actions of $\Delta \cong \mathbb{Z}^{d-1}$ on G^Δ and G'^Δ, respectively. We set $W = \pi_\Delta(X) \subset$

G^Δ, $W' = \pi_\Delta(X') \subset G'^\Delta$, and define continuous, injective homomorphism $\kappa: X \longmapsto W^{\mathbb{Z}}$, $\kappa': X' \longmapsto W'^{\mathbb{Z}}$, by $(\kappa(x))_n = \pi_\Delta(\sigma_{n\mathbf{e}(d)}(x))$, $x \in X$, and $(\kappa'(x))_n = \pi_\Delta(\sigma_{n\mathbf{e}(d)}(x))$, $x \in X'$. Note that $\kappa \cdot \sigma_{\mathbf{e}(d)} = \sigma \cdot \kappa$ and $\kappa' \cdot \sigma_{\mathbf{e}(d)} = \sigma' \cdot \kappa'$, where σ and σ' is the shifts on $Y = \kappa(X)$ and $Y' = \kappa'(X')$. By (12.2), Y and Y' are Markov subgroups of the form $Y = Y_H \subset W^{\mathbb{Z}}$ and $Y' = Y_{H'} \subset W'^{\mathbb{Z}}$, where $H = \pi_{\{0,1\}}(\kappa(X))$ and $H' = \pi_{\{0,1\}}(\kappa'(X'))$ (cf. (9.4)). Let $F_H(k), F_{H'}(k)$, $k \in \mathbb{Z}$, be given as in (10.1)–(10.3) with W, H and W', H' replacing G, H, and note that $F_H(k)$, $k \geq 1$, and $F_{H'}(k)$, $k \geq 1$, are non-decreasing sequences of closed, normal, α-invariant (resp. α'-invariant) subgroups of W and W'. From (12.1)–(12.2) we know that $F_{H'}(k)$ is the image of $F_H(k)$ under the quotient map $\theta': W \longmapsto W'$ for every $k \in \mathbb{Z}$. Lemma 12.4 shows that there exists a $K \geq 1$ with $F_{H'}(k) = F_{H'}(K)$ for all $k \geq K$, and we claim that $F_{H'}(K) = W'$.

Indeed, if $D \subset \Delta$ is a finite set, then $\pi_D(W') \subset G'^D$ is a full, connected subgroup, and hence a compact, connected Lie group with trivial centre (Proposition 2.13). Define a continuous, shift commuting homomorphism $\psi: Y_{H'} \longmapsto \pi_D(W')^{\mathbb{Z}}$ by $(\psi(y))_n = \pi_D(y_n)$, $y \in Y_{H'}, n \in \mathbb{Z}$, and note that the shift σ must be ergodic on $\psi(Y_{H'})$. For every $k \geq 0$ we put $F'(k) = \pi_D(F_{H'}(k))$, and we assert that there exists an $L \geq 1$ with $F'(L) = \pi_D(W')$: write $\pi_D(W')^{\mathbb{Z}}$ as a Markov subgroup $Y_{H''} \subset F^{\mathbb{Z}}$, where $F = \pi_{\{0,\dots,s-1\}}(\psi(Y_{H'})) \subset (G'^D)^s$ is full and hence a compact Lie group with trivial centre (cf. Lemma 9.9), apply Proposition 10.2 (6) to obtain an $L \geq 1$ such that the L-th follower set $F_{H''}(L)$ of the identity is equal to F, and use the fullness of F to see that $F'(L) = \pi_D(W)$. Hence $\pi_D(F_{H'}(K)) = \pi_D(F_{H'}(L)) = \pi_D(W')$ for every finite set $D \subset \Delta$, which implies that $F_{H'}(K) = W'$.

Since $\theta'(F_H(K)) = F_{H'}(K) = W'$ and $W' = W/C(W)$ we conclude that $W/F_H(K) \cong C(W)/(C(W) \cap F_H(K))$ is abelian. In the notation of Proposition 10.2 this means that $\boldsymbol{\eta}_K(Y)$ is abelian, and the shift on $\boldsymbol{\eta}_K(Y)$ is obviously again ergodic. Lemma 11.5 shows that the set of shift-periodic points is dense in $Y_K = \ker(\boldsymbol{\eta}_K)$, and Lemma 10.9 implies that there exists, for every $y \in Y_K$ and every $s \geq 1$, a $w \in Y_K$ with $y = \sigma^s(w)w^{-1}$. Furthermore, $\kappa^{-1}(Y_K)$ is a closed, normal, shift-invariant subgroup of X such that $X/\kappa^{-1}(Y_K)$ is abelian, $\sigma^{X/\kappa^{-1}(Y_K)}$ satisfies the d.c.c., and Theorem 11.2 shows that the set of $\sigma^{X/\kappa^{-1}(Y_K)}$-periodic points—and hence the set of $\sigma_{\mathbf{e}(d)}^{X/\kappa^{-1}(Y_K)}$-periodic points— is dense in $X/\kappa^{-1}(Y_K)$. Lemma 11.4 yields the density of $\mathrm{Per}(\sigma_{\mathbf{e}(d)})$ in X, and this completes the proof of the lemma. $\square$

PROOF OF THEOREM 12.1. By (4.10) we may assume that X is a full, shift-invariant subgroup of $G^{\mathbb{Z}^d}$, where G is a compact Lie group, and that $\alpha = \sigma$ is the shift-action of $\mathbb{Z}^d$ on X. If $\alpha_{\mathbf{n}}$ is ergodic for some $\mathbf{n} \in \mathbb{Z}^d$ then there exists a primitive element $\mathbf{n}' \in \mathbb{Z}^d$ with $l\mathbf{n}' = \mathbf{n}$ for some $l \geq 1$, $\alpha_{\mathbf{n}'}$ is again ergodic, and $\mathrm{Per}(\alpha_{\mathbf{n}'})$ is dense in X if and only if $\mathrm{Per}(\alpha_{\mathbf{n}}-)$ is dense. We may thus assume for simplicity that $\mathbf{n}$ is primitive.

By Theorem 4.11, $\alpha_{\mathbf{n}}^{X^\circ}$ is ergodic, and Proposition 4.14 allows us to assume without loss in generality that $X^\circ = X \cap (G^\circ)^{\mathbb{Z}^d}$. Lemma 12.5 shows that $\mathrm{Per}(\alpha_\mathbf{n}^{X^\circ})$ is dense in X°. Furthermore, since X/X° is zero dimensional and α^{X/X° satisfies the d.c.c., Theorem 11.2 shows that the set of α^{X/X°-periodic points—and hence the set of $\alpha_\mathbf{n}^{X/X^\circ}$-periodic points—is dense in X/X°. The ergodicity of $\alpha_\mathbf{n}^{X^\circ}$, together with Lemma 10.9, implies that the assumption of Lemma 11.4 are satisfied (with $V = X^\circ$), so that $\mathrm{Per}(\alpha_\mathbf{n})$ is dense in X. $\square$

CONCLUDING REMARK 12.6. The exposition in this section follows [45].

CHAPTER V

Entropy

13. Entropy of $\mathbb{Z}^d$-actions

In this section we establish some basic facts about entropy of $\mathbb{Z}^d$-actions. There are several ways of defining entropy, and we begin by showing that these definitions coincide.

Let Y be a compact, metrizable space, and let $T: \mathbf{m} \mapsto T_{\mathbf{m}}$ be a continuous $\mathbb{Z}^d$-action on Y, i.e. a homomorphism from $\mathbb{Z}^d$, $d \geq 1$, into the group of homeomorphisms of Y. If $\mathcal{U}$ is an open cover of Y we set $N(\mathcal{U})$ equal to the number of elements in the smallest subcover of $\mathcal{U}$. Then $\log N(\mathcal{U})$ is subadditive in the sense that $\log N(\mathcal{U} \vee \mathcal{V}) \leq \log N(\mathcal{U}) + \log N(\mathcal{V})$ for all open covers $\mathcal{U}, \mathcal{V}$ of Y. For every rectangle $Q = \prod_{j=1}^{d}\{b_j, \ldots, b_j + l_j - 1\} \subset \mathbb{Z}^d$ we set $\langle Q \rangle = \min_{j=1,\ldots,d} l_j$, and $|Q|$ is (as usual) the cardinality of Q. Put

$$h_{\text{cover}}(T) = \sup_{\mathcal{U}} h(T, \mathcal{U}), \qquad (13.1)$$

where $\mathcal{U}$ ranges over the collection of all open covers of Y, and

$$h(T, \mathcal{U}) = \lim_{\langle Q \rangle \to \infty} \frac{1}{|Q|} \log N\left(\bigvee_{\mathbf{m} \in Q} T_{-\mathbf{m}}(\mathcal{U})\right). \qquad (13.2)$$

The limit in (13.2) exists by subadditivity.

We fix a metric δ on Y. If $Q \subset \mathbb{Z}^d$ is a rectangle then a set $E \subset Y$ is (Q, δ, ε)-spanning for T if there exists, for every $y \in Y$, a $y' \in E$ with $\delta(T_{\mathbf{m}}(y), T_{\mathbf{m}}(y')) < \varepsilon$ for all $\mathbf{m} \in Q$, and E is (Q, δ, ε)-separated if there exists, for every pair $y \neq y'$ in E, an $\mathbf{m} \in Q$ with $\delta(T_{\mathbf{m}}(y), T_{\mathbf{m}}(y')) \geq \varepsilon$. Let $r_Q(\delta, \varepsilon)$ be the smallest cardinality of a (Q, δ, ε)-spanning set, $s_Q(\delta, \varepsilon)$ the largest cardinality of a (Q, δ, ε)-separated set, and put

$$
\begin{aligned}
h_{\mathrm{span}}(T) &= \lim_{\varepsilon \to 0} \limsup_{\langle Q \rangle \to \infty} \frac{1}{|Q|} \log r_Q(\delta, \varepsilon), \\
h_{\mathrm{sep}}(T) &= \lim_{\varepsilon \to 0} \limsup_{\langle Q \rangle \to \infty} \frac{1}{|Q|} \log s_Q(\delta, \varepsilon).
\end{aligned} \tag{13.3}
$$

PROPOSITION 13.1. *If T is a continuous $\mathbb{Z}^d$-action on a compact, metric space (X, δ) then*

$$
\begin{aligned}
h_{\mathrm{cover}}(T) = h_{\mathrm{span}}(T) = h_{\mathrm{sep}}(T) &= \lim_{\varepsilon \to 0} \liminf_{\langle Q \rangle \to \infty} \frac{1}{|Q|} \log r_Q(\delta, \varepsilon) \\
&= \lim_{\varepsilon \to 0} \liminf_{\langle Q \rangle \to \infty} \frac{1}{|Q|} \log s_Q(\delta, \varepsilon),
\end{aligned} \tag{13.4}
$$

and this common value (which is independent of the metric δ) is called the topological entropy $h_{\mathrm{top}}(T)$. If A is a non-singular $d \times d$-matrix with integer entries, and if T_A is the $\mathbb{Z}^d$-action $\mathbf{m} \mapsto (T_A)_{\mathbf{m}} = T_{A\mathbf{m}}$, $\mathbf{m} \in \mathbb{Z}^d$, then $h_{\mathrm{top}}(T_A) = |\det A| \cdot h_{\mathrm{top}}(T)$.

PROOF. Let $\mathcal{U}_\varepsilon$ be the open cover of X consisting of all ε-balls. Since

$$
N\left(\bigvee_{\mathbf{m} \in Q} T_{-\mathbf{m}}(\mathcal{U}_\varepsilon) \right) \le r_Q(\delta, \varepsilon) \le s_Q(\delta, \varepsilon) \le N\left(\bigvee_{\mathbf{m} \in Q} T_{-\mathbf{m}}(\mathcal{U}_{\varepsilon/2}) \right)
$$

for every $\varepsilon > 0$ and

$$
\lim_{\varepsilon \to 0} \lim_{\langle Q \rangle \to \infty} \frac{1}{|Q|} \log N\left(\bigvee_{\mathbf{m} \in Q} T_{-\mathbf{m}}(\mathcal{U}_\varepsilon) \right) = \lim_{\varepsilon \to 0} h(T, \mathcal{U}_\varepsilon) = h_{\mathrm{cover}}(T), \tag{13.5}
$$

we obtain (13.4). As the definition of h_{cover} is obviously independent of δ, the same applies to h_{top}, h_{span}, and h_{sep}.

For the final assertion we put $\bar{Q}(n) = [-n, n)^d \subset \mathbb{R}^d$, $Q(n) = \bar{Q}(n) \cap \mathbb{Z}^d$, and $Q_A(n) = A(\bar{Q}(n)) \cap \mathbb{Z}^d$ for every $n \ge 1$. For every $m, n \ge 1$, let $C(Q(m), Q_A(mn))$ be a minimal union of disjoint translates of $Q(m)$ containing $Q_A(mn)$, and let $C(Q_A(m), Q(mn))$ be a minimal union of disjoint translates of $Q_A(m)$ containing $Q(mn)$. Then $|C(Q(m), Q_A(mn))|/|Q_A(mn)| \to 1$ and $|C(Q_A(m), Q(mn))|/|Q(mn)| \to 1$ as $m \ge 1$ and $n \to \infty$.

If $\mathcal{U}$ is an open cover of X, then subadditivity implies that

$$
\begin{aligned}
h(T_A, \mathcal{U}) &= \lim_{m,n \to \infty} \frac{1}{|Q(mn)|} \log N\left(\bigvee_{\mathbf{m} \in Q(mn)} T_{-A\mathbf{m}}(\mathcal{U}) \right) \\
&\le \lim_{m \to \infty} \lim_{n \to \infty} \frac{|Q_A(mn)|}{|Q(mn)|} \frac{1}{|Q_A(mn)|} \log N\left(\bigvee_{\mathbf{m} \in Q_A(mn)} T_{-\mathbf{m}}(\mathcal{U}) \right)
\end{aligned}
$$

$$= \lim_{m \to \infty} \lim_{n \to \infty} \frac{|\det A|}{|Q_A(mn)|} \log N \left(\bigvee_{\mathbf{m} \in Q_A(mn)} T_{-\mathbf{m}}(\mathcal{U}) \right)$$

$$\leq \lim_{m \to \infty} \lim_{n \to \infty} \frac{|\det A|}{|C(Q(m), Q_A(mn))|} \log N \left(\bigvee_{\mathbf{m} \in C(Q(m), Q_A(mn))} T_{-\mathbf{m}}(\mathcal{U}) \right)$$

$$\leq \lim_{m \to \infty} \frac{|\det A|}{|Q(m)|} \log N \left(\bigvee_{\mathbf{m} \in Q(m)} T_{-\mathbf{m}}(\mathcal{U}) \right) = |\det A| \, h(T, \mathcal{U}).$$

For the reverse inequality we choose $r \geq 1$ such that $Q(r) + A\mathbb{Z}^d = \mathbb{Z}^d$, and note that $Q_A(m - 2r) \subset AQ(m) + Q(n) \subset Q_A(m + 2r)$ and $|(Q_A(m + 2r))|/|Q_A(m - 2r)| \to 1$ as $m \to \infty$. If $\mathcal{V} = \bigvee_{\mathbf{m} \in Q(r)} T_{-\mathbf{m}}(\mathcal{U})$, then

$$h(T_A, \mathcal{V}) = \lim_{m \to \infty} \frac{1}{|Q(m)|} \log N \left(\bigvee_{\mathbf{m} \in Q(m)} T_{-A\mathbf{m}}(\mathcal{V}) \right)$$

$$= \lim_{m \to \infty} \frac{1}{|Q(m)|} \log N \left(\bigvee_{\mathbf{m} \in AQ(m) + Q(r)} T_{\mathbf{m}}(\mathcal{U}) \right)$$

$$\geq \lim_{m \to \infty} \frac{1}{|Q(m)|} \log N \left(\bigvee_{\mathbf{m} \in Q_A(m-2r)} T_{\mathbf{m}}(\mathcal{U}) \right)$$

$$= \lim_{m \to \infty} \frac{|Q_A(m - 2r)|}{|Q(m)|} \frac{1}{|Q_A(m - 2r)|} \log N \left(\bigvee_{\mathbf{m} \in Q_A(m-2r)} T_{\mathbf{m}}(\mathcal{U}) \right)$$

$$\geq \lim_{m \to \infty} \lim_{n \to \infty} \frac{|\det A|}{|C(Q_A(m - 2r), Q((m - 2r)n)|}$$

$$\cdot \log N \left(\bigvee_{\mathbf{m} \in C(Q_A(m-2r), Q((m-2r)n)} T_{-\mathbf{m}}(\mathcal{U}) \right)$$

$$\geq \lim_{m \to \infty} \lim_{n \to \infty} \frac{|\det A|}{|Q((m - 2r)n)|} \log N \left(\bigvee_{\mathbf{m} \in Q((m-2r)n)} T_{-\mathbf{m}}(\mathcal{U}) \right)$$

$$= |\det A| \, h(T, \mathcal{U}).$$

By varying $\mathcal{U}$ we obtain that $h_{\text{top}}(T) = |\det A| \, h_{\text{top}}(T_A)$. $\square$

PROPOSITION 13.2. *Let T be a continuous $\mathbb{Z}^d$-action on a compact, metric space X, and let $\Delta \subset \mathbb{Z}^d$ be a subgroup of infinite index. Suppose that the restriction $T^{(\Delta)} \colon \mathbf{n} \mapsto T_{\mathbf{n}}, \mathbf{n} \in \Delta$, of T to Δ has finite topological entropy (cf. Remark 13.4). Then $h_{\text{top}}(T) = 0$.*

PROOF. We shall prove the proposition in the case where $\Delta = \{\mathbf{m} = (m_1, \dots, m_d) \in \mathbb{Z}^d : m_{k+1} = \dots = m_d = 0\}$ for some $k \in \{1, \dots, d-1\}$—the general case is only notationally more difficult. For every $m, n \geq 0$, put $Q(m, n) = \{-m, \dots, m\}^k \times \{-n, \dots, n\}^{d-k}$. Let $\mathcal{U}$ be an open cover of X, and let $\mathcal{U}_n = \bigvee_{\mathbf{m} \in Q(0,n)} T_{-\mathbf{m}}(\mathcal{U})$. Then $\lim_{m \to \infty} \frac{1}{m^k} \log N(\bigvee_{\mathbf{m} \in Q(m,0)} T_{-\mathbf{m}}(\mathcal{U}_n)) =$

$h(T^{(\Delta)}, \mathcal{U}_n) \leq h_{\text{top}}(T^{(\Delta)})$ for every $n \geq 0$, and we choose, for every $n \geq 0$, an integer $m(n) \geq n$ such that $\frac{1}{m^k} \log N(\bigvee_{\mathbf{m} \in Q(m,0)} T_{-\mathbf{m}}(\mathcal{U}_n)) \leq 2h_{\text{top}}(T^{(\Delta)})$ for every $m \geq m(n)$. Then $\langle Q(m(n), n) \rangle \to \infty$ and

$$\frac{1}{|Q(m(n), n)|} \log N\left(\bigvee_{\mathbf{m} \in Q(M(n), n)} T_{-\mathbf{m}}(\mathcal{U}) \right)$$

$$= \frac{1}{|Q(m(n), n)|} \log N\left(\bigvee_{\mathbf{m} \in Q(m(n), 0)} T_{-\mathbf{m}}(\mathcal{U}_n) \right)$$

$$\leq \frac{2}{n^{(d-k)}} h_{\text{top}}(T^{(\Delta)}) \to 0,$$

as $n \to \infty$. Since this is true for every open cover $\mathcal{U}$ of X, $h_{\text{top}}(T) = 0$. □

If $T \colon \mathbf{m} \mapsto T_{\mathbf{m}}$ is a measure preserving action of $\mathbb{Z}^d$ on a probability space $(Y, \mathfrak{T}, \mu)$ (i.e. a homomorphism from $\mathbb{Z}^d$ into the group of measure preserving automorphisms of $(Y, \mathfrak{T}, \mu)$) we define the *metric entropy* of T by

$$h_\mu(T) = \sup_{\mathcal{P}} h_\mu(T, \mathcal{P}), \tag{13.6}$$

where $\mathcal{P}$ ranges over the finite, measurable partitions of X,

$$h_\mu(T, \mathcal{P}) = \lim_{\langle Q \rangle \to \infty} \frac{1}{|Q|} H_\mu\left(\bigvee_{\mathbf{m} \in Q} T_{-\mathbf{m}}(\mathcal{P}) \right), \tag{13.7}$$

and where $H(\mathcal{Q}) = H_\mu(\mathcal{Q}) = -\sum_{P \in \mathcal{Q}} \mu(P) \log \mu(P)$ is the entropy of a finite or countable measurable partition $\mathcal{Q}$ of X. The limit in (13.7) exists by subadditivity, and is equal to $\sup_{\mathcal{Q}} h_\mu(T, \mathcal{Q}) = \sup_{\mathcal{Q}} \lim_{\langle Q \rangle \to \infty} \frac{1}{|Q|} H_\mu(\bigvee_{\mathbf{m} \in Q} T_{-\mathbf{m}}(\mathcal{Q}))$, where $\mathcal{Q}$ ranges over the countable, measurable partitions of Y with finite entropy.

We recall the notions of conditional information and conditional entropy. If $\mathcal{P}$ is a countable, measurable partition of Y and $\mathfrak{S} \subset \mathfrak{T}$ a sigma-algebra, then the *conditional information function* $I_\mu(\mathcal{P}|\mathfrak{S})$ is defined by

$$I_\mu(\mathcal{P}|\mathfrak{S}) = -\sum_{P \in \mathcal{P}} 1_P \log E_\mu(1_P|\mathfrak{S}),$$

where 1_P denotes the indicator function of P and $E_\mu(\cdot|\mathfrak{S})$ is the conditional expectation with respect to $\mathfrak{S}$, and

$$H_\mu(\mathcal{P}|\mathfrak{S}) = \int I_\mu(\mathcal{P}|\mathfrak{S}) \, d\mu$$

is the *conditional entropy* of $\mathcal{P}$ with respect to $\mathfrak{S}$. If $\mathcal{Q}$ is a second countable, measurable partition of Y we write $I_\mu(\mathcal{P}|\mathcal{Q})$ and $H_\mu(\mathcal{P}|\mathcal{Q})$ instead of $I_\mu(\mathcal{P}|\Sigma(\mathcal{Q}))$ and $H_\mu(\mathcal{P}|\Sigma(\mathcal{Q}))$, where $\Sigma(\mathcal{A})$ is the sigma-algebra generated by a family of sets $\mathcal{A}$. In the special case where $\mathfrak{S}$ is the trivial sigma-algebra $\{\varnothing, Y\}$ we use the notation $I_\mu(\mathcal{P}) = I_\mu(\mathcal{P}|\mathfrak{S})$ and note that $H_\mu(\mathcal{P}|\mathfrak{S}) = H_\mu(\mathcal{P})$.

It is sometimes convenient to use another, equivalent definition of metric entropy. Let $\prec$ be the lexicographic order on $\mathbb{Z}^d$. For any countable, measurable partition $\mathcal{P}$ of Y with finite entropy we set $\mathcal{P}_T^- = \bigvee_{\mathbf{m} \prec \mathbf{0}_{\mathbb{Z}^d}} T_{-\mathbf{m}}(\mathcal{P}) = \Sigma(\bigcup_{\mathbf{m} \prec \mathbf{0}_{\mathbb{Z}^d}} T_{-\mathbf{m}}(\mathcal{P}))$. Then

$$h_\mu(T, \mathcal{P}) = H_\mu(\mathcal{P}|\mathcal{P}_T^-) = \int_Y I_\mu(\mathcal{P}|\mathcal{P}_T^-) \, d\mu, \tag{13.8}$$

and hence

$$h_\mu(T) = \sup_{\mathcal{P}} H_\mu(\mathcal{P}|\mathcal{P}_T^-). \tag{13.9}$$

If a countable, measurable partition $\mathcal{P}$ of Y is a *generator* for T, i.e. if the sigma-algebra

$$\Sigma\left(\bigcup_{\mathbf{m} \in \mathbb{Z}^d} T_{-\mathbf{m}}(\mathcal{P})\right) = \bigvee_{\mathbf{m} \in \mathbb{Z}^d} T_{-\mathbf{m}}(\mathcal{P})$$

generated by $\bigcup_{\mathbf{m} \in \mathbb{Z}^d} T_{-\mathbf{m}}(\mathcal{P})$ is equal to $\mathfrak{T}$, then

$$h_\mu(T) = H_\mu(\mathcal{P}|\mathcal{P}_T^-) = h_\mu(T, \mathcal{P}). \tag{13.10}$$

A sequence $(\mathcal{P}_n, \, n \geq 1)$ of countable, measurable partitions of X is *increasing* if the sigma-algebras $\Sigma(\mathcal{P}_n)$ generated by $\mathcal{P}_n$ form an increasing sequence. If $(\mathcal{P}_n, \, n \geq 1)$ is such an increasing sequence, and if the sigma-algebra

$$\Sigma\left(\bigcup_{n \geq 1} \bigcup_{\mathbf{m} \in \mathbb{Z}^d} T_{-\mathbf{m}}(\mathcal{P}_n)\right)$$

is equal to $\mathfrak{T}$, then

$$\begin{aligned}
h_\mu(T) &= \lim_{n \to \infty} H_\mu(\mathcal{P}_n|(\mathcal{P}_n)_T^-) = \sup_{n \geq 1} H_\mu(\mathcal{P}_n|(\mathcal{P}_n)_T^-) \\
&= \lim_{n \to \infty} h_\mu(T, \mathcal{P}_n) = \sup_{n \geq 1} h_\mu(T, \mathcal{P}_n).
\end{aligned} \tag{13.11}$$

For background and details we refer to [84], [77], [105], [17] and [42].

Let α be a $\mathbb{Z}^d$-action by automorphisms of a compact group X with normalized Haar measure λ_X, and let δ be a metric on X. We denote by $B_\delta(\varepsilon)$ the ε-ball around $\mathbf{1}_X$ and set

$$h_{\mathrm{vol}}(\alpha) = \lim_{\varepsilon \to 0} \limsup_{\langle Q \rangle \to \infty} -\frac{1}{|Q|} \log \lambda_X(B_\delta(Q, \varepsilon)), \tag{13.12}$$

where $B_\delta(Q, \varepsilon) = \bigcap_{\mathbf{m} \in Q} \alpha_{-\mathbf{m}}(B_\delta(\varepsilon))$.

THEOREM 13.3. *Let α be a $\mathbb{Z}^d$-action by automorphisms of a compact group X. Then*

$$h_{\mathrm{top}}(\alpha) = h_{\lambda_X}(\alpha) = h_{\mathrm{vol}}(\alpha) = \lim_{\varepsilon \to 0} \liminf_{\langle Q \rangle \to \infty} -\frac{1}{|Q|} \log \lambda_X(B_\delta(Q, \varepsilon)),$$

and this common value is denoted by $h(\alpha)$.

PROOF. Let δ be a metric on X, and let E be a (Q, δ, ε)-separated set of cardinality $s_Q(\delta, \varepsilon)$. Then the sets $xB_\delta(Q, \varepsilon)$, $x \in E$, are disjoint, and hence $s_Q(\delta, \varepsilon) \leq \lambda_X(B_\delta(Q, \varepsilon))^{-1}$. By letting $\langle Q \rangle \to \infty$ and $\varepsilon \to 0$ we see that (cf. Proposition 13.1)

$$h_{\text{top}}(\alpha) = h_{\text{sep}}(\alpha) \leq \lim_{\varepsilon \to 0} \liminf_{\langle Q \rangle \to \infty} -\frac{1}{|Q|} \log \lambda_X(B_\delta(Q, \varepsilon)) \leq h_{\text{vol}}(\alpha).$$

If $\mathcal{P}$ is a finite partition of X into Borel sets of diameter $< \varepsilon$, then every set P in the partition $\mathcal{Q} = \bigvee_{\mathbf{m} \in Q} \alpha_{-\mathbf{m}}(\mathcal{P})$ is contained in a translate of $B_\delta(Q, \varepsilon)$, and hence

$$H_{\lambda_X}\left(\bigvee_{\mathbf{m} \in Q} \alpha_{-\mathbf{m}}(\mathcal{P}) \right) = -\sum_{A \in \mathcal{Q}} \lambda_X(A) \log \lambda_X(A)$$

$$\geq -\sum_{A \in \mathcal{Q}} \lambda_X(A) \log \lambda_X(B_\delta(Q, \varepsilon)) = -\log \lambda_X(B_\delta(Q, \varepsilon)).$$

This proves that $h_{\text{vol}}(\alpha) \leq h_{\lambda_X}(\alpha)$.

Finally we prove that $h_{\lambda_X}(\alpha) \leq h_{\text{top}}(\alpha)$. Let $\mathcal{P} = \{P_1, \ldots, P_k\}$ be a finite, measurable partition of X, and choose ε such that $0 < \varepsilon < 1/k \log k$. Since λ_X is regular there exist compact sets $P_i' \subset P_i$, $i = 1, \ldots, k$, with $\lambda_X(P_i \smallsetminus P_i') < \varepsilon$. Let $\mathcal{P}'$ be the partition $\{P_0', P_1', \ldots, P_k'\}$, where $P_0' = X \smallsetminus \bigcup_{i=1}^k P_i'$. Then $\lambda_X(P_0') < k\varepsilon$, and the conditional entropy $H_{\lambda_X}(\mathcal{P}|\mathcal{P}')$ satisfies that

$$H_{\lambda_X}(\mathcal{P}|\mathcal{P}') = -\sum_{i=0}^k \sum_{j=1}^k \lambda_X(P_i' \cap P_j) \log(\lambda_X(P_i' \cap P_j)/\lambda_X(P_i'))$$

$$= -\lambda_X(P_0') \sum_{j=1}^k (\lambda_X(P_0' \cap P_j)/\lambda_X(P_0')) \log(\lambda_X(P_0' \cap P_j)/\lambda_X(P_0'))$$

$$\leq \lambda_X(P_0') \log k < k\varepsilon \log k < 1.$$

In this equation we have set $0 \log 0 = 0$ and used the facts that $\lambda_X(P_i' \cap P_j)/\lambda_X(P_i') \in \{0, 1\}$ for all $i, j = 1, \ldots, k$, and that any partition of X into k measurable sets has entropy $\leq \log k$. For every $i = 1, \ldots, k$, $U_i = P_0' \cup P_i'$ is an open subset of X which contains P_i, and $\mathcal{U} = \{U_1, \ldots, U_k\}$ is an open cover of X. Let $Q \subset \mathbb{Z}^d$ be a rectangle, and let N be the number of non-empty sets in the partition $\bigvee_{\mathbf{m} \in Q} \alpha_{-\mathbf{m}}(\mathcal{P}')$. Then $H_{\lambda_X}(\bigvee_{\mathbf{m} \in Q} \alpha_{-\mathbf{m}}(\mathcal{P}')) \leq \log N$. Every set $U_{i_0} \cap \cdots \cap U_{i_{n-1}}$ in $\bigvee_{\mathbf{m} \in Q} \alpha_{-\mathbf{m}}(\mathcal{U})$ is the union of at most $2^{|Q|}$ sets in $\bigvee_{\mathbf{m} \in Q} \alpha_{-\mathbf{m}}(\mathcal{P}')$, and hence

$$H_{\lambda_X}\left(\bigvee_{\mathbf{m} \in Q} \alpha_{-\mathbf{m}}(\mathcal{P}') \right) \leq \log\left(2^{|Q|} \cdot N\left(\bigvee_{\mathbf{m} \in Q} \alpha_{-\mathbf{m}}(\mathcal{U}) \right) \right).$$

By varying Q we conclude that

$$h_{\lambda_X}(\alpha, \mathcal{P}) \leq h_{\lambda_X}(\alpha, \mathcal{P}') + H_{\lambda_X}(\mathcal{P}|\mathcal{P}')$$
$$\leq h(\alpha, \mathcal{U}) + \log 2 + 1 \leq h_{\mathrm{cover}}(\alpha) + \log 2 + 1.$$

Finally we vary $\mathcal{P}$ and obtain that

$$h_{\lambda_X}(\alpha) \leq h_{\mathrm{cover}}(\alpha) + \log 2 + 1 = h_{\mathrm{top}}(\alpha) + \log 2 + 1. \qquad (13.13)$$

Since the inequality (13.13) holds for every $\mathbb{Z}^d$-action by automorphisms of X we can apply (13.13) to the action $\alpha^{(r)} \colon \mathbf{m} \mapsto \alpha_{r\mathbf{m}}$, and obtain that

$$r^d h_{\lambda_X}(\alpha) = h_{\lambda_X}(\alpha^{(r)}) \leq h_{\mathrm{top}}(\alpha^{(r)}) + \log 2 + 1 = r^d h_{\mathrm{top}}(\alpha) + \log 2 + 1$$

for every $r \geq 1$. This proves that $h_{\lambda_X}(\alpha) \leq h_{\mathrm{top}}(\alpha)$. $\qquad \square$

REMARK 13.4. Let Δ be a countable group which is isomorphic to $\mathbb{Z}^d$ for some $d \geq 1$, and let T be a continuous Δ-action on a compact, metric space X. The last assertion in Proposition 13.1 implies that $h_{\mathrm{top}}(T)$ is well-defined, since its value is independent of the specific isomorphism $\Delta \cong \mathbb{Z}^d$. Similarly one sees that, if T is a measure preserving action of Δ on a probability space $(Y, \mathfrak{T}, \mu)$, then $h_\mu(T, \mathcal{P})$ is well-defined for any measurable partition $Q \subset \mathfrak{T}$ with finite entropy. If α is an action of Δ by automorphisms of a compact group X, then Proposition 13.1 and Theorem 13.3 together show that the (unambiguously defined) entropies $h_{\mathrm{top}}(\alpha)$ and $h_{\lambda_X}(\alpha)$ coincide.

The *variational principle for topological entropy* is the statement that, for every homeomorphism T of a compact, metrizable space X, and for every T-invariant probability measure μ on the Borel sigma-algebra $\mathfrak{B}_X$ of X, $h_\mu(T) \leq h_{\mathrm{top}}(T)$ (cf. [26]). A simple proof of the variational principle for compact group automorphisms, which can easily be generalized to $\mathbb{Z}^d$-actions by automorphisms of a compact group, appears in [10]. For a general proof we refer to [73]

PROPOSITION 13.5. *Let α be a $\mathbb{Z}^d$-action by automorphisms of a compact group X, and let μ be an α-invariant probability measure on $\mathfrak{B}_X$. Then $h_\mu(\alpha) \leq h_{\lambda_X}(\alpha) = h(\alpha)$.*

PROOF. We denote by $\pi \colon X \times X \longmapsto X$ the product map $\pi(x, y) = xy$, write $\pi_i \colon X \times X \longmapsto X$, $i = 1, 2$, for the coordinate projections, and consider the sigma-algebras $\mathfrak{S} = \pi^{-1}(\mathfrak{B}_X) \subset \mathfrak{B}_{X \times X}$ and $\mathfrak{S}_i = \pi_i^{-1}(\mathfrak{B}_X) \subset \mathfrak{B}_{X \times X}$, $i = 1, 2$. Note that $\mathfrak{S}$ and $\mathfrak{S}_i$, $i = 1, 2$, are invariant under the $\mathbb{Z}^d$-action $\alpha \times \alpha \colon \mathbf{n} \mapsto \alpha_{\mathbf{n}} \times \alpha_{\mathbf{n}}$ on $X \times X$. Furthermore,

$$\mathfrak{S} \vee \mathfrak{S}_1 = \mathfrak{S} \vee \mathfrak{S}_2 = \mathfrak{B}_{X \times X}; \qquad (13.14)$$

to prove this for $i = 1$, say, it suffices to note that the maps π_1 and $\pi_2 \colon (x, y) \mapsto y = \pi_1(x)^{-1}\pi(x, y)$ are both $\mathfrak{S} \vee \mathfrak{S}_1$-measurable.

Let $M_1(X)$ and $M_1(X)^\alpha$ be the sets of probability measures and α-invariant probability measures on $\mathfrak{B}_X$, respectively. The *convolution* $\mu * \nu$ of two measures $\mu, \nu \in M_1(X)$ is defined by $\mu * \nu(B) = \int \int 1_B(xy) \, d\mu(x) d\nu(y)$ for every $B \in \mathfrak{B}_X$. Since $\mu * \nu$ is essentially the restriction of the product measure $\mu \times \nu$ on $X \times X$ to the $\alpha \times \alpha$-invariant sigma-algebra $\mathfrak{S} \subset \mathfrak{B}_{X \times X}$, it is clear that $\mu * \nu \in M_1(X)$, and that $\mu * \nu \in M_1(X)^\alpha$ whenever $\mu, \nu \in M_1(X)^\alpha$.

We choose increasing sequences $(\mathcal{P}_n, \, n \geq 1)$ and $(\mathcal{Q}_n, \, n \geq 1)$ of finite partitions in $\mathfrak{S} \subset \mathfrak{B}_{X \times X}$ and $\mathfrak{S}_1 \subset \mathfrak{B}_{X \times X}$, respectively, such that $\Sigma(\mathcal{P}_n) \nearrow \mathfrak{S}$ and $\Sigma(\mathcal{Q}_n) \nearrow \mathfrak{S}_2$, where $\Sigma(\mathcal{C})$ is the sigma-algebra generated by a family of sets $\mathcal{C}$. From (13.8) it is obvious that $h_{\mu \times \nu}(\alpha, \mathcal{P}_n \vee \mathcal{Q}_n) \leq h_{\mu \times \nu}(\alpha, \mathcal{P}_n) + h_{\mu \times \nu}(\alpha, \mathcal{Q}_n)$ for every $n \geq 1$, and (13.11) shows that $h_{\mu \times \nu}(\alpha, \mathcal{P}_n) \to h_{\mu * \nu}(\alpha)$ and $h_{\mu \times \nu}(\alpha, \mathcal{Q}_n) \to h_\nu(\alpha)$ as $n \to \infty$. According to (13.14) and (13.11), $h_{\mu \times \nu}(\alpha, \mathcal{P}_n \vee \mathcal{Q}_n) \to h_{\mu \times \nu}(\alpha) = h_\mu(\alpha) + h_\nu(\alpha)$ as $n \to \infty$, so that

$$\begin{aligned} h_\mu(\alpha) + h_\nu(\alpha) &= \lim_n h_{\mu \times \nu}(\alpha, \mathcal{P}_n \vee \mathcal{Q}_n) \\ &\leq \lim_n h_{\mu \times \nu}(\alpha, \mathcal{P}_n) + \lim_n h_{\mu \times \nu}(\alpha, \mathcal{Q}_n) \qquad (13.15) \\ &= h_{\mu * \nu}(\alpha) + h_\nu(\alpha). \end{aligned}$$

In particular, if $h_\nu(\alpha) < \infty$, then $h_\mu(\alpha) \leq h_{\mu * \nu}(\alpha)$, and by setting $\nu = \lambda_X$ and noting that $\mu * \lambda_X = \lambda_X$ we have proved the proposition. $\square$

LEMMA 13.6. *Let α be a $\mathbb{Z}^d$-action by automorphisms of a compact group X, and let $(V_n, \, n \geq 1)$ be a decreasing sequence of closed, normal, α-invariant subgroups of X such that $\bigcap_{n \geq 1} V_n = \{1_X\}$. Then $h(\alpha) = \lim_{n \to \infty} h(\alpha^{X/V_n}) = \sup_{n \geq 1} h(\alpha^{X/V_n})$.*

PROOF. According to (13.6)–(13.7), $h(\alpha) = h_{\lambda_X}(\alpha) = \sup_{\mathcal{P}} h_{\lambda_X}(T, \mathcal{P})$, where $\mathcal{P}$ ranges over the finite, measurable partitions of X. We write $\theta_n \colon X \longmapsto X/V_n$ for the quotient map. If $\mathcal{P}$ is a finite, measurable partition of X/V_n then $h(\alpha^{X/V_n}, \mathcal{P}) = h(\alpha, \theta_n^{-1}(\mathcal{P}))$, and we conclude that $h(\alpha) \geq h(\alpha^{X/V_n})$ for all $n \geq 1$. Conversely, if $\mathcal{P}$ is a finite, measurable partition of X, the increasing martingale theorem shows that $\lim_{n \to \infty} H_{\lambda_X}(\mathcal{P}|\mathfrak{B}_n) = 0$, where $\mathfrak{B}_n = \theta_n^{-1}(\mathfrak{B}_{X/Y_n})$. Hence there exists, for every $\varepsilon > 0$, an $n \geq 1$ and a finite, measurable partition $\mathcal{P}'$ of X/V_n such that $H_{\lambda_X}(\mathcal{P}|\theta_n^{-1}(\mathcal{P}')) < \varepsilon$, and (13.7) shows that $h(\alpha, \mathcal{P}) \leq h(\alpha, \theta^{-1}(\mathcal{P}')) + \varepsilon$. Hence $h(\alpha) \leq h(\alpha^{X/V_n}) + \varepsilon$ for sufficiently large n, and the lemma is proved. $\square$

The final result shows that the topological entropy of the shift-action of $\mathbb{Z}^d$ on a closed, shift-invariant subgroup $X \subset G^{\mathbb{Z}^d}$, where G is a compact group, can be calculated in terms of a maximum metric.

Let G be a compact group, and let ϑ be a metric on G. If $X \subset G^{\mathbb{Z}^d}$ is a closed, shift-invariant subgroup, $Q \subset \mathbb{Z}^d$ a rectangle, and $\varepsilon > 0$, we call a set $E \subset X$ $[Q, \vartheta, \varepsilon]$-*spanning* for the shift-action σ of $\mathbb{Z}^d$ on X if there exist, for every $x \in X$, a $y \in E$ with $\vartheta(x_{\mathbf{m}}, y_{\mathbf{m}}) < \varepsilon$ for all $\mathbf{m} \in Q$, and E is $[Q, \delta, \varepsilon]$-*separated* if there exists, for every pair $x \neq y$ in E, an $\mathbf{m} \in Q$ with $\vartheta(x_{\mathbf{m}}, y_{\mathbf{m}}) \geq \varepsilon$. Let $\bar{r}_Q(\vartheta, \varepsilon)$ and $\bar{s}_Q(\delta, \varepsilon)$ be the smallest cardinality of a $[Q, \vartheta, \varepsilon]$-spanning set and the largest cardinality of a $[Q, \vartheta, \varepsilon]$-separated set, respectively.

PROPOSITION 13.7. *The entropy of the shift-action σ of $\mathbb{Z}^d$ on the closed, shift-invariant subgroup $X \subset G^{\mathbb{Z}^d}$ satisfies that*

$$
\begin{aligned}
h(\sigma) &= \lim_{\varepsilon \to 0} \limsup_{\langle Q \rangle \to \infty} \frac{1}{|Q|} \log \bar{r}_Q(\vartheta, \varepsilon) = \lim_{\varepsilon \to 0} \liminf_{\langle Q \rangle \to \infty} \frac{1}{|Q|} \log \bar{r}_Q(\vartheta, \varepsilon) \\
&= \lim_{\varepsilon \to 0} \limsup_{\langle Q \rangle \to \infty} \frac{1}{|Q|} \log \bar{s}_Q(\vartheta, \varepsilon) = \lim_{\varepsilon \to 0} \liminf_{\langle Q \rangle \to \infty} \frac{1}{|Q|} \log \bar{s}_Q(\vartheta, \varepsilon) \\
&\qquad\qquad\qquad\qquad\qquad\qquad\qquad\qquad\qquad\qquad\qquad (13.16) \\
&= \lim_{\varepsilon \to 0} \limsup_{\langle Q \rangle \to \infty} -\frac{1}{|Q|} \log \lambda_X(B'_\vartheta(Q, \varepsilon)) \\
&= \lim_{\varepsilon \to 0} \liminf_{\langle Q \rangle \to \infty} -\frac{1}{|Q|} \log \lambda_X(B'_\vartheta(Q, \varepsilon)),
\end{aligned}
$$

where $B'_\vartheta(\varepsilon) = \{x \in X : \vartheta(x_0, 1_G) < \varepsilon\}$ and $B'_\vartheta(Q, \varepsilon) = \bigcap_{\mathbf{m} \in Q} \sigma_{-\mathbf{m}}(B'_\vartheta(\varepsilon))$. Furthermore X is expansive if and only if there exists an $\varepsilon > 0$ such that, for every $x \neq 1_X$ in X, $\vartheta(x_{\mathbf{m}}, 1_G) > \varepsilon$ for some $\mathbf{m} \in \mathbb{Z}^d$.

PROOF. Let δ be the metric on $G^{\mathbb{Z}^d}$ defined by

$$
\delta(x, y) = \sum_{\mathbf{m} \in \mathbb{Z}^d} 2^{-|\mathbf{m}|} \vartheta(x_{\mathbf{m}}, y_{\mathbf{m}}) \qquad (13.17)
$$

for all $x, y \in G^{\mathbb{Z}^d}$, where $|\mathbf{m}| = \max\{|m_1|, \ldots, |m_d|\}$ for every $\mathbf{m} = (m_1, \ldots, m_d) \in \mathbb{Z}^d$. We claim that there exists, for every $\varepsilon > 0$, an $\varepsilon' > 0$ and an integer $b_\varepsilon \geq 0$ such that

$$
\max_{\{\mathbf{m} \in \mathbb{Z}^d : |\mathbf{m}| \leq b_\varepsilon\}} \vartheta(x_{\mathbf{m}}, y_{\mathbf{m}}) > \varepsilon \text{ whenever } \delta(x, y) > \varepsilon'.
$$

Indeed, let $K = \sum_{\mathbf{m} \in \mathbb{Z}^d} 2^{-|\mathbf{m}|} < \infty$, choose b_ε such that $\sum_{\{\mathbf{m} \in \mathbb{Z}^d : |\mathbf{m}| > b_\varepsilon\}} 2^{-|\mathbf{m}|} < \varepsilon$, and put $\varepsilon' = (K + 1)\varepsilon$. If $\vartheta(x_{\mathbf{m}}, y_{\mathbf{m}}) \leq \varepsilon$ for all $\mathbf{m} \in \mathbb{Z}^d$ with $|\mathbf{m}| \leq b_\varepsilon$ then

$$
\delta(x, y) \leq \varepsilon + \sum_{\{\mathbf{m} \in \mathbb{Z}^d : |\mathbf{m}| \leq b_\varepsilon\}} \varepsilon 2^{-|\mathbf{m}|} \leq (K + 1)\varepsilon = \varepsilon'.
$$

We fix $\varepsilon > 0$ and note that there exist $b = b_\varepsilon \geq 0$ and $\varepsilon' > 0$ such that, for every rectangle Q with sides of length l_j^Q, $j = 1, \ldots, d$, and for every (Q, δ, ε)-separated set $E \in X$ (cf. (13.3)), there exists a smaller rectangle $Q' \subset Q$ with sides of length $l_j^{Q'} = l_j - 2b$ such that F is $[Q', \vartheta, \varepsilon']$-separated. Conversely, if E is $[Q, \vartheta, \varepsilon]$-separated, then E is obviously (Q, δ, ε)-separated. This proves that

$$
h(\sigma) = \lim_{\varepsilon \to 0} \limsup_{\langle Q \rangle \to \infty} \frac{1}{|Q|} \log s_Q(\delta, \varepsilon) = \lim_{\varepsilon \to 0} \limsup_{\langle Q \rangle \to \infty} \frac{1}{|Q|} \log \bar{s}_Q(\vartheta, \varepsilon)
$$

$$= \lim_{\varepsilon \to 0} \liminf_{\langle Q \rangle \to \infty} \frac{1}{|Q|} \log s_Q(\delta, \varepsilon) = \lim_{\varepsilon \to 0} \liminf_{\langle Q \rangle \to \infty} \frac{1}{|Q|} \log \bar{s}_Q(\vartheta, \varepsilon)$$

(cf. (13.4)). The other identities are proved similarly, and the final statement about expansiveness is an immediate consequence of the assertion at the beginning of this proof. □

CONCLUDING REMARKS 13.8. (1) The exposition of entropy presented in this section follows [105] with some obvious changes necessitated by considering $\mathbb{Z}^d$-actions rather than $\mathbb{Z}$-actions (cf. [42] or [63]). The proof of Proposition 13.5 is taken from [10].

(2) In the notation of the proof of Proposition 13.5, $\mu * \nu = (\mu \times \nu)\pi^{-1}$, and $\pi \cdot (\alpha_{\mathbf{n}} \times \alpha_{\mathbf{n}}) = \alpha_{\mathbf{n}} \cdot \pi$ for every $\mathbf{n} \in \mathbb{Z}^d$. Hence $h_{\mu*\nu}(\alpha) \le h_\mu(\alpha) + h_\nu(\alpha)$ for all $\mu, \nu \in M_1(X)^\alpha$, and by combining this with (13.15) we see that $h_\mu(\alpha) + h_\nu(\alpha) \le h_{\mu*\nu}(\alpha) + h_\nu(\alpha) \le h_\mu(\alpha) + 2h_\nu(\alpha)$. In particular, if $\mu, \nu \in M_1(X)^\alpha$ and $h_\nu(\alpha) = 0$, then $h_\mu(\alpha) = h_{\mu*\nu}(\alpha)$ ([10]).

(3) If α is an expansive $\mathbb{Z}^d$-action by automorphisms of a compact group X, then $h(\alpha) < \infty$: indeed, if δ is a metric on X and $\varepsilon > 0$ is chosen so that $B_\delta(\varepsilon) = \{x \in X : \delta(x, 1_X) < \varepsilon\}$ is an expansive neighbourhood, then any finite partition $\mathcal{P}$ of X into sets of diameter $< \varepsilon$ is a generator for α, and (13.10) and (13.8) show that $h(\alpha) = h_\mu(\alpha) = \int I_\mu(\mathcal{P}|\mathcal{P}_\alpha^-) \, d\lambda \le H_\mu(\mathcal{P}) < \infty$. This result is true for every continuous, expansive $\mathbb{Z}^d$-action T on a compact, metrizable space Y, where T is *expansive* if there exists, for some metric δ on Y, an $\varepsilon > 0$ such that $\sup_{\mathbf{m} \in \mathbb{Z}^d} \delta(T_{\mathbf{m}}(x), T_{\mathbf{m}}(y)) > \varepsilon$ whenever x and y are distinct points in Y.

14. Yuzvinskii's addition formula

THEOREM 14.1. *Let α be a $\mathbb{Z}^d$-action by automorphisms of a compact group X and let $Y \subset X$ be a normal, α-invariant subgroup. Then*

$$h(\alpha) = h(\alpha^{X/Y}) + h(\alpha^Y). \tag{14.1}$$

For $d = 1$, Theorem 14.1 is due to Yuzvinskii ([111]). The proof of Yuzvinskii's addition formula (14.1) for $d > 1$ requires some preparation. Let $\theta \colon X \longmapsto X/Y = Z$ be the quotient map, and let $c \colon Z \longmapsto X$ be a Borel map with $\theta \cdot c(z) = z$ for every $z \in Z$ (cf. [78], Lemma I.5.1). We set

$$b(\mathbf{m}, z) = c(\alpha_{\mathbf{m}}^Z(z))^{-1} \alpha_{\mathbf{m}}^Z(c(z))$$

for every $z \in Z$, note that $b \colon \mathbb{Z}^d \times Z \longmapsto Y$ is a Borel map with $b(\mathbf{m} + \mathbf{n}, z) = b(\mathbf{m}, \alpha_{\mathbf{n}}^Z(z))\alpha_{\mathbf{m}}^Y(b(\mathbf{n}, z))$ for all $\mathbf{m}, \mathbf{n} \in \mathbb{Z}^d$ and $z \in Z$, and define a measure preserving $\mathbb{Z}^d$-action T on $(Z \times Y, \lambda_Z \times \lambda_Y)$ by setting

$$T_{\mathbf{m}}(z, y) = (\alpha_{\mathbf{m}}^Z(z), b(\mathbf{m}, z)\alpha_{\mathbf{m}}^Y(y)) \tag{14.2}$$

for all $\mathbf{m} \in \mathbb{Z}^d$, $z \in Z$, $y \in Y$. If $\eta \colon X \longmapsto Z \times Y$ is the map $\eta(x) = (\theta(x), c(\theta(x))^{-1}x)$ then the diagram

$$
\begin{array}{ccc}
X & \xrightarrow{\ \alpha_{\mathbf{m}}\ } & X \\[2pt]
{\scriptstyle \eta}\downarrow & & \downarrow{\scriptstyle \eta} \\[2pt]
Z \times Y & \xrightarrow[\ T_{\mathbf{m}}\ ]{} & Z \times Y
\end{array}
\tag{14.3}
$$

commutes for every $\mathbf{m} \in \mathbb{Z}^d$. This shows that the measure preserving $\mathbb{Z}^d$-actions α and T on (X, λ_X) and $(Z \times Y, \lambda_Z \times \lambda_Y)$ are measurably conjugate, and Theorem 13.3 implies that $h(\alpha) = h_{\lambda_X}(\alpha) = h_{\lambda_Z \times \lambda_Y}(T)$. Theorem 14.1 thus reduces to proving that

$$
h_{\lambda_Z \times \lambda_Y}(T) = h(\alpha^Z) + h(\alpha^Y).
\tag{14.4}
$$

The commutativity of the diagram (14.3) is preserved under quotients by closed, α-invariant subgroups of Y which are normal in X. Indeed, let $V \subset Y$ be such a subgroup, and denote by $T^{(V)}$ the $\mathbb{Z}^d$-action

$$
\begin{aligned}
T_{\mathbf{m}}^{(V)}(z, yV) &= (\alpha_{\mathbf{m}}^Z(z), b(\mathbf{m}, z)\alpha_{\mathbf{m}}^{Y/V}(yV)) \\
&= (\alpha_{\mathbf{m}}^Z(z), b(\mathbf{m}, z)\alpha_{\mathbf{m}}^Y(y)V)
\end{aligned}
\tag{14.5}
$$

on $(Z \times Y/V, \lambda_Z \times \lambda_{Y/V})$. We define $\eta^{(V)} \colon X/V \longmapsto Z \times Y/V$ by

$$
\eta^{(V)}(xV) = (\theta(x)V, c(\theta(x))^{-1}xV),
$$

and note that the diagram

$$
\begin{array}{ccc}
X/V & \xrightarrow{\ \alpha_{\mathbf{m}}^{X/V}\ } & X/V \\[2pt]
{\scriptstyle \eta^{(V)}}\downarrow & & \downarrow{\scriptstyle \eta^{(V)}} \\[2pt]
Z \times Y/V & \xrightarrow[\ T_{\mathbf{m}}^{(V)}\ ]{} & Z \times Y/V
\end{array}
\tag{14.6}
$$

again commutes for every $\mathbf{m} \in \mathbb{Z}^d$. Hence

$$
h_{\lambda_Z \times \lambda_{Y/V}}(T^{(V)}) = h(\alpha^{X/V})
\tag{14.7}
$$

for every closed, α-invariant subgroup $V \subset Y$ which is normal in X.

If $\mathcal{P}$ is a finite, measurable partition of Y and $y \in Y$, let $y \cdot \mathcal{P}$ be the partition $\{y \cdot P : P \in \mathcal{P}\}$. We choose sequences of finite, measurable partitions $(\mathcal{P}_n, n \geq 1)$ and $(\mathcal{Q}_n, n \geq 1)$ of Y and Z, respectively, which increase to the point partitions (i.e. which satisfy that $\Sigma(\bigcup_{n \geq 1} \mathcal{P}_n) = \mathcal{B}_Y$ and $\Sigma(\bigcup_{n \geq 1} \mathcal{Q}_n) = \mathcal{B}_Z$). For every rectangle $Q \subset \mathbb{Z}^d$, $z \in Z$, and $n \geq 1$, we set

$$
\mathcal{Q}_n^Q = \bigvee_{\mathbf{m} \in Q} \alpha_{-\mathbf{m}}^Z(\mathcal{Q}_n), \qquad \mathcal{P}_n^Q = \bigvee_{\mathbf{m} \in Q} \alpha_{-\mathbf{m}}^Y(\mathcal{P}_n),
$$

$$\mathcal{P}_{n,z}^{Q} = \bigvee_{\mathbf{m} \in Q} b(-\mathbf{m}, z) \cdot \alpha_{-\mathbf{m}}^{Y}(\mathcal{P}_n).$$

If

$$\mathcal{Q}_m \times \mathcal{P}_n = \{Q \times P : Q \in \mathcal{Q}_m, P \in \mathcal{P}_n\}$$

then we obtain from (13.6), (13.7) and (13.11) that

$$
\begin{aligned}
h(\alpha^Z \times \alpha^Y) &= \sup_{n \geq 1} \sup_{m \geq 1} h_{\lambda_Z \times \lambda_Y}(\alpha^Z \times \alpha^Y, \mathcal{Q}_m \times \mathcal{P}_n) \\
&= \sup_{m \geq 1} h_{\lambda_Z}(\alpha^Z, \mathcal{Q}_m) + \sup_{n \geq 1} h_{\lambda_Y}(\alpha^Y, \mathcal{P}_n) \qquad (14.8) \\
&= h(\alpha^Z) + \sup_{n \geq 1} \lim_{\langle Q \rangle \to \infty} \frac{1}{|Q|} \int_Z H_{\lambda_Y}(\mathcal{P}_n^Q) d\lambda_Z,
\end{aligned}
$$

where $\alpha^Z \times \alpha^Y$ denotes the product action $(\mathbf{m}, (z,y)) \mapsto (\alpha_{\mathbf{m}}^{Z}(z), \alpha_{\mathbf{m}}^{Y}(y))$ on $Z \times Y$. For every rectangle $Q \subset \mathbb{Z}^d$ we set

$$\overline{\mathcal{Q}}_n^Q = \{Q \times Y : Q \in \mathcal{Q}_n^Q\}, \quad \overline{\mathcal{P}}_n = \{Z \times P : P \in \mathcal{P}_n\}, \quad \overline{\mathcal{P}}_n^Q = \bigvee_{\mathbf{m} \in Q} T_{-\mathbf{m}}(\overline{\mathcal{P}}_n),$$

and note that

$$I_{\lambda_Z \times \lambda_Y}\left(\bigvee_{\mathbf{m} \in Q} T_{-\mathbf{m}}(\mathcal{Q}_n \times \mathcal{P}_n) \,\Big|\, \overline{\mathcal{Q}}_n^Q \right) = I_{\lambda_Z \times \lambda_Y}(\overline{\mathcal{P}}_n^Q | \overline{\mathcal{Q}}_n^Q).$$

Hence

$$
\begin{aligned}
H_{\lambda_Z \times \lambda_Y}&\left(\bigvee_{\mathbf{m} \in Q} T_{-\mathbf{m}}(\mathcal{Q}_m \times \mathcal{P}_n) \right) \\
&= H_{\lambda_Y}(\mathcal{Q}_m^Q) + H_{\lambda_Z \times \lambda_Y}\left(\bigvee_{\mathbf{m} \in Q} T_{-\mathbf{m}}(\mathcal{Q}_m \times \mathcal{P}_n) \,\Big|\, \overline{\mathcal{Q}}_n^Q \right) \\
&= H_{\lambda_Y}(\mathcal{Q}_n^Q) + \int_{Z \times Y} I_{\lambda_Z \times \lambda_Y}(\overline{\mathcal{P}}_n^Q | \overline{\mathcal{Q}}_n^Q) \, d\lambda_{Z \times Y},
\end{aligned}
$$

and (13.11) implies that

$$
\begin{aligned}
h_{\lambda_Z \times \lambda_Y}(T) &= \sup_{n \geq 1} \sup_{m \geq 1} h_{\lambda_Z \times \lambda_Y}(T, \mathcal{Q}_m \times \mathcal{P}_n) \\
&= h_{\lambda_Z}(\alpha^Z) + \sup_{n \geq 1} \lim_{\langle Q \rangle \to \infty} \frac{1}{|Q|} \int_{Z \times Y} I_{\lambda_Z \times \lambda_Y}(\overline{\mathcal{P}}_n^Q | \mathcal{B}_Z') \, d\lambda_{Z \times Y} \\
&= h(\alpha^Z) + \sup_{n \geq 1} \lim_{\langle Q \rangle \to \infty} \frac{1}{|Q|} \int_Z \int_Y I_{\lambda_Y}(\mathcal{P}_{n,z}^Q) \, d\lambda_Y \, d\lambda_Z(z) \quad (14.9) \\
&= h(\alpha^Z) + \sup_{n \geq 1} \lim_{\langle Q \rangle \to \infty} \frac{1}{|Q|} \int_Z H_{\lambda_Y}(\mathcal{P}_{n,z}^Q) \, d\lambda_Z(z),
\end{aligned}
$$

where $\mathcal{B}_Z' = \{B \times Y : B \in \mathcal{B}_Z\}$. The interchange of limits in (14.9) has to be justified, of course. The next lemma is motivated by a comparison of (14.8) and (14.9).

LEMMA 14.2. *Let H be a compact Lie group. Then there exist a constant $K > 0$ and a sequence $(\mathcal{P}_n, n \geq 1)$ of finite, measurable partitions of H, which increases to the point partition, such that, for every $n \geq 1$, $P \in \mathcal{P}_n$, and $h \in H$, the number of non-empty sets in $\{P \cap P' : P' \in h \cdot \mathcal{P}_n\}$ is less than or equal to K. If H is finite we may choose $K = 1$.*

PROOF. For a finite group H the assertion is trivial: put $\mathcal{P}_n$ equal to the point partition of H for every $n \geq 1$. If H is infinite, let $\mathfrak{H}$ be the Lie algebra of H. We choose and fix a positive definite, bilinear form $B(\cdot, \cdot)$ on $\mathfrak{H} \times \mathfrak{H}$ which is invariant under the adjoint representation of H on $\mathfrak{H}$, and denote by δ the Riemannian metric on H defined by $B(\cdot, \cdot)$. For notational convenience we identify $(\mathfrak{H}, B)$ with $(\mathbb{R}^m, \langle \cdot, \cdot \rangle)$, where $\langle \cdot, \cdot \rangle$ is the Euclidean inner product on $\mathbb{R}^m$. For every $\varepsilon > 0$ we set $I(\varepsilon) = [-\varepsilon, \varepsilon]^m \subset \mathbb{R}^m \cong \mathfrak{H}$, and we denote by $\mathcal{P}'_n(\varepsilon)$ the partition of $I(\varepsilon)$ into disjoint translates of $[-\varepsilon 2^{-n}, \varepsilon 2^{-n}]^m$. The exponential map $\exp \colon \mathfrak{H} \longmapsto H$ defines a homeomorphism of a neighbourhood $N(0)$ of $0 \in \mathfrak{H}$ onto a neighbourhood $N(1_H)$ of the identity in H; since the derivative of the exponential map at 1_H is the identity there exists a an ε_0 with $0 < \varepsilon_0 < \frac{1}{4m}$ such that $I(\varepsilon_0) \subset N(0)$ and, for every $n \geq 1$ and $P \subset \mathcal{P}'_n$, $\exp(P)$ contains a δ-ball of radius $\varepsilon_0 2^{-n} m^{-1}$ and is contained in a δ-ball of radius $\varepsilon_0 2^{-n} m$. Since the set $N_{\varepsilon_0} = \exp(I(\varepsilon_0))$ contains an open neighbourhood of 1_H, there exist elements $h_0 = 1_H, h_1, \ldots, h_s$ in H such that $H \subset \bigcup_{j=1}^s h_j N_{\varepsilon_0}$. Put $E_0 = N_{\varepsilon_0}$ and $E_j = h_j N_{\varepsilon_0} \smallsetminus \bigcup_{k=0}^{j-1} h_k N_{\varepsilon_0}$ for every $j = 1, \ldots, s$, denote by $\mathcal{P}_n^j$ the partition of E_j induced by $h_j \exp(\mathcal{P}'_n)$, and set $\mathcal{P}_n = \bigcup_{j=0}^s \mathcal{P}_n^j$. Then $\mathcal{P}_n$ is a finite, measurable partition of H, and $\mathcal{P}_n$ increases to the point partitions as $n \to \infty$.

There exists a constant $C > 0$ such that $C^{-1} \varepsilon^m \leq \lambda_H(B_\delta(\varepsilon)) \leq C \varepsilon^m$ for all ε with $0 < \varepsilon \leq 1$, where $B_\delta(\varepsilon)$ denotes the open ball with δ-radius ε and centre 1_H. Hence, if $0 < r \leq R < 1$, at most $C^2 2^m (R/r)^m$ disjoint open translates of $B_\delta(r)$ can intersect $B_\delta(R)$. We conclude that, for every $h \in H$ and $P \in \mathcal{P}_n^0 = \exp(\mathcal{P}'_n)$, at most $C^2 2^m m^{2m}$ sets of the form $P \cap Q$, $Q \in h \cdot \exp(\mathcal{P}'_n)$, are non-empty. From the definition of $\mathcal{P}_n$ it is now clear that every $P \in \mathcal{P}_n$ can have non-empty intersection with at most $C^2 2^m m^{2m}(s+1)$ sets in $h \cdot \mathcal{P}_n$ for every $h \in H$. □

PROOF OF THEOREM 14.1. We claim that it suffices to prove Theorem 14.1 under the additional hypotheses that $h(\alpha^Y) < \infty$ and that α satisfies the d.c.c. If $h(\alpha^Y) = \infty$, then $h(\alpha) = h_{\text{sep}}(\alpha) \geq h_{\text{sep}}(\alpha^Y) = h(\alpha^Y) = \infty$, and Theorem 14.1 holds trivially. Now suppose that Theorem 14.1 has been shown to hold for every $\mathbb{Z}^d$-action satisfying the d.c.c. If α is an arbitrary $\mathbb{Z}^d$-action by automorphisms of a compact group X with $h(\alpha) < \infty$ we apply Proposition 4.9 and choose a decreasing sequence $(V_n, n \geq 1)$ of closed, normal, α-invariant subgroups of X such that $\bigcap_{n \geq 1} V_n = \{1_X\}$ and α^{X/V_n} satisfies the d.c.c. for every $n \geq 1$. By assumption, $h(\alpha^{X/V_n}) = h(\alpha^{X/Y V_n}) + h(\alpha^{Y/V_n})$ for every

$n \geq 1$. Lemma 13.6 shows that, as $n \to \infty$, $h(\alpha^{X/V_n}) \to h(\alpha)$, $h(\alpha^{X/YV_n}) \to h(\alpha^{X/Y})$, and $h(\alpha^{Y/V_n}) \to h(\alpha^Y)$, and we obtain that $h(\alpha) = h(\alpha^{X/Y}) + h(\alpha^Y)$.

Let therefore α be a $\mathbb{Z}^d$-action by automorphisms of X which satisfies the d.c.c., and let Y be a closed, normal, α-invariant subgroup of X with $h(\alpha^Y) < \infty$. According to (4.10) we may assume that X is a full, shift-invariant subgroup of $G^{\mathbb{Z}^d}$, where G is a compact Lie group, that $Y = \{x \in X \subset G^{\mathbb{Z}^d} : x_{\mathbf{n}} \in H$ for all $\mathbf{n} \in \mathbb{Z}^d\}$, where $H = \pi_{\{0_{\mathbb{Z}^d}\}}(Y)$ is a closed, normal subgroup of G, and that α is equal to the shift-action of $\mathbb{Z}^d$ on X. By Remark 4.15 we may take it that

$$X = \{x \in G^{\mathbb{Z}^d} : \pi_{\{0,1\}^d}(\alpha_{\mathbf{n}}(x)) \in \pi_{\{0,1\}^d}(X) \text{ for every } \mathbf{n} \in \mathbb{Z}^d\},$$

$$Y = \{y \in H^{\mathbb{Z}^d} : \pi_{\{0,1\}^d}(\alpha_{\mathbf{n}}(y)) \in \pi_{\{0,1\}^d}(Y) \text{ for every } \mathbf{n} \in \mathbb{Z}^d\}. \quad (14.10)$$

Let $(\mathcal{Q}_n, n \geq 1)$ be a sequence of finite, measurable partitions of $Z = X/Y$ which increases to the point partition. We apply Lemma 14.2 and choose a sequence $(\mathcal{P}'_n, n \geq 1)$ of finite, measurable partitions of H, which increases to the point partition, such that, for every $n \geq 1$, $P \in \mathcal{P}'_n$, and $h \in H$, the number of non-empty sets in $\{P \cap P' : P' \in h \cdot \mathcal{P}'_n\}$ is $\leq K$. Since the entropy of a partition of a probability space into K sets is less than or equal to $\log K$ we obtain that

$$H_{\lambda_H}(\mathcal{P}'_n | h \cdot \mathcal{P}'_n) \leq \log K, \quad H_{\lambda_H}(h \cdot \mathcal{P}'_n | \mathcal{P}'_n) \leq \log K \quad (14.11)$$

for every $n \geq 1$ and $h \in H$. For every $n \geq 1$, let $\mathcal{P}_n = \pi^{-1}_{\{0_{\mathbb{Z}^d}\}}(\mathcal{P}'_n)$, and note that $H_{\lambda_Y}(\mathcal{P}_n | y \cdot \mathcal{P}_n) \leq \log K$ and $H_{\lambda_Y}(y \cdot \mathcal{P}_n | \mathcal{P}_n) \leq \log K$ for every $n \geq 1$. From (14.8) and (14.9) we see that, for all $n \geq 1$,

$$h(\alpha^Z \times \alpha^Y) = h(\alpha^Z) + \sup_{n \geq 1} h_{\lambda_Y}(\alpha^Y, \mathcal{P}_n)$$

$$\leq h(\alpha^Z) + \sup_{n \geq 1} \lim_{\langle Q \rangle \to \infty} \frac{1}{|Q|} \int_Z H_{\lambda_Y}(\mathcal{P}^Q_n | \mathcal{P}^Q_{n,z}) \, d\lambda_Z(z)$$

$$+ \sup_{n \geq 1} \lim_{\langle Q \rangle \to \infty} \frac{1}{|Q|} \int_Z H_{\lambda_Y}(\mathcal{P}^Q_{n,z}) \, d\lambda_Z(z)$$

$$\leq h_{\lambda_Z \times \lambda_Y}(T) + \log K,$$

$$h_{\lambda_Z \times \lambda_Y}(T) = h(\alpha^Z) + \sup_{n \geq 1} \lim_{\langle Q \rangle \to \infty} \frac{1}{|Q|} \int_Z H_{\lambda_Y}(\mathcal{P}^Q_{n,z} | \mathcal{P}^Q_n) \, d\lambda_Z(z)$$

$$+ \sup_{n \geq 1} \lim_{\langle Q \rangle \to \infty} \frac{1}{|Q|} \int_Z H_{\lambda_Y}(\mathcal{P}^Q_n) \, d\lambda_Z$$

$$\leq h(\alpha^Z \times \alpha^Y) + \log K,$$

and hence

$$h(\alpha^Z \times \alpha^Y) - \log K \leq h_{\lambda_Z \times \lambda_Y}(T) \leq h(\alpha^Z \times \alpha^Y) + \log K. \quad (14.12)$$

If Y is zero dimensional, then H is finite (Corollary 2.3), and Lemma 14.2 allows us to choose $K = 1$. This proves equation (14.4)—and hence, by (14.3), Theorem 14.1—under the additional assumption that Y is zero dimensional.

If Y is not zero dimensional then the constant $\log K$ in (14.12) will be non-zero, and we have to get rid of it. For every $k \geq 1$ we set

$$
\begin{aligned}
\Delta_k = \{&\mathbf{m} = (m_1, \ldots, m_d) \in \mathbb{Z}^d : \\
&m_j \text{ is divisible by } k \text{ for some } j \in \{1, \ldots, d\}\}, \\
S_k = \{&\mathbf{m} = (m_1, \ldots, m_d) \in \Delta_k : \\
&0 \leq m_j < k \text{ for every } j = 1, \ldots, d\}.
\end{aligned}
\tag{14.13}
$$

It may be helpful to draw a picture for $d = 2$, where the sets $S_k \subset \Delta_k$ and $\Delta_k \smallsetminus S_k$ are marked by $\otimes$ and $\times$, respectively, and where $(0,0)$ is located near the lower left hand corner of the picture:

$$
\begin{array}{ccccccccccccc}
\cdot & \times & \cdot & \cdot & \cdot & \times & \cdot & \cdot & \cdot & \times & \cdot \\
\times & \times & \times & \times & \times & \times & \times & \times & \times & \times & \times & \times \\
\cdot & \times & \cdot & \cdot & \cdot & \times & \cdot & \cdot & \cdot & \times & \cdot \\
\cdot & \times & \cdot & \cdot & \cdot & \times & \cdot & \cdot & \cdot & \times & \cdot \\
\cdot & \times & \cdot & \cdot & \cdot & \times & \cdot & \cdot & \cdot & \times & \cdot \\
\cdot & \times & \cdot & \cdot & \cdot & \times & \cdot & \cdot & \cdot & \times & \cdot \\
\times & \times & \times & \times & \times & \times & \times & \times & \times & \times & \times & \times \\
\cdot & \otimes & \cdot & \cdot & \cdot & \times & \cdot & \cdot & \cdot & \times & \cdot \\
\cdot & \otimes & \cdot & \cdot & \cdot & \times & \cdot & \cdot & \cdot & \times & \cdot \\
\cdot & \otimes & \cdot & \cdot & \cdot & \times & \cdot & \cdot & \cdot & \times & \cdot \\
\cdot & \otimes & \cdot & \cdot & \cdot & \times & \cdot & \cdot & \cdot & \times & \cdot \\
\times & \otimes & \otimes & \otimes & \otimes & \otimes & \times & \times & \times & \times & \times & \times \\
\cdot & \times & \cdot & \cdot & \cdot & \times & \cdot & \cdot & \cdot & \times & \cdot
\end{array}
\tag{14.14}
$$

We consider the projections $\pi_{\Delta_k} : Y \longmapsto H^{\Delta_k}$ and $\pi_{S_k} : Y \longmapsto H^{S_k}$, and set $H_k = \pi_{S_k}(Y) \subset H^{S_k}$, $V_k = \ker(\pi_{\Delta_k})$, $X_k = X/V_k$, and $Y_k = Y/V_k$ (note that V_k is normal in X).

The group V_k is obviously invariant under the $\mathbb{Z}^d$-action $\alpha^{(k)} : \mathbf{m} \mapsto \alpha_{\mathbf{m}}^{(k)} = \alpha_{k\mathbf{m}}$ on X, and we denote by $(\alpha^{(k)})^{V_k}$ and $(\alpha^{(k)})^{Y_k}$ the $\mathbb{Z}^d$-actions on V_k and Y_k induced by $\alpha^{(k)}$. Since $k\mathbf{m} + S_k \cap k\mathbf{n} + S_k = \varnothing$ for $\mathbf{m} \neq \mathbf{n}$ in $\mathbb{Z}^d$ and $\Delta_k = k\mathbb{Z}^d + S_k$, the homomorphism $\phi : Y \longmapsto H_k^{\mathbb{Z}^d}$, given by $\phi(y)_{\mathbf{m}} = \pi_{S_k}(\alpha_{\mathbf{m}}^{(k)}(y))$ for every $y \in Y$ and $\mathbf{m} \in \mathbb{Z}^d$, satisfies that $\ker(\phi) = V_k$. In particular, ϕ induces an isomorphism $\bar{\phi} : Y_k \longmapsto \phi(Y)$ with $\bar{\phi} \cdot (\alpha_{\mathbf{n}}^{(k)})^{Y_k} = \sigma_{\mathbf{n}} \cdot \bar{\phi}$ for every $\mathbf{n} \in \mathbb{Z}^d$, where σ denotes the shift-action on $\phi(Y) \subset H_k^{\mathbb{Z}^d}$. Since H_k is a subgroup of H^{S_k}, there exists a sequence $(\mathcal{P}_n^k, n \geq 1)$ of finite measurable partitions of H_k, which increases to the point partition, and such that, for every $n \geq 1$, $P \in \mathcal{P}_n^k$, and $h \in H_k$, the number of non-empty sets in $\{P \cap P' : P' \in h \cdot \mathcal{P}_n^k\}$ is less than or equal to $K^{|S_k|} \leq K^{dk^{d-1}}$ (such a sequence is obtained by setting $\mathcal{P}_n^{(k)}$ equal to the product partition of H^{S_k} arising from $\mathcal{P}_n'$, and by putting $\mathcal{P}_n^k = \{P \cap H_k : P \in \mathcal{P}_n^{(k)}\}$). We define a $\mathbb{Z}^d$-action $T^{(k)}$ by $T_{\mathbf{m}}^{(k)} = T_{k\mathbf{m}}$ for all $\mathbf{m} \in \mathbb{Z}^d$. Since V_k is invariant under $(\alpha^{(k)})^{Y_k}$, we can define $(T^{(k)})^{(V_k)}$ by (14.5) (with T replaced by $T^{(k)}$), and we observe exactly as in (14.7) and (14.12) that

$$h((\alpha^{(k)})^Z \times (\alpha^{(k)})^{Y_k}) - dk^{d-1} \log K$$
$$\leq h_{\lambda_Z \times \lambda_{Y_k}}((T^{(k)})^{(V_k)}) = h(\alpha^{X_k}) \qquad (14.15)$$
$$\leq h((\alpha^{(k)})^Z \times (\alpha^{(k)})^{Y_k}) + dk^{d-1} \log K.$$

Suppose that we have proved that V_k is zero dimensional for every $k \geq 1$. We add $h((\alpha^{(k)})^{V_k})$ in (14.15) and apply the first part of this proof, where we have established that Theorem 14.1 holds if Y (or, on this case, V_k) is zero dimensional. Then

$$k^d h(\alpha^Z) + k^d h(\alpha^Y) - dk^{d-1} \log K$$
$$= h((\alpha^{(k)})^Z) + h((\alpha^{(k)})^Y) - dk^{d-1} \log K$$
$$= h((\alpha^{(k)})^Z) + h((\alpha^{(k)})^{Y_k}) + h((\alpha^{(k)})^{V_k}) - dk^{d-1} \log K$$
$$= h((\alpha^{(k)})^Z \times (\alpha^{(k)})^{Y_k}) + h((\alpha^{(k)})^{V_k}) - dk^{d-1} \log K$$
$$\leq h((\alpha^{(k)})^{X_k}) + h((\alpha^{(k)})^{V_k}) = h(\alpha^{(k)}) = k^d h(\alpha)$$
$$\leq k^d h(\alpha^Z) + k^d h(\alpha^Y) + dk^{d-1} \log K.$$

This formula holds for every $k \geq 1$, and we conclude that $h(\alpha) = h(\alpha^Z) + h(\alpha^Y)$, as stated in Theorem 14.1.

We complete the proof by showing that V_k is zero dimensional. Fix $k \geq 1$ and put $R = \{0, \dots, k-1\}^d \subset \mathbb{Z}^d$ and $K = \pi_R(V_k)$. We define a continuous, injective homomorphism $\psi \colon K^{\mathbb{Z}^d} \longmapsto Y$ by $\pi_R(\alpha_{k\mathbf{m}}(\psi(w))) = w_{\mathbf{m}}$ for every $w \in K^{\mathbb{Z}^d}$ and $\mathbf{m} \in \mathbb{Z}^d$. Then $\psi(K^{\mathbb{Z}^d}) = V_k$ (an explanation follows in a moment), and $\psi \cdot \sigma_{\mathbf{m}} = (\alpha_{\mathbf{m}}^{(k)})^Y = \alpha_{k\mathbf{m}}^Y$ for every $\mathbf{m} \in \mathbb{Z}^d$, where σ is the shift-action on $K^{\mathbb{Z}^d}$. Since we are assuming that $h((\alpha^{(k)})^Y) = k^d h(\alpha^Y) < \infty$, the $\mathbb{Z}^d$-action σ on $K^{\mathbb{Z}^d}$ must have finite entropy, and this is only possible if K is finite. In particular, V_k is zero dimensional, as claimed.

In order see that $\psi(K^{\mathbb{Z}^d})$ is indeed equal to V_k it helps to have another look at (14.14): every point $k \in K$ is the image of some $v \in V_k$ under the projection π_R, and we know that $v_{\mathbf{m}} = 1_H$ for all $\mathbf{m} \in \Delta_k$. If we fill the four 'chambers' of Δ_k visible in (14.14) with arbitrary elements of K and put 1-s onto those boundaries of the chambers which have not yet been filled, then every projection of the resulting configuration onto a 2×2 square $\mathbf{m} + \{0,1\}^2$ contained in the union of the four chambers and their boundaries lies in $\pi_{\mathbf{m}+\{0,1\}^2}(Y)$; if we fill every chamber of Δ_k in this manner with an arbitrary element of K, the resulting element v of $H^{\mathbb{Z}^d}$ satisfies that $\pi_{\mathbf{m}+\{0,1\}^d}(v) \in \pi_{\mathbf{m}+\{0,1\}^d}(Y)$ for every $\mathbf{m} \in \mathbb{Z}^d$, and (14.10) implies that $v \in Y$. By construction, $v_{\mathbf{m}} = 1$ for every $\mathbf{m} \in \Delta_k$, so that $v \in \ker(\pi_{\Delta_k}) = V_k$. Although $d = 2$ in (14.14), this argument works for any $d \geq 1$. $\square$

COROLLARY 14.3. *Let α be a $\mathbb{Z}^d$-action by automorphisms of a compact group X, and let $C(X^\circ)$ denote the centre of the connected component X° of the identity in X. Then*

$$h(\alpha) = h(\alpha^{X/X^\circ}) + h(\alpha^{X^\circ/C(X^\circ)}) + h(\alpha^{C(X^\circ)}). \tag{14.16}$$

CONCLUDING REMARKS 14.4. (1) The proof of Yuzvinskii's addition formula (Theorem 14.1) is essentially based on [111] and [104], with some changes due to the transition from $\mathbb{Z}$ to $\mathbb{Z}^d$ (cf. [63]).

(2) Corollary 14.3 reduces the problem of calculating the entropy of a $\mathbb{Z}^d$-action α on a compact group to three special cases, where the group is either zero dimensional, or of the form $Y/C(Y)$ for some compact, connected group Y, or abelian. The first two cases will be dealt with fairly easily in Section 15, but the entropy formula for abelian groups turns out to be both more difficult and more interesting, and will occupy Sections 16–18.

15. $\mathbb{Z}^d$-actions on groups with zero-dimensional centres

LEMMA 15.1. *Let G be a compact, connected Lie group with trivial centre, and let $X \subset G^{\mathbb{Z}^d}$ be a full, shift-invariant subgroup. Then the shift-action σ of $\mathbb{Z}^d$ on X satisfies that $h(\sigma) \in \{0, \infty\}$.*

PROOF. According to Remark 4.15 we may assume without loss in generality that

$$X = \{x \in G^{\mathbb{Z}^d} : \pi_{\{0,1\}^d}(\sigma_{\mathbf{n}}(x)) \in \pi_{\{0,1\}^d}(X) \text{ for every } \mathbf{n} \in \mathbb{Z}^d\},$$

where G is a compact, connected Lie group with trivial centre. Let $k \geq 1$, define $\Delta_k \subset \mathbb{Z}^d$ by (14.13), and put $Y_k = \ker(\pi_{\Delta_k})$, $R_k = \{0, \ldots, k-1\}^d$, and $K_k = \pi_{R_k}(Y_k) \subset G^{R_k}$. Since $X \subset G^{\mathbb{Z}^d}$ is full, $\pi_{R_k}(X) \subset G^{R_k}$ is a compact, connected Lie group with trivial centre, and K_k is a closed, normal subgroup of $\pi_{R_k}(X)$; in particular, K_k is either trivial or uncountable. If K_k is uncountable for some $k \geq 1$, then we conclude exactly as in the last part of the proof of Theorem 14.1 that $h(\sigma) = \infty$. If K_k is trivial for every $k \geq 1$ we put $\bar{R}_k = \{0, \ldots, k\}^d$ and $B_k = \bar{R}_k \cap \Delta_k$ and observe that $\pi_{\bar{R}_k}(x)$ is determined completely by $\pi_{B_k}(x)$ for every $x \in X$ and $k \geq 1$, i.e. that the projection $\pi_{B_k}(\pi_{\bar{R}_k}(X)) \longmapsto \pi_{B_k}(X)$ is injective.

Assume that $G = G_1 \times \cdots \times G_r$, where each G_i is a compact, simple Lie group with trivial centre, and let ϑ be a metric on G which is invariant under $\text{Aut}(G)$. The invariance of ϑ under $\text{Aut}(G)$ is easily seen to imply that, for all $x, y \in X$ and $k \geq 1$,

$$r \cdot \max_{\mathbf{m} \in B_k} \vartheta(x_{\mathbf{m}}, y_{\mathbf{m}}) \geq \max_{\mathbf{m} \in R_k} \vartheta(x_{\mathbf{m}}, y_{\mathbf{m}}). \tag{15.1}$$

We fix $\varepsilon > 0$ and choose a finite $\varepsilon/2c$-spanning set $F(\varepsilon) \in G$ with cardinality $|F(\varepsilon)| = m$, say. Then there exists a set $F(k, \varepsilon) \subset \pi_{B_k}(X) \subset G^{B_k}$ with cardinality $M_k \leq m^{|B_k|} \leq m^{2dk^{d-1}}$ and with the following property: for every $y \in \pi_{B_k}(X)$ we can find a $z \in F(k,\varepsilon)$ with $\vartheta(y_{\mathbf{n}}, z_{\mathbf{n}}) < \varepsilon/c$ for all $\mathbf{n} \in B_k$. Choose a set $\bar{F}(k,\varepsilon) \subset X$ of cardinality M_k with $\pi_{B_k}(\bar{F}(k,\varepsilon)) = F(k,\varepsilon)$. Then there is, for every $x \in X$, a $y \in \bar{F}(k,\varepsilon)$ with $\vartheta(x_{\mathbf{m}}, y_{\mathbf{m}}) < \varepsilon/c$ for all $\mathbf{m} \in B_k$, and (15.1) implies that $\vartheta(x_{\mathbf{m}}, y_{\mathbf{m}}) < \varepsilon$ for all $\mathbf{m} \in \bar{R}_k$. In the notation of Proposition 13.7 this means that $\bar{r}_{\bar{R}_k}(\vartheta, \varepsilon) \leq |\bar{F}(k,\varepsilon)| \leq m^{2dk^{d-1}}$ for every $k \geq 1$; hence $\limsup_{\langle Q \rangle \to \infty} \frac{1}{|Q|} \log \bar{r}_Q(\vartheta, \varepsilon) = 0$ for every $\varepsilon > 0$, and Proposition 13.7 implies that $h(\sigma) = 0$. $\square$

PROPOSITION 15.2. *Let X be a compact, connected group, and let α be a $\mathbb{Z}^d$-action by automorphisms of X. Then $h(\alpha^{X/C(X)}) \in \{0, \infty\}$.*

PROOF. Since $X/C(X)$ has trivial centre by Proposition 2.13, the assertion follows from Lemma 15.1 and Remark 4.15. $\square$

PROPOSITION 15.3. *Let α be a $\mathbb{Z}^d$-action by automorphisms of a compact, connected group X with trivial centre. Then α is mixing if and only if it is conjugate to the shift-action of $\mathbb{Z}^d$ on $Y^{\mathbb{Z}^d}$, where $Y \neq \{1\}$ is a compact, connected group with trivial centre. If α is mixing then $h(\alpha) = \infty$.*

PROOF. Choose a decreasing sequence $(Y_n, n \geq 1)$ of closed, normal subgroups of X such that $\bigcap_{n \geq 1} Y_n = \{1\}$ and X/Y_n is a compact Lie group for every $n \geq 1$. From Proposition 2.13 we know that there exists, for every $n \geq 0$, a compact, connected Lie group G_n with trivial centre such that $Y_n \cong G_n \times Y_{n+1}$, where $Y_0 = X$. As every compact, connected Lie group with trivial centre is a cartesian product of compact, connected, simple Lie groups with trivial centres, X can be identified with a cartesian product $\prod_{k \in J} H_k$ of compact, connected, simple Lie groups H_k with trivial centres, where J is a countable set. Every $\alpha_{\mathbf{n}}$ induces a permutation $\alpha'_{\mathbf{n}}$ of J which is given by $\alpha_{\mathbf{n}}(H_j) = H_{\alpha'_{\mathbf{n}}(j)}$ for all $j \in J$, and the map $\mathbf{n} \mapsto \alpha'_{\mathbf{n}}$ is a $\mathbb{Z}^d$-action on J. Since every H_j carries a metric δ which is invariant under $\mathrm{Aut}(H_j)$, Lemma 1.2 and Theorem 1.6 together imply that the action α is mixing if and only if every orbit of α' is free, i.e. if $\alpha'_{\mathbf{n}}(j) \neq j$ for all $j \in J$ and $\mathbf{n} \neq \mathbf{0}$ in $\mathbb{Z}^d$. In this case we can choose a subset $J_0 \subset J$ which intersects each α'-orbit in exactly one point, and we write $J = J_0 \times \mathbb{Z}^d$ and set $Y = \prod_{j \in J_0} H_j$ and $X' = Y^{\mathbb{Z}^d}$. For every $\mathbf{n} \in \mathbb{Z}^d$, the automorphism $\alpha_{\mathbf{n}}$ of $X = \prod_{j \in J} H_j$ induces an isomorphism $\phi_{\mathbf{n}} \colon Y \longmapsto \prod_{j \in \alpha'_{\mathbf{n}}(J_0)} H_j$, and we denote by $\phi = \prod_{\mathbf{n} \in \mathbb{Z}^d} \phi_{\mathbf{n}} \colon X' = Y^{\mathbb{Z}^d} \longmapsto X = \prod_{\mathbf{n} \in \mathbb{Z}^d} \prod_{j \in \alpha'_{\mathbf{n}}(J_0)} H_j$ the cartesian product of these isomorphisms. Then $\phi \cdot \sigma_{\mathbf{n}} = \alpha_{\mathbf{n}} \cdot \phi$ for every $\mathbf{n} \in \mathbb{Z}^d$. Since Y is uncountable, Y has finite, measurable partitions of arbitrarily large entropy, hence the entropy of the shift-action σ of $\mathbb{Z}^d$ on $Y^{\mathbb{Z}^d}$ is infinite, and $h(\alpha) = h(\sigma) = \infty$.

Conversely, if $Y \neq \{1\}$ is a compact, connected group with trivial centre, $X = Y^{\mathbb{Z}^d}$, and α is the shift-action of $\mathbb{Z}^d$ on X, then α is obviously mixing. $\square$

REMARK 15.4. The entropy $h(\alpha)$ of the $\mathbb{Z}^d$-action α in Proposition 15.2 is infinite if and only if the $\mathbb{Z}^d$-action α' on J has a free orbit (i.e. if and only if there exists a point $j \in J$ with $\alpha'_\mathbf{n}(j) \neq j$ for all non-zero $\mathbf{n} \in \mathbb{Z}^d$). It obviously suffices to verify this in the special case where J consists of a single α'-orbit. If this orbit is free, then $h(\alpha) = \infty$ by Proposition 15.2. If this orbit is not free and $j_0 \in S$, then the stability subgroup $S = \{\mathbf{n} \in \mathbb{Z}^d : \alpha'_\mathbf{n}(j_0) = j_0\}$ is infinite. Since there exists a metric δ on $H = H_{j_0}$ which is invariant under $\mathrm{Aut}(H)$ and hence under $\{\alpha_\mathbf{n}^H : \mathbf{n} \in S\}$, we can find a metric $\bar{\delta}$ on $X \cong H^{\mathbb{Z}^d/S}$ which is invariant under $\{\alpha_\mathbf{n} : \mathbf{n} \in S\}$. If $\varepsilon > 0$ and $Q_n = \{0, \dots, n-1\}^d$, $n \geq 1$, then an elementary calculation (similar to that in the last part of the proof of Lemma 15.1) shows that the growth rate of $\log s_{Q_n}(\bar{\delta}, \varepsilon)$ is less than or equal to n^{d-1}, and this implies that $h(\alpha) = 0$.

If α is a $\mathbb{Z}^d$-action by automorphisms of a compact group X whose connected component of the identity has zero dimensional centre, and which satisfies the d.c.c., then we may assume that α is the shift-action on a closed, shift-invariant subgroup $X \subset G^{\mathbb{Z}^d}$, where G is a compact, connected Lie group such that $C(G^\circ)$ is finite (Corollary 2.3 and Remark 4.15), and Remark 4.15 also allows us to assume that

$$X = X_H = \{x \in G^{\mathbb{Z}^d} : \pi_I(\alpha_\mathbf{n}(x)) \in H \text{ for every } \mathbf{n} \in \mathbb{Z}^d\}, \qquad (15.2)$$

where $I = \{0,1\}^d \subset \mathbb{Z}^d$ and $H = \pi_I(X)$ is a full subgroup of G^I. The next lemma enables us to characterize the entropy of the shift-action σ of $\mathbb{Z}^d$ on X_H in terms of the group H.

LEMMA 15.5. *For every $\mathbf{m} \in I$, let $I_\mathbf{m} = I \smallsetminus \{\mathbf{m}\}$, and let $H_\mathbf{m} \subset H$ be the kernel of the projection $\pi_{I_\mathbf{m}} : H \longmapsto G^{I_\mathbf{m}}$. Then the entropy of the shift-action σ of $\mathbb{Z}^d$ on the group $X = X_H$ in (15.2) is equal to $\log |H_\mathbf{m}|$ for every $\mathbf{m} \in I$, where $|H_\mathbf{m}|$ is the cardinality of $H_\mathbf{m}$.*

PROOF. We assume for simplicity (and without loss in generality) that $\mathbf{m} = (1, \dots, 1) \subset \mathbb{Z}^d$, fix a metric ϑ on G which is invariant under $\mathrm{Aut}(G)$, and use the notation introduced in the proof of Lemma 15.1.

If $|H_\mathbf{m}| = n < \infty$, then we have to prove that $h(\sigma) = \log n$. In order to do this we choose $\varepsilon > 0$ such that $\vartheta(g, g') > 2\varepsilon$ for all $g \neq g'$ in $G' = \pi_\mathbf{m}(H_\mathbf{m}) \subset G$. Let $k \geq 1$, set $S_k = R_k \cap \Delta_k$, $X_k = \pi_{R_k}(X) \subset G^{R_k}$, and consider the projection $\pi_{S_k} : X_k \longmapsto G^{S_k}$ of the points in X_k onto their coordinates in S_k. The kernel of the homomorphism π_{S_k} of X_k has cardinality $n^{(k-1)^d}$, and any two distinct points in this kernel have distance $> \varepsilon$ in at least one coordinate in R_k. Hence there exists a $[R_k, \vartheta, \varepsilon]$-separated set in X of cardinality $n^{(k-1)^d}$, and by letting $k \to \infty$ we see from Proposition 13.7 that $h(\sigma) \geq \log n$.

If $H_\mathbf{m}$ is infinite (i.e. uncountable), then we can find, for every $M \geq 1$, an $\varepsilon > 0$ such that the group $H' = \pi_\mathbf{m}(H_\mathbf{m}) \subset G$ has an ε-separated set of cardinality $> M$. By applying the argument in the preceding paragraph we obtain an $[R_k, \vartheta, \varepsilon]$-separated set of cardinality $> M^{(k-1)^d}$, and by letting $k \to \infty$ and $\varepsilon \to 0$ we conclude that $h(\sigma) = \infty$.

We return to the assumption that $|H_{\mathbf{m}}| = n < \infty$ and claim that $h(\sigma) \leq \log n$. If $E \subset G$ is a finite, normal subgroup of G, then $E \cap G^\circ \subset C(G^\circ)$ (Proposition 2.13). Hence there exists a unique, maximal, finite, normal subgroup $E' \subset G$, and $E' \cap G^\circ = C(G^\circ)$. Since $H_{\mathbf{m}}$ is finite, the groups $Y_k = \ker(\pi_{\Delta_k})$ and K_k in the proof of Lemma 15.1 are zero dimensional and finite, respectively, and $\pi_{\mathbf{m}}(Y_k) = \pi_{\mathbf{m}}(K_k) \subset E'$ for every $\mathbf{m} \in R_k$. In particular, the coordinates in B_k of a point $x \in X$ determine completely the coset $x_{\mathbf{m}} E'$ for every $\mathbf{m} \in R_k$.

We write $\eta \colon G \longmapsto G'$ for the quotient map, define $\boldsymbol{\eta} \colon X \longmapsto G'^{\mathbb{Z}^d}$ by $(\boldsymbol{\eta}(x))_{\mathbf{m}} = \eta(x_{\mathbf{m}})$ for all $\mathbf{m} \in \mathbb{Z}^d$, and set $X' = \boldsymbol{\eta}(X) \subset G'^{\mathbb{Z}^d}$. As we have just seen, the projection $\pi_{R_k}(y)$ of a point $y \in X'$ onto its coordinates in R_k is completely determined by $\pi_{B_k}(y)$. From (15.1) (applied to G') we show that there exists a constant $c > 0$ such that, for all x, y in X' and $k \geq 1$, $c \cdot \max_{\mathbf{m} \in B_k} \vartheta'(x_{\mathbf{m}}, y_{\mathbf{m}}) \geq \max_{\mathbf{m} \in \bar{R}_k} \vartheta'(x_{\mathbf{m}}, y_{\mathbf{m}})$, where ϑ' is the invariant metric on G' induced by ϑ, and conclude as in the proof of Lemma 15.1 that the entropy of the shift-action σ' of $\mathbb{Z}^d$ on X' is zero.

Put $B'_{\vartheta'}(\varepsilon) = \{x \in X' : \vartheta'(x_0, \mathbf{1}_{G'}) < \varepsilon\}$, and let, for every rectangle $Q \subset \mathbb{Z}^d$, $B'_{\vartheta'}(Q, \varepsilon) = \bigcap_{\mathbf{m} \in Q} \sigma'_{-\mathbf{m}}(B'_{\vartheta'}(\varepsilon))$. From Proposition 13.7 we know that

$$\lim_{\varepsilon \to 0} \liminf_{k \to \infty} -(k-1)^{-d} \log \lambda_{X'}(B'_{\vartheta'}(R_k, \varepsilon)) = 0.$$

If ε is small enough, then $\boldsymbol{\eta}^{-1}(B'_{\vartheta'}(R_k, \varepsilon))$ consists of at most $|E'|^{2dk^{d-1}} n^{(k-1)^d}$ disjoint translates of $B'_\vartheta(Q, \varepsilon) = \bigcap_{\mathbf{m} \in R_k} \sigma_{-\mathbf{m}}(B'_\vartheta(\varepsilon))$, where $B'_\vartheta(\varepsilon) = \{x \in X : \vartheta(x_0, \mathbf{1}_G) < \varepsilon\}$. Hence $\lambda_X(B'_\vartheta(R_k, \varepsilon)) \geq \lambda_{X'}(B'_{\vartheta'}(R_k, \varepsilon))/|E'|^{2dk^{d-1}} n^{(k-1)^d}$, and

$$h(\sigma) = \lim_{\varepsilon \to 0} \liminf_{k \to \infty} -(k-1)^{-d} \log \lambda_X(B'_\vartheta(R_k, \varepsilon)) \leq \log n. \quad \square$$

THEOREM 15.6. *Let α be a $\mathbb{Z}^d$-action by automorphisms of a compact group X whose connected component of the identity has zero dimensional centre. If $h(\alpha) < \infty$ then $h(\alpha) = \log k$ for some integer $k \geq 1$.*

PROOF. Let $(V_n, n \geq 1)$ be a decreasing sequence of closed, normal, α-invariant subgroups of X such that $\bigcap_{n \geq 1} V_n = \{\mathbf{1}_X\}$ and α^{X/V_n} satisfies the d.c.c. for every $n \geq 1$ (Proposition 4.9). By Lemma 13.6, $h(\alpha^{X/V_n}) \to h(\alpha)$ as $n \to \infty$, so that it suffices to prove the proposition under the assumption that α satisfies the d.c.c. The proof is completed by applying Lemma 15.5, together with the brief discussion preceding its statement. $\square$

COROLLARY 15.7. *Let α be a $\mathbb{Z}^d$-action by automorphisms of a zero dimensional compact group X. If $h(\alpha) < \infty$ then $h(\alpha) = \log k$ for some integer $k \geq 1$.*

CONCLUDING REMARKS 15.8. (1) Lemma 15.5 implies that $|H_{\mathbf{m}}| = |H_{\mathbf{n}}|$ for all $\mathbf{m}, \mathbf{n} \in I$. In order to prove this directly we assume for simplicity that $d = 2$, and consider the sets $P = \{(m, n) \in \mathbb{Z}^2 : n < 0\}$ and $L = \{(m, 0) : m \in \mathbb{Z}\} \subset \mathbb{Z}^2$. Then $\pi_L(\ker(\pi_P))$ is a closed, shift-invariant subgroup of G^L, and by identifying $L = \mathbb{Z} \times \{0\}$ with $\mathbb{Z}$ we may assume that $Y = \pi_L(\ker(\pi_P)) \subset G^{\mathbb{Z}}$. From the definition of X_H in (15.2) we see that $Y = Y_K$, where Y_K is the Markov subgroup arising from $K = \{(k_1, k_2) \in G \times G : \left(\begin{smallmatrix} k_1 & k_1 \\ 1_G & 1_G \end{smallmatrix}\right) \in H\}$ as in (9.4). We set $G' = \pi_{\{0\}}(Y) \subset G$ and note that the groups $F_K(1) = \{g \in G' : \left(\begin{smallmatrix} 1_G & g \\ 1_G & 1_G \end{smallmatrix}\right) \in H\}$ and $F_K(-1) = \{g \in G' : \left(\begin{smallmatrix} g & 1_G \\ 1_G & 1_G \end{smallmatrix}\right) \in H\}$ are isomorphic to $H_{(1,1)}$ and $H_{(0,1)}$, respectively, and that $G'/F_K(1) \cong G'/F_K(-1)$ by (10.1)–(10.4). Since G' is normal in G, G' has finite centre, and the isomorphism of $G'/F_K(1)$ and $G'/F_K(-1)$ implies that $|H_{(1,1)}| = |F_K(1)| = |F_K(-1)| = |H_{(0,1)}|$. Similarly one proves that $|H_{(0,1)}| = |H_{(0,0)}| = |H_{(0,1)}|$, and the proof in higher dimensions is completely analogous.

(2) Corollary 14.3 reduces the study of $\mathbb{Z}^d$-actions by automorphisms of compact groups with zero entropy to the analysis of zero entropy $\mathbb{Z}^d$-actions on zero dimensional groups, on connected groups with trivial centres, and on abelian groups, and Lemma 13.6 allows us to restrict our attention to actions satisfying the d.c.c. $\mathbb{Z}^d$-actions on compact, abelian groups will be discussed in Section 18, and the other two cases can be dealt with by investigating the closed, shift-invariant subgroups $X \subset G^{\mathbb{Z}^d}$ with zero entropy, where G is a compact Lie group such that G° has finite centre. By using both the notation and the arguments employed in the proofs of the Lemmas 15.1 and 15.5 one can easily show the following: the shift-action by $\mathbb{Z}^d$ on a closed, shift-invariant subgroup $X \subset G^{\mathbb{Z}^d}$ satisfying (15.2) has zero entropy if and only if, for every $k \geq 1$, the projection of any point $x \in X$ onto its coordinates in $\bar{R}_k = \{0, \ldots, k\}^d$ is completely determined by its projection onto the *boundary* $B_k = \bar{R}_k \cap \Delta_k$ of $\bar{R}_k$. The analogous statement for closed, shift-invariant subgroups $X \subset (\mathbb{T}^n)^{\mathbb{Z}^d}$ is obviously incorrect (Remark 3.10 (2)).

16. Mahler measure

In this section we discuss a quantity introduced by Mahler in [67], [68], which will yield the last remaining ingredient of equation (14.16) by providing an entropy formula for $\mathbb{Z}^d$-actions by automorphisms of compact, abelian groups. The *Mahler measure* of a polynomial $f \in \mathfrak{R}_d$, $d \geq 1$, is defined as

$$\mathbb{M}(f) = \begin{cases} \exp\left(\int_{\mathbb{S}^d} \log|f(\mathbf{s})| \, d\mathbf{s}\right) & \text{if } f \neq 0, \\ 0 & \text{if } f = 0, \end{cases} \tag{16.1}$$

where $d\mathbf{s}$ denotes integration with respect to the normalized Haar measure on the multiplicative subgroup $\mathbb{S}^d \subset \mathbb{C}^d$, and where f is regarded as a function on $\mathbb{C}^d$.

PROPOSITION 16.1. *Let $f = a_0 + \cdots + a_s u^s$ be a polynomial with complex coefficients, with $a_0 a_s \neq 0$, and with roots $\xi_1, \ldots, \xi_s$. Then*

$$\int_0^1 \log |f(e^{2\pi it})|\, dt = \sum_{j=1}^s \log^+ |\xi_j| + \log |a_s|, \tag{16.2}$$

where $\log^+ t = \log (\max\{1, t\})$ for every $t \geq 0$.

PROOF. According to Jensen's formula ([2], p. 208),

$$\log |a_0| = \log |f(0)| = -\sum_{j=1}^s \log^+ |\xi_j^{-1}| + \int_0^1 \log |f(e^{2\pi it})|\, dt.$$

Since $a_0/a_s = \prod_{j=1}^s \xi_j$ we have that

$$\int_0^1 \log |f(e^{2\pi it})|\, dt = \log |a_0| + \sum_{j=1}^s \log^+ |\xi_j^{-1}|$$

$$= \log |a_s| + \sum_{j=1}^s \log |\xi_j| - \sum_{\{j:1 \leq j \leq s \text{ and } |\xi_j| < 1\}} \log |\xi_j|$$

$$= \log |a_s| + \sum_{j=1}^s \log^+ |\xi_j|. \quad \square$$

Now assume that $d \geq 1$, and that $f \in \mathfrak{R}_d$. For $\mathbf{r} = (r_1, \ldots, r_d)$, $\mathbf{m} = (m_1, \ldots, m_d)$ in $\mathbb{Z}^d$ we put

$$|\mathbf{m}| = \max\{|m_1|, \ldots, |m_d|\},$$
$$\langle \mathbf{r}, \mathbf{m} \rangle = r_1 m_1 + \cdots + r_d m_d, \tag{16.3}$$
$$\langle \mathbf{r} \rangle = \min\{|\mathbf{n}| : \mathbf{0} \neq \mathbf{n} \in \mathbb{Z}^d \text{ and } \langle \mathbf{r}, \mathbf{n} \rangle = 0\},$$

and

$$f_{\mathbf{r}}(u) = f(u^{r_1}, \ldots, u^{r_d}) = \sum_{\mathbf{m} \in \mathbb{Z}^d} c_f(\mathbf{m}) u^{\langle \mathbf{r}, \mathbf{m} \rangle} \in \mathfrak{R}_1 = \mathbb{Z}[u^{\pm 1}].$$

Then $\mathbb{M}(f)$ has the following asymptotic expression.

PROPOSITION 16.2. *For every $f \in \mathfrak{R}_d$, $\lim_{\langle \mathbf{r} \rangle \to \infty} \mathbb{M}(f_{\mathbf{r}}) = \mathbb{M}(f)$.*

For the proof of Proposition 16.2 we introduce the following terminology. An *interval* $I \subset \mathbb{S}$ is a connected subset, and a continuous function $f : \mathbb{S} \longmapsto \mathbb{C}$ is *uniform* on an interval $I \subset \mathbb{S}$ if $f(I)$ is contained in one of the four quadrants of $\mathbb{C}$, i.e. $f(I) \subset \{z \in \mathbb{C} : \text{Re}(z) \geq 0 \text{ and } \text{Im}(z) \geq 0\}$, $f(I) \subset \{z \in \mathbb{C} : \text{Re}(z) \geq 0 \text{ and } \text{Im}(z) \leq 0\}$, $f(I) \subset \{z \in \mathbb{C} : \text{Re}(z) \leq 0 \text{ and } \text{Im}(z) \geq 0\}$, or $f(I) \subset \{z \in \mathbb{C} : \text{Re}(z) \leq 0 \text{ and } \text{Im}(z) \leq 0\}$. The normalized Haar measure on $\mathbb{S}$ is denoted by $\lambda = \lambda_{\mathbb{S}}$.

LEMMA 16.3. *Let $m \geq 1$, and let $g(u) = c_g(0) + \cdots + c_g(m-1)u^{m-1} + u^m$ be a monic polynomial with real coefficients. If $a > 0$, and if $A = \{z \in \mathbb{S} : |g(z)| \geq a\}$, $B = \{z \in \mathbb{S} : |g(z)| \leq a\}$, then A and B each have at most m connected components. Furthermore, $\mathbb{S}$ is the union of at most $4m$ closed intervals I_j such that g is uniform on each I_j.*

PROOF. Every element in the set $S = \{z \in \mathbb{S} : |g(z)| = a\}$ is a root of the polynomial $h(u) = u^m(g(u^{-1})g(u) - a^2)$; since the degree of h is equal to $2m$, S has at most $2m$ elements, and the sets A and B each have at most m connected components. The second assertion is proved in the same manner by taking $h = u^m(g(u) \pm g(u^{-1}))$. $\square$

LEMMA 16.4. *For every $k \geq 2$ there exists a constant $C_k > 0$ with the following property: if $g(u) = c_g(0) + \cdots + c_g(m-1)u^{m-1} + u^m$ is a monic polynomial with real coefficients, and if g has exactly $k \geq 2$ non-zero coefficients, then*

$$\lambda(\{z \in \mathbb{S} : |g(z)| \leq a\}) \leq C_k a^{\frac{1}{k-1}}$$

for every $a > 0$.

PROOF. We assume without loss in generality that g has a non-zero constant term. Let $A(a) = \{z \in \mathbb{S} : |g(z)| \leq a\}$. If $k = 2$ then $\lambda(A(a)) = \lambda(\{z \in \mathbb{S} : |c_g(0) + z| \leq a\}) \leq 2a\pi$, and we set $C_2 = 2\pi$. Suppose that the lemma has been proved for $k \leq K$, where $K \geq 2$, and that g has $K + 1$ non-zero coefficients. Then $g' = \frac{dg}{du}$ has K non-zero coefficients, and our induction hypothesis implies that

$$\lambda(\{z \in \mathbb{S} : |g'(z)/m| \leq a\}) \leq C_k a^{\frac{1}{k-1}}$$

for every $a > 0$. We set $B(b) = \{z \in \mathbb{C} : |g'(z)| \geq b\}$ and obtain that

$$\lambda(A(a)) \leq \lambda(\{z \in \mathbb{S} : |g'(z)| \leq b\}) + \lambda(A(a) \cap B(b))$$
$$\leq C_k \left(\frac{b}{m}\right)^{\frac{1}{k-1}} + \lambda(A(a) \cap B(b))$$

for all $a, b > 0$. Since both g and g' have degree $\leq m$, Lemma 16.3 shows that $A(a) \cap B(b)$ can be expressed as the union of at most $m + m + 4m = 6m$ closed intervals, on each of which g' is uniform. If $I = [c, d]$ is one of these intervals, then the uniformity of g' implies that $|\mathrm{Re}(g(d)) - \mathrm{Re}(g(c))| = \int_I |\mathrm{Re}(g')| \, d\lambda$ and $|\mathrm{Im}(g(d)) - \mathrm{Im}(g(c))| = \int_I |\mathrm{Im}(g')| \, d\lambda$, and hence $2a \geq |g(d) - g(c)| \geq b\lambda(I)/\sqrt{2}$. We obtain that $\lambda(A(a) \cap B(b)) \leq 12m\sqrt{2}\frac{a}{b}$, and $\lambda(A(a)) \leq C_k(\frac{b}{m})^{\frac{1}{k-1}} + 12m\sqrt{2}\frac{a}{b}$. This inequality holds for every $b > 0$, and the induction step is proved by setting $\frac{b}{m} = a^{\frac{k-1}{k}}$ and $C_{k+1} = C_k + 12\sqrt{2}$. $\square$

LEMMA 16.5. *Let $g \in \mathfrak{R}_1$, and assume that g has $k \geq 2$ non-zero coefficients. For every $y \in \mathbb{R}$ with $0 < y \leq 1$ we put $A_g(y) = \{z \in \mathbb{S} : |g(z)| < y\}$. Then*

$$0 \geq \int_{A_g(y)} \log|g| \, d\lambda \geq C_k((1 - k + \log y)y^{\frac{1}{k-1}}).$$

PROOF. Let $\nu(y) = \lambda(A_g(y))$, $0 \leq y < 1$. We assume without loss in generality that $g(y) = c_g(0) + \cdots + c_g(m)u^m$ with $c_g(m) \neq 0$, apply Lemma 16.4 to $g/c_g(m)$, and obtain that $\nu(y) \leq C_k(y/c_g(m))^{\frac{1}{k-1}} \leq C_k y^{\frac{1}{k-1}}$. Integration by parts yields that

$$\int_{A_g(y)} \log|g| \, d\lambda = \int_0^y \log x \, d\nu(x) = \nu(y) \log y - \int_0^y C_k x^{\frac{1}{k-1}} \frac{1}{x} \, dx$$

$$\geq C_k y^{\frac{1}{k-1}} \log y - \int_0^y C_k x^{\frac{1}{k-1}} \frac{1}{x} \, dx$$

$$= C_k((1 - k + \log y)y^{\frac{1}{k-1}}). \quad \square$$

PROOF OF PROPOSITION 16.2. For every non-zero $\mathbf{r} \in \mathbb{Z}^d$ we define a homomorphism $\eta_{\mathbf{r}} \colon \mathbb{Z}^d \longmapsto \mathbb{Z}$ by $\eta_{\mathbf{r}}(\mathbf{m}) = \langle \mathbf{r}, \mathbf{m} \rangle$. The group $Z = \eta_{\mathbf{r}}(\mathbb{Z}^d) \subset \mathbb{Z}$ is isomorphic to $\mathbb{Z}$, and by dualizing we obtain an injective homomorphism $\hat{\eta}_{\mathbf{r}} \colon \mathbb{S} \cong \hat{Z} \longmapsto \mathbb{S}^d$ and set $\mu_{\mathbf{r}} = \lambda_{\mathbb{S}} \hat{\eta}_{\mathbf{r}}^{-1}$. Then $\lim_{\langle \mathbf{r} \rangle \to \infty} \int_{\mathbb{S}^d} \chi \, d\mu_{\mathbf{r}} = 0$ for every non-trivial character χ of $\mathbb{S}^d$, and Weierstrass' approximation theorem implies that $\lim_{\langle \mathbf{r} \rangle \to \infty} \int_{\mathbb{S}^d} g \, d\mu_{\mathbf{r}} = \int g \, d\lambda_{\mathbb{S}^d}$ for all continuous functions $g \colon \mathbb{S}^d \longmapsto \mathbb{C}$. Hence $\int_{\mathbb{S}} \log|f_{\mathbf{r}}| \, d\lambda = \int_{\mathbb{S}^d} \log|f| \, d\mu_{\mathbf{r}} \to \int_{\mathbb{S}^d} \log|f| \, d\lambda_{\mathbb{S}^d}$ for every $f \in \mathfrak{R}_d$ with $f(\mathbf{z}) \neq 0$ for all $\mathbf{z} \in \mathbb{S}^d$.

If $f = 0$, the assertion is obvious, since all terms are equal to 0 by definition. If $f \neq 0$, but $f(\mathbf{z}) = 0$ for some $\mathbf{z} \in \mathbb{S}^d$, then the function $\log|f|$ is integrable on $\mathbb{S}^d$, and

$$\int_{\mathbb{S}} \log|f_{\mathbf{r}}| \, d\lambda = \int_{\mathbb{S}^d} \log|f| \, d\mu_{\mathbf{r}} \to \int_{\mathbb{S}^d} f \, d\lambda_{\mathbb{S}^d}$$

provided that there exists, for every $\varepsilon > 0$, a $y > 0$ in $\mathbb{R}$ with

$$\int_{A_{f_{\mathbf{r}}}(y)} \log|f_{\mathbf{r}}| \, d\lambda_{\mathbb{S}} = \int_{\{\mathbf{z} \in \mathbb{S}^d : |f(\mathbf{z})| \leq y\}} \log|f| \, d\mu_{\mathbf{r}} > -\varepsilon$$

whenever $\langle \mathbf{r} \rangle$ is sufficiently large. However, since $f_{\mathbf{r}}$ has at most as many non-zero coefficients as f, this is precisely the statement of Lemma 16.5, so that the proposition is proved. $\square$

COROLLARY 16.6. *For every non-zero polynomial $f \in \mathfrak{R}_d$, $1 \leq \mathrm{M}(f) < \infty$.*

PROOF. From Proposition 16.1 we know that $\mathbb{M}(h) \geq 1$ for every non-zero polynomial $h \in \mathfrak{R}_1$, and Proposition 16.2 implies the same inequality for every $f \in \mathfrak{R}_d$. The assertion that $M(f) < \infty$ is obvious, since $|\log f|$ is integrable on $\mathbb{S}^d$. $\square$

CONCLUDING REMARK 16.7. Proposition 16.2 is taken from [53], [54].

17. Mahler measure and entropy of group automorphisms

Any reader familiar with the entropy formula for toral automorphisms will have recognized a connection between entropy and the expression for the Mahler measure of a polynomial $f \in \mathfrak{R}_1$ in (16.2). However, before we can enter into a discussion of Mahler measure and entropy, we have to prove a lemma about the uniform convergence of the entropy of linear maps on $\mathbb{C}^K$, where $K \geq 1$. The corresponding point-wise result can be found in [105], Theorem 8.14, and is sufficient for determining the entropy of $\mathbb{Z}$-actions on compact, abelian groups. The full strength of Lemma 17.1 will only be required when we discuss $\mathbb{Z}^d$-actions with $d > 1$.

For every $z = (z_0, \ldots, z_{K-1}) \in \mathbb{C}^K$ we set $\|z\|_\infty = \max_{k=0,\ldots,K-1} |z_k|$, and we put $C(\varepsilon) = \{z \in \mathbb{C}^K : \|z\|_\infty < \varepsilon\}$ for every $\varepsilon > 0$. The symbol λ will denote the Lebesgue measure on $\mathbb{C}^K \cong \mathbb{R}^{2K}$.

LEMMA 17.1. *Let $M_K(\mathbb{C}) \cong \mathbb{C}^{K^2}$ be the set of $K \times K$-matrices with complex entries. For every $A \in M_K(\mathbb{C})$ and $m \geq 1$ we set*

$$\bar{h}_m(A) = -\frac{1}{m} \log \lambda \left(\bigcap_{j=0}^{m-1} A^{-j} C(1) \right),$$

and

$$\bar{h}(A) = \sum_{k=1}^{K} \log^+ |\zeta_k|^2 = 2 \int_0^1 \log |\chi_A(e^{2\pi i t})| \, dt,$$

where χ_A is the characteristic polynomial and $\zeta_1, \ldots, \zeta_K$ are the eigenvalues of A. Then the following is true.

(1) *The functions $A \mapsto \bar{h}(A)$ and $A \mapsto \bar{h}_m(A)$, $m \geq 1$, are continuous on $M_K(\mathbb{C})$;*
(2) *As $m \to \infty$, $\bar{h}_m(A) \to \bar{h}(A)$ uniformly on compact subsets of $M_K(\mathbb{C})$.*

PROOF. Fix $A \in M_K(\mathbb{C})$, $m \geq 1$, and $0 < \theta < 1$, and note that

$$\theta \bigcap_{j=0}^{m-1} B^{-j} C(1) = \bigcap_{j=0}^{m-1} B^{-j} C(\theta)$$

for every $B \in M_K(\mathbb{C})$. If $(B_n, n \geq 1)$ is a sequence in $M_K(\mathbb{C})$ which converges to A then $\bigcap_{j=0}^{m-1} B_n^{-j} C(\theta) \subset \bigcap_{j=0}^{m-1} A^{-j} C(1)$ for all sufficiently large n, so that

$$\theta^{2K} \bar{h}_m(B_n) = \theta^{2K} \lambda \left(\bigcap_{j=0}^{m-1} B_n^{-j} C(1) \right) = \lambda \left(\bigcap_{j=0}^{m-1} B_n^{-j} C(\theta) \right)$$

$$\leq \lambda \left(\bigcap_{j=0}^{m-1} A^{-j} C(1) \right) = \bar{h}_m(A)$$

for all sufficiently large n. By letting $\theta \to 1$ we see that $\bar{h}_m$ is lower semicontinuous on $M_K(\mathbb{C})$. Upper semicontinuity is proved similarly by noting that, for $0 < \theta < 1$, $\bigcap_{j=0}^{m-1} B_n^{-j} C(1) \supset \bigcap_{j=0}^{m-1} A^{-j} C(\theta)$ for all sufficiently large n. Hence $\bar{h}_m$ is continuous for every $m \geq 1$, and the continuity of $\bar{h}$ will follow from (2).

To prove (2) it suffices to show that there exists, for every $A \in M_K(\mathbb{C})$ and every $\varepsilon > 0$, a neighbourhood $N(A) \subset M_K(\mathbb{C})$ and an $m_0 \geq 1$ such that $|\bar{h}_m(B) - \bar{h}(B)| < \varepsilon$ for all $B \in N(A)$ and $m \geq m_0$. We fix A, write the eigenvalues of A without multiplicity as $\xi_{-r}, \ldots, \xi_s$ with $|\xi_k| \leq 1$ for $k = -r, \ldots, 0$ and $|\xi_k| > 1$ for $k = 1, \ldots, s$, and denote the multiplicity of ξ_k by N_k. Fix $\varepsilon > 0$ and choose $\gamma > 0$ such that the circles C_k of radius γ around ξ_k are disjoint, $|\xi_k| - \gamma > 1$ for all $k = 1, \ldots, s$, and $4K \log(1+\gamma) < \varepsilon$. According to §I.5.3 in [36],

$$P_k = P_k(A) = \frac{1}{2\pi i} \int_{C_k} (\xi I - A)^{-1} \, d\xi,$$

where I is the $K \times K$ identity matrix, is the projection onto the generalized eigenspace V_k of A for ξ_k along the sum of the other generalized eigenspaces, and $\dim_{\mathbb{C}} V_k = N_k$. We fix k for the moment and set, for every $y \in V_k$,

$$\|y\|_k = \sum_{n=0}^{\infty} \frac{\|A^n y\|_\infty}{(|\xi_k| + \gamma)^n}.$$

The spectral radius formula shows that this series converges geometrically. If A_k denotes the restriction of A to V_k, then $\|A_k y\|_k < (|\xi_k| + \gamma) \|y\|_k$ for $0 \neq y \in V_k$. We denote by E_k the unit ball in V_k in the norm $\| \cdot \|_k$ and observe that $(|\xi_k| + \gamma)^{-1} E_k \Subset A_k^{-1} E_k$, where $F \Subset F'$ indicates that the closure of F is contained in F'. Since $E = \bigoplus_{k=-r}^s E_k \subset \bigoplus_{k=-r}^s V_k = \mathbb{C}^K$ is a bounded neighbourhood of 0 there exists a $\theta > 0$ with $C(\theta) \Subset E \Subset C(\theta^{-1})$.

We have to investigate how this picture changes as A varies. There exists a small neighbourhood $N(A)$ of the matrix A such that, for every $\xi \in C_k$, the map $B \mapsto (\xi I - B)^{-1}$ is analytic in $N(A)$ and the dimension of the projection $P_k(B) = \frac{1}{2\pi i} \int_{C_k} (\xi I - B)^{-1} \, d\xi$ is constant and equal to N_k in this neighbourhood. Since the projection $P_k(B)$ is close to $P_k(A)$ if B is close to A, we may further reduce the size of $N(A)$, if necessary, and assume that there exists a continuous map $B \mapsto U(B)$ from $N(A)$ into $\mathrm{GL}(K, \mathbb{C})$ such that $P_k(B)U(B) = U(B)P_k(A)$ for every $B \in N(A)$ (cf. [36], Section 1.6). Write

$V_k(B) = P_k(B)\mathbb{C}^K$ for the range of $P_k(B)$, and put $E_k(B) = U(B)E_k$. Finally we vary k and set $E(B) = \bigoplus_{k=-r}^{s} E_k(B)$. The restriction of B to $V_k(B)$ will be denoted by B_k. After restricting $N(A)$ still further we can assume that $(|\xi_k| + \gamma)^{-1}E_k(B) \Subset B_k^{-1}E_k(B)$ and

$$C(\theta^2) \Subset \theta E \Subset E(B) \Subset \theta^{-1}E \Subset C(\theta^{-2}). \tag{17.1}$$

We shall first prove that $\bar{h}_m(B) > \bar{h}(B) - \varepsilon$ for all large enough m and all $B \in N(A)$. Since

$$\bigcap_{j=0}^{m-1} B^{-j}E(B) \subset \bigoplus_{k=-r}^{0} E_k(B) \oplus \bigoplus_{k=1}^{s} B_k^{-m+1}E_k(B),$$

we have that

$$-\frac{1}{m} \log \lambda \left(\bigcap_{j=0}^{m-1} B^{-j}E(B) \right)$$

$$\geq -\frac{1}{m} \log \lambda(E(B)) + \frac{m-1}{m} \sum_{k=1}^{s} \log |\det B_k|^2, \tag{17.2}$$

where the square in the last term is a consequence of working over $\mathbb{C}$ and not over $\mathbb{R}$. For every $k \geq 1$ every eigenvalue of B_k lies inside C_k, and therefore has modulus > 1. As there are at most K other eigenvalues of B, each with modulus $< 1 + \gamma$, we obtain that

$$\sum_{k=1}^{s} \log |\det B_k|^2 \geq \bar{h}(B) - 2K \log(1 + \gamma). \tag{17.3}$$

Since $\theta E \subset E(B)$ we have that $\theta \bigcap_{j=0}^{m-1} B^{-j}E \subset \bigcap_{j=0}^{m-1} B^{-j}E(B)$, and (17.1)–(17.3) imply that

$$\bar{h}_m(B) - \frac{4K \log \theta}{m} \geq -\frac{1}{m} \log \lambda \left(\bigcap_{j=0}^{m-1} B^{-j}E(B) \right)$$

$$\geq (1 - m^{-1})\bar{h}(B) - \frac{1}{m} \log \lambda(C(\theta^{-2})) - 2K \log(1 + \gamma),$$

so that $\bar{h}_m(B) \geq \bar{h}(B) - \varepsilon$ for all sufficiently large m and all $B \in N(A)$. For the opposite inequality we recall that $(|\xi_k| + \gamma)^{-j}E_k(B) \Subset B_k^{-j}E_k(B)$, so that

$$\bigcap_{j=0}^{m-1} B^{-j}E(B) \supset \bigoplus_{k=-r}^{0} (1 + \gamma)^{-m+1}E_k(B) \oplus \bigoplus_{k=1}^{s} (|\xi_k| + \gamma)^{-m+1}E_k(B)$$

and

$$\lambda \left(\bigcap_{j=0}^{m-1} B^{-j}E(B) \right) \geq \lambda(E(B))(1 + \gamma)^{-2K(m-1)} \prod_{k=1}^{s} (|\xi_k| + \gamma)^{-2N_k(m-1)}. \tag{17.4}$$

Since $E(B) \subset \theta^{-1}E$ we see as above that

$$-\frac{1}{m}\log\lambda\left(\bigcap_{j=0}^{m-1}B^{-j}E(B)\right) \geq \frac{2K\log\theta}{m} + \bar{h}_m(B), \qquad (17.5)$$

and (17.4) and (17.5) imply that

$$\bar{h}_m(B) + \frac{2K\log\theta}{m} \leq \frac{m-1}{m}\sum_{k=1}^{s}2N_k\log(|\xi_k|+\gamma) - \frac{1}{m}\lambda(E(B))$$

$$+ \frac{m-1}{m}2K\log(1+\gamma)$$

$$\leq \sum_{k=1}^{s}2N_k\log(|\xi_k|+\gamma) - \frac{1}{m}\log\lambda(C(\theta^2)) + 2K\log(1+\gamma).$$

As B has exactly N_k eigenvalues within γ of ξ_k for $k = 1,\ldots,s$, and as the remaining eigenvalues all have modulus $< 1 + \gamma$, our choice of γ implies that the sum $\sum_{k=1}^{s}2N_k\log(|\xi_k|+\gamma)$ is within $\varepsilon/2$ of $\bar{h}(B)$. It follows that $\bar{h}_m(B) < \bar{h}(B) + \varepsilon$ for all sufficiently large m and $B \in N(A)$. $\square$

PROPOSITION 17.2. *For every* $f \in \mathfrak{R}_1$,

$$h(\alpha^{\mathfrak{R}_1/(f)}) = |\log \mathrm{M}(f)| = \begin{cases} \log \mathrm{M}(f) & \text{if } f \neq 0, \\ \infty & \text{if } f = 0. \end{cases} \qquad (17.6)$$

PROOF. If $f = 0$ then $\mathfrak{R}_1/(f) = \mathfrak{R}_1$, $\alpha^{\mathfrak{R}_1}$ is the shift-action of $\mathbb{Z}$ on $\mathbb{T}^{\mathbb{Z}}$, and $h(\alpha^{\mathfrak{R}_1}) = \infty$ (Example 5.2 (2)). If $f = cu_1^m$ for some $c, m \in \mathbb{Z}$ then $(f) = c\mathfrak{R}_1$, $\alpha^{\mathfrak{R}_1/(f)}$ is isomorphic to the shift on $(\mathbb{Z}_{/|c|})^{\mathbb{Z}}$ (cf. Example 5.2 (2)), and $h(\alpha^{\mathfrak{R}_1/(f)}) = \log|c| = \log \mathrm{M}(f)$. Assume therefore that f has at least two non-zero terms. After multiplying f by a monomial we may take it that $f(u_1) = c_s u_1^s + c_{s-1}u_1^{s-1} + \cdots + c_1 u_1 + c_0$ with $c_s c_0 \neq 0$, and we set $a_j = c_j/c_s$ for $j = 1,\ldots,s-1$. According to Example 5.2 (2), $X = X^{\mathfrak{R}_1/(f)} = \{(x_n) \in \mathbb{T}^{\mathbb{Z}} : \sum_{k=0}^{s}c_k x_{n+k} = 0 \pmod 1 \text{ for all } n \in \mathbb{Z}\}$, and $\alpha = \alpha^{\mathfrak{R}_1/(f)}$ is the restriction to X of the shift-action σ of $\mathbb{Z}$ on $\mathbb{T}^{\mathbb{Z}}$.

In order to calculate the entropy $h(\alpha) = h_{\mathrm{vol}}(\alpha)$ (cf. Proposition 13.7) we write ϑ for the usual metric

$$\vartheta(s,t) = \|t - s\| = \min\{|s - t - m| : m \in \mathbb{Z}\} \qquad (17.7)$$

on $\mathbb{T}$ and denote by $\zeta_1,\ldots,\zeta_s$ the roots of the polynomial f, counted with multiplicity. Fix $0 < \varepsilon < \frac{1}{2}$, and consider the sets $B(N,\varepsilon) = \{x \in X : |x_j| < \varepsilon \text{ for all } j = 0,\ldots,N-1\}$, where $N \geq 1$. For every $x \in B(N,\varepsilon)$ there exists a unique point $\phi_N(x) = y = (y_0,\ldots,y_{N-1}) \in \mathbb{R}^N$ with $|y_j| < \varepsilon$ and $y_j \pmod 1 = x_j$ for $j = 0,\ldots,N-1$, and we set $\phi(x) = (y_0,\ldots,y_{s-1})$. If $\varepsilon < (4|c_s|(\max_{k=1,\ldots,s}|\zeta_k|+1))^{-1}$ and $N > s$ then $\phi_N(B(N,\varepsilon)) = \{y = (y_0,\ldots,y_{N-1}) \in \mathbb{R}^N : a_0 y_k + \cdots + a_{s-1}y_{k+s-1} + y_{k+s} = 0 \text{ for } k = 0,\ldots,N-s-$

1}. We fix $y = (y_0, \ldots, y_{N-1}) \in \phi_N(B(N, \varepsilon))$, set $z_k = (y_k, \ldots, y_{k+s-1}) \in \mathbb{R}^s$, and observe that $z_{k+1} = Az_k$ for $k = 0, \ldots, N - s - 2$, where

$$A = \begin{pmatrix} 0 & 1 & 0 & \cdots & 0 \\ 0 & 0 & 1 & \cdots & 0 \\ \vdots & \vdots & \vdots & \ddots & \vdots \\ 0 & 0 & 0 & \cdots & 1 \\ -a_0 & -a_1 & -a_2 & \cdots & -a_{s-1} \end{pmatrix}.$$

Hence

$$\phi(B(N, \varepsilon)) = \{t \in \mathbb{R}^s : \|A^k t\|_\infty < \varepsilon \text{ for } k = 0, \ldots, N - s - 1\},$$

where $\| \cdot \|_\infty$ denotes the maximum norm both on $\mathbb{R}^s$ and $\mathbb{C}^s$. Let $\lambda_{\mathbb{R}^s}$ and $\lambda_{\mathbb{C}^s}$ denote the Lebesgue measures on $\mathbb{R}^s$ and $\mathbb{C}^s \cong \mathbb{R}^{2s}$, respectively, and note that

$$\lambda_X(B(N-1, \varepsilon))/\lambda_X(B(N, \varepsilon)) = |c_s|\lambda_{\mathbb{R}^s}(\phi(B(N-1, \varepsilon)))/\lambda_{\mathbb{R}^s}(\phi(B(N, \varepsilon))).$$

Since $\lambda_X(B(s, \varepsilon)) = \lambda_{\mathbb{R}^s}(\phi(B(s, \varepsilon))) = (2\varepsilon)^s$ we obtain that

$$\lambda_X(B(N, \varepsilon)) = |c_s|^{N-s}\lambda_{\mathbb{R}^s}(\phi(B(N, \varepsilon))) \tag{17.8}$$

for every $N \geq s$, and

$$\lambda_{\mathbb{R}^s}(\phi(B(N, \varepsilon)))^2 > \lambda_{\mathbb{C}^s}(\{t \in \mathbb{C}^s : \|A^k t\|_\infty < \varepsilon \text{ for } k = 0, \ldots, N - s - 1\}),$$

where we regard A as a linear map on both $\mathbb{R}^s$ and $\mathbb{C}^s$. According to Lemma 17.1,

$$\begin{aligned} &-\frac{2}{N} \log \lambda_{\mathbb{R}^s}(\phi(B(N, \varepsilon))) \\ &= -\frac{2}{N} \log \lambda_{\mathbb{R}^s}(\{t \in \mathbb{R}^s : \|A^k t\|_\infty < \varepsilon \text{ for } 0 \leq k < N - s\}) \\ &< -\frac{1}{N} \log \lambda_{\mathbb{C}^s}(\{t \in \mathbb{C}^s : \|A^k t\|_\infty < \varepsilon \text{ for } 0 \leq k < N - s\}) \\ &= \frac{N - s}{N} \bar{h}_{N-s}(A) \rightarrow 2 \sum_{j=1}^{s} \log^+ |\zeta_j| \end{aligned} \tag{17.9}$$

as $N \rightarrow \infty$. By combining (17.8), (17.9), and Proposition 16.1, we obtain that

$$h(\alpha) = h_{\text{vol}}(\alpha) = \sup_{\varepsilon > 0} \limsup_{N \rightarrow \infty} -\frac{1}{N} \log \lambda_X(B(N, \varepsilon))$$

$$= \log |c_s| + \sum_{j=1}^{s} \log^+ |\zeta_j| = \log M(f). \quad \square$$

We present a second proof of Proposition 17.2, taken from [64], which gives a very detailed picture of the dynamics of $\alpha^{\Re_1/(f)}$, but which requires familiarity with the arithmetical ideas introduced in Section 7.

SECOND PROOF OF PROPOSITION 17.2. As in the first proof we can imme-
diately dismiss the cases where $f = 0$, or where $f = cu_1^m$ for some $c, m \in \mathbb{Z}$.
Let us first prove the proposition under the additional assumption that $f \in \mathfrak{R}_1$
is an irreducible polynomial of the form $f(u_1) = c_s u_1^s + c_{s-1} u_1^{s-1} + \cdots + c_0$
with $c_s c_0 \neq 0$, and with roots $\zeta_1, \ldots, \zeta_s$. We use the terminology of Section
7 and set $c = \zeta_1$, $j_c = \{f \in \mathfrak{R}_1 : f(c) = 0\} = (f)$, $\mathbb{K} = \mathbb{Q}(c)$, and define
$P^{\mathbb{K}}, P_f^{\mathbb{K}}, P_\infty^{\mathbb{K}}, F(c), P(c)$ and R_c by (7.1)–(7.8). Put $\Omega = \prod_{v \in P(c)} \mathbb{K}_v$, write a
typical element of Ω as $\omega = (\omega_v) = (\omega_v, v \in P(c))$, and consider the diag-
onal embedding $i_F \colon R_c \longmapsto \Omega$ of R_c as a discrete, co-compact subgroup of
Ω. We denote by α the automorphism of Ω defined by $\alpha(\omega)_v = c\omega_v$ for all
$\omega = (\omega_v) \in \Omega$. Then $\alpha(i_F(R_c)) = i_F(R_c)$, so that α induces an automorphism
$\alpha^{(c)}$ of $Y^{(c)} = \Omega/i_F(R_c)$ as in (7.9)–(7.10). Since the subgroup $i_F(R_c) \subset \Omega$ is
discrete, every neighbourhood N of the identity in $Y^{(c)}$ contains a neighbour-
hood of the form $N' = \prod_{v \in P(c)} N_v + i_F(R_c)$, where N_v is a neighbourhood
of 0 in $\mathbb{K}_v$ for every $v \in P(c)$, and where $\prod_{v \in P(c)} N_v \cap i_F(R_c) = \varnothing$. If N' is
sufficiently small then

$$\lambda_{Y(c)}\left(\bigcap_{k=0}^{n-1} \alpha_k^{(c)}(N')\right) = \prod_{\{v \in P(c):|c|_v > 1\}} |c|_v^{-n+1} \lambda_{Y(c)}(N'),$$

and from (7.10), Theorem 7.1 and Theorem 13.3 we see that

$$h(\alpha^{\mathfrak{R}_1/(f)}) = h(\alpha^{\mathfrak{R}_1/j_c}) = h(\alpha^{(c)}) = h_{\mathrm{vol}}(\alpha^{(c)})$$

$$= - \sum_{\{v \in P(c):|c|_v > 1\}} \log |c|_v^{-1} \tag{17.10}$$

$$= \sum_{\{v \in P(c):|c|_v > 1\}} \log |c|_v = \sum_{v \in P(c)} \log^+ |c|_v.$$

In order to reconcile (17.10) with (17.6) we have to prove that

$$\sum_{j=1}^{s} \log^+ |\zeta_j| + \log |c_s| = \sum_{v \in P(c)} \log^+ |c|_v = \sum_{v \in P^{\mathbb{K}}} \log^+ |c|_v \tag{17.11}$$

(cf. Proposition 16.1). Let $f' = f/c_s = a_0 + \cdots + a_{s-1} u_1^{s-1} + u_1^s$ with $a_j = c_j/c_s$.
If v is an infinite place of $\mathbb{K}$ then it arises from one of the irreducible factors of
f' over $\mathbb{R}$: if v is real, this factor is of the form $u_1 - \zeta_j$ for some $j = 1, \ldots, s$,
$\mathbb{K}_v = \mathbb{R}$, and $|c|_v = |\zeta_j|$; if v is complex the factor is given by $(u_1 - \zeta_j)(u_1 - \bar\zeta_j)$
for some $j = 1, \ldots, n$, and $|c|_v = |\zeta_j|^2$. Hence

$$\sum_{j=1}^{s} \log^+ |\zeta_j| = \sum_{v \in P_\infty^{\mathbb{K}}} \log^+ |c|_v, \tag{17.12}$$

and we claim that

$$\log|c_s| = \sum_{v\in F(c)} \log^+ |c|_v = \sum_{v\in P_f^{\mathbb{K}}} \log^+ |c|_v. \qquad (17.13)$$

Let p be a rational prime, and let $f' = f_1 \cdots f_r$ be the factorization of f' into irreducible, monic polynomials f_j of degree s_j over $\mathbb{Q}_p$. Each of the polynomials f_j defines a place v_j of $\mathbb{K}$ above p, as explained at the beginning of Section 7. We choose a minimal splitting field $\mathbb{L} \supset \mathbb{Q}_p$ for f' of degree $n = [\mathbb{L} : \mathbb{Q}_p]$ over $\mathbb{Q}_p$, write $\mathrm{mod}_\mathbb{L}(a)$ for the module of an element a in $\mathbb{L}$ (cf. Section 7), and assume that $f'(u_1) = \prod_{j=1}^s (u_1 - \xi_j)$ with $\xi_j \in \mathbb{L}$ and

$$\mathrm{mod}_\mathbb{L}(\xi_1) \ge \mathrm{mod}_\mathbb{L}(\xi_2) \ge \cdots \ge \mathrm{mod}_\mathbb{L}(\xi_k) > 1 \ge \mathrm{mod}_\mathbb{L}(\xi_{k+1}) \ge \cdots \ge \mathrm{mod}_\mathbb{L}(\xi_s),$$

where $k = 0$ if $\mathrm{mod}_\mathbb{L}(\xi_i) \le 1$ for all $i = 1, \ldots, s$. By expressing each a_i as a symmetric polynomial in the ξ_j and by using the ultrametric inequality

$$\mathrm{mod}_\mathbb{L}(a + b) \le \max\{\mathrm{mod}_\mathbb{L}(a), \mathrm{mod}_\mathbb{L}(b)\}$$

we see that, if $k \ge 1$, then

$$\max_{j=1,\ldots,s} \mathrm{mod}_\mathbb{L}(a_j) = \mathrm{mod}_\mathbb{L}(a_k) = \prod_{i=1}^k \mathrm{mod}_\mathbb{L}(\xi_i)$$

and

$$\max_{j=1,\ldots,s} \log^+ \mathrm{mod}_\mathbb{L}(a_j) = \log^+ \mathrm{mod}_\mathbb{L}(a_k) = \sum_{i=1}^s \log^+ \mathrm{mod}_\mathbb{L}(\xi_i),$$

and the last equation remains correct if $k = 0$. Fix temporarily $j \in \{1, \ldots, r\}$ and consider the roots $\{\xi_{i_1}, \ldots, \xi_{i_{s_j}}\}$ of f_j in $\mathbb{L}$. Since $\mathbb{L}$ is a vector space of dimension of dimension n/s_j over $\mathbb{L}_j = \mathbb{Q}_p[u_1]/(f_j)$, and since $|\xi_{i_l}|_{v_j} = \mathrm{mod}_{\mathbb{L}_j}(\xi_{i_l})$, we see that

$$\mathrm{mod}_\mathbb{L}(a_j) = \mathrm{mod}_{\mathbb{Q}_p}(a_j)^n, \quad \mathrm{mod}_\mathbb{L}(\xi_{i_l}) = \mathrm{mod}_{\mathbb{L}_j}(\xi_{i_l})^{n/s_j} = |c|_{v_j}^{n/s_j}$$

for $j = 1, \ldots, s$, $l = 1, \ldots, s_j$, and that

$$\sum_{l=1}^{s_j} \log^+ \mathrm{mod}_\mathbb{L}(\xi_{i_l}) = n \log^+ |c|_{v_j}.$$

The polynomial f is prime and hence primitive, so that

$$n \log^+ |c_s^{-1}|_p = n \max_{i=1,\ldots,s} \log^+ |a_i|_p = \log^+ |a_k|_p^n$$

$$= \log^+ \mathrm{mod}_\mathbb{L}(a_k) = \sum_{i=1}^s \log^+ \mathrm{mod}_\mathbb{L}(\xi_i) = n \sum_{j=1}^s \log^+ |c|_{v_j}.$$

By summing over all primes p and remembering that $|c_s| \cdot \prod_p |c_s|_p = 1$ one obtains (17.13), and (17.11) follows from (17.12)–(17.13). This proves the assertion if f is prime.

Finally we turn to the general case, where f is not necessarily prime, choose a prime factorization $f = f_1 \cdots f_s$ of f in $\mathfrak{R}_1$, and set $\mathfrak{M}_s = \mathfrak{M} = \mathfrak{R}_1/(f)$ and $\mathfrak{M}_j = (\prod_{i=j+1}^{s} f_i)\mathfrak{M}$ for $j = 0, \ldots, s-1$. Then $\mathfrak{M} = \mathfrak{M}_s \supset \cdots \supset \mathfrak{M}_0 = \{0\}$ is a prime filtration of $\mathfrak{M}$ with $\mathfrak{M}_j/\mathfrak{M}_{j-1} \cong \mathfrak{R}_1/(f_j)$ for all j. Put $X_j = \mathfrak{M}_j^{\perp} \subset X^{\mathfrak{R}_1/(f)} = X$ and note that $X = X_0 \supset \cdots \supset X_s = \{0\}$, each X_j is a closed, subgroup of X which is invariant under $\alpha = \alpha^{\mathfrak{R}_1/(f)}$, and α^{X_{j-1}/X_j} is conjugate to $\alpha^{\mathfrak{R}_1/(f_j)}$ for all $j = 1, \ldots, s$. The addition formula (14.1) shows that $h(\alpha^{\mathfrak{R}_1/(f)}) = \sum_{j=1}^{s} h(\alpha^{\mathfrak{R}_1/(f_j)})$. Since $\mathrm{M}(f) = \prod_{i=1}^{s} \mathrm{M}(f_i)$ we have proved (17.6). $\square$

EXAMPLES 17.3. (1) Consider the automorphism $\alpha = \alpha^{\mathfrak{R}_1/(2u_1-3)}$ in Examples 6.17 (2) and 7. (1). According to (17.6) and (16.2), $h(\alpha) = \log 2 + \log^+ \frac{3}{2} = \log 3$, and the second proof of Proposition 17.2 shows how this comes about: locally the space $X = X^{\mathfrak{R}_1/(2u_1-3)}$ is a cartesian product of $\mathbb{S}$, $\mathbb{Z}_2$, and $\mathbb{Z}_3$, and α acts by multiplying each component locally by $\frac{3}{2}$. This leads to an expansion of Haar measure by $\frac{3}{2}$ on the first component, by 2 on the second component (since multiplication by 3 is an invertible automorphism of the group $\mathbb{Z}_2$ of dyadic integers), and by $\frac{1}{3}$ on the last component (for similar reasons). The entropy of α is equal to $\log^+ \frac{3}{2} + \log^+ 2 + \log^+ \frac{1}{3} = \log 3$. The entropy of α^{-1} is—of course—again equal to $\log 3$, but this time the value arises as $\log^+ \frac{2}{3} + \log^+ \frac{1}{2} + \log^+ 3$.

(2) Let $f \in \mathfrak{R}_1$ be a non-zero, irreducible polynomial of the form $f(u_1) = c_0 + \cdots + c_s u_1^s$ with $s \geq 1$ and $c_0 c_s \neq 0$, and with roots $\zeta_1, \ldots, \zeta_s$. As in the second proof of Proposition 17.2 we set $c = \zeta_1$, $\mathbb{K} = \mathbb{Q}(c)$, and denote by $P^{\mathbb{K}}$ the set of places of $\mathbb{K}$. From (16.2) and (17.12)–(17.13) we know that

$$\log \mathrm{M}(f) = \sum_{v \in P^{\mathbb{K}}} \log^+ |c|_v. \tag{17.14}$$

Furthermore, if we set, for every $a \in \mathbb{K}$,

$$|a|_v^+ = \begin{cases} \max\{|a|_v, |a^{-1}|_v\} & \text{if } a \neq 0, \\ 0 & \text{if } a = 0, \end{cases}$$

then the product formula

$$\prod_{v \in P^{\mathbb{K}}} |a|_v = \begin{cases} 1 & \text{if } a \neq 0, \\ 0 & \text{if } a = 0 \end{cases} \tag{17.15}$$

(cf. [16] or [109]) implies that

$$\sum_{v \in P^{\mathbb{K}}} \log |a|_v^+ = \sum_{v \in P^{\mathbb{K}}} \log^+ |a|_v + \sum_{v \in P^{\mathbb{K}}} \log^+ |a^{-1}|_v = 2 \sum_{v \in P^{\mathbb{K}}} \log^+ |a|_v$$

for every non-zero $a \in \mathbb{K}$. By setting $a = c$ we obtain in particular that

$$\prod_{v \in P^{\mathbb{K}}} |c|_v^+ = \mathbb{M}(f)^2. \tag{17.16}$$

(3) Let $d \geq 1$, $c = (c_1, \dots, c_d) \in (\overline{\mathbb{Q}}^{\times})^d$, and let $(X^{\mathfrak{R}_d/\mathfrak{j}_c}, \alpha^{\mathfrak{R}_d/\mathfrak{j}_c})$ and $(Y^{(c)}, \alpha^{(c)})$ be defined as in Theorem 7.1. For every $\mathbf{n} = (n_1, \dots, n_d) \in \mathbb{Z}^d$ we have that $h(\alpha_{\mathbf{n}}^{\mathfrak{R}_d/\mathfrak{j}_c}) = h(\alpha_{\mathbf{n}}^{(c)})$, since the map ϕ in (7.11) is finite-to-one and $h(\alpha_{\mathbf{n}}^{\mathfrak{R}_d/\mathfrak{j}_c}) = h_{\text{vol}}(\alpha_{\mathbf{n}}^{\mathfrak{R}_d/\mathfrak{j}_c}) = h_{\text{vol}}(\alpha_{\mathbf{n}}^{(c)}) = h(\alpha_{\mathbf{n}}^{(c)})$. In order to calculate $h(\alpha_{\mathbf{n}}^{(c)})$ we proceed as in Section 7, denote by $\mathbb{K} = \mathbb{Q}(c)$ the algebraic number field generated by $c_1, \dots, c_d$, and define $F(c)$ and $P(c)$ by (7.7) and (7.8). Exactly as in the second proof of Proposition 17.2 we obtain that

$$h(\alpha_{\mathbf{n}}^{(c)}) = \sum_{v \in P(c)} \log^+ |c^{\mathbf{n}}|_v = \sum_{v \in P^{\mathbb{K}}} \log^+ |c^{\mathbf{n}}|_v,$$

where $c^{\mathbf{n}} = c_1^{n_1} \cdots c_d^{n_d}$. Then quantity $\sum_{v \in P^{\mathbb{K}}} \log^+ |c^{\mathbf{n}}|_v$ is also called the *logarithmic height* of $c^{\mathbf{n}}$ in $\mathbb{K}$. $\square$

LEMMA 17.4. *Let $d \geq 1$, and let $\mathfrak{p} \subsetneq \mathfrak{q} \subset \mathfrak{R}_d$ be prime ideals. If $h(\alpha^{\mathfrak{R}_d/\mathfrak{p}}) < \infty$ then $h(\alpha^{\mathfrak{R}_d/\mathfrak{q}}) = 0$.*

PROOF. Since $\mathfrak{p} \neq \mathfrak{q}$ there exists a prime polynomial $g \in \mathfrak{q} \smallsetminus \mathfrak{p}$. The map $\psi \colon \mathfrak{R}_d/\mathfrak{p} \longmapsto \mathfrak{R}_d/\mathfrak{p}$ consisting of multiplication by g is injective, since g is prime and $g \notin \mathfrak{p}$, and $\psi(\mathfrak{R}_d/\mathfrak{p}) = (g)/\mathfrak{p} = \mathfrak{p} + (g)/\mathfrak{p}$. The dual of the exact sequence

$$0 \longrightarrow \mathfrak{R}_d/\mathfrak{p} \xrightarrow{\psi} \mathfrak{R}_d/\mathfrak{p} \longrightarrow \mathfrak{R}_d/(\mathfrak{p} + (g)) \longrightarrow 0$$

is the exact sequence

$$0 \longrightarrow X^{\mathfrak{R}_d/(\mathfrak{p}+(g))} \longrightarrow X^{\mathfrak{R}_d/\mathfrak{p}} \xrightarrow{\hat{\psi}} X^{\mathfrak{R}_d/\mathfrak{p}} \longrightarrow 0,$$

and the addition formula (14.1) yields that

$$h(\alpha^{\mathfrak{R}_d/\mathfrak{p}}) = h(\alpha^{\mathfrak{R}_d/\mathfrak{p}}) + h(\alpha^{\mathfrak{R}_d/(\mathfrak{p}+(g))}).$$

Since $\mathfrak{p} + (g) \subset \mathfrak{q}$, the group $X^{\mathfrak{R}_d/\mathfrak{q}}$ is a closed, $\alpha^{\mathfrak{R}_d/\mathfrak{p}+(g)}$-invariant subgroup of $X^{\mathfrak{R}_d/\mathfrak{p}+(g)}$, and $h(\alpha^{\mathfrak{R}_d/\mathfrak{q}}) = h_{\text{sep}}(\alpha^{\mathfrak{R}_d/\mathfrak{q}}) \leq h_{\text{sep}}(\alpha^{\mathfrak{R}_d/\mathfrak{p}+(g)}) = h(\alpha^{\mathfrak{R}_d/\mathfrak{p}+(g)})$. As $h(\alpha^{\mathfrak{R}_d/\mathfrak{p}}) < \infty$ we obtain that $h(\alpha^{\mathfrak{R}_d/\mathfrak{q}}) \leq h(\alpha^{\mathfrak{R}_d/(\mathfrak{p}+(g))}) = 0$. $\square$

PROPOSITION 17.5. *Let $\mathfrak{p} \subset \mathfrak{R}_1$ be a prime ideal. Then*

$$h(\alpha^{\mathfrak{R}_1/\mathfrak{p}}) = \begin{cases} |\log \mathbb{M}(f)| & \text{if } \mathfrak{p} = (f) \text{ is principal,} \\ 0 & \text{if } \mathfrak{p} \text{ is non-principal.} \end{cases}$$

PROOF. If $\mathfrak{p}$ is principal the statement follows from Proposition 17.2. If $\mathfrak{p}$ is non-principal we apply Lemma 17.4. $\square$

The next lemma provides the tool for proving the familiar entropy formulae for automorphisms of finite-dimensional tori and solenoids.

LEMMA 17.6. *Let $\mathfrak{M}$ be a Noetherian $\mathfrak{R}_1$-module which is torsion-free as an additive group. Put $\mathfrak{R}_1^{(\mathbb{Q})} = \mathbb{Q}[u_1^{\pm 1}]$, $\mathfrak{M}^{(\mathbb{Q})} = \mathbb{Q} \otimes_{\mathbb{Z}} \mathfrak{M}$, and choose primitive elements $f_1, \ldots, f_r$ in $\mathfrak{R}_1$ such that f_j divides f_{j+1} for all $j = 1, \ldots, r-1$, and $\mathfrak{M}^{(\mathbb{Q})}$ is isomorphic (as an $\mathfrak{R}_1^{(\mathbb{Q})}$-module) to the direct sum $\mathfrak{R}_1^{(\mathbb{Q})}/f_1\mathfrak{R}_1^{(\mathbb{Q})} \oplus \cdots \oplus \mathfrak{R}_1^{(\mathbb{Q})}/f_r\mathfrak{R}_1^{(\mathbb{Q})}$ (cf. Lemma 8.1). Then*

$$h(\alpha^{\mathfrak{M}}) = \left| \log \mathbb{M}\left(\prod_{j=1}^{r} f_j \right) \right|. \tag{17.17}$$

Furthermore, if we regard $\mathfrak{M}^{(\mathbb{Q})}$ as an $\mathfrak{R}_1$-module and denote by $\alpha^{\mathfrak{M}^{(\mathbb{Q})}}$ the automorphism of the group $X^{\mathfrak{M}^{(\mathbb{Q})}} = \widehat{\mathfrak{M}^{(\mathbb{Q})}}$ defined by (5.5)–(5.6), then $h(\alpha^{\mathfrak{M}^{(\mathbb{Q})}}) = h(\alpha^{\mathfrak{M}})$.

PROOF. The existence of the polynomials $f_1, \ldots, f_r$ and the $\mathfrak{R}_1^{(\mathbb{Q})}$-module isomorphism $\theta' : \mathfrak{M}^{(\mathbb{Q})} \longmapsto \mathfrak{R}_1^{(\mathbb{Q})}/f_1\mathfrak{R}_1^{(\mathbb{Q})} \oplus \cdots \oplus \mathfrak{R}_1^{(\mathbb{Q})}/f_r\mathfrak{R}_1^{(\mathbb{Q})} = \mathfrak{N}^{(\mathbb{Q})}$ was established in the proof of Lemma 9.1, and (14.1), Theorem 9.2, and Proposition 17.2, together imply (17.17). As in the proof of Lemma 9.1 we set $\mathfrak{N} = \mathfrak{R}_1/(f_1) \oplus \cdots \oplus \mathfrak{R}_1/(f_r)$ and consider the Noetherian $\mathfrak{R}_1$-modules $\frac{1}{k}\mathfrak{N} \subset \mathfrak{N}^{(\mathbb{Q})}$. Then $h(\alpha^{\frac{1}{k}\mathfrak{N}}) = h(\alpha)$ for every $k \geq 1$, and the last assertion follows from Lemma 13.6 and duality by letting $k \to \infty$. $\square$

THEOREM 17.7. *Let $A \in \mathrm{GL}(n, \mathbb{Q})$, and let $X = \widehat{\mathbb{Q}^n}$ and $\alpha^X = A = \widehat{A^\top} \in \mathrm{Aut}(X)$, where $A^\top$ is the transpose matrix of A. If $N \in \mathbb{Q}^n$ is any subgroup which contains $\mathbb{Z}^n$ and is invariant under multiplication by $A^\top$, and if α^V denotes the automorphism of $V = \hat{N} = \widehat{\mathbb{Q}^n}/N^\perp$ induced by α, then*

$$h(\alpha^V) = \sum_{j=1}^{n} \log^+ |\zeta_j| + \log |a|$$

$$= \int_0^1 \log |\chi_A(e^{2\pi i s})|\, ds = \log \mathbb{M}(\chi_A), \tag{17.18}$$

where $\chi_A(u) = u_1^n + a_{n-1}u_1^{n-1} + \cdots + a_0$ is the characteristic polynomial of A with roots $\zeta_1, \ldots, \zeta_n$, and where a is the lowest common multiple of the denominators of the coefficients a_i, $i = 0, \ldots, n-1$. Furthermore, $h(\alpha^V) = h(\alpha^A)$, where α^A is the automorphism of X^A defined in Example 9.5 (3).

PROOF. Consider the automorphism α^A of X^A and the dual $\mathfrak{R}_1$-module $\mathfrak{M}^A$ in Example 9.5 (3), and define $\mathfrak{R}_1^{(\mathbb{Q})}$, $\mathfrak{M}^* = \mathbb{Q}^n$, and $f_1, \ldots, f_r$ as in Lemma 17.6. Lemma 17.6 and Proposition 16.1 together imply that $h(\alpha^A)$ is given by (17.18). If $N \subset \mathbb{Q}^n$ is any subgroup which is invariant under multiplication by $A^\top$ and contains $\mathbb{Z}^n$ and hence $\mathfrak{M}^A$, and if $V = \hat{N}$, then $\mathfrak{M}^A \subset N \subset \mathfrak{M}^*$, and by duality we obtain surjective homomorphisms $X = \widehat{\mathfrak{M}^*} \longmapsto V = \hat{N} \longmapsto X^A$. According to Lemma 17.6, $h(\alpha^X) = h(\alpha^A)$, and (14.1) implies that $h(\alpha^V) = h(\alpha^X) = h(\alpha^A)$. $\square$

CONCLUDING REMARK 17.8. Lemmas 17.1 and 17.4 are taken from [63], as is the method used in the first proof of Proposition 17.2. If $A^\top \in \mathrm{GL}(n, \mathbb{Z}) = \mathrm{Aut}(\mathbb{Z}^n)$ and $\alpha = A \in \mathrm{Aut}(\mathbb{T}^n)$, then (17.18) is the well known entropy formula for toral automorphisms due to Sinai and Rokhlin [85], and for $A^\top \in \mathrm{GL}(n, \mathbb{Q})$ and $V = X = \mathbb{Q}^n$, (17.18) is Yuzvinskii's formula in [111]. The reason for deriving (17.18) by the particular route used in this section is that the argument presented here anticipates the proof of the entropy formula for $\mathbb{Z}^d$-actions by automorphisms of compact, abelian groups in Section 18.

18. Mahler measure and entropy of $\mathbb{Z}^d$-actions

We begin our discussion of $\mathbb{Z}^d$-actions by generalizing Proposition 17.2 to higher dimensions.

THEOREM 18.1. *For every $f \in \mathfrak{R}_d$, $h(\alpha^{\mathfrak{R}_d/(f)}) = |\log \mathbb{M}(f)|$.*

If $f = 0$, then Theorem 18.1 is trivially true, since

$$h(\alpha^{\mathfrak{R}_d/(f)}) = |\log \mathbb{M}(f)| = \infty.$$

If $f = cu^{\mathbf{m}}$ for some $c \in \mathbb{Z}$ and $\mathbf{m} \in \mathbb{Z}^d$, then $\alpha^{\mathfrak{R}_d/(f)}$ is isomorphic to the shift-action of $\mathbb{Z}^d$ on $(\mathbb{Z}_{/|c|})^{\mathbb{Z}^d}$ (cf. Example 5.2 (2)), and $h(\alpha^{\mathfrak{R}_d/(f)}) = \log|c| = \log \mathbb{M}(f) = |\log \mathbb{M}(f)|$, as claimed in Theorem 18.1. Assume therefore from now on that f has at least two non-zero terms. We start by estimating $h(\alpha^{\mathfrak{R}_d/(f)})$ from below.

LEMMA 18.2. $h(\alpha^{\mathfrak{R}_d/(f)}) \geq \log \mathbb{M}(f)$.

PROOF. Let $\mathbf{r} = (r_1, \ldots, r_d) \in \mathbb{Z}^d$ be primitive, i.e. assume that the highest common factor of $\{r_1, \ldots, r_d\}$ is equal to 1. As in the proof of Proposition 16.2 we define a homomorphism $\eta_{\mathbf{r}} \colon \mathbb{Z}^d \longmapsto \mathbb{Z}$ by $\eta_{\mathbf{r}}(\mathbf{m}) = \langle \mathbf{r}, \mathbf{m} \rangle$, and we note that $\eta_{\mathbf{r}}$ is surjective since $\mathbf{r}$ is primitive. The map $\eta_{\mathbf{r}}$ induces an injective homomorphism $\psi_{\mathbf{r}} \colon \mathbb{T}^{\mathbb{Z}} \longmapsto \mathbb{T}^{\mathbb{Z}^d}$ by $\psi_{\mathbf{r}}(x)_{\mathbf{m}} = x_{\eta_{\mathbf{r}}(\mathbf{m})}$, and we claim that $\psi_{\mathbf{r}}(X^{\mathfrak{R}_1/(f_{\mathbf{r}})}) \subset X^{\mathfrak{R}_d/(f)}$: indeed, if $f = \sum_{\mathbf{n} \in \mathbb{Z}^d} c_f(\mathbf{n}) u^{\mathbf{n}}$, then $f_{\mathbf{r}} = \sum_{\mathbf{n} \in \mathbb{Z}^d} c_f(\mathbf{n}) u^{\eta_{\mathbf{r}}(\mathbf{n})}$, $x \in \mathbb{T}^{\mathbb{Z}}$ lies in $X^{\mathfrak{R}_1/(f_{\mathbf{r}})}$ if and only if $\sum_{\mathbf{n} \in \mathbb{Z}^d} c_f(\mathbf{n}) x_{m + \eta_{\mathbf{r}}(\mathbf{n})} = 0 \pmod 1$ for all $m \in \mathbb{Z}$, and the last condition implies that $\psi_{\mathbf{r}}(x) \in X^{\mathfrak{R}_d/(f)}$.

For every $n \geq 1$ we set $\mathbf{r}_n = (1, n, \ldots, n^{d-1})$. Then $\mathbf{r}_n$ is primitive, $\langle \mathbf{r}_n \rangle = n$, and $\eta_{\mathbf{r}_n}$ is a bijection of the rectangle $Q_{n,m} = \{0, \ldots, n-1\}^{d-1} \times \{0, \ldots, m-1\} \subset \mathbb{Z}^d$ onto $Q'_{n,m} = \{0, \ldots, n^{d-1}m - 1\}$. In particular, if $\vartheta(s,t) = |s - t|$ is the metric on $\mathbb{T}$ defined in (17.6), $\varepsilon > 0$, and if $F \subset X^{\mathfrak{R}_1/(f_{\mathbf{r}_n})}$ is $[Q'_{n,m}, \vartheta, \varepsilon]$-separated (we are using the notation of Proposition 13.7), then $\psi_{\mathbf{r}_n}(F)$ is a

$[Q_{n,m}, \vartheta, \varepsilon]$-separated subset of $X^{\Re_d/(f)}$. From Proposition 13.7 it is now clear that

$$h(\alpha^{\Re_1/(f_{r_n})}) = \lim_{\varepsilon \to 0} \liminf_{m,n \to \infty} \frac{1}{|Q'_{n,m}|} \log \bar{s}_{Q'_{n,m}}(\vartheta, \varepsilon)$$

$$\leq \lim_{\varepsilon \to 0} \liminf_{m,n \to \infty} \frac{1}{|Q_{n,m}|} \log \bar{s}_{Q_{n,m}}(\vartheta, \varepsilon) \leq h(\alpha^{\Re_d/(f)})$$

(note that $\bar{s}_{Q'_{n,m}}(\vartheta, \varepsilon)$ and $\bar{s}_{Q_{n,m}}(\vartheta, \varepsilon)$ denote cardinalities of sets in different spaces!). According to the Propositions 16.2 and 17.2, $h(\alpha^{\Re_1/(f_{r_n})}) = \log M(f_{r_n}) \to \log M(f)$, so that $\log M(f) \leq h(\alpha^{\Re_d/(f)})$. $\square$

The proof of the opposite inequality

$$h(\alpha^{\Re_d/(f)}) \leq \log M(f) \tag{18.1}$$

is much more involved, and it will be helpful to present it first for a specific polynomial of a particularly simple form.

LEMMA 18.3. *The inequality* (18.1) *holds for* $f = 1 + u_1 + u_2 \in \Re_2$.

PROOF. According to Example 5.2 (2) we may assume that

$$X = X^{\Re_2/(f)} = \{x \in \mathbb{T}^{\mathbb{Z}^2} : x_{(m,n)} + x_{(m+1,n)} + x_{(m,n+1)} = 0 \ (\mathrm{mod}\ 1)$$
$$\text{for all } (m,n) \in \mathbb{Z}^2\},$$

and that $\alpha = \alpha^{\Re_2/(f)}$ is to the shift-action σ of $\mathbb{Z}^2$ on X. We define $\vartheta(s,t) = \|s - t\|$ on $\mathbb{T}$ by (17.7), fix $\varepsilon > 0$ and a rectangle $Q = Q(M, N) = \{0, \ldots, M - 1\} \times \{0, \ldots, N - 1\} \subset \mathbb{Z}^2$, and construct explicitly an $[Q, \vartheta, \varepsilon]$-spanning set $F \subset X$ of the desired cardinality (for notation we refer to Proposition 13.7).

Let $x \in X$. If $y \in X$ satisfies that $\pi_Q(x) = \pi_Q(y)$, then $y_\mathbf{m} = x_\mathbf{m}$ for all $\mathbf{m} \in \{(m_1, m_2) \in \mathbb{Z}^2 : 0 \leq m_1 \leq M + N - 2, 0 \leq m_2 \leq M + N - 2 - m_1\}$. Apart from this constraint we have considerable freedom in choosing y: for example, the coordinates $y_{(k,0)}$, $k \notin \{0, \ldots, M + N - 2\}$, are completely at our disposal, and we can choose them so that $y_{(k+L,0)} = y_{(k,0)}$ for all $k \in \mathbb{Z}$, where $L = M + N - 1$. It follows that there exists a point $y \in X$ with $\pi_Q(y) = \pi_Q(x)$ and $\sigma_{(L,0)}(y) = y$.

Denote by $z \in \ell^\infty(\mathbb{Z})$ the point with $-\frac{1}{2} < z_m \leq \frac{1}{2}$ and $z_m \ (\mathrm{mod}\ 1) = y_{(m,0)}$ for all $m \in \mathbb{Z}$. There exist unique complex coefficients c_k, $k = 0, \ldots, L-1$, such that $z = \sum_{k=0}^{L-1} c_k v^{(k)}$, where $v^{(k)} \in \ell^\infty(\mathbb{Z})$ and $v^{(k)}(m) = e^{2\pi i k m/L}$ for all $k = 0, \ldots, L-1$ and $m \in \mathbb{Z}$. It is clear that $|c_k| \leq \frac{1}{2}$ for all $k = 0, \ldots, L-1$, and that $c_k = \bar{c}_{L-k}$ for $k = 1, \ldots, L-1$. For every $k = 0, \ldots, L-1$, we set $\zeta_k = -1 - e^{2\pi i k/L}$ and define a map $w^{(k)} : \mathbb{Z}^2 \longmapsto \mathbb{C}$ by $w^{(k)}(m,n) = \zeta_k^n v^{(k)}(m)$ for all $(m,n) \in \mathbb{Z}^2$ if $\zeta_k \neq 0$, and $w^{(k)} \equiv 0$ otherwise. Then $\sum_{k=0}^{L-1} c_k w^{(k)}(\mathbf{m})$ is real and $\sum_{k=0}^{L-1} c_k w^{(k)}(\mathbf{m}) \ (\mathrm{mod}\ 1) = y_\mathbf{m}$ for all $\mathbf{m} \in \mathbb{Z}^2$. If we replace each coefficient c_k by a complex number of the form $c'_k = r_k/Lt_k + is_k/Lt_k$, where t_k is a positive integer, r_k and s_k are integers in the interval $[-Lt_k/2, Lt_k/2]$, and $c'_k = \bar{c}'_{L-k}$ for $k = 1, \ldots, L-1$, then we obviously cannot expect $y'_{\mathbf{m}} = \sum_{k=0}^{L-1} c'_k w^{(k)}(\mathbf{m})$ (mod 1) to be equal to $y_{\mathbf{m}}$ for every $\mathbf{m} \in \mathbb{Z}^2$. Suppose, however, that we only wish to achieve that $|y'_{\mathbf{m}} - y_{\mathbf{m}}| < \varepsilon$ for all $\mathbf{m} \in Q' = Q(M,N)' = \{0, \ldots, L-1\} \times \{0, \ldots, N-1\}$. If we choose $t_k = \text{Int}((1+|\zeta_k|^{N-1})/\varepsilon)$ for every $k = 0, \ldots, L-1$, where Int denotes the integral part, then we can indeed find such coefficients $c'_k = r_k/Lt_k + is_k/Lt_k$ such that $y' = \sum_{k=0}^{L-1} c'_k w^{(k)}$ (mod 1) satisfies that $|y'_{\mathbf{m}} - y_{\mathbf{m}}| < \varepsilon$ for all $\mathbf{m} \in Q'$. The total number of all y' which can be written in this form is less than or equal to $K(M,N,\varepsilon) = \prod_{k=0}^{(L+1)/2}[\frac{L+1}{\varepsilon} \cdot (1+|\zeta_k|^{N-1})]^2$. Since $Q = Q(M,N) \subset Q' = Q(M,N)'$ we have found an $[Q(M,N), \vartheta, \varepsilon]$-spanning set of cardinality $K(M,N,\varepsilon)$, and Proposition 13.7 shows that

$$h(\alpha) = h(\sigma) \leq \sup_{\varepsilon > 0} \liminf_{M,N \to \infty} \frac{1}{MN} \log K(M,N,\varepsilon).$$

Furthermore,

$$\liminf_{M,N \to \infty} \frac{1}{MN} \log K(M,N,\varepsilon)$$

$$= \liminf_{M,N \to \infty} \frac{2}{MN} \sum_{k=0}^{(M+N)/2} (\log(M+N) - \log \varepsilon + \log^+(1+|\zeta_k|^{N-1}))$$

$$\leq \liminf_{M,N \to \infty} \frac{1}{MN} \sum_{k=0}^{M+N-2} (\log(M+N) - \log \varepsilon + (N-1)\log^+ |\zeta_k|),$$

and by setting $N = \text{Int}(\log M)$ and letting $M \to \infty$ we obtain that

$$h(\alpha) \leq \lim_{L \to \infty} \frac{1}{L} \sum_{k=0}^{L-1} \log^+ |1 + e^{2\pi ik/L}| = \int_0^1 \log^+ |1 + e^{2\pi is}| \, ds,$$

where $L = M + N - 1$. From Proposition 16.1 we know that

$$\log^+ |1 + e^{2\pi is}| = \int_0^1 \log |1 + e^{2\pi is} + e^{2\pi it}| \, dt,$$

so that

$$h(\alpha) \leq \int_0^1 \int_0^1 \log |1 + e^{2\pi is} + e^{2\pi it}| \, ds \, dt = \log \mathbb{M}(f). \qquad \square$$

Although this is probably the most direct proof of Lemma 18.3, its extension to a general polynomial $f \in \mathfrak{R}_d$ is slightly awkward. For this reason we present a second proof of Lemma 18.3, based on h_{vol} rather than h_{span}.

SECOND PROOF OF LEMMA 18.3. We use the same notation as in the first proof of this lemma. Let $Q = Q(M,N) = \{0, \ldots, M-1\} \times \{0, \ldots, N-1\}$, $L = M+N-1$, $Q' = Q'(M,N) = \{0, \ldots, L-1\} \times \{0, \ldots, N-1\}$, and let $Y \subset X$ be the subgroup of points with $\sigma_{(L,0)}(y) = y$. Then $\pi_Q(Y) = \pi_Q(X)$, and we write $\pi'_Q : Y \longmapsto \mathbb{T}^Q$ for the restriction to Y of $\pi_Q : X \longmapsto \mathbb{T}^Q$. As in Proposition

13.7 we set, for every $\varepsilon > 0$, $B(Q, \varepsilon) = \{x \in X : |x_{\mathbf{m}}| < \varepsilon$ for all $\mathbf{m} \in Q\}$ and note that ${\pi'_Q}^{-1}(\pi_Q(B(Q, \varepsilon))) \supset \{y \in Y : |y_{\mathbf{m}}| < \varepsilon$ for all $\mathbf{m} \in Q'\} = C(Q', \varepsilon)$, say. Since $\pi_Q^{-1}(\pi_Q(B(Q, \varepsilon)) = B(Q, \varepsilon)$ we have that

$$\lambda_X(B(Q, \varepsilon)) = \lambda_{\pi_Q(X)}(\pi_Q(B(Q, \varepsilon)) \geq \lambda_{\pi'_Q(Y)}(\pi'_Q(C(Q', \varepsilon))) \geq \lambda_Y(C(Q', \varepsilon)),$$

and Proposition 13.7 shows that

$$h(\alpha) = h(\sigma) = \lim_{\varepsilon \to 0} \liminf_{M,N \to \infty} -\frac{1}{MN} \log \lambda_X(B(Q(M, N), \varepsilon))$$

$$\leq \lim_{\varepsilon \to 0} \liminf_{M,N \to \infty} -\frac{1}{MN} \log \lambda_Y(C(Q(M, N)', \varepsilon)).$$

In order to estimate $\lambda_Y(C(Q(M, N)', \varepsilon))$ we consider the surjective homomorphism

$$\eta = \pi_{\{0,\dots,L-1\} \times \{0\}} : Y \longmapsto \mathbb{T}^L$$

and note that the diagram

$$
\begin{array}{ccc}
Y & \xrightarrow{\sigma_{(0,1)}} & Y \\
\eta \downarrow & & \downarrow \eta \\
\mathbb{T}^L & \xrightarrow[A]{} & \mathbb{T}^L
\end{array}
$$

commutes, where

$$A = \begin{pmatrix} -1 & -1 & 0 & \dots & 0 \\ 0 & -1 & -1 & \dots & 0 \\ \vdots & \vdots & \ddots & \ddots & \vdots \\ 0 & 0 & \dots & -1 & -1 \\ -1 & 0 & \dots & 0 & -1 \end{pmatrix} \in \mathrm{GL}(L, \mathbb{Z}) = \mathrm{Aut}(\mathbb{T}^L).$$

The matrix A induces a linear map on $\mathbb{C}^L$, which will again be denoted by A. For every $k = 0, \dots, M - 1$ we set $v(k) = L^{-\frac{1}{2}}(1, e^{2\pi i k/L}, \dots, e^{2\pi i k(L-1)/L}) \in \mathbb{C}^L$, note that $Av(k) = \zeta_k v(k)$ with $\zeta_k = -1 - e^{2\pi i k/L}$, and define a unitary map V of $\mathbb{C}^L$ by $Vz = \sum_{k=0}^{L-1} z_k v(k)$ for every $z = (z_0, \dots, z_{L-1}) \in \mathbb{C}^L$.

Let $\|z\|_\infty = \max_{j=0,\dots,L-1} |z_j|$ for all $z = (z_0, \dots, z_{L-1}) \in \mathbb{C}^L$, denote by $\lambda = \lambda_{\mathbb{C}^L}$ the Lebesgue measure on $\mathbb{C}^L \cong \mathbb{R}^{2L}$, and put $C(\varepsilon) = \{z \in \mathbb{C}^L : \|z\|_\infty < \varepsilon\}$ for every $\varepsilon > 0$. Then

$$\lambda(D(N, \varepsilon)) \leq \lambda_Y(C(Q', \varepsilon))^2,$$

where

$$D(N, \varepsilon) = \{z \in \mathbb{C}^L : \max_{k=0,\dots,N-1} \|A^k z\|_\infty \leq \varepsilon\}.$$

Since $\lambda \cdot V = \lambda$ and $V(C(1)) \supset C(L^{-\frac{1}{2}})$ we have that

$$V(D(N, \varepsilon)) \supset E(N, L^{-\frac{1}{2}}\varepsilon),$$

where

$$E(N, \varepsilon) = \{z = (z_0, \ldots, z_{L-1}) \in \mathbb{C}^L : |z_j| \leq \varepsilon \max\{1, |\zeta_j|\}^{-N+1}$$
$$\text{for } 0 \leq j < L\}.$$

Hence

$$\lambda_Y(C(Q(M, N)', \varepsilon))^2 \geq \lambda(D(N, \varepsilon)) \geq \lambda(E(N, L^{-\frac{1}{2}}\varepsilon))$$
$$= L^{-\frac{L}{2}} \varepsilon^L \prod_{j=0}^{L-1} \max\{1, |\zeta_j|\}^{-N+1},$$

and

$$2h(\alpha) \leq \lim_{\varepsilon \to 0} \liminf_{M,N \to \infty} -\frac{2}{MN} \log \lambda_Y(C(Q(M, N)', \varepsilon))$$

$$\leq \lim_{\varepsilon \to 0} \liminf_{M,N \to \infty} -\frac{1}{MN} \log \lambda(E(N, \varepsilon L^{-\frac{1}{2}}))$$

$$= \lim_{L \to \infty} -\frac{2}{L} \sum_{j=0}^{L-1} \log^+ |\zeta_j| = 2 \int_0^1 \log^+ |1 + e^{2\pi i s}| ds,$$

where $N = \text{Int}(\log M)$, $L = M + N - 1$, and $M \to \infty$. As in the first proof of this lemma we conclude that $h(\alpha) \leq \log \text{M}(f)$. $\square$

In order to prove the inequality (18.1) for an arbitrary polynomial $f \in \mathfrak{R}_d$ it helps to transform f into a more convenient form. If $h \in \mathfrak{R}_d$ and $A \in \text{GL}(d, \mathbb{Z})$ we put $h^A = \sum_{\mathbf{n} \in \mathbb{Z}^d} c_h(\mathbf{n}) u^{A\mathbf{n}}$ (cf. (5.2)).

LEMMA 18.4. *For every $A \in \text{GL}(d, \mathbb{Z})$ and $\mathbf{m} \in \mathbb{Z}^d$,*

$$h(\alpha^{\mathfrak{R}_d/(f)}) = h(\alpha^{\mathfrak{R}_d/(u^{\mathbf{m}} f^A)}) \quad \text{and} \quad \text{M}(f) = \text{M}(u^{\mathbf{m}} f^A).$$

PROOF. Since $(g) = (u^{\mathbf{m}} g)$ and $\text{M}(g) = \text{M}(u^{\mathbf{m}} g)$ for every $g \in \mathfrak{R}_d$ we may assume that $\mathbf{m} = \mathbf{0}$. Example 5.2 (2) allows us to realize $X^{\mathfrak{R}_d/(f)}$ and $X^{\mathfrak{R}_d/(f^A)}$ as closed, shift-invariant subgroups of $\mathbb{T}^{\mathbb{Z}^d}$, and to assume that $\alpha^{\mathfrak{R}_d/(f)}$ and $\alpha^{\mathfrak{R}_d/(f^A)}$ are the restrictions of the shift-action σ of $\mathbb{Z}^d$ on $\mathbb{T}^{\mathbb{Z}^d}$ to $X^{\mathfrak{R}_d/(f)}$ and $X^{\mathfrak{R}_d/(f^A)}$, respectively. We define an isomorphism $\psi_A \colon \mathbb{T}^{\mathbb{Z}^d} \longrightarrow \mathbb{T}^{\mathbb{Z}^d}$ by $(\psi_A(x))_{\mathbf{n}} = x_{A\mathbf{n}}$ for every $x = (x_{\mathbf{n}}) \in \mathbb{T}^{\mathbb{Z}^d}$ and note that $\psi_A(X^{\mathfrak{R}_d/(f)}) = X^{\mathfrak{R}_d/(f^A)}$, and that $\psi_A \cdot \sigma_{\mathbf{n}}(x) = \psi_A \cdot \alpha_{\mathbf{n}}^{\mathfrak{R}_d/(f)}(x) = \alpha_{A\mathbf{n}}^{\mathfrak{R}_d/(f^A)} \cdot \psi_A(x) = \sigma_{A\mathbf{n}} \cdot \psi_A(x)$ for every $x \in X^{\mathfrak{R}_d/(f)}$. In other words, ψ_A is a conjugacy of the $\mathbb{Z}^d$-actions $\alpha_A^{\mathfrak{R}_d/(f)}$ and $\alpha_A^{\mathfrak{R}_d/(f^A)} \colon \mathbf{n} \mapsto \alpha_{A\mathbf{n}}^{\mathfrak{R}_d/(f^A)}$, and Proposition 13.1 shows that $h(\alpha^{\mathfrak{R}_d/(f)}) = h(\alpha_A^{\mathfrak{R}_d/(f^A)}) = h(\alpha^{\mathfrak{R}_d/(f^A)})$. The second assertion follows from the fact that $A \in \text{GL}(d, \mathbb{Z}) = \text{Aut}(\mathbb{T}^d)$ induces a (Haar measure preserving) automorphism A' of $\mathbb{S}^d \cong \mathbb{T}^d$ with $f^A(\mathbf{z}) = f(A'\mathbf{z})$ for every $\mathbf{z} \in \mathbb{S}^d$. $\square$

PROOF OF THEOREM 18.1. Let $d > 1$, and let $f \in \mathfrak{R}_d$ be a polynomial with at least two non-zero terms (the other possibilities have already been taken care of). From Lemma 18.2 we know that $h(\alpha^{\mathfrak{R}_d/(f)}) \geq \log \mathrm{M}(f)$, but we still have to prove the opposite inequality (18.1). In view of Lemma 18.4 we can replace f by any polynomial $u^{\mathbf{n}} f^A$ with $\mathbf{n} \in \mathbb{Z}^d$ and $A \in \mathrm{GL}(d, \mathbb{Z})$, without affecting the assertion of the theorem. In particular we may assume without loss in generality that f has the form

$$f = a u_d^K + f_{K-1} u_d^{K-1} + \cdots + f_0, \tag{18.2}$$

where $a \geq 1$, $K \geq 1$, and $f_k \in \mathbb{Z}[u_1, \ldots, u_{d-1}]$ for every $k = 0, \ldots, K - 1$. According to Example 5.2 (2) we may take it that $X = X^{\mathfrak{R}_d/(f)} \subset \mathbb{T}^{\mathbb{Z}^d}$, and that $\alpha = \alpha^{\mathfrak{R}_d/(f)}$ is equal to the shift-action of $\mathbb{Z}^d$ on X. From (5.9) and (18.2) it is clear that the projection $\pi_Z \colon X \longmapsto \mathbb{T}^Z$ is surjective, where Z is the strip $\mathbb{Z}^{d-1} \times \{0, \ldots, K - 1\} \subset \mathbb{Z}^d$.

We proceed as in the first proof of Lemma 18.3, consider the rectangle $Q = Q(M, N) = \{0, \ldots, M - 1\}^{d-1} \times \{0, \ldots, N - 1\}$ with $N > K$, and fix a point $x \in X$. If $y \in X$ satisfies that $\pi_Q(y) = \pi_Q(x)$ then y need no longer be *completely* determined on a set of coordinates which strictly contains Q, but there exists an integer $q > 0$ which is independent of M and N such that the coordinates of y are *completely undetermined* in $Z \smallsetminus Q'$, where $Q' = Q(M, N)' = \{0, \ldots, M + qN - 1)\}^{d-1} \times \{0, \ldots, N - 1\}$ (this is equivalent to saying that $\pi_{Z \smallsetminus Q'} \cdot \pi_Q^{-1}(\{x\}) = \mathbb{T}^{Z \smallsetminus Q'}$). The integer q is given by $q = \min\{p : \mathcal{S}(f) \subset \Delta(p)\}$, where $\mathcal{S}(f) = \{\mathbf{m} \in \mathbb{Z}^d : c_f(\mathbf{m}) \neq 0\}$ is the support of f and $\Delta(p) = \{\mathbf{m} = (m_1, \ldots, m_d) \in \mathbb{Z}^d : 0 \leq m_d \leq K$ and $0 \leq m_j \leq p(K - m_d)$ for all $j = 1, \ldots, d - 1\}$.

Let $Y_n \subset X$ be the subgroup of points which have period n in the first $d - 1$ coordinates, i.e. which satisfy that $\alpha_{n\mathbf{e}^{(j)}}(x) = x$ for $j = 1, \ldots, d - 1$, where $\mathbf{e}^{(j)}$ denotes the j-th unit vector in $\mathbb{Z}^d$. Exactly as in the second proof of Lemma 18.3 we obtain that $\pi_Q(Y) = \pi_Q(X)$ with $L = M + qN$ and $Y = Y_L$, and observe that

$$\begin{aligned} h(\alpha) &= \lim_{\varepsilon \to 0} \liminf_{M, N \to \infty} -\frac{1}{M^{d-1}N} \log \lambda_X(B(Q(M, N), \varepsilon)) \\ &\leq \lim_{\varepsilon \to 0} \liminf_{M, N \to \infty} -\frac{1}{M^{d-1}N} \log \lambda_Y(C(Q(M, N)', \varepsilon)), \end{aligned} \tag{18.3}$$

where

$$B(Q(M, N), \varepsilon) = \{x \in X : |x_{\mathbf{m}}| < \varepsilon \text{ for all } \mathbf{m} \in Q(M, N)\},$$
$$C(Q(M, N)', \varepsilon) = \{y \in Y : |y_{\mathbf{m}}| < \varepsilon \text{ for all } \mathbf{m} \in Q(M, N)'\},$$

and where $\| \cdot \|$ is given by (17.7). We put $R = R(M, N) = \{0, \ldots, L - 1\}^{d-1}$, $S = S(M, N) = R(M, N) \times \{0, \ldots, K - 1\}$, and note that the projection $\pi_S \colon Y \longmapsto \mathbb{T}^S$ is surjective. For every $0 < \varepsilon < \frac{1}{2}$ and every $y \in C(Q', \varepsilon) =$

$C(Q(M, N)', \varepsilon)$ there exists a unique point $\phi(y) = z = (z_\mathbf{n}) \in \mathbb{R}^{Q'}$ with $|z_\mathbf{n}| < \varepsilon$ and $z_\mathbf{n}$ (mod 1) $= y_\mathbf{n}$ for all $\mathbf{n} \in Q'$, and we set $D(Q', \varepsilon) = \phi(C(Q', \varepsilon))$.

Write a typical element $v \in \mathbb{R}^R$ as $(v_\mathbf{m}) = (v_\mathbf{m}, \mathbf{m} \in R)$, and define automorphisms $\tau_j \in \mathrm{GL}(L^{d-1}, \mathbb{R}) = \mathrm{Aut}(\mathbb{R}^R)$ by $(\tau_j v)_\mathbf{m} = v_{\mathbf{m}+\mathbf{e}^{(j)} \pmod L}$ for every $v \in \mathbb{R}^R$, where (mod L) indicates reduction modulo L in every coordinate. For every $\mathbf{k} = (k_1, \ldots, k_{d-1}) \in \mathbb{Z}^{d-1}$ and $h \in \mathfrak{R}_{d-1}$ we put $\tau^\mathbf{k} = \tau_1^{k_1} \cdots \tau_{d-1}^{k_{d-1}} \in \mathrm{GL}(L^{d-1}, \mathbb{R})$ and $h(\tau) = \sum_{\mathbf{k} \in \mathbb{Z}^{d-1}} c_h(\mathbf{k}) \tau^\mathbf{k} \in M_{L^{d-1}}(\mathbb{R})$, where $M_s(\mathbb{R})$ is the set of $s \times s$-matrices with real entries for every $s \geq 1$, and where $c_h(\mathbf{k})$ is defined as in (5.2). The matrix

$$A = A(M, N) = \begin{pmatrix} 0 & I & 0 & \cdots & 0 \\ 0 & 0 & I & \cdots & 0 \\ \vdots & \vdots & \vdots & \ddots & \vdots \\ 0 & 0 & 0 & \cdots & I \\ -a^{-1}f_0(\tau) & -a^{-1}f_1(\tau) & -a^{-1}f_2(\tau) & \cdots & -a^{-1}f_{K-1}(\tau) \end{pmatrix} \in M_{|S|}(\mathbb{R})$$

induces linear maps both on $\mathbb{R}^S = (\mathbb{R}^R)^K$ and $\mathbb{C}^S = (\mathbb{C}^R)^K$, which will again be denoted by $A = A(M, N)$.

Let $\omega = e^{2\pi i/L}$ and define, for every $\mathbf{n} \in R$, a vector $v(\mathbf{n}) \in \mathbb{C}^R$ by $v(\mathbf{n})_\mathbf{m} = \omega^{\langle \mathbf{m}, \mathbf{n} \rangle}/\sqrt{L^{d-1}}$ for every $\mathbf{m} \in R$ (cf. (16.3)). The set $\{v(\mathbf{n}) : \mathbf{n} \in R\}$ is a basis of $\mathbb{C}^R$ which is orthonormal in the Euclidean norm, and $h(\tau)v(\mathbf{n}) = h(\omega)v(\mathbf{n})$ for all $\mathbf{n} \in R$ and $h \in \mathfrak{R}_{d-1}$, so that the $v(\mathbf{n})$ are simultaneous eigenvectors for all $h(\tau)$, $h \in \mathfrak{R}_{d-1}$. Consider the bases $\{\mathbf{e}(\mathbf{m}) : \mathbf{m} \in S\}$ and $\{w(\mathbf{m}) : \mathbf{m} \in S\}$ of $\mathbb{C}^S$ defined by $\mathbf{e}(\mathbf{m})_\mathbf{n} = \delta_{\mathbf{m},\mathbf{n}}$ and $w(\mathbf{m})_\mathbf{n} = \delta_{\mathbf{m},\mathbf{n}} v(\mathbf{m}')$ for all $\mathbf{m}, \mathbf{n} \in S$, where $\mathbf{m}' = (m_1, \ldots, m_{d-1}) \in \mathbb{Z}^{d-1}$ for every $\mathbf{m} = (m_1, \ldots, m_d) \in \mathbb{Z}^d$, and $\delta_{\mathbf{m},\mathbf{n}} = 1$ if $\mathbf{m} = \mathbf{n}$, and $= 0$ otherwise. For every $\mathbf{k} \in \mathbb{Z}^{d-1}$ we consider the linear span $W_\mathbf{k}$ of $\{w(\mathbf{n}) : \mathbf{n} \in \mathbb{Z}^d$ and $\mathbf{n}' = \mathbf{k}\}$ in $\mathbb{C}^S$, write

$$\|t\|_\mathbf{k} = \max_{\{\mathbf{n} \in \mathbb{Z}^d : \mathbf{n}' = \mathbf{k}\}} |c_\mathbf{n}|$$

for the maximum norm of an element

$$t = \sum_{\{\mathbf{n} \in \mathbb{Z}^d : \mathbf{n}' = \mathbf{k}\}} c_\mathbf{n} w_\mathbf{n} \in W_\mathbf{k}$$

in the basis $\{w(\mathbf{n}) : \mathbf{n} \in \mathbb{Z}^d$ and $\mathbf{n}' = \mathbf{k}\}$, set $S(1)_\mathbf{k} = \{t \in W_\mathbf{k} : \|t\|_\mathbf{k} < 1\}$, and denote by $\lambda_\mathbf{k}$ the Lebesgue measure on $W_\mathbf{k}$, i.e. the Haar measure on $W_\mathbf{k}$ with $\lambda_\mathbf{k}(S(1)_\mathbf{k}) = (2\pi)^K$. Then $\mathbb{C}^S = \bigoplus_{\mathbf{k} \in R} W_\mathbf{k}$, each $W_\mathbf{k}$ is invariant under A, and the restriction $A_\mathbf{k} = A(M, N)_\mathbf{k}$ of A to $W_\mathbf{k}$ has the form

$$A_\mathbf{k} = \begin{pmatrix} 0 & 1 & 0 & \cdots & 0 \\ 0 & 0 & 1 & \cdots & 0 \\ \vdots & \vdots & \vdots & \ddots & \vdots \\ 0 & 0 & 0 & \cdots & 1 \\ -a^{-1}f_0(\omega(\mathbf{k})) & -a^{-1}f_1(\omega(\mathbf{k})) & -a^{-1}f_2(\omega(\mathbf{k})) & \cdots & -a^{-1}f_{K-1}(\omega(\mathbf{k})) \end{pmatrix},$$

where $\omega(\mathbf{k}) = (\omega^{k_1}, \ldots, \omega^{k_{d-1}})$ for every $\mathbf{k} = (k_1, \ldots, k_{d-1}) \in \mathbb{Z}^{d-1}$. More generally we define, for every $\mathbf{s} = (s_1, \ldots, s_{d-1}) \in \mathbb{S}^{d-1} \subset \mathbb{C}^{d-1}$,

$$
A_{\mathbf{s}} = \begin{pmatrix} 0 & 1 & 0 & \cdots & 0 \\ 0 & 0 & 1 & \cdots & 0 \\ \vdots & \vdots & \vdots & \ddots & \vdots \\ 0 & 0 & 0 & \cdots & 1 \\ -a^{-1} f_0(\mathbf{s}) & -a^{-1} f_1(\mathbf{s}) & -a^{-1} f_2(\mathbf{s}) & \cdots & -a^{-1} f_{K-1}(\mathbf{s}) \end{pmatrix},
$$

and denote by $\zeta_{\mathbf{s},1}, \ldots, \zeta_{\mathbf{s},K}$ the eigenvalues of $A_{\mathbf{s}}$, i.e. the roots of the polynomial

$$
f_{\mathbf{s}}(u_d) = \sum_{\mathbf{m} = (m_1, \ldots, m_d) \in \mathbb{Z}^d} c_f(\mathbf{m}) s_1^{m_1} \cdots s_{d-1}^{m_{d-1}} u_d^{m_d}.
$$

In particular we obtain that

$$
\xi = \max_{\mathbf{s} \in \mathbb{S}^{d-1}} \max_{j=1,\ldots,k} |\zeta_{\mathbf{s},j}| < \infty, \tag{18.4}
$$

and since $A_{\mathbf{k}} = A_{\omega(\mathbf{k})}$ for every $\mathbf{k} \in R$, every eigenvalue ζ of $A(M, N)$ satisfies that $\zeta \leq \xi$, uniformly in M and N.

Lemma 17.1, applied to the compact set $\{A_{\mathbf{s}} : \mathbf{s} \in \mathbb{S}^{d-1}\} \subset M_K(\mathbb{C})$, implies that there exists, for every $\varepsilon' > 0$, an $N(\varepsilon') \geq 1$ with

$$
\left| -\frac{1}{N} \log \lambda_{\mathbf{k}} \left(\bigcap_{j=0}^{N-1} A_{\mathbf{k}} C(1)_{\mathbf{k}} \right) - 2 \sum_{j=1}^{K} \log^+ |\zeta_{\mathbf{k},j}| \right| < \varepsilon' \tag{18.5}
$$

for all $M \geq 1$, $N \geq N(\varepsilon')$, and $\mathbf{k} \in R(M, N)$. We define a linear map $W : \mathbb{C}^S \longmapsto \mathbb{C}^S$ by $W\mathbf{e}(\mathbf{n}) = w(\mathbf{n})$ for all $\mathbf{n} \in S$, and note that W is unitary with respect to the Euclidean inner product on $\mathbb{C}^S$, that $\lambda_{\mathbb{C}^S} W = \lambda_{\mathbb{C}^S}$ and $\{z \in \mathbb{C}^S : \|z\|_\infty < 1\} \supset W\{z \in \mathbb{C}^S : \|z\|_\infty < |S|^{-\frac{1}{2}}\}$, where $\|v\|_\infty = \max_{\mathbf{n} \in S} |v_{\mathbf{n}}|$ is, as usual, the maximum norm both on $\mathbb{R}^S$ and on $\mathbb{C}^S$, and that

$$
\left\{ z \in \mathbb{C}^S : \max_{j=0,\ldots,N-1} \|A^j z\|_\infty < \varepsilon \right\} \supset \varepsilon |S|^{-\frac{1}{2}} \bigoplus_{\mathbf{k} \in R} \left[\bigcap_{j=0}^{N-1} A_{\mathbf{k}}^j C(1)_{\mathbf{k}} \right].
$$

Since $\lambda_{\mathbb{C}^S}$ is the product of the measures $\lambda_{\mathbf{k}}$, $\mathbf{k} \in R$, we obtain that

$$
-\frac{1}{N} \log \lambda_{\mathbb{C}^S}(\{z \in \mathbb{C}^S : \|A^j z\|_\infty < \varepsilon \text{ for } j = 0 \ldots, N-1\})
$$
$$
\leq -\frac{1}{N} \log(\varepsilon |S|^{-\frac{1}{2}}) - \frac{1}{N} \sum_{\mathbf{k} \in R} \log \lambda_{\mathbf{k}} \left(\bigcap_{j=0}^{N-1} A_{\mathbf{k}}^j C(1)_{\mathbf{k}} \right), \tag{18.6}
$$

and (18.5) shows that

$$
\left| -\frac{1}{N} \sum_{\mathbf{k} \in R(M,N)} \log \lambda_{\mathbf{k}} \left(\bigcap_{j=0}^{N-1} A_{\mathbf{k}}^j C(1)_{\mathbf{k}} \right) - 2 \sum_{\mathbf{k} \in R(M,N)} \sum_{j=1}^{K} \log^+ |\zeta_{\omega(\mathbf{k}),j}| \right| < L^{d-1} \varepsilon' \tag{18.7}
$$

for all $M \geq 1$ and $N \geq N(\varepsilon')$.

We can now determine $\lambda_Y(C(Q',\varepsilon))$ exactly as in the first proof of Proposition 17.2. Let $\varepsilon < (4|a|(\xi+1))^{-1}$, where ξ is given by (18.4), and let $v = (v_{\mathbf{n}}) \in \mathbb{R}^{Q'}$. For every $k = 0,\ldots,N-K$ we define a point $\pi_k(v) = v(k) \in \mathbb{R}^S$ by setting $v(k)_{\mathbf{n}} = v_{\mathbf{n}+k\mathbf{e}^{(d)}}$ for all $\mathbf{n} \in S$. The maps $\pi_k : \mathbb{R}^{Q'} \longmapsto \mathbb{R}^S$ are linear, and $\pi_{k+1}(z) = z(k+1) = Az(k) = A\pi_k(z)$ for every $z \in D(Q',\varepsilon)$. Hence

$$\pi_0(D(Q',\varepsilon)) = \{v \in \mathbb{R}^S : \max_{k=0,\ldots,N-K} \|A^k v\|_\infty < \varepsilon\},$$
$$\lambda_Y(C(Q',\varepsilon)) = |a|^{-(N-K)|S|}\lambda_{\mathbb{R}^S}(\pi_0(D(Q',\varepsilon))), \tag{18.8}$$

and

$$\lambda_{\mathbb{R}^S}(\pi_0(D(Q',\varepsilon)))^2 = \lambda_{\mathbb{R}^S}\left(\{v \in \mathbb{R}^S : \max_{k=0,\ldots,N-K} \|A^k v\|_\infty < \varepsilon\}\right)^2$$
$$> \lambda_{\mathbb{C}^S}\left(\{v \in \mathbb{C}^S : \max_{k=0,\ldots,N-K} \|A^k v\|_\infty < \varepsilon\}\right). \tag{18.9}$$

It follows that

$$2h(\alpha) = \lim_{\varepsilon \to 0} \liminf_{M,N \to \infty} -\frac{2}{M^{d-1}N} \log \lambda_X(B(Q(M,N),\varepsilon))$$
$$\text{(by (18.3))}$$
$$\leq \lim_{\varepsilon \to 0} \liminf_{M,N \to \infty} -\frac{2}{M^{d-1}N} \log \lambda_{Y_L}(C(Q(M,N)',\varepsilon))$$
$$\text{(by (18.8))}$$
$$= \lim_{\varepsilon \to 0} \liminf_{M,N \to \infty} -\frac{2}{M^{d-1}N} \log[|a|^{-(N-K)|S|}\lambda_{\mathbb{R}^{S(M,N)}}(\pi_0(D(Q(M,N)',\varepsilon)))]$$
$$\text{(by (18.9))}$$
$$\leq 2\log|a| + \lim_{\varepsilon \to 0} \liminf_{M,N \to \infty} \frac{1}{M^{d-1}N}$$
$$\cdot \log \lambda_{\mathbb{C}^{S(M,N)}}\left(\{z \in \mathbb{C}^{S(M,N)} : \max_{j=0,\ldots,N-1} \|A^j z\|_\infty < \varepsilon\}\right)$$
$$\text{(by (18.6)–(18.7))}$$
$$= 2\log|a| + \liminf_{M,N \to \infty} \frac{2}{M^{d-1}} \sum_{\mathbf{k} \in R(M,N)} \sum_{j=1}^K \log^+ |\zeta_{\omega(\mathbf{k}),j}|$$
$$\text{(by letting } M \to \infty \text{ and setting } N = \mathrm{Int}(\log M), L = M + qN)$$
$$\leq 2\log|a| + \lim_{L \to \infty} \frac{2}{L^{d-1}} \sum_{\mathbf{k} \in R(M,N)} \sum_{j=1}^K \log^+ |\zeta_{\omega(\mathbf{k}),j}|$$
$$\text{(by Lemma 17.1)}$$
$$= 2\log|a| + 2\int_{\mathbb{S}^{d-1}} \log^+ |\zeta_{\mathbf{s},j}| \, d\lambda_{\mathbb{S}^{d-1}}(\mathbf{s})$$
$$\text{(by Proposition 16.1)}$$
$$= 2\int_{\mathbb{S}^{d-1}} \int_{\mathbb{S}} \log|f_{\mathbf{s}}| \, d\lambda_{\mathbb{S}} \, d\lambda_{\mathbb{S}^{d-1}} = 2\log \mathrm{M}(f). \quad \square$$

We state some consequences of Theorem 18.1. For notation we refer to Lemma 5.1.

COROLLARY 18.5. *Let* $d \geq 1$, *and let* $\mathfrak{p} \subset \mathfrak{R}_d$ *be a prime ideal. Then*

$$h(\alpha^{\mathfrak{R}_d/\mathfrak{p}}) = \begin{cases} |\log \mathbb{M}(f)| & \text{if } \mathfrak{p} = (f) \text{ is principal,} \\ 0 & \text{if } \mathfrak{p} \text{ is not principal.} \end{cases}$$

PROOF. Use Lemma 17.4. $\square$

PROPOSITION 18.6. *Let* $d \geq 1$, *and let* $\mathfrak{M}$ *be a Noetherian* $\mathfrak{R}_d$-*module. If* $\mathfrak{M} = \mathfrak{N}_s \supset \cdots \supset \mathfrak{N}_0 = \{0\}$ *is a prime filtration of* $\mathfrak{M}$, *and if* $\mathfrak{q}_i \subset \mathfrak{R}_d$ *is the prime ideal satisfying that* $\mathfrak{N}_i/\mathfrak{N}_{i-1} \cong \mathfrak{R}_d/\mathfrak{q}_i$, $i = 1, \ldots, s$, *then*

$$h(\alpha^{\mathfrak{M}}) = \sum_{i=1}^{s} h(\alpha^{\mathfrak{R}_d/\mathfrak{q}_i}), \qquad (18.10)$$

where $h(\alpha^{\mathfrak{R}_d/\mathfrak{q}_i})$ *is given by* Corollary 18.5 *for every* $i = 1, \ldots, s$.

If $\mathfrak{L}$ *is an arbitrary countable* $\mathfrak{R}_d$-*module, there exists a increasing sequence* $(\mathfrak{L}_n, n \geq 1)$ *of Noetherian submodules of* $\mathfrak{L}$ *such that* $\bigcup_{n \geq 1} \mathfrak{L}_n = \mathfrak{L}$. *Then* $h(\alpha^{\mathfrak{L}}) = \lim_{n \to \infty} h(\alpha^{\mathfrak{L}_n})$, *and* $h(\alpha^{\mathfrak{L}_n})$ *is determined by* (18.10) *for every* $n \geq 1$.

PROOF. The first assertion is clear from the addition formula (14.1), and the second follows from Lemma 13.6 by setting $V_n = \mathfrak{L}_n^{\perp} \subset X^{\mathfrak{L}} = \widehat{\mathfrak{L}}$ for all $n \geq 1$. $\square$

EXAMPLES 18.7. (1) Let $d \geq 1$, denote by $M_n(\mathfrak{R}_d)$ the ring of $n \times n$ matrices with entries in $\mathfrak{R}_d$, and let $A \in M_n(\mathfrak{R}_d)$. We claim that

$$h(\alpha^{\mathfrak{R}_d^n/A\mathfrak{R}_d^n}) = |\log \mathbb{M}(\det A)|. \qquad (18.11)$$

In order to prove (18.11) we set $X = X^{\mathfrak{R}_d^n/A\mathfrak{R}_d^n} = \widehat{\mathfrak{R}_d^n/A\mathfrak{R}_d^n}$ and $\alpha = \alpha^{\mathfrak{R}_d^n/A\mathfrak{R}_d^n}$, and assume first that $\det A = 0$.

Put $\mathfrak{N} = \{w \in \mathfrak{R}_d^n : A^k w = 0 \text{ for some } k > 0\}$. Since $\mathfrak{R}_d^n$ is a Noetherian $\mathfrak{R}_d$-module, there exists a $K \geq 1$ with $\mathfrak{N} \supsetneq A\mathfrak{N} \supsetneq \cdots \supsetneq A^K \mathfrak{N} = \{0\}$, and we choose an element $w \in \mathfrak{N}$ with $A^{K-1}w \neq 0$. Then $f A^{K-1} w = A^{k-1} f w \neq 0$ for every non-zero $f \in \mathfrak{R}_d$. It follows that $fw \in \mathfrak{N} \smallsetminus A\mathfrak{R}_d^n$ for every non-zero $f \in \mathfrak{R}_d$, and that $\mathfrak{R}_d \cong \mathfrak{L} = \mathfrak{R}_d w/A\mathfrak{R}_d^n \subset \mathfrak{R}_d^n/A\mathfrak{R}_d^n$. We set $Y = \mathfrak{L}^{\perp} \subset \widehat{\mathfrak{R}_d^n/A\mathfrak{R}_d^n} = X$ and obtain that $h(\alpha) \geq h(\alpha^{X/Y}) = h(\alpha^{\mathfrak{L}}) = h(\alpha^{\mathfrak{R}_d/(0)}) = |\log \mathbb{M}(0)| = \infty$ by Theorem 18.1.

Now assume that $B, C \in M_n(\mathfrak{R}_d)$ have non-zero determinant, and that (18.11) holds for B and C. Consider the exact sequence

$$0 \longrightarrow B\mathfrak{R}_d^n/BC\mathfrak{R}_d^n \longrightarrow \mathfrak{R}_d^n/BC\mathfrak{R}_d^n \longrightarrow \mathfrak{R}_d^n/B\mathfrak{R}_d^n \longrightarrow 0.$$

Since $\det B \neq 0$, the map B is injective on $\mathfrak{R}_d^n$, so that $B\mathfrak{R}_d^n/BC\mathfrak{R}_d^n \cong \mathfrak{R}_d^n/C\mathfrak{R}_d^n$, and the addition formula (14.1) shows that

$$h(\alpha^{\mathfrak{R}_d^n/BC\mathfrak{R}_d^n}) = h(\alpha^{\mathfrak{R}_d^n/B\mathfrak{R}_d^n}) + h(\alpha^{\mathfrak{R}_d^n/C\mathfrak{R}_d^n})$$
$$= \log \mathrm{M}\,(\det B) + \log \mathrm{M}\,(\det C) \qquad (18.12)$$
$$= \log \mathrm{M}\,(\det BC),$$

so that (18.11) holds for BC.

We regard A as a matrix over the field of fractions $\mathfrak{R}_d$ and note that there exists a non-zero polynomial $f \in \mathfrak{R}_d$ such that fA can be written as a product of elementary matrices of the form

$$\begin{pmatrix} 1 \\ & \ddots \\ & & 1 \\ & & & g \\ & & & & 1 \\ & & & & & \ddots \\ & & & & & & 1 \end{pmatrix}, \quad \begin{pmatrix} 1 \\ & \ddots \\ & & 0 & & 1 \\ & & & \ddots \\ & & 1 & & 0 \\ & & & & & \ddots \\ & & & & & & 1 \end{pmatrix}, \quad \text{and} \quad \begin{pmatrix} 1 & & 0 \\ & \ddots \\ 1 & & 1 \\ & & & \ddots \\ & & & & 1 \end{pmatrix}, (18.13)$$

where $0 \neq g \in \mathfrak{R}_d$. It is clear that (18.11) holds for any of the elementary matrices in (18.13), and hence for their product fA. We conclude that $h(\alpha^{\mathfrak{R}_d^n/fA\mathfrak{R}_d^n}) = \log \mathrm{M}\,(\det fA) = n\log \mathrm{M}\,(f) + \log \mathrm{M}\,(\det A)$, and (18.12) (with $B = fI$ and $C = A$) shows that

$$h(\alpha^{\mathfrak{R}_d^n/fA\mathfrak{R}_d^n}) = h(\alpha^{\mathfrak{R}_d^n/f\mathfrak{R}_d^n}) + h(\alpha^{\mathfrak{R}_d^n/A\mathfrak{R}_d^n}) = n\log \mathrm{M}\,(f) + h(\alpha^{\mathfrak{R}_d^n/A\mathfrak{R}_d^n}).$$

Since $\mathrm{M}\,(f) < \infty$ we have verified (18.11).

(2) If $d \geq 1$ and $f \in \mathfrak{R}_d$, then $f \in \mathfrak{R}_{d+n}$ for every $n \geq 0$. However, $\mathrm{M}\,(f)$ is independent of n, since the extra variables integrate to 1. To see this dynamically, let $\alpha = \alpha^{\mathfrak{R}_d/(f)} = \alpha^{\mathfrak{R}_d/f\mathfrak{R}_d}$. Then $\alpha^{\mathfrak{R}_{d+n}/f\mathfrak{R}_{d+n}}$ is isomorphic to an action $\bar{\alpha}$ of $\mathbb{Z}^d \times \mathbb{Z}^n$ on $X^{\mathbb{Z}^n}$, where $X = X^{\mathfrak{R}_d/(f)}$, and where $(\bar{\alpha}_{(\mathbf{m},\mathbf{n})}(x))_{\mathbf{k}} = (\alpha_{\mathbf{m}}(x))_{\mathbf{k}+\mathbf{n}}$ for all $x = (x_{\mathbf{k}}) \in X^{\mathbb{Z}^n}$, $\mathbf{m} \in \mathbb{Z}^d$, and $\mathbf{k} \in \mathbb{Z}^n$. Since the action of $\mathbb{Z}^n$ on $X^{\mathbb{Z}^n}$ is the shift-action, it follows from the last equality in Proposition 13.7 that $h(\bar{\alpha}) = h(\alpha)$.

(4) Let $d > 1$, $c \in (\overline{\mathbb{Q}}^\times)^d$, and consider the $\mathbb{Z}^d$-actions $\alpha^{\mathfrak{R}_d/\mathfrak{j}_c}$ and $\alpha^{(c)}$ in Theorem 5.1. Since the homomorphism ϕ in (5.8) is finite-to-one, we know that $h(\alpha^{(c)}) = h(\alpha^{\mathfrak{R}_d/\mathfrak{j}_c})$, and Corollary 18.5 implies that $h(\alpha^{\mathfrak{R}_d/\mathfrak{j}_c}) = 0$, since the ideal $\mathfrak{j}_c$ is prime and non-principal. An alternative proof of this fact is based on Example 17.3 (2), where we saw that $h(\alpha_{\mathbf{n}}^{(c)}) = h(\alpha_{\mathbf{n}}^{\mathfrak{R}_d/\mathfrak{j}_c}) < \infty$ for all $\mathbf{n} \in \mathbb{Z}^d$; a glance at the definition of $h_{\mathrm{cover}}(\cdot) = h(\cdot)$ now shows that $h(\alpha^{(c)}) = h(\alpha^{\mathfrak{R}_d/\mathfrak{j}_c}) = 0$. $\boxdot$

CONCLUDING REMARK 18.8. All the material in this section is based on [63].

CHAPTER VI

Positive entropy

19. Positive entropy

We begin this section with a brief discussion of entropy for a single automorphism α of a compact group X, in which we prove that every ergodic automorphism of an infinite, compact group has positive entropy, and that automorphisms with zero entropy have a very degenerate structure.

LEMMA 19.1. *Let $f \in \mathfrak{R}_1$ be a Laurent polynomial with $\mathbb{M}(f) = 1$. Then there exist integers $r \geq 1$, $t \in \mathbb{Z}$, and cyclotomic polynomials $c_1, \ldots, c_r$, such that $f(u) = \pm u^t \prod_{i=1}^{r} c_r(u)$. In other words, $\pm f$ is a finite product of generalized cyclotomic polynomials (cf. Corollary 6.11).*

PROOF. It is obviously enough to prove the lemma under the assumption that f is irreducible. We choose an integer t such that $u^t f(u) = a_0 + a_1 u + \cdots + a_s u^s$ with $a_0 a_s \neq 0$. By Proposition 16.1, $|a_s| = 1$, and the roots $\xi_1, \ldots, \xi_s$ (counted with multiplicity) must all have modulus ≤ 1. Hence $a_0 = \xi_1 \cdots \xi_s = 1$, and $|\xi_j| = 1$ for $j = 1, \ldots, s$, so that $\xi_j^{-1} = \bar{\xi}_j \in \{\xi_1, \ldots, \xi_s\}$ for every $j = 1, \ldots, s$. Let $\mathbb{K} = \mathbb{Q}(\xi_1)$, and let $P^{\mathbb{K}}$ be the set of places of $\mathbb{K}$ (we are using the same notation as in Section 7). Since both ξ_1 and ξ_1^{-1} are algebraic integers, we have that $|\xi_j|_v = 1$ for every $v \in P^{\mathbb{K}}$. As in Section 7 we write $\mathbb{K}_{\mathbb{A}}$ for the adele ring of $\mathbb{K}$ and consider the diagonal embedding $i \colon \mathbb{K} \longmapsto \mathbb{A}_{\mathbb{K}}$. Then the elements $i(\xi_j^k)$, $k \geq 1, j = 1, \ldots, s$, lie in the compact subset $\prod_{v \in P^{\mathbb{K}}} \{\omega \in \mathbb{K}_v : |\omega|_v \leq 1\}$ of $\mathbb{K}_{\mathbb{A}}$. In particular the set $\{i(\xi_j^k) : k \geq 1, j = 1, \ldots, s\}$ is finite, and there exists an integer $k \geq 1$ such that $i(\xi_1^k) = i(1)$. It follows that $\xi_1^k = 1$ and that either f or $-f$ is a generalized cyclotomic polynomial. $\square$

THEOREM 19.2. *Let α be an automorphism of a compact group X with zero entropy. Then there exists a decreasing sequence of closed, normal, α-invariant subgroups $X = Y_0 \supset Y_1 \supset \cdots \supset Y_n \supset \cdots$ such that $\bigcap_{i \geq 1} Y_i = \{1_X\}$, and Y_{i-1}/Y_i is a compact Lie group with an α^{Y_{i-1}/Y_i}-invariant metric for every*

$i \geq 1$. In particular, there exists no infinite, closed, α-invariant subgroup $Y \subset X$ such that α^Y is ergodic. Finally, X is zero-dimensional and α satisfies the d.c.c., then X is finite.

PROOF. First assume that α satisfies the d.c.c. If α is ergodic, we apply Theorem 10.6 and conclude that, in the notation of that theorem, $\Lambda_i = \{1\}$ for $i = 1, \ldots, K$, and that $\log M(\chi_A) = 0$ (Theorem 17.7), where A is the $n \times n$-matrix appearing in Theorem 10.6 and χ_A is the characteristic polynomial of A. From (17.18) we see in particular that $\chi_A \in \mathfrak{R}_1$, and Lemma 19.1 implies that every eigenvalue of A is a root of unity, contrary to Example 9.5 (3). This shows that α must be non-ergodic.

Since $0 = h(\alpha) \geq h(\alpha^V) = 0$ for every closed, α-invariant subgroup $V \subset X$, α^V will again be non-ergodic by the first part of this proof, and Theorem 1.4, combined with the d.c.c., yields that there exist closed, normal, α-invariant subgroups $X = V_0 \supset \cdots \supset V_r = \{1\}$ such that V_{i-1}/V_i is a compact Lie group with an α-invariant metric for every $i = 1, \ldots, r$. In particular, if X is zero-dimensional, then V_{i-1}/V_i must be finite for $i = 1, \ldots, r$, so that X is finite.

For an arbitrary automorphism α of a compact group X with $h(\alpha) = 0$ we choose a decreasing sequence $(V_n, n \geq 1)$ of closed, normal, α-invariant subgroups such that $\bigcap_{n \geq 1} V_n = \{1_X\}$ and α^{X/V_n} satisfies the d.c.c. for every $n \geq 1$ (Example 4.8 (1) and Proposition 4.9). Then $h(\alpha^{V_{n-1}/V_n}) = 0$ for every $n \geq 1$, where $V_0 = X$, and by applying the preceding paragraph to each α^{V_{n-1}/V_n} we obtain the promised sequence $(Y_n, n \geq 1)$. $\square$

If $d > 1$, and if α is a $\mathbb{Z}^d$-action by automorphisms of a compact, abelian group X, we assume without loss in generality that $\alpha = \alpha^{\mathfrak{M}}$ and $X = X^{\mathfrak{M}} = \widehat{\mathfrak{M}}$ for some countable $\mathfrak{R}_d$-module $\mathfrak{M}$ (Lemma 5.1). We already know how to characterize ergodicity of $\alpha^{\mathfrak{M}}$ in terms of the associated primes of $\mathfrak{M}$ (Proposition 6.6), and we now prove that positive entropy is again determined by the prime ideals associated with $\mathfrak{M}$.

DEFINITION 19.3. A prime ideal $\mathfrak{q} \subset \mathfrak{R}_d$, $d \geq 1$, is positive if $h(\alpha^{\mathfrak{R}_d/\mathfrak{q}}) > 0$, and null otherwise.

PROPOSITION 19.4. Let $\mathfrak{M}$ be a countable $\mathfrak{R}_d$-module, and let $\alpha = \alpha^{\mathfrak{M}}$ and $X = X^{\mathfrak{M}} = \widehat{\mathfrak{M}}$.

(1) $h(\alpha^{\mathfrak{M}}) = 0$ if and only if every prime ideal $\mathfrak{p}$ associated with $\mathfrak{M}$ is null;
(2) If $\mathfrak{M}$ is Noetherian, then $h(\alpha^{\mathfrak{M}}) < \infty$ if and only if every prime ideal associated with $\mathfrak{M}$ is non-zero.

PROOF. For every prime ideal $\mathfrak{p}$ associated with $\mathfrak{M}$ there exists an element $a \in \mathfrak{M}$ with $\mathrm{ann}(a) = \mathfrak{p}$. Put $\mathfrak{N} = \mathfrak{R}_d \cdot a$ and $Y = \mathfrak{N}^\perp \subset X$, and note that $\widehat{\mathfrak{N}} = X/Y$ and $\alpha^{\mathfrak{N}} = \alpha^{X/Y}$. In particular, $h(\alpha) \geq h(\alpha^{X/Y}) = h(\alpha^{\mathfrak{R}_d/\mathfrak{p}})$, so that $\mathfrak{p}$ must be null if $h(\alpha) = 0$, and $\mathfrak{p}$ is non-zero if $h(\alpha) < \infty$ (otherwise $\infty = h(\alpha^{\mathfrak{R}_d}) = h(\alpha^{\mathfrak{R}_d/\mathfrak{p}}) = h(\alpha^{X/Y}) \leq h(\alpha) < \infty$).

Conversely, if every prime ideal $\mathfrak{p} \subset \mathfrak{R}_d$ associated with $\mathfrak{M}$ is null, we choose an increasing sequence $(\mathfrak{M}_n, n \geq 1)$ of Noetherian submodules of $\mathfrak{M}$ with $\mathfrak{M} = \bigcup_{n \geq 1} \mathfrak{M}_n$. Fix $n \geq 1$ for the moment and choose a prime filtration $\mathfrak{M}_n = \mathfrak{N}_s \supset \cdots \supset \mathfrak{N}_0 = \{0\}$ of $\mathfrak{M}$. Then $\mathfrak{N}_j/\mathfrak{N}_{j-1} \cong \mathfrak{R}_d/\mathfrak{q}_j$ for some prime ideal $\mathfrak{q}_j \subset \mathfrak{R}_d$ containing one of the associated primes of $\mathfrak{M}_n$, and hence one of the prime ideals associated with $\mathfrak{M}$, and Lemma 17.4 shows that $\mathfrak{q}_j$ is null for $j = 1, \ldots, s$. According to Proposition 18.6, $h(\alpha^{\mathfrak{M}_n}) = 0$, and Lemma 13.6 implies that $h(\alpha) = 0$, as claimed in (1).

In order to complete the proof of (2) we observe that, if every prime ideal $\mathfrak{p} \subset \mathfrak{R}_d$ associated with $\mathfrak{M}$ is non-zero, then $h(\alpha^{\mathfrak{R}_d/\mathfrak{p}}) < \infty$ (Corollary 18.5), and Proposition 18.6, Lemma 17.4, and Corollary 6.2 together imply that $h(\alpha) < \infty$. $\square$

By Corollary 18.5, every non-principal prime ideal $\mathfrak{p} \subset \mathfrak{R}_d$ is null, and the characterization of all null prime ideals $\mathfrak{p} \subset \mathfrak{R}_d$ which are *principal* is equivalent to the problem of determining the polynomials $f \in \mathfrak{R}_d$ with $\mathrm{M}(f) = 1$.

THEOREM 19.5. *Let $f \in \mathfrak{R}_d$, $d \geq 1$. Then $\mathrm{M}(f) = 1$ if and only if $\pm f$ is a product of generalized cyclotomic polynomials (cf. Corollary 6.12).*

For $d = 1$, Theorem 19.5 is Lemma 19.1. For $d > 1$, the proof of Theorem 19.5 depends on an inequality in [100], which compares the Mahler measure of a non-zero polynomial $f \in \mathfrak{R}_d$ with that of certain polynomials associated with the faces of the convex hull of the support of f. In order to formulate Smyth's inequality we have to introduce a little bit of terminology. Let $C \subset \mathbb{R}^d$ be a closed, convex set. A hyperplane $H \subset \mathbb{R}^d$ supports C if $C \cap H \neq \varnothing$, and C is contained in one of the two closed half-spaces defined by H. A face Φ of C is an intersection of the form $\Phi = C \cap H$, where H is a supporting hyperplane of C (in particular, a face may consist of a single extremal point). Now assume that $0 \neq f = \sum_{\mathbf{m} \in \mathbb{Z}^d} c_f(\mathbf{m}) u^{\mathbf{m}} \in \mathfrak{R}_d$, denote by

$$\mathcal{S}(f) = \{\mathbf{m} \in \mathbb{Z}^d : c_f(\mathbf{m}) \neq 0\}, \tag{19.1}$$

the support of f, and write

$$\mathcal{C}(f) \subset \mathbb{R}^d \tag{19.2}$$

for the closed, convex hull of $\mathcal{S}(f) \subset \mathbb{Z}^d \subset \mathbb{R}^d$. If Φ is a face of $\mathcal{C}(f)$ we set

$$f_\Phi = \sum_{\mathbf{m} \in \Phi} c_f(\mathbf{m}) u^{\mathbf{m}}. \tag{19.3}$$

LEMMA 19.6. *For every face Φ of $\mathcal{C}(f)$, $\mathrm{M}(f) \geq \mathrm{M}(f_\Phi)$.*

PROOF. We may assume without loss in generality that $\Phi = \mathcal{C}(f) \cap H$, where $H = \{\mathbf{m} \in \mathbb{Z}^d : \langle \mathbf{m}, \mathbf{v} \rangle = k\}$ for some non-zero, primitive element $\mathbf{v} \in \mathbb{Z}^d$ and some $k \in \mathbb{Z}$, and the Lemmas 18.4 and 11.3 allow us to assume furthermore that $\mathbf{v} = \mathbf{e}^{(d)} = (0, \ldots, 0, 1) \in \mathbb{Z}^d$, $k = 0$, and that $\mathcal{C}(f) \subset \{\mathbf{m} = (m_1, \ldots, m_d) \in \mathbb{Z}^d : m_d \leq 0\}$ (this amounts to replacing the polynomial f by $u^{\mathbf{n}} f^A$ for some $\mathbf{n} \in \mathbb{Z}^d$ and $A \in \mathrm{GL}(d, \mathbb{Z})$, where f^A has the same meaning as in Lemma 18.4). Then $f_\Phi = \sum_{\mathbf{m}=(m_1,\ldots,m_{d-1},0) \in \mathbb{Z}^d} c_f(\mathbf{m}) u^{\mathbf{m}}$.

We write $X = X^{\mathfrak{R}_d/(f)}$ in the form (5.9) (with $\mathfrak{a} = (f)$) and consider the shift-action $\alpha = \alpha^{\mathfrak{R}_d/(f)}$ of $\mathbb{Z}^d$ on X. If $x = (x_{\mathbf{m}}) \in X \subset \mathbb{T}^{\mathbb{Z}^d}$ satisfies that $x_{\mathbf{m}} = 0$ for all $\mathbf{m} = (m_1, \ldots, m_d) \in \mathbb{Z}^d$ with $m_d < 0$, and if we set $y_{\mathbf{n}} = x_{(n_1,\ldots,n_{d-1},0)}$ for all $\mathbf{n} = (n_1, \ldots, n_{d-1}) \in \mathbb{Z}^{d-1}$, then $y = (y_{\mathbf{n}}) \in Y = X^{\mathfrak{R}_{d-1}/f_\Phi} \subset \mathbb{T}^{\mathbb{Z}^{d-1}}$. Conversely there exists, for every $x \in X$ and $y \in Y$, an element $z \in X$ such that $z_{\mathbf{m}} = x_{\mathbf{m}}$ whenever $\mathbf{m} = (m_1, \ldots, m_d)$ with $m_d < 0$, and $z_{(m_1,\ldots,m_{d-1},0)} = x_{(m_1,\ldots,m_{d-1},0)} + y_{(m_1,\ldots,m_{d-1})} \pmod 1$ for all $(m_1, \ldots, m_{d-1}) \in \mathbb{Z}^{d-1}$.

Let $Q(m) = \{-m, \ldots, m\}^{d-1} \subset \mathbb{Z}^{d-1}$, and let $Q(m, n) = Q(m) \times \{0, \ldots, n\} \subset \mathbb{Z}^d$. We consider the metric ϑ on $\mathbb{T}$ defined in (17.7), fix $\varepsilon > 0$, and choose a maximal $[Q(m), \vartheta, \varepsilon]$-separated set $F \subset Y$ (for notation see Proposition 13.7). The discussion in the preceding paragraph shows that there exists a $[Q(m, n), \vartheta, \varepsilon]$-separated set in X whose cardinality is $\geq |F|^n$, where $|F|$ is the cardinality of F. By letting first $n \to \infty$ and then $m \to \infty$ we see from Proposition 13.7 and Theorem 18.1 that $\log \mathrm{M}(f) = h(\alpha) \geq h(\alpha^{\mathfrak{R}_{d-1}/(f_\Phi)}) = \log \mathrm{M}(f_\Phi)$, as claimed. $\square$

PROOF OF THEOREM 19.5. The case where $d = 1$ is dealt with by Lemma 19.1. Assume therefore that $d > 1$. If $f \in \mathfrak{R}_d$ is a product of generalized cyclotomic polynomials, then the same is true for every polynomial of the form $u^{\mathbf{m}} f^A$ with $\mathbf{m} \in \mathbb{Z}^d$ and $A \in \mathrm{GL}(d, \mathbb{Z})$ (the notation is taken from Lemma 18.4). Hence the assertion of the theorem is unaffected if we replace f by $u^{\mathbf{m}} f^A$.

If f is a generalized cyclotomic polynomial, then $f = u^{\mathbf{m}} c(u^{\mathbf{n}})$, where $\mathbf{m}, \mathbf{n} \in \mathbb{Z}^d$, $\mathbf{n} \neq \mathbf{0}$, and where c is a cyclotomic polynomial in a single variable. From (16.1) and a change of variable it is clear that $\mathrm{M}(f) = \mathrm{M}(c)$, and Proposition 16.1 shows that $\mathrm{M}(f) = \mathrm{M}(c) = 1$. If f is a product of generalized cyclotomic polynomials $f = h_1 \cdots h_s$, then $\mathrm{M}(f) = \prod_{j=1}^s \mathrm{M}(h_j) = 1$, which completes the proof of one of the implications in Theorem 19.5.

In order to prove the reverse implication we have to show that every polynomial $f \in \mathfrak{R}_d$ with $\mathrm{M}(f) = 1$ is a product of generalized cyclotomic polynomials. We use double induction on the number of variables d and the number of irreducible factors r of f in $\mathbb{C}[u_1^{\pm 1} \cdots u_d^{\pm 1}]$ (excluding monomials). The result is clearly true if $d = 1$, or if $r = 0$, in which case $f = a u^{\mathbf{m}}$ for some $a \in \mathbb{Z}$ and $\mathbf{m} \in \mathbb{Z}^d$. Now suppose that the assertion is true if either $d < D$ or $r < R$, where $D \geq 2$ and $R \geq 1$. Let $f \in \mathfrak{R}_D$, and assume that $\mathrm{M}(f) = 1$ and that f has R irreducible factors in $\mathbb{C}[u_1^{\pm 1} \cdots u_d^{\pm 1}]$. If $\mathcal{S}(f)$ lies on a straight line we can find elements $\mathbf{m} \in \mathbb{Z}^D$ and $A \in \mathrm{GL}(D, \mathbb{Z})$ such that $g = u^{\mathbf{m}} f^A$ is a polynomial of the single variable u_1. Since $\mathrm{M}(f) = \mathrm{M}(g)$, Lemma 19.1 implies that $\pm g$ is a product of cyclotomic polynomials, and hence that $\pm f$ is a product of generalized cyclotomic polynomials.

Assume therefore that $\mathcal{S}(f)$ does not lie on a straight line or, equivalently, that $\mathcal{C}(f)$ has a face $\Phi \subsetneq \mathcal{C}(f)$ which consists of more than one point. As in the proof of Lemma 19.6 we may replace f by $u^{\mathbf{m}} f^A$ for some $\mathbf{m} \in \mathbb{Z}^D$ and $A \in \mathrm{GL}(D, \mathbb{Z})$ and assume that $\mathcal{S}(f) \subset \{\mathbf{m} = (m_1, \ldots, m_D) \in \mathbb{Z}^D : m_D \leq 0\}$, and that $\Phi \subset \{\mathbf{m} = (m_1, \ldots, m_D) \in \mathbb{Z}^D : m_D = 0\}$. After multiplying f by a monomial we may thus take it that $f = \sum_{j=0}^{L} u_D^j f_j$, where $f_j \in \mathfrak{R}_{D-1}$ for $j = 0, \ldots, L$, $f_0 f_L \neq 0$, and $\mathcal{S}(f_L)$ consists of more than one point. Lemma 19.6 implies that $\mathrm{M}(f_L) = 1$ and, since $f_L \in \mathfrak{R}_{D-1}$, our induction hypothesis implies that f_L is a (necessarily non-trivial) product of generalized cyclotomic polynomials in the variables $u_1, \ldots, u_{D-1}$. In particular, f_L can be written—in $\mathbb{C}[u_1^{\pm 1}, \ldots, u_{D-1}^{\pm 1}]$—as a product of the form $f_L = c u^{\mathbf{m}} \prod_{j=1}^{t} (u^{\mathbf{a}_j} - \alpha_j)$, where $0 \neq c \in \mathbb{C}$, $\alpha_1, \ldots, \alpha_t \in \mathbb{S} \subset \mathbb{C}$, $\mathbf{n} \in \mathbb{Z}^{D-1}$, and where $\mathbf{a}_1, \ldots, \mathbf{a}_t$ are primitive elements in $\mathbb{Z}^{D-1}$ (as usual, $u^{\mathbf{a}} = u_1^{a_1} \cdots u_{D-1}^{a_{D-1}}$ for all $\mathbf{a} = (a_1, \ldots, a_{D-1}) \in \mathbb{Z}^{D-1}$).

We claim that f_L divides f_j for $j = 0, \ldots, L - 1$. Indeed, if f_L does not divide f_k for some $k \in \{0, \ldots, L - 1\}$, then f_k is not divisible by some $(u^{\mathbf{a}_j} - \alpha_j)$. Since $\mathbf{a}_j$ is primitive, we can apply Lemma 11.3 to find a matrix $B \in \mathrm{GL}(D, \mathbb{Z})$ such that $B\mathbf{e}^{(D)} = \mathbf{e}^{(D)}$ and $B\mathbf{a}_j = \mathbf{e}^{(D-1)}$, where $\mathbf{e}^{(i)}$ is the i-th unit vector in $\mathbb{Z}^D$. After replacing f with f^B, if necessary, we can write f_k as $f_k = (u_{D-1} - \alpha_j)g + h$, where h is a non-zero complex polynomial in the variables $u_1^{\pm 1}, \ldots, u_{D-2}^{\pm 1}$. There exists a point $\mathbf{c} = (c_1, \ldots, c_{D-2}) \in \mathbb{S}^{D-2}$ with $h(\mathbf{c}) \neq 0$, and a neighbourhood $N(c_1, \ldots, c_{D-2}, \alpha_j) \subset \mathbb{S}^{D-1}$ of $(c_1, \ldots, c_{D-2}, \alpha_j)$ such that

$$|f_k(\mathbf{t})/f_L(\mathbf{t})| > 1 + \binom{L-2}{k-2} \tag{19.4}$$

for all $\mathbf{t} \in N(c_1, \ldots, c_{D-2}, \alpha_j)$. We fix $\mathbf{t} = (t_1, \ldots, t_{D-1}) \in \mathbb{S}^{D-1}$ and regard

$$F_{\mathbf{t}}(u_D) = f(t_1, \ldots, t_{D-1}, u_D)$$

as a function of the single variable u_D. If every root of $F_{\mathbf{t}}$ has modulus ≤ 1, then the coefficient of u_D^k in $F_{\mathbf{t}}$ must have modulus $\leq \binom{L-2}{k-2}$, and by comparing this with (19.4) and (16.2) we see that $\int_{\mathbb{S}} \log |F_{\mathbf{t}}| \, d\lambda_{\mathbb{S}} > |f_L(\mathbf{t})|$ for every $\mathbf{t} \in N(c_1, \ldots, c_{D-1}, \alpha_j)$. According to (16.2), $\int_{\mathbb{S}} \log |F_{\mathbf{t}}| \, d\lambda_{\mathbb{S}} \geq |f_L(\mathbf{t})|$ for every $\mathbf{t} \in \mathbb{S}^{D-1}$, and we obtain that

$$0 = \log \mathrm{M}(f) = \int_{\mathbb{S}^{D-1}} \int_{\mathbb{S}} |f_{\mathbf{t}}| \, d\lambda_{\mathbb{S}} \, d\lambda_{\mathbb{S}^{D-1}}$$
$$> \int_{\mathbb{S}^{D-1}} |f_L| \, d\lambda_{\mathbb{S}^{D-1}} = \log \mathrm{M}(f_L) = 0,$$

which is absurd. This contradiction implies that f_L divides each f_j, $j = 0, \ldots,$ $L-1$, and hence f.

We can write f as a product $f = gf_L$, where f_L is a non-trivial product of generalized cyclotomic polynomials. Hence g has fewer than R irreducible factors, and $1 \le \mathrm{M}(g) \le \mathrm{M}(f) = 1$. Our induction hypothesis (on polynomials with fewer than R irreducible factors) allows us to write g as a product of generalized cyclotomic polynomials, so that f is a product of generalized cyclotomic polynomials. $\square$

Having determined the principal prime ideals $\mathfrak{p} \subset \mathfrak{R}_d$ which are null, we turn to the problem of calculating the Mahler measure $\mathrm{M}(f)$ of a polynomial $f \in \mathfrak{R}_d$. An explicit calculation is, of course, possible only in some special cases, such as when $d = 1$ (Proposition 16.1), or for certain polynomials discussed in [13] and [101]. We present some of Smyth's results, for which we shall use the following notation: a *character* (mod q) is a homomorphism $\chi \colon \mathbb{Z} \longmapsto \mathbb{C}$ with $\chi(0) = 0$, $\chi(1) = 1$, $\chi(m + q) = \chi(m)$, and $\chi(mm') = \chi(m)\chi(m')$ for all $m, m' \in \mathbb{Z}$. The symbols χ_q, $q = 3, 4$, will denote the unique non-trivial characters (mod q) given by

$$
\chi_3(m) = \begin{cases} 0 & \text{if } m \equiv 0 \pmod{3}, \\ 1 & \text{if } m \equiv 1 \pmod{3}, \\ -1 & \text{if } m \equiv 2 \pmod{3}, \end{cases}
$$
$$
\chi_4(m) = \begin{cases} 0 & \text{if } m \equiv 0 \pmod{2}, \\ 1 & \text{if } m \equiv 1 \pmod{4}, \\ -1 & \text{if } m \equiv 3 \pmod{4}. \end{cases}
\tag{19.5}
$$

The *L-function* $L(s, \chi)$ associated with a character χ is defined by

$$
L(s, \chi) = \sum_{n=1}^{\infty} \frac{\chi(n)}{n^s} = \prod_{p \text{ prime}} \left(1 - \frac{\chi(p)}{p^s}\right).
\tag{19.6}
$$

PROPOSITION 19.7. *Let $k \in \mathbb{Z}$, and let $f_k = u_1 + u_2 + k \in \mathfrak{R}_2$. Then*

$$
h(\alpha^{\mathfrak{R}_2/(f_k)}) = \log \mathrm{M}(f_k) = \begin{cases} 0 & \text{if } k = 0, \\ \frac{3\sqrt{3}}{4\pi} L(2, \chi_3) & \text{if } |k| = 1, \\ \log |k| & \text{if } |k| \ge 2, \end{cases}
\tag{19.7}
$$

PROOF. According to (16.2), $\mathrm{M}(f_0) = 1$. If $k \ne 0$ we also apply (16.2) and obtain that

$$
\int_0^1 \int_0^1 \log \left| e^{2\pi i u_1} + e^{2\pi i k u_2} + k \right| du_1 du_2
$$
$$
= \int_0^1 \int_0^1 \log \left| 1 + e^{2\pi i (u_2 - u_1)} + k e^{-2\pi i k u_1} \right| du_1 du_2
$$

$$= \int_0^1 \log |k| + \log^+ \frac{|1 + e^{2\pi i t}|}{|k|} \, dt$$

$$= \int_0^1 \max\{\log |k|, \log |1 + e^{2\pi i t}|\} \, dt.$$

Hence, if $|k| \geq 2$, $\log \mathrm{M}(f_k) = \int_0^1 \log |k| \, dt = \log |k|$. If $|k| = 1$, then

$$\log \mathrm{M}(f_k) = \int_0^1 \log^+ |1 + e^{2\pi i t}| \, dt = \int_{-1/3}^{1/3} \log |1 + e^{2\pi i t}| \, dt.$$

As

$$\log |1 + e^{2\pi i t}| = \mathrm{Re}\left(\sum_{n=1}^{\infty} \frac{(-1)^{n-1}}{n} e^{2\pi i n t}\right), \tag{19.8}$$

where Re stands for real part, and

$$\int_{-1/3}^{1/3} e^{2\pi i n t} \, dt = \frac{1}{\pi n} \sin \frac{2\pi n}{3} = \frac{\sqrt{3}}{2\pi n} \chi_3(n),$$

it follows that

$$\log \mathrm{M}(f_k) = \frac{\sqrt{3}}{2\pi} \sum_{n=1}^{\infty} \frac{(-1)^{n-1}\chi_3(n)}{n^2} = \frac{\sqrt{3}}{2\pi}\left(\sum_{n=1}^{\infty} \frac{\chi_3(n)}{n^2} - 2\sum_{n=1}^{\infty} \frac{\chi_3(2n)}{(2n)^2}\right).$$

The proof is completed by noting that $\chi_3(2n) = \chi_3(2)\chi_3(n) = -\chi_3(n)$ for every $n \geq 1$. $\square$

LEMMA 19.8. *For every $f \in \mathfrak{R}_d$,*

$$\mathrm{M}(u_{d+1}f(u_1, \ldots, u_d) + f(u_1^{-1}, \ldots, u_d^{-1})) = \mathrm{M}(f).$$

PROOF. If $f = 0$, the statement is trivial, and if $f \neq 0$, it is a consequence of (16.2):

$$\log \mathrm{M}(u_{d+1}f(u_1, \ldots, u_d) + f(u_1^{-1}, \ldots, u_d^{-1}))$$

$$= \int_{\mathbb{S}} \cdots \int_{\mathbb{S}} \int_{\mathbb{S}} \log |u_{d+1}f(u_1, \ldots, u_d) + f(u_1^{-1}, \ldots, u_d^{-1})|$$
$$d\lambda_{\mathbb{S}}(u_{d+1})d\lambda_{\mathbb{S}}(u_1) \cdots d\lambda_{\mathbb{S}}(u_d)$$

$$= \log \mathrm{M}(f),$$

since $\frac{|f(u_1, \ldots, u_d)|}{|f(u_1^{-1}, \ldots, u_d^{-1})|} = 1$ for all $(u_1, \ldots, u_d) \in \mathbb{S}^d$. $\square$

EXAMPLE 19.9. Let $f \in \mathfrak{R}_3$ be the *octahedron* $f = (u_1 + u_2 + u_3 + u_1^{-1} + u_2^{-1} + u_3^{-1})$. Then $\mathrm{M}(f) = \mathrm{M}(1 + u_1 + u_2)$ (cf. (19.7)). Indeed, $\mathrm{M}(f) = \mathrm{M}(u_1u_3^{-1} + u_2u_3^{-1} + 1 + u_3^{-2}(u_1^{-1}u_3 + u_2^{-1}u_3 + 1)) = \mathrm{M}(z_1 + z_2 + 1 + z_3(z_1^{-1} + z_2^{-1} + 1)) = \mathrm{M}(z_1 + z_2 + 1)$ by Lemma 19.8, where $z_1 = u_1u_3^{-1}$, $z_2 = u_2u_3^{-1}$, and $z_3 = u_3^{-2}$.

Note that $\mathrm{M}(f) = \mathrm{M}(f_\Phi)$ for every two dimensional face Φ of $\mathcal{S}(f)$ (cf. Lemma 19.6). $\boxdot$

PROPOSITION 19.10. *Let $k \in \mathbb{Z}$, and let $f_k = (u_1 + u_2)^2 + k \in \mathfrak{R}_2$. Then*

$$h(\alpha^{\mathfrak{R}_2/(f_k)}) = \log M(f_k) = \begin{cases} 0 & \text{if } k = 0, \\ \frac{3\sqrt{3}}{2\pi} L(2, \chi_3) & \text{if } |k| = 1, \\ \frac{1}{2} \log 2 + \frac{2}{\pi} L(2, \chi_4) & \text{if } |k| = 2, \\ \frac{2}{3} \log 3 + \frac{\sqrt{3}}{\pi} L(2, \chi_3) & \text{if } |k| = 3, \\ \log |k| & \text{if } |k| \geq 4, \end{cases} \quad (19.9)$$

where the characters χ_3, χ_4 are given by (19.5), and where $L(m, \chi)$ is defined in (19.6).

PROOF. It is clear that $M(f_k) = M(f_{-k})$ for every $k \in \mathbb{Z}$, so that we may assume that $k \leq 0$. By (16.2), $\log M(f_0) = 2 \log M(1 + u_1) = 0$, and

$$\log M(f_1) = \log M(u_1 + u_2 + 1) + \log M(u_1 + u_2 - 1)$$

$$= \frac{3\sqrt{3}}{2\pi} \sum_{n=1}^{\infty} \frac{\chi(n)}{n^2} = \frac{3\sqrt{3}}{2\pi} L(2, \chi_3)$$

by (19.7). In order to check the remaining cases, fix $\omega \in [0, 1) \setminus \{\frac{1}{4}, \frac{3}{4}\}$ and consider the integral

$$\int_0^1 \int_0^1 \log |e^{2\pi i u_1} + e^{2\pi i u_2} + 2 \cos 2\pi \omega| \, du_1 du_2$$

$$= \int_0^1 \int_0^1 \log |e^{4\pi i u_1} + e^{4\pi i u_2} + 2 \cos 2\pi \omega| \, du_1 du_2$$

$$= \int_0^1 \int_0^1 \log |e^{2\pi i (-u_1 + u_2)} + e^{2\pi i (-u_1 - u_2)} + 2 \cos 2\pi \omega| \, du_1 du_2$$

$$= \log |2 \cos 2\pi \omega| + \int_0^1 \int_0^1 \log \left| e^{2\pi i u_1} + \frac{2 \cos 2\pi u_2}{2 \cos 2\pi \omega} \right| \, du_1 du_2$$

$$= \log |2 \cos 2\pi \omega| + \int_0^1 \log^+ \left| \frac{2 \cos 2\pi u_2}{2 \cos 2\pi \omega} \right| \, du_2 \qquad \text{(by (16.2))}$$

$$= (1 - 4\omega') \log(2 \cos 2\pi \omega') + 4 \int_0^{\omega'} \log |2 \cos 2\pi u_2| \, du_2$$

$$= (1 - 4\omega') \log(2 \cos 2\pi \omega') + 4 \int_0^{\omega'} \log |1 + e^{4\pi i u_2}| \, du_2,$$

where

$$\omega' = \begin{cases} \omega & \text{if } 0 \leq \omega < \frac{1}{4}, \\ \frac{1}{2} - \omega & \text{if } \frac{1}{4} < \omega \leq \frac{1}{2}, \\ \omega - \frac{1}{2} & \text{if } \frac{1}{2} \leq \omega < \frac{3}{4}, \\ 1 - \omega & \text{if } \frac{3}{4} < \omega < 1. \end{cases} \quad (19.10)$$

According to (19.8), $\log|1+e^{4\pi it}| = \mathrm{Re}(\sum_{n=1}^{\infty}\frac{(-1)^{n-1}}{n}e^{4\pi int})$, and term-by-term integration yields that

$$\int_0^1\int_0^1 \log|e^{2\pi iu_1} + e^{2\pi iu_2} + 2\cos 2\pi w|\,du_1du_2$$

$$= (1 - 4\omega')\log(2\cos 2\pi w') + \frac{1}{\pi}\sum_{n=1}^{\infty}\frac{(-1)^{n-1}}{n^2}\sin 4\pi n\omega', \qquad (19.11)$$

with ω' given by (19.10). Since

$$f_{-2} = ((u_1 + u_2) + \sqrt{2})((u_1 + u_2) - \sqrt{2})$$

$$= ((u_1 + u_2) + 2\cos\frac{\pi}{4})((u_1 + u_2) + 2\cos\frac{3\pi}{4}),$$

(19.11) implies that

$$\log \mathrm{M}(f_{\pm 2}) = \frac{1}{2}\log 2 + \frac{2}{\pi}\sum_{n=1}^{\infty}\frac{(-1)^{n-1}}{n^2}\sin\frac{\pi n}{2} = \frac{1}{2}\log 2 + \frac{2}{\pi}L(2,\chi_4).$$

Similarly,

$$f_{-3} = ((u_1 + u_2) + \sqrt{3})((u_1 + u_2) - \sqrt{3})$$

$$= ((u_1 + u_2) + 2\cos\frac{\pi}{6})((u_1 + u_2) + 2\cos\frac{5\pi}{6}),$$

and (19.12) yields that

$$\log \mathrm{M}(f_{\pm 3}) = \frac{2}{3}\log 3 + \frac{2}{\pi}\sum_{n=1}^{\infty}\frac{(-1)^{n-1}}{n^2}\sin\frac{\pi n}{3} = \frac{2}{3}\log 3 + \frac{\sqrt{3}}{\pi}L(2,\chi_3).$$

For $k \le -4$, $f_k = ((u_1+u_2)+\sqrt{k})((u_1+u_2)-\sqrt{k})$, and the proof of Proposition 19.7 shows that $\log \mathrm{M}(f_{\pm k}) = 0$. $\square$

We conclude our list of examples with another interesting formula, taken from [13].

PROPOSITION 19.11. *Let $f = 1 + u_1 + u_2 + u_3 \in \Re_3$. Then*

$$h(\alpha^{\Re_3/(f)}) = \log \mathrm{M}(f) = \frac{7}{2\pi^2}\zeta(3),$$

where $\zeta(3) = \sum_{n=1}^{\infty}n^{-3}$.

PROOF. As in the proof of Proposition 19.7 we note that $\mathrm{M}(au_3 + b) = \max\{|a|, |b|\}$. We set $u = u_2u_3^{-1}$ and see that

$$\log \mathrm{M}(1 + u_1 + u_2 + u_3) = \log \mathrm{M}(1 + u_1 + u_3(1 + u))$$

$$= \int_0^1\int_0^1 \log\max\{|1 + e^{2\pi iu_1}|, |1 + e^{2\pi iu}|\}\,du_1du$$

$$= \int_0^1 \int_0^1 \max\{\log|1 + e^{2\pi i u_1}|, \log|1 + e^{2\pi i u}|\}\, du_1\, du$$

$$= 8 \int_0^{1/2} (\tfrac{1}{2} - u_1) \log|1 + e^{2\pi i u_1}|\, du_1$$

$$= -8 \int_0^{1/2} u_1 \log|1 + e^{2\pi i u_1}|\, du_1,$$

since

$$\int_0^{1/2} \log|1 + e^{2\pi i u_1}|\, du_1 = \frac{1}{2}\int_0^1 \log|1 + e^{2\pi i u_1}|\, du_1 = \log \mathrm{M}\,(1 + u_1) = 0.$$

As $\mathrm{Re} \int_0^{1/2} t e^{2\pi i n t}\, dt = -\frac{(-1)^n - 1}{(2\pi n)^2}$ for every $n \geq 1$, we obtain from (19.8) that

$$\log \mathrm{M}\,(1 + u_1 + u_2 + u_3) = -8 \int_0^{1/2} t \log|1 + e^{2\pi i t}|\, du_1$$

$$= -8 \int_0^{1/2} \mathrm{Re}\left(\sum_{n=1}^{\infty} \frac{(-1)^{n-1}}{n} t e^{2\pi i n t}\right) dt$$

$$= \frac{4}{\pi^2} \sum_{k=0}^{\infty} (2k+1)^{-3} = \frac{4}{\pi^2}\left(\zeta(3) - \frac{1}{8}\zeta(3)\right)$$

$$= \frac{7}{2\pi^2}\zeta(3). \quad \square$$

CONCLUDING REMARKS 19.12. (1) Lemma 19.1 is due to [49], and Theorem 19.2 is an immediate consequence of [71] and [72]. Proposition 19.4 is from [63], and Theorem 19.5 is due to [54], [12], and [100]; our proof follows [100] very closely. The dynamical proof of Lemma 19.6 given here is an expansion of the outline presented in [63], Remark 5.5. An algebraic proof of Lemma 19.6 appears in [100]. Lemma 19.8, Example 19.9, and Proposition 19.11 are taken from [13], and Propositions 19.7 and 19.10 from [101]. We refer to [101] for further explicit calculations of Mahler measures of polynomials $f \in \mathfrak{R}_2$ with arithmetically intriguing values. Are the formulae in Propositions 19.7, 19.10, and 19.11 indications of a deeper link between dynamical and arithmetical properties of these systems?

(2) Equation (14.16) shows that, if α satisfies the d.c.c., then $h(\alpha) = |\log \mathrm{M}\,(f)|$ for some polynomial $f \in \mathfrak{R}_d$, since the contributions to entropy arising from the groups X/X° and $X^{\circ}/C(X^{\circ})$ are equal to ∞ or $\log n$ for some $n \geq 1$ (Theorem 15.6). If α does not satisfy the d.c.c., and if $h(\alpha) < \infty$, then Proposition 18.6 implies that there exists a sequence $(f_n, n \geq 1)$ of non-zero polynomials in $\mathfrak{R}_d$ such that $h(\alpha) = \sum_{n=1}^{\infty} \log \mathrm{M}\,(f_n)$. The question whether this sum can only have finitely many non-zero terms (or whether $h(\alpha)$ has to be equal to $\log \mathrm{M}\,(f)$ for some $f \in \mathfrak{R}_d$) is equivalent to the question whether 1 is an isolated point in $\{\mathrm{M}\,(f) : f \in \mathfrak{R}_1\}$ (cf. Proposition 16.2), which was posed

in [57] in 1933 and is still unresolved. The smallest known value of $\mathrm{M}(f) > 1$ can be found in [57]: if

$$f(u) = u^{10} + u^9 - u^7 - u^6 - u^5 - u^4 - u^3 + u + 1 \in \mathfrak{R}_1,$$

then $\mathrm{M}(f) = 1.1762\ldots$. For a discussion of Lehmer's problem we refer to [13].

Lind has pointed out that a negative answer to Lehmer's question is equivalent to the existence of ergodic automorphisms of $\mathbb{T}^\infty$ with finite entropy. Indeed, if α is an ergodic automorphism of $X = \mathbb{T}^\infty$ with finite entropy, then every prime ideal $\mathfrak{p} \subset \mathfrak{R}_1$ associated with the module $\mathfrak{M} = \hat{X}$ arising via Lemma 5.1 must be principal by Example 6.17 (4). If $\mathfrak{p} = (f)$ for some $f \in \mathfrak{R}_1$, then f must be non-zero by Proposition 17.5, since $h(\alpha^{\mathfrak{R}_1 \cdot a}) \leq h(\alpha) < \infty$ for every $a \in \mathfrak{M}$, f cannot be constant, since X is connected, and f cannot be a generalized cyclotomic polynomial by Example 6.17 (4).

If 1 is an isolated point in $\{\mathrm{M}(h) : h \in \mathfrak{R}_1\}$, then $\mathfrak{M}$ can only have finitely many associated prime ideals: indeed, if $\mathfrak{p}_1, \mathfrak{p}_2, \ldots$ is an infinite sequence of distinct prime ideals associated with $\mathfrak{M}$, we can find elements $a_1, a_2, \ldots$ in $\mathfrak{M}$ such that $\mathfrak{R}_1 \cdot a_j \cong \mathfrak{R}_1/\mathfrak{p}_j$, and the submodules $\mathfrak{N}_j = \mathfrak{R}_1 \cdot a_j$ satisfy that $\mathfrak{N}_j \cap \mathfrak{N}_k = \{0\}$ whenever $1 \leq j < k$ (otherwise we could find a non-principal prime ideal associated with $\mathfrak{M}$). Hence $\infty > h(\alpha) = h(\alpha^{\mathfrak{M}}) \geq h(\alpha^{\sum_{j=1}^\infty \mathfrak{N}_j}) = \sum_{i=1}^\infty h(\alpha^{\mathfrak{N}_j}) = \sum_{j=1}^\infty \log \mathrm{M}(f_j) = \infty$, which is impossible. We denote by $(f_1), \ldots, (f_m)$ the prime ideals associated with $\mathfrak{M}$. Since $X^{\mathfrak{R}_1/(f_j)} = \widehat{\mathfrak{R}_1/(f_j)}$ is a quotient group of $X = \mathbb{T}^\infty$, $\widehat{\mathfrak{R}_1/(f_j)}$ must be locally connected, hence a finite-dimensional torus, and Example 6.17 (3) shows that every f_j may be assumed to be of the form $f_j = c_0^{(j)} + \cdots + c_{s_j}^{(j)} u^{s_j}$ with $|c_0^{(j)} c_{s_j}^{(j)}| = 1$.

We set $\mathfrak{W}_j = \{a \in \mathfrak{M} : f_1^l \cdots f_{j-1}^l f_{j+1}^l \cdots f_m^l \cdot a = 0$ for some $l \geq 1\}$, $j = 1, \ldots, m$, and note that $\{\mathfrak{W}_1, \ldots, \mathfrak{W}_m\}$ is a reduced primary decomposition of $\mathfrak{M}$ (cf. (6.5)). Furthermore, $\mathfrak{M}/\mathfrak{W}_j$ is a torsion-free additive group for every $j = 1, \ldots, m$, and there exists a $j_0 \in \{1, \ldots, m\}$ with $X' = \widehat{\mathfrak{N}} \cong \mathbb{T}^\infty$, where $\mathfrak{N} = \mathfrak{M}/\mathfrak{W}_{j_0}$. Since $\mathfrak{N}$ is associated with (f_{j_0}) and $h(\alpha^{\mathfrak{N}}) \leq h(\alpha) < \infty$, the addition formula (14.1) implies that we can find a $k \geq 1$ with $f^k \cdot a = 0$ for every $a \in \mathfrak{N}$. Hence $\mathfrak{N}$ contains an infinite direct sum of $\mathfrak{R}_1$-modules $\mathfrak{K}_1 \oplus \mathfrak{K}_2 \oplus \cdots$ with $\mathfrak{K}_i \cong \mathfrak{R}_1/(f) \cong \mathbb{Z}^s$ for every $i \geq 1$, and $\infty > h(\alpha) \geq h(\alpha^{\mathfrak{N}}) \geq h(\alpha^{\bigoplus_{i \geq 1} \mathfrak{K}_i}) = \infty$, which is absurd.

We have proved that the existence of an ergodic automorphism α of $\mathbb{T}^\infty$ with finite entropy implies that 1 cannot be isolated in $\{\mathrm{M}(f) : f \in \mathfrak{R}_1\}$. Conversely, if 1 is not an isolated point in $\{\mathrm{M}(f) : f \in \mathfrak{R}_1\}$, then we can choose a sequence $(f_n, n \geq 1)$ of irreducible polynomials in $\mathfrak{R}_1$ with $0 < \log \mathrm{M}(f_n) < 2^{-n} \log 2$ for every $n \geq 1$. According to Lemma 19.1, f_n is not a generalized cyclotomic polynomial, and Example 6.17 (3) shows that $\alpha_n = \alpha^{\mathfrak{R}_1/(f_n)}$ is ergodic—and hence mixing—for every $n \geq 1$ (Theorem 1.6). Furthermore, since $\log \mathrm{M}(f_n) < \log 2$, f_n is monic by Proposition 16.1, and the product of the roots of f_n must have modulus 1 (since it is an integer). After multiplying f_n by a power of the variable u we may assume that $\pm f(u) = c_0 + c_1 u + \cdots + u^s$ for some $s \geq 1$, where $|c_0| = 1$, and Example 6.17 (3) shows that $X_n = X^{\Re_1/(f_n)} \cong \mathbb{T}^{m_n}$ for some $m_n \geq 2$. The automorphism $\alpha = \alpha_1 \times \alpha_2 \times \cdots$ on $\mathbb{T}^\infty = X_1 \times X_2 \times \cdots$ is ergodic and has entropy $h(\alpha) = \sum_{n \geq 1} h(\alpha_n) < \infty$.

20. Completely positive entropy

Let $T \colon \mathbf{m} \mapsto T_\mathbf{m}$ be a measure preserving $\mathbb{Z}^d$-action on a probability space $(Y, \mathfrak{T}, \mu)$. The *Pinsker algebra* $\mathfrak{P}(T)$ is the smallest sigma-algebra containing all finite partitions $\mathcal{P} \subset \mathfrak{T}$ with $h_\mu(T, \mathcal{P}) = 0$. It is clear that $\mathfrak{P}(T)$ is T-invariant. The action T has *completely positive entropy* if $\mathfrak{P}(T)$ is the trivial sigma-algebra $\{\varnothing, X\}$ or, equivalently, if $h_\mu(T, \mathcal{P}) > 0$ for every non-trivial, finite, measurable partition $\mathcal{P} \subset \mathfrak{T}$. If $\mathfrak{S} \subset \mathfrak{T}$ is a T-invariant sigma-algebra then T has *completely positive entropy on* $\mathfrak{S}$ if $h_\mu(T, \mathcal{P}) > 0$ for every non-trivial, finite, measurable partition $\mathcal{P} \subset \mathfrak{S}$.

If $\mathcal{P} \subset \mathfrak{T}$ is a countable partition we set

$$\mathcal{P}_T = \bigvee_{\mathbf{n} \in \mathbb{Z}^d} T_{-\mathbf{n}}(\mathcal{P}) = \Sigma\left(\bigcup_{\mathbf{n} \in \mathbb{Z}^d} T_{-\mathbf{n}}(\mathcal{P})\right),$$

$$\mathcal{P}_T^- = \bigvee_{\mathbf{n} \prec 0_{\mathbb{Z}^d}} T_{-\mathbf{n}}(\mathcal{P}) = \Sigma\left(\bigcup_{\mathbf{n} \prec 0_{\mathbb{Z}^d}} T_{-\mathbf{n}}(\mathcal{P})\right),$$

where $\prec$ is the lexicographic order on $\mathbb{Z}^d$ and $\Sigma(\mathcal{C})$ is the sigma-algebra generated by a collection of sets $\mathcal{C} \subset \mathfrak{T}$ (for notation we refer to Section 12).

LEMMA 20.1. *Let* $\mathcal{P}, \mathcal{Q}$ *be countable partitions in* $\mathfrak{T}$ *such that* $H_\mu(\mathcal{P}) + H_\mu(\mathcal{Q}) < \infty$. *Then*

$$h_\mu(T, \mathcal{P} \vee \mathcal{Q}) = h_\mu(T, \mathcal{P}) + H_\mu(\mathcal{Q} | \mathcal{Q}_T^- \vee \mathcal{P}_T). \qquad (20.1)$$

PROOF. For every $n \geq 0$ we set $Q(n) = \{-n, \ldots, n\}^d$ and obtain that, for $m, n \geq 0$,

$$H_\mu\left(\bigvee_{\mathbf{n} \in Q(m)} T_{-\mathbf{n}}(\mathcal{P})\right) \leq H_\mu\left(\bigvee_{\mathbf{n} \in Q(m)} T_{-\mathbf{n}}(\mathcal{P}) \vee \bigvee_{\mathbf{n} \in Q(m+n)} T_{-\mathbf{n}}(\mathcal{Q})\right)$$

$$= H_\mu\left(\bigvee_{\mathbf{n} \in Q(m+n)} T_{-\mathbf{n}}(\mathcal{Q})\right) + H_\mu\left(\bigvee_{\mathbf{n} \in Q(m)} T_{-\mathbf{n}}(\mathcal{P}) \,\middle|\, \bigvee_{\mathbf{n} \in Q(m+n)} T_{-\mathbf{n}}(\mathcal{Q})\right)$$

$$\leq H_\mu\left(\bigvee_{\mathbf{n} \in Q(m+n)} T_{-\mathbf{n}}(\mathcal{Q})\right) + (2m+1)^d H_\mu\left(\mathcal{P} \,\middle|\, \bigvee_{\mathbf{n} \in Q(n)} T_{-\mathbf{n}}(\mathcal{Q})\right).$$

We divide by $|Q(m)| = (2m+1)^d$ and let $m \to \infty$. Then

$$h_\mu(T, \mathcal{P}) \leq h_\mu(T, \mathcal{Q}) + H_\mu\left(\mathcal{P} \,\middle|\, \bigvee_{\mathbf{n} \in Q(n)} T_{-\mathbf{n}}(\mathcal{Q})\right),$$

and by letting $n \to \infty$ we see that

$$h_\mu(T, \mathcal{P}) \leq h_\mu(T, \mathcal{Q}) + H_\mu(\mathcal{P}|\mathcal{Q}_T). \tag{20.2}$$

If we replace $\mathcal{P}$ by $\mathcal{P} \vee T_{-\mathbf{n}}(\mathcal{Q})$, $\mathbf{n} \in \mathbb{Z}^d$, and $\mathcal{Q}$ by $\mathcal{P}$, then (20.2) becomes

$$h_\mu(T, \mathcal{P} \vee \mathcal{Q}) \leq h_\mu(T, \mathcal{P}) + H_\mu(T_{-\mathbf{n}}(\mathcal{Q})|\mathcal{P}_T). \tag{20.3}$$

We average (20.3) over $\mathbf{n} \in Q(m)$ and obtain that, as $m \to \infty$,

$$h_\mu(T, \mathcal{P} \vee \mathcal{Q}) \leq h_\mu(T, \mathcal{P}) + \frac{1}{|Q(m)|} \sum_{\mathbf{n} \in Q(m)} H_\mu(T_{-\mathbf{n}}(\mathcal{Q})|\mathcal{P}_T) \tag{20.4}$$

$$= H_\mu(\mathcal{Q}|\mathcal{Q}_T^- \vee \mathcal{P}_T),$$

where the last equality in (20.4) is proved in the same way as the equivalence of (13.7) and (13.8). On the other hand,

$$H_\mu\left(\bigvee_{\mathbf{n} \in Q(m)} T_{-\mathbf{n}}(\mathcal{P} \vee \mathcal{Q}) \right)$$

$$= H_\mu\left(\bigvee_{\mathbf{n} \in Q(m)} T_{-\mathbf{n}}(\mathcal{P}) \right) + H_\mu\left(\bigvee_{\mathbf{n} \in Q(m)} T_{-\mathbf{n}}(\mathcal{Q}) \,\middle|\, \bigvee_{\mathbf{n} \in Q(m)} T_{-\mathbf{n}}(\mathcal{P}) \right)$$

$$\geq H_\mu\left(\bigvee_{\mathbf{n} \in Q(m)} T_{-\mathbf{n}}(\mathcal{P}) \right) + H_\mu\left(\bigvee_{\mathbf{n} \in Q(m)} T_{-\mathbf{n}}(\mathcal{Q}) \,\middle|\, \mathcal{P}_T \right).$$

We divide by $|Q(m)|$, let $m \to \infty$, and obtain as in (20.4) that

$$h_\mu(T, \mathcal{P} \vee \mathcal{Q}) \geq h_\mu(T, \mathcal{P}) + \lim_{m \to \infty} \frac{1}{|Q(m)|} \sum_{\mathbf{n} \in Q(m)} H_\mu(T_{-\mathbf{n}}(\mathcal{Q})|\mathcal{P}_T)$$

$$= H_\mu(\mathcal{Q}|\mathcal{Q}_T^- \vee \mathcal{P}_T),$$

which completes the proof of (20.1). $\square$

PROPOSITION 20.2. *Let $\mathcal{P}$ and $\mathcal{Q}$ be countable partitions in $\mathfrak{T}$ such that $H_\mu(\mathcal{P}) + H_\mu(\mathcal{Q}) < \infty$, and let $\mathcal{Q}_T^{-\infty} = \bigcap_{\mathbf{n} \in \mathbb{Z}^d} T_{-\mathbf{n}}(\mathcal{Q}_T^-)$. Then $H_\mu(\mathcal{P}|\mathcal{P}_T^- \vee \mathcal{Q}_T^{-\infty}) = H_\mu(\mathcal{P}|\mathcal{P}_T^-)$, and $\mathcal{Q}_T^{-\infty} \subset \mathfrak{P}(T)$. Furthermore, $\mathfrak{P}(T)$ is the smallest sigma-algebra containing $\mathcal{Q}_T^{-\infty}$ for every finite partition $\mathcal{Q} \subset \mathfrak{T}$.*

PROOF. For every $\mathbf{m} \in \mathbb{Z}^d$,

$$h_\mu(T, \mathcal{P} \vee \mathcal{Q}) = h_\mu(T, \mathcal{P} \vee T_{-\mathbf{m}}(\mathcal{Q})) = H_\mu(\mathcal{P} \vee T_{-\mathbf{m}}(\mathcal{Q})|\mathcal{P}_T^- \vee T_{-\mathbf{m}}(\mathcal{Q}_T^-))$$

$$= H_\mu(\mathcal{P}|\mathcal{P}_T^- \vee T_{-\mathbf{m}}(\mathcal{Q}_T^-))$$

$$\qquad\qquad + H_\mu(T_{-\mathbf{m}}(\mathcal{Q})|\mathcal{P} \vee \mathcal{P}_T^- \vee T_{-\mathbf{m}}(\mathcal{Q}_T^-)) \tag{20.5}$$

$$= H_\mu(\mathcal{P}|\mathcal{P}_T^- \vee T_{-\mathbf{m}}(\mathcal{Q}_T^-)) + H_\mu(\mathcal{Q}|T_{\mathbf{m}}(\mathcal{P} \vee \mathcal{P}_T^-) \vee \mathcal{Q}_T^-).$$

By letting $\mathbf{m} \nearrow (\infty, \ldots, \infty)$ lexicographically in (20.5) we see from (20.1) that

$$h_\mu(T, \mathcal{P}) + H_\mu(\mathcal{Q}|\mathcal{Q}_T^- \vee \mathcal{P}_T) = h_\mu(T, \mathcal{P} \vee \mathcal{Q})$$
$$= H_\mu(\mathcal{P}|\mathcal{P}_T^- \vee \mathcal{Q}_T^{-\infty}) + H_\mu(\mathcal{Q}|\mathcal{Q}_T^- \vee \mathcal{P}_T),$$

which proves our first assertion. For the second assertion we choose a finite partition $\mathcal{Q}' \subset \mathcal{Q}_T^{-\infty}$ and note that, according to (20.1), $h_\mu(T, \mathcal{Q}') = H_\mu(\mathcal{Q}'|\mathcal{Q}'_T^-) = H_\mu(\mathcal{Q}'|\mathcal{Q}'_T^- \vee \mathcal{Q}_T^{-\infty}) = 0$.

Conversely, let $\mathcal{Q}$ be a finite partition in $\mathfrak{P}(T)$, and consider the action $T^{(k)}: \mathbf{n} \mapsto T_\mathbf{n}^{(k)} = T_{k\mathbf{n}}$, $\mathbf{n} \in \mathbb{Z}^d$. For every finite partition $\mathcal{P} \subset \mathfrak{X}$ we set

$$\mathcal{P}_k = \bigvee_{\mathbf{n} \in Q_k} T_{-\mathbf{n}}(\mathcal{P}), \tag{20.6}$$

where $Q_k = \{0, \ldots, k-1\}^d$, and note that, according to (13.7),

$$h_\mu(T^{(k)}, \mathcal{P}_k) = k^d h_\mu(T, \mathcal{P}). \tag{20.7}$$

Hence $h_\mu(T^{(k)}, \mathcal{Q}) \le h_\mu(T^{(k)}, \mathcal{Q}_k) = k^d h_\mu(T, \mathcal{Q}) = 0$, i.e. $\mathcal{Q} \subset \mathfrak{P}(T^{(k)})$. Equation (20.1) shows that, for any finite partition $\mathcal{P} \subset \mathfrak{X}$, $h_\mu(T^{(k)}, \mathcal{P}) + H_\mu(\mathcal{Q}|\mathcal{Q}_{T^{(k)}}^- \vee \mathcal{P}_{T^{(k)}}) = h_\mu(T^{(k)}, \mathcal{P} \vee \mathcal{Q}) = h_\mu(T^{(k)}, \mathcal{Q}) + H_\mu(\mathcal{P}|\mathcal{P}_{T^{(k)}}^- \vee \mathcal{Q}_{T^{(k)}})$, which implies that $h_\mu(T^{(k)}, \mathcal{P}) = H_\mu(\mathcal{P}|\mathcal{P}_{T^{(k)}}^-) = H_\mu(\mathcal{P}|\mathcal{P}_{T^{(k)}}^- \vee \mathcal{Q}_{T^{(k)}}) = H_\mu(\mathcal{P}|\mathcal{P}_{T^{(k)}}^- \vee \mathcal{Q}) \le H_\mu(\mathcal{P}|\mathcal{Q})$ for every $k \ge 1$. By letting $k \to \infty$ we see that

$$H_\mu(\mathcal{P}|\mathcal{Q}) \ge H_\mu(\mathcal{P}|\mathcal{P}_{T^{(k)}}^{-\infty}) \ge H_\mu(\mathcal{P}|\mathcal{P}_T^{-\infty}), \tag{20.8}$$

and by setting $\mathcal{P} = \mathcal{Q}$ we obtain that $0 = H_\mu(\mathcal{Q}|\mathcal{Q}) = H_\mu(\mathcal{Q}|\mathcal{Q}_T^{-\infty})$, i.e. that $\mathcal{Q} \subset \mathcal{Q}_T^{-\infty}$. As $\mathcal{Q}$ was an arbitrary finite partition in $\mathfrak{P}(T)$, the proposition is proved. $\square$

COROLLARY 20.3. *If $\mathfrak{S} \subset \mathfrak{X}$ is a sigma-algebra such that T has completely positive entropy on $\mathfrak{S}$, then $H_\mu(\mathcal{P}|\mathcal{Q}) = H_\mu(\mathcal{P})$ for all finite partitions $\mathcal{P} \subset \mathfrak{S}$ and $\mathcal{Q} \subset \mathfrak{P}(T)$. Furthermore, if there exists an increasing sequence $(\mathfrak{S}_n, n \ge 1)$ of T-invariant sigma-algebras in $\mathfrak{X}$ such that $\bigvee_{n \ge 1} \mathfrak{S}_n = \mathfrak{X}$ and T has completely positive entropy on $\mathfrak{S}_n$ for every $n \ge 1$, then T has completely positive entropy.*

PROOF. Let $\mathcal{P} \subset \mathfrak{S}$ and $\mathcal{Q} \subset \mathfrak{P}(T)$ be finite partitions. According to (20.8), $H_\mu(\mathcal{P}|\mathcal{Q}) \ge H_\mu(\mathcal{P}|\mathcal{P}_T^{-\infty}) = H_\mu(\mathcal{P})$, since $\mathcal{P}_T^{-\infty} \subset \mathfrak{S}$ must be trivial by Proposition 20.2. If $\mathfrak{S}_n \nearrow \mathfrak{X}$, and if T has completely positive entropy on every $\mathfrak{S}_n$, then there exists, for every $A \in \mathfrak{X}$, a sequence of sets $(A_n, n \ge 1)$ such that $A_n \in \mathfrak{S}_n$ for every n and $\lim_{n \to \infty} \mu(A \triangle A_n) = 0$. We set $\mathcal{P} = \{A, Y \smallsetminus A\}$, $\mathcal{P}_n = \{A_n, Y \smallsetminus A_n\}$, and note that $H_\mu(\mathcal{P}|\mathcal{Q}) = \lim_{n \to \infty} H_\mu(\mathcal{P}_n|\mathcal{Q}) = \lim_{n \to \infty} H_\mu(\mathcal{P}_n) = H_\mu(\mathcal{P})$ for every finite partition $\mathcal{Q} \subset \mathfrak{P}(T)$. If $\mathfrak{P}(T) \ne \{\varnothing, Y\}$ we obtain a contradiction by choosing $A \in \mathfrak{P}(T)$. Hence $\mathfrak{P}(T)$ must be trivial, i.e. T has completely positive entropy. $\square$

LEMMA 20.4. *Let X be a compact group, and let $\mathfrak{S} \subset \mathfrak{B}_X$ be a left and right translation invariant sigma-algebra (i.e. $xB = \{xy : y \in B\} \in \mathfrak{S}$ and $Bx = \{yx : y \in B\} \in \mathfrak{S}$ for every $B \in \mathfrak{S}$ and $x \in X$). Then there exists a closed, normal subgroup $Y \subset X$ such that $\mathfrak{S} = \mathfrak{B}_{X/Y}$ (mod λ_X), where $\mathfrak{B}_{X/Y} = \{BY : B \in \mathfrak{B}_X\}$ is the sigma-algebra of Borel subsets of X/Y. If $\mathfrak{S}$ is invariant under an action α of a countable group Γ by automorphisms of X, then Y is α-invariant.*

PROOF. For every $f \in L^2(X, \mathfrak{S}, \lambda_X)$ and every continuous function $h\colon X \longmapsto \mathbb{C}$, the convolution $x \mapsto f * h(x) = \int_X f(y^{-1}x)h(y)\, d\lambda_X(y)$ is continuous and lies in $L^2(X, \mathfrak{S}, \lambda_X)$. By varying h we see that $L^2(X, \mathfrak{S}, \lambda_X)$ is spanned by its continuous elements. We write $\mathcal{A}$ for the set of continuous, $\mathfrak{S}$-measurable, complex valued functions on X, and note that $\mathcal{A}$ is invariant under left translation: for every $f \in \mathcal{A}$ and $x' \in X$, the map $x \mapsto f(x'x)$ is again an element of $\mathcal{A}$. We define an equivalence relation $\sim$ on X by setting $x \sim x'$ if and only if $f(x) = f(x')$ for all $f \in \mathcal{A}$, denote by $[x]$ the equivalence class of a point $x \in X$, and consider the space $Z = \{[x] : x \in X\}$ of equivalence classes. Then Z is compact and metrizable in the smallest topology in which every element of $\mathcal{A}$, regarded as a function on Z, is continuous.

As $\mathcal{A}$ is invariant under left translation we obtain a continuous action $x \mapsto \phi_x$ of X on Z by $\phi_x([x']) = [xx']$ for every $x, x' \in X$. This action is obviously transitive, i.e. Z consist of a single ϕ-orbit. We set $Y = \{x \in X : \phi_x([\mathbf{1}_X]) = [x] = [\mathbf{1}_X]\}$ and note that Y is a closed subgroup of X, and that the continuous, surjective map $x \mapsto [x]$ induces a homeomorphism $X/Y \longmapsto Z$ which allows us to set $Z = X/Y$. Since $\mathcal{A}$ is dense in $L^2(X, \mathfrak{S}, \lambda_X)$, the collection of open subsets of $Z = X/Y$ generates $\mathfrak{S}$. The normality of Y follows from the fact that, for every $f \in \mathcal{A}$ and $x \in X$, the map $x' \mapsto f(x^{-1}x'x)$, $x' \in X$ is also an element of A. Hence the map $[x'] \mapsto \psi_x([x']) = [xx'x^{-1}]$ is a homeomorphism of Z for every $x \in X$, and $xYx^{-1} = \psi_x([\mathbf{1}_X]) = [\mathbf{1}_X] = Y$ for every $x \in X$. This proves the first assertion of the lemma.

The second assertion follows from the observation that $f \cdot \alpha_\gamma \in \mathcal{A}$ for every $f \in \mathcal{A}$ and $\gamma \in \Gamma$, so that α induces a continuous Γ-action $\bar{\alpha}$ on Z. Since $\bar{\alpha}_\gamma([\mathbf{1}_X]) = [\mathbf{1}_X] = Y$ for every $\gamma \in \Gamma$, the group Y is α-invariant. $\square$

PROPOSITION 20.5. *Let α be a $\mathbb{Z}^d$-action by automorphisms of a compact group X such that the set of α-periodic points is dense in X (Definition 5.5). Then there exists a closed, normal, α-invariant subgroup $Y \subset X$ such that the Pinsker algebra $\mathfrak{P}(\alpha)$ is equal to $\mathfrak{B}_{X/Y}$ (mod λ_X). In particular, Y is the unique minimal, closed, normal, α-invariant subgroup of X such that $h(\alpha^{X/Y}) = 0$, and α has completely positive entropy if and only if $Y = \{\mathbf{1}_X\}$.*

PROOF. If $\Lambda \subset \mathbb{Z}^d$ is a subgroup of finite index we denote by $\alpha^{(\Lambda)}$ the restriction of α to Λ. A glance at the definition of $h_{\lambda_X}(\alpha, \mathcal{P})$ for any finite, measurable partition $\mathcal{P}$ of X in (13.7) shows that $\mathfrak{P}(\alpha^{(\Lambda)}) = \mathfrak{P}(\alpha)$ for every subgroup $\Lambda \subset \mathbb{Z}^d$ of finite index. Furthermore, if Λ is such a subgroup, and if

$\mathrm{Fix}_\Lambda(\alpha) = \{x \in X : \alpha_\mathbf{n}(x) = x \text{ for every } \mathbf{n} \in \Lambda\}$, then $\alpha_\mathbf{n}$ commutes both with left and right translation by x for every $\mathbf{n} \in \Lambda$ and $x \in \mathrm{Fix}_\Lambda(\alpha)$, and $\mathfrak{P}(\alpha) = \mathfrak{P}(\alpha^{(\Lambda)})$ is invariant under left and right translation by every $x \in \mathrm{Fix}_\Lambda(\alpha)$. By varying Λ we see that $\mathfrak{P}(\alpha)$ is invariant under left and right translation by every α-periodic point in X, hence under left and right translation by every element in a dense subset of X, and therefore under all translations. An application of Lemma 20.4 yields that $\mathfrak{P}(\alpha) = \mathfrak{B}_{X/Y}$ for some closed, normal, α-invariant subgroup $Y \subset X$. In particular, $h(\alpha^{X/Y}) = 0$. If Y' is a second closed, normal, α-invariant subgroup with $h(\alpha^{X/Y'}) = 0$ and $Y'' = Y \cap Y''$, then $h(\alpha^{X/Y''}) = 0$ by (14.1), which proves that Y is the unique minimal, closed, normal, α-invariant subgroup of X with $h(\alpha^{X/Y}) = 0$. The last assertion is obvious. $\square$

COROLLARY 20.6. *Let α be an ergodic automorphism of a compact group X. Then α has completely positive entropy.*

PROOF. Corollary 20.3 and Proposition 4.9 allow us to assume without loss in generality that α satisfies the d.c.c. In this case Theorem 11.1 implies that α has a dense set of periodic points, and Proposition 20.5 and Theorem 19.2 complete the proof. $\square$

COROLLARY 20.7. *Let α be a $\mathbb{Z}^d$-action by automorphisms of a compact group X, and assume that X° is abelian. Then there exists a closed, α-invariant subgroup $Y \subset X$ such that the Pinsker algebra $\mathfrak{P}(\alpha)$ is equal to $\mathfrak{B}_{X/Y}$ (mod λ_X), and α^Y has completely positive entropy whenever $Y = \{\mathbf{1}_X\}$.*

PROOF. Choose a decreasing sequence $(V_n, n \geq 1)$ of closed, normal, α-invariant subgroups of X such that $\bigcap_{n \geq 1} V_n = \{\mathbf{1}_X\}$ and α^{X/V_n} satisfies the d.c.c. for every $n \geq 1$ (Proposition 4.9). By Theorem 10.2, the set of α^{X/V_n}-periodic points is dense in X/V_n for every $n \geq 1$, and Proposition 20.5 implies the existence of closed, normal, α-invariant subgroups $Y_n \supset V_n$ such that $\mathfrak{P}(\alpha^{X/V_n}) = \mathfrak{B}_{X/Y_n}$ for every $n \geq 1$. The sequence $(Y_n, n \geq 1)$ is obviously non-increasing, and we set $Y = \bigcap_{n \geq 1} Y_n$ and note that $Y_n/V_n = Y/V_n$ for every $n \geq 1$, and that $\mathfrak{P}(\alpha) \supset \mathfrak{B}_{X/Y}$. Proposition 20.5 implies that Y is the unique minimal closed, normal, α-invariant subgroup of X such that $h(\alpha^{X/Y}) = 0$.

Now assume that α^Y does not have completely positive entropy. Since $Y^\circ \subset X^\circ$ is abelian, the set of α^Y-periodic points is dense in Y by Theorem 10.2, and Proposition 20.5 implies the existence of a unique minimal, closed, normal (in Y), α^Y-invariant subgroup $Z \subsetneq Y$ with $h(\alpha^{Y/Z}) = 0$. As $h(\alpha^{Y/xZx^{-1}}) = 0$ for every $x \in X$, the uniqueness of Z implies that Z is normal in X, and (14.1) yields that $h(\alpha^{X/Z}) = 0$, in violation of the minimality of Y. This shows that α^Y has completely positive entropy. $\square$

If the group X is abelian, our ability to translate the dynamical properties of a $\mathbb{Z}^d$-action α by automorphisms of X into algebraic properties of the $\mathfrak{R}_d$-module $\mathfrak{M} = \hat{X}$ gives us a characterization of the Pinsker algebra $\mathfrak{P}(\alpha)$ in terms of the prime ideals associated with $\mathfrak{M}$.

THEOREM 20.8. *Let α be a $\mathbb{Z}^d$-action by automorphisms of a compact, abelian group X, and let $\mathfrak{M} = \hat{X}$ be the $\mathfrak{R}_d$-module defined in Lemma 5.1. Then the Pinsker algebra $\mathfrak{P}(\alpha)$ is given by $\mathfrak{P}(\alpha) = \mathfrak{B}_{X/\mathfrak{N}^\perp}$, where $\mathfrak{N} \subset \mathfrak{M}$ is the unique maximal submodule such that every prime ideal $\mathfrak{p} \subset \mathfrak{R}_d$ associated with $\mathfrak{N}$ is null. In particular, α^Y has completely positive entropy, where $Y = \mathfrak{N}^\perp \subset X$, $h(\alpha^{X/Y}) = 0$, and α has completely positive entropy if and only if every prime ideal $\mathfrak{p} \subset \mathfrak{R}_d$ associated with $\mathfrak{M}$ is positive.*

PROOF. Consider the collection $\mathcal{N}$ of submodules $\mathfrak{M}' \subset \mathfrak{M}$ with $h(\alpha^{\mathfrak{M}'}) = h(\alpha^{X/\mathfrak{M}'^\perp}) = 0$, partially ordered by inclusion. Then $\{0\} \in \mathcal{N} \neq \varnothing$ and, for every totally ordered subset $\mathcal{N}' \subset \mathcal{N}$, $\mathfrak{M}'' = \bigcup_{\mathfrak{M}' \in \mathcal{N}'} \mathfrak{M}' \in \mathcal{N}$, since $h(\alpha^{\mathfrak{M}''}) = \sup_{\mathfrak{M}' \in \mathcal{N}'} h(\alpha^{\mathfrak{M}'}) = 0$ by Lemma 13.6 (as $\mathfrak{M}$ is countable, $\mathcal{N}'$ is a countable set). According to Zorn's lemma, $\mathcal{N}$ has a maximal element.

For every pair $\mathfrak{N}, \mathfrak{N}' \in \mathcal{N}$, the injective homomorphism $X^{\mathfrak{N}+\mathfrak{N}'} = \widehat{\mathfrak{N}+\mathfrak{N}'} \longmapsto X^{\mathfrak{N}} \times X^{\mathfrak{N}'}$ dual to the addition map $\mathfrak{N} \times \mathfrak{N}' \longmapsto \mathfrak{N} + \mathfrak{N}'$ carries $\alpha^{\mathfrak{N}+\mathfrak{N}'}$ to the restriction of the zero entropy $\mathbb{Z}^d$-action $\alpha^{\mathfrak{N}} \times \alpha^{\mathfrak{N}'}$ to a closed, invariant subgroup, so that $h(\alpha^{\mathfrak{N}+\mathfrak{N}'}) = 0$ and $\mathfrak{N}+\mathfrak{N}' \in \mathcal{N}$. If $\mathfrak{N}$ and $\mathfrak{N}'$ are both maximal elements in $\mathcal{N}$, then $\mathfrak{N}+\mathfrak{N}' \in \mathcal{N}$, which implies that $\mathfrak{N} = \mathfrak{N}'$, i.e. that $\mathcal{N}$ contains a unique maximal element $\mathfrak{N}$. Proposition 19.4 shows that every prime ideal $\mathfrak{p} \subset \mathfrak{R}_d$ associated with $\mathfrak{N}$ is null, and that $\mathfrak{N}$ is the maximal submodule of $\mathfrak{M}$ with this property.

By Corollary 20.7 there exists a unique minimal, closed, α-invariant subgroup $Y \subset X$ such that $h(\alpha^{X/Y}) = 0$ and α^Y has completely positive entropy, and it is clear that $\mathfrak{N} = Y^\perp$.

The last assertion follows from the trivial observation that $\mathcal{N} \neq \{\{0\}\}$ if some prime ideal $\mathfrak{p} \subset \mathfrak{R}_d$ associated with $\mathfrak{M}$ is null, since there exists an $a \in \mathfrak{M}$ with $\mathfrak{R}_d/\mathfrak{p} \cong \mathfrak{R}_d \cdot a = \mathfrak{N}' \in \mathcal{N}$. $\square$

COROLLARY 20.9. *Let α be a $\mathbb{Z}^d$-action by automorphisms of a compact, abelian group, and let $Y \subset X$ be a closed, α-invariant subgroup. If both α^Y and $\alpha^{X/Y}$ have completely positive entropy then α has completely positive entropy.*

PROOF. Let $\mathfrak{M} = \hat{X}$ be the $\mathfrak{R}_d$-module defined in Lemma 5.1, and consider the submodule $Y^\perp = \mathfrak{N} \subset \mathfrak{M}$. If $\mathfrak{p} \subset \mathfrak{R}_d$ is a prime ideal associated with $\mathfrak{M}$ there exists an element $a \in \mathfrak{M}$ such that $\mathfrak{p} = \{f \in \mathfrak{R}_d : f \cdot a = 0\}$. For $a \in \mathfrak{N}$, $h(\alpha^{\mathfrak{R}_d/\mathfrak{p}}) > 0$ by Theorem 20.8 (applied to $\alpha^{\mathfrak{N}}$); for $a \notin \mathfrak{N}$ we set $\bar{a} = a + \mathfrak{N} \in \mathfrak{M}/\mathfrak{N}$ and consider the submodule $\mathfrak{L} = \mathfrak{R}_d \cdot \bar{a} \in \mathfrak{M}/\mathfrak{N}$. Every prime ideal $\mathfrak{q}$ associated with $\mathfrak{L}$ is also associated with $\mathfrak{M}/\mathfrak{N}$, and must contain $\mathfrak{p}$. According to Lemma 17.4 and Theorem 20.8 (applied to $\alpha^{\mathfrak{M}/\mathfrak{N}}$), $0 < h(\alpha^{\mathfrak{R}_d/\mathfrak{q}}) \leq h(\alpha^{\mathfrak{R}_d/\mathfrak{p}})$. $\square$

In order to derive a further corollary of Theorem 20.8 we need a lemma. A much more general version of this result, due to Kamiński, will be proved below (Theorem 20.14).

LEMMA 20.10. *Let α be a $\mathbb{Z}^d$-action by automorphisms of a compact group X. If α is not mixing, then there exists a closed, normal, α-invariant subgroup $Y \subsetneq X$ such that $h(\alpha^{X/Y}) = 0$. In particular, if α has completely positive entropy, then it is mixing.*

PROOF. Suppose that α is not mixing. Then $\alpha_{\mathbf{n}}$ is non-ergodic for some $\mathbf{0} \neq \mathbf{n} \in \mathbb{Z}^d$ (Theorem 1.6 (2)), and we may assume that $\mathbf{n}$ is primitive. Lemma 11.3 allows us to find a matrix $A \in \mathrm{GL}(d, \mathbb{Z})$ such that $A\mathbf{n} = \mathbf{e}^{(d)} = (0, \dots, 0, 1) \in \mathbb{Z}^d$, and by replacing the $\mathbb{Z}^d$-action α by the action $\mathbf{m} \mapsto \alpha_{A\mathbf{m}}$, if necessary, we can ensure that $\mathbf{n} = \mathbf{e}^{(d)}$. Lemma 1.2 implies that there exists a closed, normal, $\alpha_{\mathbf{e}^{(d)}}$-invariant subgroup $V \subsetneq X$ such that $G = X/V$ is a compact Lie group and $\alpha_{\mathbf{e}^{(d)}}^G$ preserves a metric δ on G. We write $\eta \colon X \longmapsto G$ for the quotient map and define a homomorphism $\boldsymbol{\eta} \colon X \longmapsto G^{\mathbb{Z}^{d-1}}$ by setting $(\boldsymbol{\eta}(x))_{\mathbf{m}} = \eta \cdot \alpha_{(m_1, \dots, m_{d-1}, 0)}(x)$ for every $\mathbf{m} = (m_1, \dots, m_{d-1}) \in \mathbb{Z}^{d-1}$. Let $\zeta \in \mathrm{Aut}(\boldsymbol{\eta}(X))$ be given by $(\zeta(y))_{\mathbf{m}} = \alpha_{\mathbf{e}^{(d)}}(y(\mathbf{m}))$ for every $\mathbf{m} \in \mathbb{Z}^{d-1}$, and define a $\mathbb{Z}^d$-action β by automorphisms of $\boldsymbol{\eta}(X)$ by setting $(\beta_{\mathbf{m}}(y))_{\mathbf{n}} = \zeta_{m_d}(y_{(m_1 + n_1, \dots, m_{d-1} + n_{d-1})})$ for every $y \in \boldsymbol{\eta}(X)$, $\mathbf{m} = (m_1, \dots, m_d) \in \mathbb{Z}^d$, and $\mathbf{n} = (n_1, \dots, n_{d-1}) \in \mathbb{Z}^{d-1}$. Then $\boldsymbol{\eta} \cdot \alpha_{\mathbf{m}} = \beta_{\mathbf{m}} \cdot \boldsymbol{\eta}$ for every $\mathbf{m} \in \mathbb{Z}^d$, and $\beta_{\mathbf{e}^{(d)}}$ preserves a metric on $\boldsymbol{\eta}(X)$. Therefore $h_{\lambda_{\boldsymbol{\eta}(X)}}(\beta_{\mathbf{e}^{(d)}}) = 0$ by (13.12) and Theorem 13.3, and $h(\beta) = 0$ by Proposition 13.2. If $Y = \ker(\boldsymbol{\eta})$ then $\{\varnothing, X\} \neq \mathfrak{B}_{X/Y} \subset \mathfrak{P}(\alpha)$, so that α cannot have completely positive entropy. $\square$

COROLLARY 20.11. *For $f \in \mathfrak{R}_d$, $\alpha^{\mathfrak{R}_d/(f)}$ is mixing if and only if it has completely positive entropy.*

PROOF. The prime ideals associated with $\mathfrak{M} = \mathfrak{R}_d/(f)$ are the principal ideals arising from the irreducible factors of f. If $\alpha^{\mathfrak{R}_d/(f)}$ is mixing, Theorem 6.5 (2) implies that f cannot be divisible by any generalized cyclotomic polynomial, so that every prime ideal associated with $\mathfrak{M}$ is positive by Theorem 19.5. An application of Theorem 20.8 shows that $\alpha^{\mathfrak{R}_d/(f)}$ has completely positive entropy. The reverse implication follows from Lemma 20.10. $\square$

To illustrate the difference between the prime ideals associated with a Noetherian $\mathfrak{R}_d$-module $\mathfrak{M}$ and the prime ideals occurring in a given prime filtration of $\mathfrak{M}$, we present an example of two Noetherian $\mathfrak{R}_d$-modules $\mathfrak{M}$ and $\mathfrak{M}'$ which have prime filtrations with term-wise isomorphic quotients, but which satisfy that $\alpha^{\mathfrak{M}}$ has completely positive entropy, but not $\alpha^{\mathfrak{M}'}$.

EXAMPLE 20.12. Let $d = 2$, and let $f(u_1) \in \mathbb{Z}[u_1] \subset \mathfrak{R}_2$ be irreducible and non-cyclotomic. We set $g = u_2 - 1$ and define ideals in $\mathfrak{R}_2$ by

$$\mathfrak{p} = (f), \quad \mathfrak{q} = (f^2, fg),$$

$$\mathfrak{a}_1 = (f, g^2), \quad \mathfrak{b}_1 = (f^2, g), \quad \mathfrak{a}_2 = (f, g).$$

Let $\mathfrak{M} = \mathfrak{R}_2/\mathfrak{p}$ and $\mathfrak{M}' = \mathfrak{R}_2/\mathfrak{q}$, and consider the filtrations

$$0 \subset \mathfrak{a}_1/\mathfrak{p} \subset \mathfrak{a}_2/\mathfrak{p} \subset \mathfrak{R}_2/\mathfrak{p} = \mathfrak{M},$$

$$0 \subset \mathfrak{b}_1/\mathfrak{q} \subset \mathfrak{a}_2/\mathfrak{q} \subset \mathfrak{R}_2/\mathfrak{q} = \mathfrak{M}'.$$

The first quotients $\mathfrak{a}_1/\mathfrak{p}$ and $\mathfrak{b}_1/\mathfrak{q}$ are both isomorphic to $\mathfrak{R}_2/\mathfrak{p}$, and all the other quotients are isomorphic to $\mathfrak{R}/\mathfrak{a}_2$. Since $\mathfrak{p}$ and $\mathfrak{a}_2$ are both prime, these are prime filtrations with term-wise isomorphic quotients. However, $\mathfrak{p}$ is the only prime ideal associated with $\mathfrak{M}$, so that $\alpha^{\mathfrak{M}}$ has completely positive entropy by Theorem 20.8, but the set of associated primes of $\mathfrak{M}'$ is equal to $\{\mathfrak{p}, \mathfrak{a}_2\}$, and Corollary 18.5 and Theorem 20.8 together imply that $\alpha^{\mathfrak{M}'}$ does not have completely positive entropy. $\boxdot$

REMARK 20.13. Example 20.12 indicates that prime filtrations are an unreliable guide to completely positive entropy. However, Corollary 20.9 *does* allow us to conclude that $\alpha^{\mathfrak{M}}$ has completely positive entropy if $\mathfrak{M}$ is a Noetherian $\mathfrak{R}_d$-module which has a prime filtration $\mathfrak{M} = \mathfrak{N}_s \supset \cdots \supset \mathfrak{N}_0 = \{0\}$ in which every quotient $\mathfrak{N}_j/\mathfrak{N}_{j-1} \cong \mathfrak{R}_d/\mathfrak{q}_j$ for some positive prime ideal $\mathfrak{q}_j \subset \mathfrak{R}_d$.

We return to the connection between completely positive entropy and mixing first mentioned in Lemma 20.10. A measure preserving $\mathbb{Z}^d$-action T on a probability space $(Y, \mathfrak{T}, \mu)$ is *mixing of order n* (or *n-mixing*) if, for all sets $B_1, \ldots, B_n$ in $\mathfrak{T}$,

$$\lim_{\substack{\mathbf{m}_i \in \mathbb{Z}^d \text{ and } \mathbf{m}_i - \mathbf{m}_j \to \infty \\ \text{for all } i,j=1,\ldots,n,\, i \neq j}} \mu(T_{-\mathbf{m}_1}(B_1) \cap \cdots \cap T_{-\mathbf{m}_n}(B_n)) = \mu(B_1) \cdots \mu(B_n). \tag{20.9}$$

For $n = 2$, (20.9) is equivalent to the usual definition of mixing.

THEOREM 20.14. *Let T be a measure preserving $\mathbb{Z}^d$-action on a probability space $(Y, \mathfrak{T}, \mu)$. If T has completely positive entropy, then T is n-mixing for every $n \geq 2$.*

PROOF. We have to prove (20.9) for every $n \geq 1$. For $n = 1$, (20.9) is obvious, and we assume that (20.9) holds for all $n < k$ with $k \geq 2$. Let $B_1, \ldots, B_k$ be sets in $\mathfrak{T}$, and let $\mathcal{P} \subset \mathfrak{T}$ be the finite partition generated by $B_2, \ldots, B_k$. Suppose that $((\mathbf{m}(r,1), \ldots, \mathbf{m}(r,k)), r \geq 1)$ is a sequence in $(\mathbb{Z}^d)^k$ with $\lim_{r \to \infty} \mathbf{m}(r,i) - \mathbf{m}(r,j) = \infty$ whenever $1 \leq i < j \leq k$, and that

$$\liminf_{r \to \infty} \left| \mu(T_{-\mathbf{m}(r,1)}(B_1) \cap \cdots \cap T_{-\mathbf{m}(r,k)}(B_k)) - \prod_{i=1}^k \mu(B_i) \right| > 0.$$

After permuting the indices $1, \ldots, k$, replacing $\mathbf{m}(r,i)$ by $A\mathbf{m}(r,i)$ for some matrix $A \in \mathrm{GL}(d, \mathbb{Z})$ and all $r \geq 1$, $1 \leq i \leq k$ (cf. Lemma 11.3), and after replacing the sequence $((A\mathbf{m}(r,1), \ldots, A\mathbf{m}(r,k)), r \geq 1)$ by a subsequence, if necessary, we may assume that $\lim_{r \to \infty} m(r,1)^{(1)} = \infty$, and that $m(r,i)^{(1)} \geq m(r,2)^{(1)} \geq m(r,1)^{(1)}$ and $\lim_{r \to \infty}(m(r,i)^{(1)} - m(r,1)^{(1)}) = \infty$ for $i = 2, \ldots, k$, with $\mathbf{m}(r,i) = (m(r,i)^{(1)}, \ldots, m(r,i)^{(d)}) \in \mathbb{Z}^d$ for every r, i. Then

$$A_r = T_{-(\mathbf{m}(r,2)-\mathbf{m}(r,1))}(B_2) \cap \cdots \cap T_{-(\mathbf{m}(r,k)-\mathbf{m}(r,1))}(B_k)$$

$$\in T_{-(m(r,2)^{(1)}-m(r,1)^{(1)}-1)\mathbf{e}^{(1)}}(\mathcal{P}_T^-),$$

where $\mathbf{e}^{(1)} = (1,0,\dots,0) \in \mathbb{Z}^d$, and where $\mathcal{P}_T^-$ was defined at the beginning of this section. It follows that

$$\left| \mu(T_{-\mathbf{m}(r,1)}(B_1) \cap \dots \cap T_{-\mathbf{m}(r,k)}(B_k)) - \prod_{i=1}^{k} \mu(B_i) \right|$$

$$= \left| \mu(B_1 \cap A_r) - \prod_{i=1}^{k} \mu(B_i) \right|$$

$$= \left| \int_{A_r} E_\mu(1_{B_1} | T_{-(m(r,2)^{(1)}-m(r,1)^{(1)}-1)\mathbf{e}^{(1)}}(\mathcal{P}_T^-)) \, d\mu - \prod_{i=1}^{k} \mu(B_i) \right|$$

$$\le \int_Y \left| E_\mu(1_{B_1} | T_{-(m(r,2)^{(1)}-m(r,1)^{(1)}-1)\mathbf{e}^{(1)}}(\mathcal{P}_T^-)) - \mu(B_1) \right| d\mu$$

$$+ \mu(B_1) \left| \mu(A_r) - \prod_{i=2}^{k} \mu(B_i) \right|.$$

By Proposition 20.2, $T_{-m\mathbf{e}^{(1)}}(\mathcal{P}_T^-) \searrow \{\varnothing, Y\} \pmod{\mu}$ as $m \nearrow \infty$, and the decreasing martingale theorem shows that

$$\int_Y \left| E_\mu(1_{B_1} | T_{-(m(r,2)^{(1)}-m(r,1)^{(1)}-1)\mathbf{e}^{(1)}}(\mathcal{P}_T^-)) - \mu(B_1) \right| d\mu \to 0$$

as $r \to \infty$. Our induction hypothesis implies that

$$\left| \mu(A_r) - \prod_{i=2}^{k} \mu(B_i) \right| = \left| \mu(T_{-\mathbf{m}(r,2)}(B_2) \cap \dots \cap T_{-\mathbf{m}(r,k)}(B_k)) - \prod_{i=2}^{k} \mu(B_i) \right| \to 0$$

as $r \to \infty$, so that $\lim_{r \to \infty} \mu(T_{-\mathbf{m}(r,1)}(B_1) \cap \dots \cap T_{-\mathbf{m}(r,k)}(B_k)) = \prod_{i=1}^{k} \mu(B_i)$, contrary to our choice of $((\mathbf{m}(r,1),\dots,\mathbf{m}(r,k)), r \ge 1)$ and $B_1,\dots,B_k$. It follows that (20.9) holds for $n = k$ and, by induction, for every $n \ge 1$, which proves the theorem. $\square$

We conclude this section with another consequence of completely positive entropy. If α is a $\mathbb{Z}^d$-action by automorphisms of a compact group X, Proposition 13.5 shows that $h(\alpha) = h_{\lambda_X}(\alpha) \ge h_\mu(\alpha)$ for every $\mu \in M_1(X)^\alpha$, the set of α-invariant probability measures on $\mathfrak{B}_X$. If α has completely positive entropy and $h(\alpha) < \infty$, then this inequality is strict.

THEOREM 20.15. *Let α be a $\mathbb{Z}^d$-action by automorphisms of a compact group X with $h(\alpha) < \infty$. The following conditions are equivalent.*

(1) *α has completely positive entropy;*
(2) *$h_\mu(\alpha) < h_{\lambda_X}(\alpha)$ whenever $\lambda_X \ne \mu \in M_1(X)^\alpha$.*

For $d = 1$, Theorem 20.15 was established in [10], and we follow Berg's proof with the modifications described in [63]. Let $T\colon \mathbf{m} \mapsto T_{\mathbf{m}}$ be a measure preserving $\mathbb{Z}^d$-action on a probability space $(Y, \mathfrak{T}, \mu)$.

LEMMA 20.16. *Assume that* $\mathcal{P}, \mathcal{Q}$ *are countable partitions in* $\mathfrak{T}$ *such that* $H_\mu(\mathcal{P}) + H_\mu(\mathcal{Q}) < \infty$, *that* T *has completely positive entropy on* $\mathcal{P}_T$, *and that*

$$h_\mu(T, \mathcal{P} \vee \mathcal{Q}) = h_\mu(T, \mathcal{P}) + h_\mu(T, \mathcal{Q}). \qquad (20.10)$$

Then the sigma-algebras $\mathcal{P}_T$ *and* $\mathcal{Q}_T$ *are independent.*

PROOF. Equation (20.1) shows that (20.10) is equivalent to

$$H_\mu(\mathcal{Q}|\mathcal{Q}_T^-) = H_\mu(\mathcal{Q}|\mathcal{Q}_T^- \vee \mathcal{P}_T). \qquad (20.11)$$

We set $Q_k = \{0, \dots, k-1\}^d$, define $\mathcal{P}_k$ and $\mathcal{Q}_k$ as in (20.6), and consider the $\mathbb{Z}^d$-action $T^{(k)}\colon \mathbf{n} \mapsto T_{\mathbf{n}}^{(k)} = T_{k\mathbf{n}}$. According to (20.7),

$$\begin{aligned} h_\mu(T^{(k)}, \mathcal{P}_k \vee \mathcal{Q}_k) &= k^d h_\mu(T, \mathcal{P} \vee \mathcal{Q}) = k^d(h_\mu(T, \mathcal{P}) + h_\mu(T, \mathcal{Q})) \\ &= h_\mu(T^{(k)}, \mathcal{P}_k) + h_\mu(T^{(k)}, \mathcal{Q}_k), \end{aligned}$$

and the equivalence of (20.10) and (20.11) yields that $H_\mu(\mathcal{Q}_k|(\mathcal{Q}_k)_{T^{(k)}}^-) = H_\mu(\mathcal{Q}_k|\mathcal{P}_{T^{(k)}} \vee (\mathcal{Q}_k)_{T^{(k)}}^-) = H_\mu(\mathcal{Q}_k|(\mathcal{P}_k)_{T^{(k)}} \vee (\mathcal{Q}_k)_{T^{(k)}}^-)$. By applying the equivalence of (20.10) and (20.11) twice more we see that $h_\mu(T^{(k)}, \mathcal{P} \vee \mathcal{Q}_k) = h_\mu(T^{(k)}, \mathcal{P}) + h_\mu(T^{(k)}, \mathcal{Q}_k)$ and $H_\mu(\mathcal{P}|\mathcal{P}_{T^{(k)}}^-) = H_\mu(\mathcal{P}|\mathcal{P}_{T^{(k)}}^- \vee (\mathcal{Q}_k)_{T^{(k)}})$. Since $\bigcap_{k \geq 1} \mathcal{P}_{T^{(k)}}^- \subset \mathcal{P}_T^{-\infty} = \{\varnothing, X\}$ by Proposition 20.2 and the assumption of completely positive entropy of T on $\mathcal{P}_T$,

$$\begin{aligned} H_\mu(\mathcal{P}|\mathcal{Q}) &\geq H_\mu(\mathcal{P}|\mathcal{P}_{T^{(k)}}^- \vee \mathcal{Q}) \\ &\geq H_\mu(\mathcal{P}|\mathcal{P}_{T^{(k)}}^- \vee (\mathcal{Q}_k)_{T^{(k)}}) = H_\mu(\mathcal{P}|\mathcal{P}_{T^{(k)}}^-) \to H_\mu(\mathcal{P}) \end{aligned}$$

as $k \to \infty$, so that $\mathcal{P}$ and $\mathcal{Q}$ are independent.

If we replace $\mathcal{P}$ and $\mathcal{Q}$ by $\mathcal{P}_k$ and $\mathcal{Q}_k$, then $h_\mu(T, \mathcal{P}_k \vee \mathcal{Q}_k) = h_\mu(T, \mathcal{P} \vee \mathcal{Q}) = h_\mu(T, \mathcal{P}) + h_\mu(T, \mathcal{Q}) = h_\mu(T, \mathcal{P}_k) + h_\mu(T, \mathcal{Q}_k)$, and the first part of this proof shows that $\mathcal{P}_k$ and $\mathcal{Q}_k$ are independent for every $k \geq 1$. Hence $\mathcal{P}_T$ and $\mathcal{Q}_T$ are independent, as claimed. $\square$

Now assume that α is a $\mathbb{Z}^d$-action by automorphisms of a compact group X, and that μ is an α-invariant probability measure on $\mathfrak{B}_X$. The sigma-algebras $\mathfrak{S}, \mathfrak{S}_1 \subset \mathfrak{B}_{X \times X}$ are defined as in the proof of Proposition 13.5.

LEMMA 20.17. *The sigma-algebras* $\mathfrak{S}_1$ *and* $\mathfrak{S}$ *are independent with respect to the measure* $\lambda_X \times \mu$ *if and only if* $\mu = \lambda_X$.

PROOF. We use the same notation as in the proof of Proposition 13.5. If $\mu = \lambda_X$, then

$$
(\lambda_X \times \lambda_X)(\pi_1^{-1}(E) \cap \pi^{-1}(F))
$$
$$
= (\lambda_X \times \lambda_X)(\{(x,y) \in X \times X : x \in E,\, xy \in F\})
$$
$$
= \int \int 1_E 1_{x^{-1}F}\, d\lambda_X d\lambda_X = \lambda_X(E)\lambda_X(F)
$$

for all $E, F \in \mathfrak{S}$, so that $\mathfrak{S}$ and $\mathfrak{S}_1$ are independent. Conversely, if $\mathfrak{S}_1$ and $\mathfrak{S}$ are independent with respect to $\lambda_x \times \mu$, then

$$
(\lambda_X \times \mu)(\pi_1^{-1}(E) \cap \pi^{-1}(F)) = (\lambda_X \times \mu)(\pi_1^{-1}(E))(\lambda_X \times \mu)(\pi^{-1}(F))
$$
$$
= \lambda_X(E)\lambda_X * \mu(F) = \lambda_X(E)\lambda_X(F),
$$

so that $\lambda_X \times \mu$ and $\lambda_X \times \lambda_X$ coincide on the sigma-algebra $\mathfrak{S}_1 \vee \mathfrak{S} = \mathfrak{B}_X \times \mathfrak{B}_X$ (cf. (13.15)). Hence $\mu(F) = (\lambda_X \times \mu)(\pi_2^{-1}(F)) = (\lambda_X \times \lambda_X)(\pi_2^{-1}(F)) = \lambda_X(F)$ for every $F \in \mathfrak{B}_X$. $\square$

PROOF OF THEOREM 20.15. Suppose that we have proved the implication (1)$\Rightarrow$(2) in Theorem 20.15 under the additional condition that μ is ergodic under α. If $\nu \in M_1(X)^\alpha$ is non-ergodic and satisfies that $h_\nu(\alpha) = h(\alpha)$ (for notation we refer to the proof of Proposition 13.5), then we can write ν as an integral of the form $\nu = \int_X \mu_x\, d\rho(x)$, where ρ is a probability measure on $\mathfrak{B}_X$, $\mu_x \in M_1(X)^\alpha$ is ergodic for every $x \in X$, and the map $x \mapsto \int f\, d\mu_x$ is an α-invariant Borel map from X to $\mathbb{R}$ for every continuous function $f\colon X \longmapsto \mathbb{R}$. Then $h_\nu(\alpha) = \int_X h_{\mu_x}(\alpha)\, d\rho(x) = h_{\lambda_X}(\alpha)$, which implies that $\mu_x = \lambda_X$ ρ−a.e., so that $\nu = \lambda_X$.

Assume therefore that α has completely positive entropy, and that $\mu \in M_1(X)^\alpha$ is ergodic. We claim that there exists a countable partition $\mathcal{P} \subset \mathfrak{B}_X$ such that $h_\mu(\mathcal{P}) < \infty$ and $\mathcal{P}_\alpha = \mathfrak{B}_X$ (mod μ), so that $h_\mu(\alpha, \mathcal{P}) = h_\mu(\alpha)$ (cf. (13.10)). If $d = 1$ this assertion is proved in [84]; if $d > 1$ it follows from Théorème 3.2 in [17], provided that

$$
\mu(\{x \in X : \alpha_{\mathbf{n}}(x) = x\}) = 0 \tag{20.12}
$$

for every non-zero $\mathbf{n} \in \mathbb{Z}^d$. In order to verify (20.12) we argue by contradiction and assume that (20.12) is not satisfied for some $0 \neq \mathbf{n} \in \mathbb{Z}^d$. Then $Y = \{x \in X : \alpha_{\mathbf{n}}(x) = x\}$ is a closed, α-invariant subgroup with $\mu(Y) > 0$, and hence, by ergodicity, with $\mu(Y) = 1$, and we conclude that $h_\mu(\alpha_{\mathbf{n}}) = 0$. Propositions 13.2 and 13.5 imply that $h_\mu(\alpha) \leq h_{\lambda_Y}(\alpha^Y) = h(\alpha^Y) = 0$, which is absurd. Hence (20.12) is satisfied, and there exists a partition $\mathcal{P}$ with the required properties.

Since α has completely positive entropy with respect to λ_X, α is mixing by Lemma 20.10, hence α satisfies (20.12) with μ replaced by λ_X, and we conclude as in the last paragraph that there exists a countable partition $\mathcal{Q} \subset \mathfrak{B}_X$ with $\mathcal{Q}_\alpha = \mathfrak{B}_X$ (mod λ_X), $h_{\lambda_X}(\mathcal{Q}) < \infty$, and $h_{\lambda_X}(\alpha) = h_{\lambda_X}(\alpha, \mathcal{Q})$.

We set $\mathcal{Q}' = \pi_1^{-1}(\mathcal{Q}) \subset \mathfrak{S}_1$ and $\mathcal{P}' = \pi^{-1}(\mathcal{Q}) \subset \mathfrak{S}$, and note that

$$
\begin{aligned}
h_{\lambda_X}(\alpha, \mathcal{Q}) + h_\mu(\alpha, \mathcal{P}) &= h_{\lambda_X}(\alpha) + h_\mu(\alpha) = h_{\lambda_X \times \mu}(\alpha \times \alpha) \\
&\leq h_{\lambda_X \times \mu}(\alpha \times \alpha, \mathfrak{S}_1 \vee \mathfrak{S}) \\
&\leq h_{\lambda_X \times \mu}(\alpha \times \alpha, \mathfrak{S}_1) + h_{\lambda_X \times \mu}(\alpha \times \alpha, \mathfrak{S}) \\
&= h_{\lambda_X}(\alpha) + h_{\lambda_X * \mu}(\alpha) = 2h_{\lambda_X}(\alpha).
\end{aligned}
\tag{20.13}
$$

Here $h_\nu(T, \mathfrak{U})$ denotes the entropy of a measure preserving $\mathbb{Z}^d$-action T on a probability space $(Z, 3, \nu)$, restricted to a T-invariant sigma-algebra $\mathfrak{U} \subset 3$, i.e. $h_\nu(T, \mathfrak{U}) = \sup_{\mathcal{U}} h_\nu(T, \mathcal{U})$, where $\mathcal{U}$ varies over all finite partitions of Z in $\mathfrak{U}$. If $h_\mu(\alpha) = h_{\lambda_X}(\alpha)$, then (20.13) shows that

$$
\begin{aligned}
h_{\lambda_X \times \mu}(\alpha \times \alpha, \mathcal{P}' \vee \mathcal{Q}') &= h_{\lambda_X \times \mu}(\alpha \times \alpha, \mathfrak{S}_1 \vee \mathfrak{S}) \\
&= h_{\lambda_X \times \mu}(\alpha \times \alpha, \mathfrak{S}_1) + h_{\lambda_X \times \mu}(\alpha \times \alpha, \mathfrak{S}) \\
&= h_{\lambda_X \times \mu}(\alpha \times \alpha, \mathcal{Q}') + h_{\lambda_X \times \mu}(\alpha \times \alpha, \mathcal{P}').
\end{aligned}
$$

According to Lemma 20.16, the sigma-algebras $\mathcal{P}'_{\alpha \times \alpha} = \mathfrak{S} \pmod{\lambda_X \times \mu}$ and $\mathcal{Q}'_{\alpha \times \alpha} = \mathfrak{S}_1 \pmod{\lambda_X \times \mu}$ are independent with respect to $\lambda_X \times \mu$, so that $\mu = \lambda_X$ by Lemma 20.17.

Conversely, if (2) is satisfied, then α must be mixing by Lemma 20.10; otherwise $h_{\lambda_X}(\alpha) = h_{\lambda_Y}(\alpha)$, but $\lambda_X \neq \lambda_Y$. According to Corollary 4.12, α^{X° and $\alpha^{X^\circ/C(X^\circ)}$ are again mixing, and $h(\alpha^{X^\circ/C(X^\circ)}) < h(\alpha^{X^\circ}) < h(\alpha) < \infty$. Propositions 2.13 and 15.3 imply that $X^\circ/C(X^\circ)$ is trivial, i.e. that X° is abelian. If α does not have completely positive entropy, then Corollary 20.7 shows that here exists a closed, normal, α-invariant subgroup $Y \subsetneqq X$ such that $h(\alpha^{X/Y}) = 0$ and $h_{\lambda_X}(\alpha) = h_{\lambda_Y}(\alpha)$, contrary to (2). This shows that (2)$\Rightarrow$(1). $\square$

CONCLUDING REMARKS 20.18. (1) The material on completely positive entropy is due to [85] and [111], and the modifications needed for the transition from $\mathbb{Z}$-actions to $\mathbb{Z}^d$-actions are taken from [17]: Lemma 20.1 and Proposition 20.2 follow [17], Corollary 20.3 [85], and Lemma 20.4, Proposition 20.5, and Corollary 20.6 [111]. Theorem 20.8 and Corollary 20.9 were proved in [63]. Theorem 20.14 and its proof are taken from [33]; we shall return to the higher order mixing properties of $\mathbb{Z}^d$-actions by automorphisms of compact groups in Sections 27–28. Example 20.12 is due to Paul Smith (cf. [63]). For Theorem 20.15 and its proof we follow [10] with the modifications outlined in [63], where X was assumed to be abelian.

(2) Following [84] and [17] we call a measure preserving $\mathbb{Z}^d$-action T on a probability space $(Y, \mathfrak{T}, \mu))$ aperiodic if $\mu(\{y \in Y : T_\mathbf{n}(y) = y\}) = 0$ whenever $0 \neq \mathbf{n} \in \mathbb{Z}^d$. From the proof of Theorem 20.15 it is clear that an ergodic $\mathbb{Z}^d$-action α by automorphisms of a compact group X is aperiodic if and only if it is faithful, i.e. if and only if $\alpha_\mathbf{n} \neq \mathrm{id}_X$ for all non-zero $\mathbf{n} \in \mathbb{Z}^d$. If α is ergodic,

faithful, and $h(\alpha) < \infty$, then α has a countable generator with finite entropy by [17].

21. Entropy and periodic points

Let α be a $\mathbb{Z}^d$-action by automorphisms of a compact group X. For every subgroup $\Lambda \subset \mathbb{Z}^d$ of finite index we set

$$\langle \Lambda \rangle = \min\{|\mathbf{m}| : \mathbf{0} \neq \mathbf{m} \in \Lambda\} \tag{21.1}$$

where $|\mathbf{m}|$, $\mathbf{m} \in \mathbb{Z}^d$, is given by (16.3), and define the fixed point set $\mathrm{Fix}_\Lambda(\alpha)$ as in Theorem 6.5 (3). We wish to investigate the growth rate of $\mathrm{Fix}_\Lambda(\alpha)$ as $\langle \Lambda \rangle \to \infty$, and to relate this growth rate to the topological entropy of α.

If α is an expansive automorphism of a finite-dimensional torus $X = \mathbb{T}^n$, then it is well known that

$$\lim_{n \to \infty} \frac{1}{n} \log |\mathrm{Fix}_{n\mathbb{Z}}(\alpha)| = h(\alpha). \tag{21.2}$$

If the toral automorphism α in (21.2) is only ergodic, but not expansive, then (21.2) remains true, but its proof becomes much more subtle, and requires a deep diophantine result on rational approximation of logarithms of algebraic numbers due to Gelfond (cf. [24], [60], and Lemma 21.8).

If α is an expansive $\mathbb{Z}^d$-action by automorphisms of a compact group X, then $\mathrm{Fix}_\Lambda(\alpha)$ is obviously finite for every subgroup $\Lambda \subset \mathbb{Z}^d$ of finite index, and it is natural to ask whether

$$\lim_{\langle \Lambda \rangle \to \infty} \frac{1}{|\mathbb{Z}^d/\Lambda|} \log |\mathrm{Fix}_\Lambda(\alpha)| = h(\alpha). \tag{21.3}$$

We shall prove in this section that (21.3) does indeed hold if α is expansive. However, if α is non-expansive, the cardinality of $\mathrm{Fix}_\Lambda(\alpha)$ may be equal to 1 for every $\Lambda \subset \mathbb{Z}^d$ (Examples 5.6), or it may be infinite for every Λ (Examples 10.1 (1)–(2), and Example 6.18 (3)), so that (21.3) cannot hold in complete generality. The connection between periodic points and the d.c.c. exhibited in Section 5 (Examples 5.6 and Theorem 5.7) shows that any generalization of (21.3) to non-expansive actions will—in general—require the d.c.c. The problem of having *too many* periodic points (Examples 10.1) can be overcome by counting not the periodic points, but the number of connected components in the set of all points with a given period. Even with these natural adaptations of (21.3) some problems remain, and we can prove (an analogue of) (21.3) only with $\lim_{\langle \Lambda \rangle \to \infty}$ replaced by $\limsup_{\langle \Lambda \rangle \to \infty}$.

We begin with a few definitions. Let α be a $\mathbb{Z}^d$-action by automorphisms of a compact, abelian group X, $\Lambda \subset \mathbb{Z}^d$ a subgroup with finite index, and let $\mathrm{Fix}_\Lambda(\alpha)^\circ$ be the connected component of the identity in $\mathrm{Fix}_\Lambda(\alpha)$. Consider the index

$$P_\Lambda(\alpha) = |\mathrm{Fix}_\Lambda(\alpha)/\mathrm{Fix}_\Lambda(\alpha)^\circ| < \infty \tag{21.4}$$

of $\mathrm{Fix}_\Lambda(\alpha)^\circ$ in $\mathrm{Fix}_\Lambda(\alpha)$, and note that

$$P_\Lambda(\alpha) = |\mathrm{Fix}_\Lambda(\alpha)| \text{ if and only if } |\mathrm{Fix}_\Lambda(\alpha)| < \infty \tag{21.5}$$

(cf. Theorem 6.5 (3)). By

$$
\begin{aligned}
p^-(\alpha) &= \liminf_{\langle\Lambda\rangle\to\infty} \frac{1}{|\mathbb{Z}^d/\Lambda|} \log P_\Lambda(\alpha), \\
p^+(\alpha) &= \limsup_{\langle\Lambda\rangle\to\infty} \frac{1}{|\mathbb{Z}^d/\Lambda|} \log P_\Lambda(\alpha).
\end{aligned}
\tag{21.6}
$$

we denote the upper and lower growth rates of $P_\Lambda(\alpha)$ as $\langle\Lambda\rangle \to \infty$. In this section we prove the following result.

THEOREM 21.1. *Let $d \geq 1$, and let α be a $\mathbb{Z}^d$-action by automorphisms of a compact, abelian group X which satisfies the d.c.c. Then the following is true.*

(1) $p^+(\alpha) < \infty$;
(2) *If $h(\alpha) < \infty$ then $p^+(\alpha) = h(\alpha)$;*
(3) *If $d = 1$, or if α is expansive, then $p^-(\alpha) = p^+(\alpha) = h(\alpha)$.*

The proof of Theorem 21.1 requires some preparation.

LEMMA 21.2. *Let $\mathfrak{M}$ be a Noetherian $\mathfrak{R}_d$-module, and let $\alpha = \alpha^{\mathfrak{M}}$ and $X = X^{\mathfrak{M}}$ be defined as in Lemma 5.1. If δ is a metric on X, then there exists an $\varepsilon > 0$ with the following property: for every $x \in X$ with $\delta(\alpha_\mathbf{n}(x), 1_X) < \varepsilon$ for all $\mathbf{n} \in \mathbb{Z}^d$ there is a unique continuous homomorphism $t \mapsto x^{(t)}$ from $\mathbb{R}$ into X such that*

(1) $\delta(\alpha_\mathbf{n}(x^{(t)}), 1_X) < \varepsilon$ *for every $t \in [-1, 1]$ and $\mathbf{n} \in \mathbb{Z}^d$;*
(2) $x^{(1)} = x$;
(3) *If $x \in \mathrm{Fix}_\Lambda(\alpha)$ for some subgroup $\Lambda \subset \mathbb{Z}^d$, then $x^{(t)} \in \mathrm{Fix}_\Lambda(\alpha)$ for every $t \in \mathbb{R}$.*

PROOF. We begin by assuming that $\mathfrak{M} = \mathfrak{R}_d/\mathfrak{a}$ for some ideal $\mathfrak{a} \subset \mathfrak{R}_d$; the general proof is very similar and is given below. If $\mathfrak{a}$ contains a non-zero constant $m > 1$ the assertion is trivial, since $x_\mathbf{m} \in \frac{1}{m}\mathbb{Z} \pmod 1$ for every $x \in X$ and $\mathbf{m} \in \mathbb{Z}^d$. If $\mathfrak{a} \cap \mathbb{Z} = \{0\}$ we choose $\{f_1, \dots, f_k\} \subset \mathfrak{R}_d$ such that $\mathfrak{a} = (f_1, \dots, f_k) = \sum_{j=1}^k f_j \mathfrak{R}_d$ and set

$$\varepsilon = \left(10 \sum_{j=1}^k \|f_j\| + 4\right)^{-1}, \tag{21.7}$$

where $\|f_j\| = \sum_{\mathbf{n}\in\mathbb{Z}^d} |c_{f_j}(\mathbf{n})|$ for every $j = 1, \dots, k$ (cf. (5.2)). Suppose that $x \in X$ satisfies that $\|x_\mathbf{n}\| < \varepsilon$ for every $\mathbf{n} \in \mathbb{Z}^d$, where $\|s - t\| = \vartheta(s, t)$ is the metric (17.7) on $\mathbb{T}$. We proceed exactly as in the proof of Lemma 6.8 and denote by $y \in \mathbf{B} = \ell^\infty(\mathbb{Z}^d)$ the unique point with $|y_\mathbf{n}| < \varepsilon$ and $y_\mathbf{n} \pmod 1 = x_\mathbf{n}$ for every $\mathbf{n} \in \mathbb{Z}^d$. The choice of ε in (21.7) implies that $y \in \mathbf{S}$, where $\mathbf{S} \subset \mathbf{B}$

is the subspace defined in (6.9), and hence that $ty \in \mathbf{S}$ for every $t \in \mathbb{R}$. We set $x_{\mathbf{n}}^{(t)} = ty_{\mathbf{n}} \pmod{1}$ for every $t \in \mathbb{R}$ and $\mathbf{n} \in \mathbb{Z}^d$. The map $t \mapsto x^{(t)}$ is a continuous homomorphism from $\mathbb{R}$ into X with $\|x_{\mathbf{n}}^{(t)}\| < |t|\varepsilon$ for all $|t| \le 1$ and $\mathbf{n} \in \mathbb{Z}^d$, and (2)–(3) are obvious from the definition of $x^{(t)}$. Finally, if δ is an arbitrary metric on X, then there exists an $\varepsilon' > 0$ such that $\|x_0\| < \varepsilon$ for every $x = (x_{\mathbf{n}}) \in X$ with $\delta(x, \mathbf{1}_X) < \varepsilon'$, which proves (1) and completes the proof of the lemma for $\mathfrak{M} = \mathfrak{R}_d/\mathfrak{a}$.

If $\mathfrak{M}$ is a general Noetherian $\mathfrak{R}_d$-module, we can find an integer $l \ge 1$ and a submodule $\mathfrak{L} \subset \mathfrak{R}_d^l$ such that $\mathfrak{M} \cong \mathfrak{R}_d^l/\mathfrak{L}$. Choose a set of generators $\{\mathbf{f}^{(1)}, \dots, \mathbf{f}^{(k)}\}$ for $\mathfrak{L}$, where $\mathbf{f}^{(j)} = (f_1^{(j)}, \dots, f_l^{(j)}) \in \mathfrak{R}_d^l$ for $j = 1, \dots, k$. Since $\widehat{\mathfrak{R}_d^l} \cong (\mathbb{T}^l)^{\mathbb{Z}^d}$, X can be regarded as a closed, shift-invariant subgroup of $(\mathbb{T}^l)^{\mathbb{Z}^d}$, and α as the shift-action of $\mathbb{Z}^d$ on $X \subset (\mathbb{T}^l)^{\mathbb{Z}^d}$ (Example 5.2 (4)). We put $\|\mathbf{t}\| = \max_{i=1,\dots,l} \|t_i\|$ for all $\mathbf{t} = (t_1, \dots, t_l)$ in $\mathbb{T}^l$, denote by $\mathbf{B} = \ell^\infty(\mathbb{Z}^d, \mathbb{C}^l)$ the Banach space of all functions $\mathbf{v} \colon \mathbb{Z}^d \longmapsto \mathbb{C}^l$ which are bounded with respect to the maximum norm on $\mathbb{C}^l$, and write each $\mathbf{v} \in \mathbf{B}$ as $\mathbf{v} = (v_1, \dots, v_l)$ with $v_i \in \ell^\infty(\mathbb{Z}^d)$, $i = 1, \dots, l$. Consider the subspace

$$\mathbf{S} = \{v \in \mathbf{B} : \alpha_{\mathbf{f}^{(j)}}(\mathbf{v}) = 0 \text{ for } j = 1, \dots, k\},$$

where $(\alpha_{\mathbf{f}^{(j)}}(\mathbf{v}))_i = \sum_{\mathbf{m} \in \mathbb{Z}^d} c_{f_i^{(j)}}(\mathbf{m})\alpha_{\mathbf{m}}(v_i)$ for every $\mathbf{v} = (v_1, \dots, v_l) \in \mathbf{B}$ and $i = 1, \dots, l$. If

$$\varepsilon = \left(10 \sum_{i=1}^{l} \sum_{j=1}^{k} \|f_i^{(j)}\| + 4\right)^{-1},$$

then exactly the same argument as above shows that, for every $x \in X \subset (\mathbb{T}^l)^{\mathbb{Z}^d}$ with $\|x_{\mathbf{m}}\| < \varepsilon$ for all $\mathbf{m} \in \mathbb{Z}^d$, there exists a continuous homomorphism $t \mapsto x^{(t)}$ from $\mathbb{R}$ into X such that $\|x_{\mathbf{n}}^{(t)}\| < |t|\varepsilon$ for all $|t| \le 1$ and $\mathbf{n} \in \mathbb{Z}^d$, and such that (2)–(3) are satisfied. The proof is completed as in the case where $\mathfrak{M} = \mathfrak{R}_d/\mathfrak{a}$. $\square$

LEMMA 21.3. *Let α be a $\mathbb{Z}^d$-action by automorphisms of a compact, abelian group X which satisfies the d.c.c. Then $p^+(\alpha) < \infty$, and $p^+(\alpha) \le h(\alpha)$ whenever $h(\alpha) < \infty$.*

PROOF. Let α be a $\mathbb{Z}^d$-action by automorphisms of a compact group X which satisfies the d.c.c, and let $\mathfrak{M} = \hat{X}$ be the $\mathfrak{R}_d$-module defined by Lemma 5.1. We choose a metric δ on X and fix an $\varepsilon > 0$ which satisfies the conditions (1)–(3) in Lemma 21.2. For every subgroup $\Lambda \subset \mathbb{Z}^d$ of finite index we select a set of generators $\{\mathbf{m}_\Lambda^{(1)}, \dots, \mathbf{m}_\Lambda^{(d)}\}$ of Λ such that $|\mathbf{m}_\Lambda^{(1)}| + \cdots + |\mathbf{m}_\Lambda^{(d)}|$ is minimal, where $|\mathbf{m}|$, $\mathbf{m} \in \mathbb{Z}^d$, is defined in (16.3), and set

$$F_\Lambda = \mathbb{Z}^d \cap \left\{ \sum_{j=1}^{d} t_j \mathbf{m}_\Lambda^{(j)} : (t_1, \dots, t_d) \in [0, 1)^d \right\}.$$

Let $Q_n = \{0, \dots, n-1\}^d$, and let $D(Q_n, F_\Lambda)$ be a minimal union of disjoint translates of Q_n containing F_Λ. Then $\frac{|F_\Lambda|}{|D(Q_n,F_\Lambda)|} = \frac{|\mathbb{Z}^d/\Lambda|}{|D(Q_n,F_\Lambda)|} \to 1$ as $\langle \Lambda \rangle \to \infty$, and subadditivity shows that

$$
\frac{1}{|F_\Lambda|} \log N\left(\bigvee_{\mathbf{m} \in F_\Lambda} \alpha_{-\mathbf{m}}(\mathcal{U}) \right)
$$

$$
\leq \frac{|D(Q_n, F_\Lambda)|}{|F_\Lambda|} \frac{1}{|D(Q_n, F_\Lambda)|} \log N\left(\bigvee_{\mathbf{m} \in D(Q_n,F_\Lambda)} \alpha_{-\mathbf{m}}(\mathcal{U}) \right) \quad (21.8)
$$

$$
\leq \frac{|D(Q_n, F_\Lambda)|}{|F_\Lambda|} \frac{1}{|Q_n|} \log N\left(\bigvee_{\mathbf{m} \in Q_n} \alpha_{-\mathbf{m}}(\mathcal{U}) \right)
$$

for every $n \geq 1$, $\Lambda \in \mathbb{Z}^d$, and every open cover $\mathcal{U}$ of X. If we let first $\langle \Lambda \rangle \to \infty$ and then $n \to \infty$ we see that

$$
\lim_{\langle \Lambda \rangle \to \infty} \frac{1}{|F_\Lambda|} \log N\left(\bigvee_{\mathbf{m} \in F_\Lambda} \alpha_{-\mathbf{m}}(\mathcal{U}) \right) \leq h(\alpha). \quad (21.9)
$$

We specialize (21.9) by setting $\mathcal{U} = \mathcal{U}_{\varepsilon'}$, where $\varepsilon' < \varepsilon/2$ and $\mathcal{U}_{\varepsilon'}$ is the open cover of X consisting of all ε'-balls in the metric δ, and claim that every $U \in \bigvee_{\mathbf{m} \in F_\Lambda} \alpha_{-\mathbf{m}}(\mathcal{U}_{\varepsilon'})$ intersects at most one coset of $\mathrm{Fix}_\Lambda(\alpha)^\circ$ in $\mathrm{Fix}_\Lambda(\alpha)$. Indeed, if there exist $x, x' \in \mathrm{Fix}_\Lambda(\alpha)$ such that $U \cap (\mathrm{Fix}_\Lambda(\alpha)^\circ + x) \neq \varnothing$ and $U \cap (\mathrm{Fix}_\Lambda(\alpha)^\circ + x') \neq \varnothing$, then $\delta(\alpha_\mathbf{n}(x - x'), 1_X) < \varepsilon$ for every $\mathbf{n} \in \mathbb{Z}^d$, and Lemma 21.2 implies that $x - x' \in \mathrm{Fix}_\Lambda(\alpha)^\circ$, so that $\mathrm{Fix}_\Lambda(\alpha)^\circ + x = \mathrm{Fix}_\Lambda(\alpha)^\circ + x'$. Hence

$$
P_\Lambda(\alpha) \leq N\left(\bigvee_{\mathbf{m} \in F_\Lambda} \alpha_{-\mathbf{m}}(\mathcal{U}_{\varepsilon'}) \right) \quad (21.10)
$$

for every subgroup $\Lambda \subset \mathbb{Z}^d$ of finite index and every $\varepsilon' \leq \varepsilon/2$, and

$$
p^+(\alpha) = \limsup_{\langle \Lambda \rangle \to \infty} \frac{1}{|\mathbb{Z}^d/\Lambda|} \log P_\Lambda(\alpha)
$$

$$
\leq \lim_{\langle \Lambda \rangle \to \infty} \frac{1}{|F_\Lambda|} \log N\left(\bigvee_{\mathbf{m} \in F_\Lambda} \alpha_{-\mathbf{m}}(\mathcal{U}_{\varepsilon/2}) \right) \quad (21.11)
$$

$$
\leq \log N(\mathcal{U}_{\varepsilon/2}) < \infty.
$$

If $h(\alpha) < \infty$, Proposition 13.1, Theorem 13.3, (13.5), and (21.8)–(21.11) together imply that

$$
p^+(\alpha) = \limsup_{\langle \Lambda \rangle \to \infty} \frac{1}{|\mathbb{Z}^d/\Lambda|} \log P_\Lambda(\alpha)
$$

$$
\leq \lim_{\varepsilon' \to 0} \lim_{\langle \Lambda \rangle \to \infty} \frac{1}{|F_\Lambda|} \log N\left(\bigvee_{\mathbf{m} \in F_\Lambda} \alpha_{-\mathbf{m}}(\mathcal{U}_{\varepsilon'}) \right) = h(\alpha). \quad \square \quad (21.12)
$$

PROOF OF THEOREM 21.1 (1). Apply Lemma 21.3. $\quad \Box$

Motivated by Theorem 6.5 (3) we put, for every subgroup $\Lambda \subset \mathbb{Z}^d$ of finite index,

$$\mathfrak{b}(\Lambda) = \sum_{\mathbf{m} \in \Lambda} (u^{\mathbf{m}} - 1) \mathfrak{R}_d,$$

$$\Omega(\Lambda) = V_{\mathbb{C}}(\mathfrak{b}(\Lambda)) = \{\omega = (\omega_1, \ldots, \omega_d) \in \mathbb{S}^d : \tag{21.13}$$

$$\omega^{\mathbf{n}} = \omega_1^{n_1} \cdot \ldots \cdot \omega_d^{n_d} = 1$$

$$\text{for every } \mathbf{n} = (n_1, \ldots, n_d) \in \Lambda\} \subset \mathbb{S}^d.$$

An element $\mathbf{p} = (p_1, \ldots, p_d) \in \mathbb{Z}^d$ is *positive* if $p_i > 0$ for $i = 1, \ldots, d$, and *prime* if each coordinate $p_1, \ldots, p_d$ is a rational prime. For every positive element $\mathbf{p} = (p_1, \ldots, p_d) \in \mathbb{Z}^d$ we set

$$\Lambda_{\mathbf{p}} = \{(n_1 p_1, \ldots, n_d p_d) : (n_1, \ldots, n_d) \in \mathbb{Z}^d\},$$

$$\Omega_{\mathbf{p}} = \Omega(\Lambda_{\mathbf{p}}), \tag{21.14}$$

$$\mathfrak{b}_{\mathbf{p}} = \mathfrak{b}(\Lambda_{\mathbf{p}}),$$

$$\mathfrak{j}_{\mathbf{p}} = (\phi_{p_1}(u_1), \ldots, \phi_{p_d}(u_d)) = \phi_{p_1}(u_1) \mathfrak{R}_d + \cdots + \phi_{p_d}(u_d) \mathfrak{R}_d,$$

where ϕ_q denotes the q-th cyclotomic polynomial for every $q \geq 1$. Then

$$V_{\mathbb{C}}(\mathfrak{j}_{\mathbf{p}}) = \{\omega = (\omega_1, \ldots, \omega_d) \in V_{\mathbb{C}}(\mathfrak{b}_{\mathbf{p}}) : \omega_i \neq 1 \text{ for } i = 1, \ldots, d\} \quad (21.15)$$

and

$$\langle \Lambda_{\mathbf{p}} \rangle = \min_{i=1,\ldots,d} p_i$$

for every prime element $\mathbf{p} \in \mathbb{Z}^d$. In particular, if $(\mathbf{p}^{(n)} = (p_1^{(n)}, \ldots, p_d^{(n)}), n \geq 1)$ is a sequence of prime elements of $\mathbb{Z}^d$, and if $\mathbf{p}^{(n)} \to \infty$ indicates that $\lim_{n \to \infty} p_i^{(n)} = \infty$ for $i = 1, \ldots, d$, then

$$\langle \Lambda_{\mathbf{p}^{(n)}} \rangle \to \infty \text{ as } \mathbf{p}^{(n)} \to \infty. \tag{21.16}$$

LEMMA 21.4. *For every subgroup* $\Lambda \subset \mathbb{Z}^d$ *of finite index, the ideal* $\mathfrak{b}(\Lambda)$ *in* (21.13) *is radical, i.e.* $\mathfrak{b}(\Lambda) = \sqrt{\mathfrak{b}(\Lambda)} = \{f \in \mathfrak{R}_d : f^k \in \mathfrak{b}(\Lambda) \text{ for some } k \geq 1\} = \{f \in \mathfrak{R}_d : f(c) = 0 \text{ for every } c \in V_{\mathbb{C}}(\mathfrak{b}(\Lambda))\}$ *(cf. Example* 7.5 (3)).

PROOF. According to (21.13), $\Omega(\Lambda) = V_{\mathbb{C}}(\mathfrak{b}(\Lambda))$ is a multiplicative subgroup of $\mathbb{S}^d \subset \mathbb{C}^d$, which is isomorphic to $\widehat{\mathbb{Z}^d/\Lambda}$. By duality, $\mathbb{Z}^d/\Lambda \cong \widehat{\Omega(\Lambda)}$, and this isomorphism extends to the integral group rings $\mathfrak{R}_d/\mathfrak{b}(\Lambda) \cong \mathbb{Z}[\mathbb{Z}^d/\Lambda] \cong \mathbb{Z}[\widehat{\Omega(\Lambda)}]$; note that the composition of these isomorphisms is nothing but the evaluation of a Laurent polynomial $f \in \mathfrak{R}_d$ (or, more precisely, of $f + \mathfrak{b}(\Lambda)$) on $\Omega(\Lambda)$. In particular, a Laurent polynomial $f \in \mathfrak{R}_d$ lies in $\mathfrak{b}(\Lambda)$ if and only if its restriction to $\Omega(\Lambda)$ is equal to zero, which is what we claimed. $\quad \Box$

LEMMA 21.5. *For every prime element* $\mathbf{p} = (p_1, \ldots, p_d) \in \mathbb{Z}^d$, *the ideal* $\mathfrak{j}_\mathbf{p} \subset \mathfrak{R}_d$ *defined by* (21.14) *is radical.*

PROOF. Define $\widehat{\mathfrak{R}_d/\mathfrak{j}_\mathbf{p}} = X^{\mathfrak{R}_d/\mathfrak{j}_\mathbf{p}} \subset \mathbb{T}^{\mathbb{Z}^d}$ by (5.9) and set $Q_\mathbf{p} = \prod_{i=1}^{d}\{1, \ldots, p_i - 1\} \subset \mathbb{Z}^d$. Since $\phi_{p_i}(u) = 1 + u + \cdots + u^{p_i - 1}$ for $i = 1, \ldots, d$, the projection map $\pi_{Q_\mathbf{p}} \colon X^{\mathfrak{R}_d/\mathfrak{j}_\mathbf{p}} \longmapsto \mathbb{T}^{Q_\mathbf{p}}$ is a continuous group isomorphism, and we conclude that $X^{\mathfrak{R}_d/\mathfrak{j}_\mathbf{p}} \cong \mathbb{T}^{Q_\mathbf{p}}$ is connected. Hence $\mathfrak{R}_d/\mathfrak{j}_\mathbf{p}$ is a torsion-free $\mathbb{Z}$-module. We denote by $\mathfrak{R}_d^{(\mathbb{Q})} = \mathbb{Q}[u_1^{\pm 1}, \ldots, u_d^{\pm 1}]$ the ring of Laurent polynomials in $u_1, \ldots, u_d$ with coefficients in $\mathbb{Q}$, write $\mathfrak{j}_\mathbf{p}^{(\mathbb{Q})} \subset \mathfrak{R}_d^{(\mathbb{Q})}$ for the ideal $\phi_{p_1}(u_1)\mathfrak{R}_d^{(\mathbb{Q})} + \cdots + \phi_{p_d}(u_d)\mathfrak{R}_d^{(\mathbb{Q})}$ generated by $\{\phi_{p_1}(u_1), \ldots, \phi_{p_d}(u_d)\}$, and obtain that the embedding $\iota\colon a \mapsto 1 \otimes a$ of $\mathfrak{R}_d/\mathfrak{j}_\mathbf{p}$ in $\mathbb{Q} \otimes_\mathbb{Z} \mathfrak{R}_d/\mathfrak{j}_\mathbf{p} = \mathfrak{R}_d^{(\mathbb{Q})}/\mathfrak{j}_\mathbf{p}^{(\mathbb{Q})}$ is injective.

The ideal $\mathfrak{j}_\mathbf{p}$ is radical if and only if the ring $\mathfrak{R}_d/\mathfrak{j}_\mathbf{p}$ has no non-zero nilpotent elements. Since $\iota\colon \mathfrak{R}_d/\mathfrak{j}_\mathbf{p} \longmapsto \mathfrak{R}_d^{(\mathbb{Q})}/\mathfrak{j}_\mathbf{p}^{(\mathbb{Q})}$ is injective, it will be sufficient to prove that the quotient ring $\mathfrak{R}_d^{(\mathbb{Q})}/\mathfrak{j}_\mathbf{p}^{(\mathbb{Q})}$ has no non-zero nilpotent elements or, equivalently, that the ideal $\mathfrak{j}_\mathbf{p}^{(\mathbb{Q})} \subset \mathfrak{R}_d^{(\mathbb{Q})}$ is radical. Let therefore $f \in \sqrt{\mathfrak{j}_\mathbf{p}^{(\mathbb{Q})}} = \{h \in \mathfrak{R}_d^{(\mathbb{Q})} : h(c) = 0 \text{ for every } c \in V\}$, where $V = V_\mathbb{C}(\mathfrak{j}_\mathbf{p}) = V_\mathbb{C}(\mathfrak{j}_\mathbf{p}^{(\mathbb{Q})})$ is given by (21.15). After multiplying f by an integer we may take it that $f \in \mathfrak{R}_d$. Then

$$g(u_1, \ldots, u_d) = f(u_1, \ldots, u_d) \cdot \prod_{i=1}^{d}\left(1 - \frac{1}{p_i}(1 + u_i + \ldots u_i^{p_i - 1})\right)$$

vanishes on $V_\mathbb{C}(\mathfrak{b}_\mathbf{p})$, and Lemma 21.4 implies that there exist Laurent polynomials $a_i \in \mathfrak{R}_d$ such that

$$(p_1 \cdot \ldots \cdot p_d)g(u_1, \ldots, u_d) = \sum_{i=1}^{d} a_i(u_1, \ldots, u_d)(u_i^{p_i} - 1)$$

$$= \sum_{i=1}^{d} a_i(u_1, \ldots, u_d)(u_i - 1)\phi_{p_i}(u_i).$$

From the definition of g it is clear that we can find Laurent polynomials $h_i \in \mathfrak{R}_d$ such that

$$(p_1 \cdot \ldots \cdot p_d)g(u_1, \ldots, u_d) + \sum_{i=1}^{d} h_i(u_1, \ldots, u_d)\phi_{p_i}(u_i)$$

$$= (p_1 \cdot \ldots \cdot p_d)f(u_1, \ldots, u_d),$$

and we conclude the existence of Laurent polynomials $b_1, \ldots, b_d$ in $\mathfrak{R}_d$ such that

$$(p_1 \cdot \ldots \cdot p_d)f(u_1, \ldots, u_d) = \sum_{i=1}^{d} b_i(u_1, \ldots, u_d)\phi_{p_i}(u_i).$$

Hence $f \in j_{\mathbf{p}}^{(\mathbb{Q})} \cap \mathfrak{R}_d$, which implies that $j_{\mathbf{p}}^{(\mathbb{Q})}$ is radical. As explained above, this shows that $j_{\mathbf{p}}$ is radical. $\square$

LEMMA 21.6. *Let $f \in \mathfrak{R}_d$ a non-zero polynomial, and let*

$$\deg(f) = \max\{|\mathbf{m}| : \mathbf{m} \in \mathcal{S}(f)\}$$

be the maximum degree of f in each of the variables $u_1, \ldots, u_d$ (cf. (16.3) and (19.1)). If $p_1, \ldots, p_d$ are rational primes such that

$$p_1 > \deg(f) + 1 \quad \text{and} \quad p_j > (\deg(f) + 1) \cdot \prod_{i=1}^{j-1} p_i \ \text{for } 2 \leq j \leq d, \quad (21.17)$$

then the prime element $\mathbf{p} = (p_1, \ldots, p_d) \in \mathbb{Z}^d$ satisfies that

$$\begin{aligned} V_{\mathbb{C}}(f) \cap V_{\mathbb{C}}(\mathfrak{b}_{\mathbf{p}}) &\subset \{(1, \ldots, 1)\}, \\ V_{\mathbb{C}}(f) \cap V_{\mathbb{C}}(j_{\mathbf{p}}) &= \varnothing. \end{aligned} \quad (21.18)$$

PROOF. If we can prove that $V_{\mathbb{C}}(f) \cap V_{\mathbb{C}}(\mathfrak{b}_{\mathbf{p}}) \subset \{(1, \ldots, 1)\}$ for every prime element $\mathbf{p} \in \mathbb{Z}^d$ satisfying (21.17), then the second equation in (21.18) follows from (21.15). We shall verify (21.18) by induction on d. If $d = 1$, and if $f(\omega) = 0$ for any primitive p-th unit root ω, where $p > \deg(f) + 1$ is a rational prime, then $f(\omega^k) = 0$ for every $k = 1, \ldots, p - 1$, which is impossible. This proves our assertion for $d = 1$. Now assume that the lemma has been established for every $d \in \{1, \ldots, D - 1\}$, where $D \geq 2$. Let $f \in \mathfrak{R}_D$ be a non-zero polynomial, $\mathbf{p} = (p_1, \ldots, p_D) \in \mathbb{Z}^D$ a prime element satisfying (21.17) with D replacing d, and define $\mathfrak{b}_{\mathbf{p}'} \subset \mathfrak{R}_D$ as in (21.14). If there exists an integer $d' < D$ and an element $\omega = (\omega_1, \ldots, \omega_D) \in V_{\mathbb{C}}(\mathfrak{b}_{\mathbf{p}}) \smallsetminus \{(1, \ldots, 1)\}$ such that $f(\omega) = 0$ and $\omega_j = 1$ for $j = d' + 1, \ldots, D$, then we replace f by the polynomial $f'(u_1, \ldots, u_{d'}) = f(u_1, \ldots, u_{d'}, 1, \ldots, 1) \in \mathfrak{R}_{d'}$ and obtain that $V_{\mathbb{C}}(f') \cap V_{\mathbb{C}}(\mathfrak{b}_{(p_1, \ldots, p_{d'})}) \not\subset \{(1, \ldots, 1)\}$, in violation of our induction hypothesis (note that $\deg(f') \leq \deg(f)$). We conclude that any $\omega \in (V_{\mathbb{C}}(f) \cap V_{\mathbb{C}}(\mathfrak{b}_{\mathbf{p}})) \smallsetminus \{(1, \ldots, 1)\}$ must be of the form $\omega = (\omega_1, \ldots, \omega_D)$, where ω_D is a primitive p_D-th root of unity. We set $\mathbf{p}' = (p_1, \ldots, p_{D-1})$ and obtain a polynomial

$$F(x) = \prod_{(\omega_1, \ldots, \omega_{D-1}) \in V_{\mathbb{C}}(\mathfrak{b}_{\mathbf{p}'})} f(\omega_1, \ldots, \omega_{D-1}, x)$$

with integral coefficients and degree $p_1 \cdot \ldots \cdot p_{D-1} \cdot \deg(f)$. Since ω_D is primitive, and since $F(\omega_D) = 0$, we know that $F(\omega_D^q) = 0$ for every $q \in \{1, \ldots, p_D - 1\}$, which implies that the degree of F is $\geq p_D - 1$. Since $p_D - 1 > \deg(F)$, we obtain a contradiction which proves (21.18) for $d = D$, and hence for every $d \geq 1$. $\square$

LEMMA 21.7. *Let $\mathfrak{M}$ be a Noetherian torsion $\mathfrak{R}_d$-module with associated primes $\{\mathfrak{p}_1, \ldots, \mathfrak{p}_m\}$, and let $\alpha = \alpha^{\mathfrak{M}}$ be the $\mathbb{Z}^d$-action on $X = X^{\mathfrak{M}}$ defined by Lemma 5.1. Then there exists an integer $M = M(\mathfrak{p}_1, \ldots, \mathfrak{p}_m) \geq 1$, which depends only on the prime ideals $\mathfrak{p}_1, \ldots, \mathfrak{p}_d$ associated with $\mathfrak{M}$, but not on the module $\mathfrak{M}$, with the following property. If $\mathbf{p} = (p_1, \ldots, p_d) \in \mathbb{Z}^d$ is a prime element with*

$$p_1 > M, \quad and \quad p_j > M \cdot \prod_{i=1}^{j-1} p_i \quad for \ j = 2, \ldots, d, \tag{21.19}$$

then the group

$$Y_{\mathbf{p}}^{\mathfrak{M}} = \widehat{\mathfrak{M}/j_{\mathbf{p}} \cdot \mathfrak{M}} = (j_{\mathbf{p}} \cdot \mathfrak{M})^{\perp} \subset X \tag{21.20}$$

satisfies that

$$Y_{\mathbf{p}}^{\mathfrak{M}} \cap \mathrm{Fix}_{\Lambda_{\mathbf{p}}}(\alpha)^{\circ} = \{0\}, \tag{21.21}$$

and hence that

$$|Y_{\mathbf{p}}^{\mathfrak{M}}| \leq P_{\Lambda_{\mathbf{p}}}(\alpha). \tag{21.22}$$

PROOF. First we prove the lemma under the assumption that $\mathfrak{M} = \mathfrak{R}_d/\mathfrak{p}$ for a non-zero prime ideal $\mathfrak{p} \subset \mathfrak{R}_d$. If $\mathfrak{p} \cap \mathbb{Z} \neq \{0\}$, then (21.21) is trivially satisfied. If $\mathfrak{p} \cap \mathbb{Z} = \{0\}$ we realize $X^{\mathfrak{R}_d/\mathfrak{p}}$ as a closed, shift-invariant subgroup of $\mathbb{T}^{\mathbb{Z}^d}$, choose a non-zero polynomial $f \in \mathfrak{p}$, and let $\mathbf{p} \in \mathbb{Z}^d$ be a prime element satisfying (21.19) with $M = \deg(f) + 1$. Let $\mathbf{B} = \ell^{\infty}(\mathbb{Z}^d)$ be the Banach space of all bounded, complex valued functions on $\mathbb{Z}^d$, define the closed, shift-invariant subspace $\mathbf{S} \subset \mathbf{B}$ by (6.9), and let, for every prime element $\mathbf{p} \in \mathbb{Z}^d$, $\mathbf{S}_{\mathbf{p}} = \{z = (z_{\mathbf{n}}) \in \mathbf{S} : z_{\mathbf{n}} = z_{\mathbf{n}+\mathbf{m}} \text{ for all } \mathbf{n} \in \mathbb{Z}^d, \mathbf{m} \in \Lambda_{\mathbf{p}}\}$. Then $\mathbf{S}_{\mathbf{p}}$ is finite-dimensional. We denote by $\mathbf{S}_{\mathbf{p}}^{\mathbb{R}}$ the subspace of all real valued function in $\mathbf{S}_{\mathbf{p}}$, define a map $\zeta : \mathbf{S}_{\mathbf{p}}^{\mathbb{R}} \longmapsto \mathrm{Fix}_{\Lambda_{\mathbf{p}}}(\alpha^{\mathfrak{R}_d/\mathfrak{p}})$ by $(\zeta(z))_{\mathbf{m}} = z_{\mathbf{m}} \pmod 1$ for every $z \in \mathbf{S}_{\mathbf{p}}^{\mathbb{R}}$ and $\mathbf{m} \in \mathbb{Z}^d$, and claim that $\zeta(\mathbf{S}_{\mathbf{p}}^{\mathbb{R}}) = \mathrm{Fix}_{\Lambda_{\mathbf{p}}}(\alpha^{\mathfrak{R}_d/\mathfrak{p}})^{\circ}$.

In order to prove this assertion we denote by $\|s - t\| = \vartheta(s, t)$ the metric (17.7) on $\mathbb{T}$ and recall the proof of Lemma 21.2, which shows that there exists an $\varepsilon > 0$ such that every $x \in \mathrm{Fix}_{\Lambda_{\mathbf{p}}}(\alpha)$ with $\|x_{\mathbf{n}}\| < \varepsilon$ for every $\mathbf{n} \in \mathbb{Z}^d$, or equivalently, for every $\mathbf{n} \in Q_{\mathbf{p}} = \{(n_1, \ldots, n_d) \in \mathbb{Z}^d : 0 \leq n_i < p_i \text{ for } i = 1, \ldots, d\}$, lies in $\zeta(\mathbf{S}_{\mathbf{p}}^{\mathbb{R}})$. Hence the connected set $\zeta(\mathbf{S}_{\mathbf{p}}^{\mathbb{R}})$ contains an open subset of $\mathrm{Fix}_{\Lambda_{\mathbf{p}}}(\alpha^{\mathfrak{R}_d/\mathfrak{p}})$. Since ζ is a continuous group homomorphism with respect to addition on $\mathbf{S}_{\mathbf{p}}^{\mathbb{R}}$, this shows that $\zeta(\mathbf{S}_{\mathbf{p}}^{\mathbb{R}}) = \mathrm{Fix}_{\Lambda_{\mathbf{p}}}(\alpha^{\mathfrak{R}_d/\mathfrak{p}})^{\circ}$.

If $\mathbf{S}_{\mathbf{p}} = \{0\}$, the group $\mathrm{Fix}_{\Lambda}(\alpha)$ is discrete, and (21.21)–(21.22) are obvious. If $\mathbf{S}_{\mathbf{p}} \neq \{0\}$, then it is finite-dimensional, as mentioned above, and we define an inner product $[\cdot, \cdot]$ on $\mathbf{S}_{\mathbf{p}}$ by setting $[z, z'] = \sum_{\mathbf{n} \in Q_{\mathbf{p}}} z_{\mathbf{n}} \overline{z'_{\mathbf{n}}}$ for all $z, z' \in \mathbf{S}_{\mathbf{p}}$. For every $\mathbf{m} \in \mathbb{Z}^d$ we denote by $V_{\mathbf{m}}$ the restriction to $\mathbf{S}_{\mathbf{p}}$ of the shift operator $(U_{\mathbf{m}}z)_{\mathbf{n}} = z_{\mathbf{m}+\mathbf{n}}, z \in \mathbf{B}, \mathbf{n} \in \mathbb{Z}^d$, on $\mathbf{B}$. The map $\mathbf{m} \mapsto V_{\mathbf{m}}, \mathbf{m} \in \mathbb{Z}^d$, is a unitary representation of $\mathbb{Z}^d$ on the finite-dimensional Hilbert space $(\mathbf{S}_{\mathbf{p}}, [\cdot, \cdot])$, and elementary spectral theory allows us to choose an orthonormal basis $\{w_1, \ldots, w_r\}$

of $\mathbf{S_p}$ and characters $\mathbf{t}^{(1)}, \ldots, \mathbf{t}^{(r)} \in \mathbb{T}^d = \widehat{\mathbb{Z}^d}$, such that $V_{\mathbf{m}} w_j = \langle \mathbf{m}, \mathbf{t} \rangle w_j$ for every $\mathbf{m} \in \mathbb{Z}^d$ and $j = 1, \ldots, r$. Here $\langle \mathbf{m}, \mathbf{t} \rangle$ denotes, as usual, the value of the character $\mathbf{t} \in \mathbb{T}^d = \widehat{\mathbb{Z}^d}$ at an element $\mathbf{n} \in \mathbb{Z}^d$. We write $\mathbf{e}^{(i)}$ for the i-th unit vector in $\mathbb{Z}^d$, put $c^{(j)} = (c_1^{(j)}, \ldots, c_d^{(j)}) \in \mathbb{C}$ for every $j = 1, \ldots, k$, where $c_i^{(j)} = \langle \mathbf{e}^{(i)}, \mathbf{t}_j \rangle$ for $i = 1, \ldots, d$, and set $(c^{(j)})^{\mathbf{n}} = (c_1^{(j)})^{n_1} \cdot \ldots \cdot (c_d^{(j)})^{n_d}$ for every $\mathbf{n} = (n_1, \ldots, n_d) \in \mathbb{Z}^d$. Since $\sum_{\mathbf{m} \in \mathbb{Z}^d} c_{f_j}(\mathbf{m}) V_{\mathbf{m}} = 0$ for $j = 1, \ldots, k$, and $V_{\mathbf{m}}$ is the identity operator on $\mathbf{S_p}$ for every $\mathbf{m} \in \Lambda_{\mathbf{p}}$, we obtain that $\sum_{\mathbf{m} \in \mathbb{Z}^d} c_{f_j}(\mathbf{m})(c^{(j)})^{\mathbf{m}} = 0$ and $(c^{(j)})^{\mathbf{n}} = 1$ for every $j = 1, \ldots, k$ and $\mathbf{n} \in \Lambda_{\mathbf{p}}$. According to Lemma 21.6, $c^{(j)} \in V_{\mathbb{C}}(\mathfrak{p}) \cap V_{\mathbb{C}}(\mathfrak{b_p}) \subset V_{\mathbb{C}}(f) \cap V_{\mathbb{C}}(\mathfrak{b_p}) \subset \{(1, \ldots, 1)\}$ for $j = 1, \ldots, r$, so that every $z = (z_{\mathbf{n}}) \in \mathbf{S_p}$ is fixed under every $V_{\mathbf{m}}, \mathbf{m} \in \mathbb{Z}^d$. This shows that $\mathrm{Fix}_{\Lambda}(\alpha)^{\circ} = \mathrm{Fix}_{\mathbb{Z}^d}(\alpha)^{\circ}$.

Now consider an element $x = (x_{\mathbf{n}}) \in Y_{\mathbf{p}} \cap \mathrm{Fix}_{\Lambda}(\alpha)^{\circ} = Y_{\mathbf{p}} \cap \mathrm{Fix}_{\mathbb{Z}^d}(\alpha)^{\circ}$. Then there exists an element $t \in \mathbb{T}$ with $x_{\mathbf{n}} = t$ for every $\mathbf{n} \in \mathbb{Z}^d$. As $x \in Y_{\mathbf{p}}$, (5.7) implies that $\langle x, \phi_{p_i}(u_i) \rangle = 1$ for $i = 1, \ldots, d$, which is equivalent to saying that $p_i t = 0 \pmod 1$ for $i = 1, \ldots, d$. However, since the p_i are distinct primes, this implies that $t = \mathbf{0}_{\mathbb{T}}$ and $x = \mathbf{0}_X$, as claimed in (21.21), and (21.22) is obvious from (21.21).

Having established the lemma for $\mathfrak{R}_d$-modules of the form $\mathfrak{R}_d/\mathfrak{p}$, where $\mathfrak{p} \subset \mathfrak{R}_d$ is a non-zero prime ideal, we assume that $\mathfrak{M}$ is an arbitrary Noetherian torsion $\mathfrak{R}_d$-module with (non-zero) associated primes $\{\mathfrak{p}_1, \ldots, \mathfrak{p}_m\}$, set $X = X^{\mathfrak{M}}$ and $\alpha = \alpha^{\mathfrak{M}}$, use Corollary 6.2 to find a prime filtration $\mathfrak{M} = \mathfrak{N}_s \supset \cdots \supset \mathfrak{N}_0 = \{0\}$ with the properties described there, and put $Y_j = \mathfrak{N}_j^{\perp}$ for $j = 0, \ldots, s$. From Corollary 6.2 we know that, for $j = 1, \ldots, s$, $\mathfrak{N}_j/\mathfrak{N}_{j-1} \cong \mathfrak{R}_d/\mathfrak{q}_j$ for a prime ideal $\mathfrak{q}_j \subset \mathfrak{R}_d$ which contains one of the associated primes of $\mathfrak{M}$ and is therefore non-zero.

For each prime ideal $\mathfrak{p}_i$ associated with $\mathfrak{M}$ we choose a non-zero polynomial $f_i \in \mathfrak{p}_i$ and set $M = M(\mathfrak{p}_1, \ldots, \mathfrak{p}_m) = \max_{i=1,\ldots,m} \deg(f_i) + 1$ (cf. Lemma 21.6). According to (21.17)–(21.18),

$$
\left(\bigcup_{j=1}^{s} V_{\mathbb{C}}(\mathfrak{q}_j) \right) \cap V_{\mathbb{C}}(\mathfrak{b_p}) \subset \left(\bigcup_{i=1}^{m} V_{\mathbb{C}}(\mathfrak{p}_i) \right) \cap V_{\mathbb{C}}(\mathfrak{b_p})
$$
$$
\subset \left(\bigcup_{i=1}^{m} V_{\mathbb{C}}(f_i) \right) \cap V_{\mathbb{C}}(\mathfrak{b_p}) \subset \{(1, \ldots, 1)\} \tag{21.23}
$$

whenever $\mathbf{p} \in \mathbb{Z}^d$ is a prime element satisfying (21.19). We fix such a prime element $\mathbf{p} \in \mathbb{Z}^d$ and consider the quotient $Y_{\mathbf{p}}^{\mathfrak{M}}/Y_1 \subset X/Y_1 \cong X^{\mathfrak{R}_d/\mathfrak{q}_1}$. As $\widehat{Y_{\mathbf{p}}^{\mathfrak{M}}/Y_1} = \mathfrak{N}_1/\mathfrak{j_p} \cdot \mathfrak{M}$ is a quotient of $\mathfrak{N}_1/\mathfrak{j_p} \cdot \mathfrak{N}_1 \cong \mathfrak{R}_d/(\mathfrak{q}_1 + \mathfrak{j_p})$, the quotient map $\eta \colon X \longmapsto X^{\mathfrak{R}_d/\mathfrak{q}_1} \cong X/Y_1$ carries $Y_{\mathbf{p}}^{\mathfrak{M}}$ to a subgroup of $Y_{\mathbf{p}}^{\mathfrak{R}_d/\mathfrak{q}_1}$. Similarly we see that $\eta(\mathrm{Fix}_{\Lambda_{\mathbf{p}}}(\alpha)^{\circ})$ is a subgroup of $\mathrm{Fix}_{\Lambda_{\mathbf{p}}}(\alpha^{\mathfrak{R}_d/\mathfrak{q}_1})^{\circ}$. The inclusion (21.23) allows us to apply the first part of this proof to $\mathfrak{R}_d/\mathfrak{q}_1$. We conclude that

$(Y_{\mathbf{p}}^{\mathfrak{M}}/Y_1) \cap (\mathrm{Fix}_{\Lambda_{\mathbf{p}}}(\alpha)^\circ/Y_1) = \{0_{X/Y_1}\}$ or, equivalently, that

$$Y_{\mathbf{p}}^{\mathfrak{M}} \cap \mathrm{Fix}_{\Lambda_{\mathbf{p}}}(\alpha)^\circ \subset Y_1.$$

Hence

$$Y_{\mathbf{p}}^{\mathfrak{M}} \cap \mathrm{Fix}_{\Lambda_{\mathbf{p}}}(\alpha)^\circ \subset Y_{\mathbf{p}}^{\mathfrak{M}/\mathfrak{N}_1} \cap \mathrm{Fix}_{\Lambda_{\mathbf{p}}}(\alpha^{\mathfrak{M}/\mathfrak{N}_1})^\circ,$$

and by applying the above argument to the module $\mathfrak{M}/\mathfrak{N}_1$ instead of $\mathfrak{M}$ we find that

$$Y_{\mathbf{p}}^{\mathfrak{M}} \cap \mathrm{Fix}_{\Lambda_{\mathbf{p}}}(\alpha)^\circ \subset Y_{\mathbf{p}}^{\mathfrak{M}/\mathfrak{N}_1} \cap \mathrm{Fix}_{\Lambda_{\mathbf{p}}}(\alpha^{\mathfrak{M}/\mathfrak{N}_1})^\circ$$
$$\subset Y_{\mathbf{p}}^{\mathfrak{M}/\mathfrak{N}_2} \cap \mathrm{Fix}_{\Lambda_{\mathbf{p}}}(\alpha^{\mathfrak{M}/\mathfrak{N}_2})^\circ.$$

After repeating this argument s times we obtain that

$$Y_{\mathbf{p}}^{\mathfrak{M}} \cap \mathrm{Fix}_{\Lambda_{\mathbf{p}}}(\alpha)^\circ \subset Y_{\mathbf{p}}^{\mathfrak{M}/\mathfrak{N}_1} \cap \mathrm{Fix}_{\Lambda_{\mathbf{p}}}(\alpha^{\mathfrak{M}/\mathfrak{N}_1})^\circ$$

$$\vdots$$

$$\subset Y_{\mathbf{p}}^{\mathfrak{M}/\mathfrak{N}_{s-1}} \cap \mathrm{Fix}_{\Lambda_{\mathbf{p}}}(\alpha^{\mathfrak{M}/\mathfrak{N}_{s-1}})^\circ$$
$$\cong Y_{\mathbf{p}}^{\mathfrak{R}_d/\mathfrak{q}_s} \cap \mathrm{Fix}_{\Lambda_{\mathbf{p}}}(\alpha^{\mathfrak{R}_d/\mathfrak{q}_s})^\circ = \{0\}.$$

This proves (21.21) and hence (21.22). □

Let α be a $\mathbb{Z}^d$-action by automorphisms of a compact, abelian group X with finite entropy, and satisfying the d.c.c. If $\mathfrak{M} = \hat{X}$ is the Noetherian $\mathfrak{R}_d$-module arising from Lemma 5.1, then the Lemmas 21.3 and 21.7 show that

$$\limsup_{\mathbf{p}} \frac{1}{|\mathbb{Z}^d/\Lambda_{\mathbf{p}}|} \log |Y_{\mathbf{p}}^{\mathfrak{M}}| \leq p^+(\alpha) \leq h(\alpha), \qquad (21.24)$$

where $Y_{\mathbf{p}}^{\mathfrak{M}}$ is defined in (21.20), and where the $\limsup_{\mathbf{p}}$ is taken over all sequences of prime elements $(\mathbf{p}^{(n)} = (p_1^{(n)}, \ldots, p_d^{(n)}), n \geq 1)$ in $\mathbb{Z}^d$ with

$$\lim_{n\to\infty} p_1^{(n)} = \lim_{n\to\infty} p_j^{(n)} \Big/ \prod_{i=1}^{j-1} p_i^{(n)} = \infty \qquad (21.25)$$

for $j = 2, \ldots, d$. Our next task will be to prove that all the terms in (21.24) coincide.

LEMMA 21.8. *Let $k \geq 1$, and let $h = c_0 + c_1 t + \cdots + c_k t^k$ be a polynomial with $c_0 c_k \neq 0$, whose coefficients are complex algebraic numbers. Then*

$$\lim_{m\to\infty} \frac{1}{m} \sum_{\{k:0\leq k<m \text{ and } h(e^{2\pi i k/m})\neq 0\}} \log |h(e^{2\pi i k/m})| = \int_0^1 \log |h(e^{2\pi i t})|\, dt. \qquad (21.26)$$

PROOF. If $h(c) \neq 0$ for every $c \in \mathbb{S}$, then $\log|h|$ is continuous and hence integrable on $\mathbb{S}$, and the sum in (21.26) is a Riemann sum approximation to $\int_0^1 \log|h(e^{2\pi it})|\, dt$. This proves (21.26) if f does not vanish on $\mathbb{S}$.

If $h(c) = 0$ for some $c \in \mathbb{S}$, the convergence in (21.26) is much more subtle, and depends on a result in [24]: if $\zeta \in \mathbb{S}$ is an algebraic number, but not a root of unity, then there exists, for every $\varepsilon > 0$, a constant $C(\zeta, \varepsilon) > 0$ such that $|\zeta^k - 1| > C(\zeta, \varepsilon)e^{-\varepsilon k}$ for every $k \geq 1$. Suppose that $h(e^{2\pi it_0}) = 0$ for some irrational $t_0 \in \mathbb{R}$, and that the derivatives $h^{(j)}$ of h satisfy that $0 = h(e^{2\pi it_0}) = h^{(1)}(e^{2\pi it_0}) = \cdots = h^{(l-1)}(e^{2\pi it_0}) \neq h^{(l)}(e^{2\pi it_0})$ for some $l \geq 1$. Gelfond's result shows that there exists, for every $\varepsilon > 0$, a constant $C(t_0, \varepsilon)$ such that $\|t_0 - \frac{k}{m}\| > \frac{1}{m}C(t_0, \varepsilon)e^{-\varepsilon m}$ for every $m \geq 1$ and $k \in \mathbb{Z}$, where $\|\cdot\|$ is defined in (17.7). It follows that $|h(e^{2\pi ik_m/m})| > \frac{|h^{(l)}(e^{2\pi it_0})|}{2m^l l!}C(t_0,\varepsilon)^l e^{-l\varepsilon m}$ for all sufficiently large m, where k_m is the unique integer with $0 \leq k < m$ and $k_m/m \in [t_0 - \frac{1}{2m}, t_0 + \frac{1}{2m}]$. By applying this to each root of h in $\mathbb{S}$ which is not a root of unity, and by letting $m \to \infty$ and $\varepsilon \to 0$ (in this order), we obtain (21.26). $\square$

LEMMA 21.9. *Let $d \geq 1$, and let $\mathfrak{p} \subset \mathfrak{R}_d$ be a non-zero prime ideal. Then*

$$\limsup_{\mathfrak{p}} \frac{1}{|\mathbb{Z}^d/\Lambda_{\mathfrak{p}}|} \log |Y_{\mathfrak{p}}^{\mathfrak{R}_d/\mathfrak{p}}| = p^+(\alpha^{\mathfrak{R}_d/\mathfrak{p}}) = h(\alpha^{\mathfrak{R}_d/\mathfrak{p}}), \qquad (21.27)$$

where $\limsup_{\mathfrak{p}}$ is interpreted as in (21.25).

PROOF. If $\mathfrak{p}$ is not principal, or if $\mathfrak{p} = (f)$ for some generalized cyclotomic polynomial $f \in \mathfrak{R}_d$, then Lemma 21.3, Lemma 21.7, Corollary 18.5, and Theorem 19.5, show that

$$\limsup_{\mathfrak{p}} \frac{1}{|\mathbb{Z}^d/\Lambda_{\mathfrak{p}}|} \log |Y_{\mathfrak{p}}^{\mathfrak{R}_d/\mathfrak{p}}| = p^{\pm}(\alpha) = h(\alpha) = 0.$$

Assume therefore that $\mathfrak{p} = (f)$ for some non-zero polynomial $f \in \mathfrak{R}_d$ which is not generalized cyclotomic. We set $\alpha = \alpha^{\mathfrak{R}_d/(f)}$, choose a prime element $\mathbf{p} = (p_1, \ldots, p_d) \in \mathbb{Z}^d$ satisfying (21.17) and hence (21.18), and define $j_{\mathbf{p}} \subset \mathfrak{R}_d$ by (21.14). If $\mathfrak{N}_{\mathbf{p}} = \mathfrak{R}_d/j_{\mathbf{p}}$ then

$$|Y_{\mathfrak{p}}^{\mathfrak{R}_d/\mathfrak{p}}| = |\mathfrak{R}_d/((f) + j_{\mathbf{p}})| = |\mathfrak{N}_{\mathbf{p}}/f \cdot \mathfrak{N}_{\mathbf{p}}|. \qquad (21.28)$$

In order to calculate $|\mathfrak{N}_{\mathbf{p}}/f \cdot \mathfrak{N}_{\mathbf{p}}|$ in (21.28) we choose and fix an element $\omega = (\omega_1, \ldots, \omega_d) \in V_{\mathbb{C}}(j_{\mathbf{p}})$. Since $V = V_{\mathbb{C}}(j_{\mathbf{p}})$ is equal to the orbit of ω in $\overline{\mathbb{Q}}^d \subset \mathbb{C}^d$ under the action of the Galois group, and since $j_{\mathbf{p}}$ is radical by Lemma 21.5, the evaluation map $\eta_\omega \colon g \mapsto g(\omega)$, $g \in \mathfrak{R}_d$, has kernel $j_{\mathbf{p}}$. Every $c = (c_1, \ldots, c_d) \in V$ is of the form $c = (\omega_1^{k_1(c)}, \ldots, \omega_d^{k_d(c)})$ for a unique element $\mathbf{k}(c) = (k_1(c), \ldots, k_d(c)) \in Q_{\mathbf{p}} = \prod_{i=1}^d \{1, \ldots, p_i - 1\} \subset \mathbb{Z}^d$, and we define a $\mathbb{Z}$-module isomorphism $\psi \colon \mathbb{Z}^V \longmapsto \mathfrak{N}_{\mathbf{p}}$ by setting, for every $w \colon V \longmapsto \mathbb{Z}$, $\psi(w) = \sum_{c \in V} w(c)u^{\mathbf{k}(c)} + j_{\mathbf{p}}$. Let $\beta_f \colon \mathfrak{N}_{\mathbf{p}} \longmapsto \mathfrak{N}_{\mathbf{p}}$ be multiplication by f, and let $\beta_f' = \psi \cdot \beta_f \cdot \psi^{-1} \colon \mathbb{Z}^V \longmapsto \mathbb{Z}^V$. We embed $\mathbb{Z}^V$ linearly in $\mathbb{C}^V$ and consider the linear map $\beta_f'' : \mathbb{C}^V \longmapsto \mathbb{C}^V$ induced by β_f'. For every $c \in V$, the indicator function $1_{\{c\}} \in \mathbb{C}^V$ is an eigenvector of β_f'' with eigenvalue $f(c)$, and the set of all such eigenvectors is a basis of $\mathbb{C}^V$. According to (21.18), $f(c) \neq 0$ for every $c \in V$, so that β_f'' is non-singular, and

$$|Y_{\mathbf{p}}^{\mathfrak{R}_d/\mathfrak{p}}| = |\mathfrak{N}_{\mathbf{p}}/f \cdot \mathfrak{N}_{\mathbf{p}}| = |\mathbb{Z}^V / \beta_f' \mathbb{Z}^V| = |\det \beta_f''| = \prod_{c \in V} |f(c)| \quad (21.29)$$

by (21.28). If we combine (21.29) with (21.24) we see that

$$\frac{1}{|\mathbb{Z}^d / \Lambda_{\mathbf{p}}|} \log P_{\Lambda_{\mathbf{p}}}(\alpha) \geq \frac{1}{|\mathbb{Z}^d / \Lambda_{\mathbf{p}}|} \log |Y_{\mathbf{p}}^{\mathfrak{R}_d/\mathfrak{p}}|$$

$$= \frac{1}{|\mathbb{Z}^d / \Lambda_{\mathbf{p}}|} \sum_{c \in V_{\mathbb{C}}(\mathfrak{j}_{\mathbf{p}})} \log |f(c)| \quad (21.30)$$

$$= \frac{1}{p_1 \cdots p_d} \sum_{(k_1, \ldots, k_d) \in Q_{\mathbf{p}}} \log |f(\omega_1^{k_1}, \ldots, \omega_d^{k_d})|$$

for every prime element $\mathbf{p} = (p_1, \ldots, p_d) \in \mathbb{Z}^d$ satisfying (21.17). We fix the rational primes $p_1, \ldots, p_{d-1}$ and consider the polynomial

$$F(t) = \prod_{k_1=1}^{p_1-1} \cdots \prod_{k_{d-1}=1}^{p_{d-1}-1} f(\omega_1^{k_1}, \ldots, \omega_{d-1}^{k_{d-1}}, t).$$

According to Lemma 21.8,

$$\lim_{m \to \infty} \frac{1}{m} \sum_{\{k : 0 \leq k < m \text{ and } F(e^{2\pi ik/m}) \neq 0\}} \log |F(e^{2\pi ik/m})|$$

$$= \lim_{p_d \to \infty} \frac{1}{p_d} \sum_{k=1}^{p_d-1} \log |F(e^{2\pi ik/p_d})| \quad (21.31)$$

$$= \int_0^1 \log |F(e^{2\pi it})| \, dt,$$

where the second limit in (21.31) is taken over the sequence of rational primes $p_d \to \infty$.

For every $\mathbf{s} \in \mathbb{S}^{d-1}$ we set $f_{\mathbf{s}}(t) = f(s_1, \ldots, s_{d-1}, t)$. According to Proposition 16.1 and Lemma 17.1, applied to the companion matrix of $f_{\mathbf{s}}$, $\mathbf{s} \in \mathbb{S}^{d-1}$, the map

$$\mathbf{s} = (s_1, \ldots, s_{d-1}) \mapsto \int_0^1 \log |f_{\mathbf{s}}(e^{2\pi it})| \, dt = \int_{\mathbb{S}} \log |f(s_1, \ldots, s_{d-1}, s_d)| \, d\lambda_{\mathbb{S}}(s_d)$$

from $\mathbb{S}^{d-1}$ to $\mathbb{R}$ is continuous. By (21.31) and (21.17)–(21.18) we can find, for every $\varepsilon > 0$, an integer $M > \deg(f)$ such that

$$\left| \left(\lim_{p_d \to \infty} \frac{1}{p_1 \cdots p_d} \sum_{c \in V_{\mathbb{C}}(\mathfrak{j}_{(p_1, \ldots, p_d)})} \log |f(c)| \right) - \log \mathbb{M}(f) \right| < \varepsilon \quad (21.32)$$

whenever $(p_1, \ldots, p_{d-1}) \in \mathbb{Z}^{d-1}$ is prime with $p_1 > M$ and $p_j > M \prod_{i=1}^{j-1} p_i$ for $j = 2, \ldots, d-1$, where the limit in (21.32) is taken over the sequence of primes $p_d \to \infty$. From (21.30)–(21.32) and Theorem 18.1 we conclude that

$$\log M(f) = \limsup_{\mathbf{p}} \frac{1}{|\mathbb{Z}^d/\Lambda_{\mathbf{p}}|} \log |Y_{\mathbf{p}}^{\mathfrak{R}_d/\mathfrak{p}}| \le p^+(\alpha) \le h(\alpha) = \log \mathrm{M}(f),$$

which proves (21.27). $\square$

In order to extend (21.27) to an arbitrary Noetherian torsion $\mathfrak{R}_d$-module $\mathfrak{M}$ we have to investigate how the quantity $|\mathfrak{M}/j_{\mathbf{p}} \cdot \mathfrak{M}|$ is related to the corresponding quantity for the modules $\mathfrak{R}_d/\mathfrak{q}_j$ occurring in a prime filtration of $\mathfrak{M}$. We shall derive some of the necessary results in a slightly more general form than is currently required in order to avoid duplication at a later stage.

LEMMA 21.10. *Assume that* $0 \ne f \in \mathfrak{R}_d$ *is an irreducible polynomial and* $\mathfrak{a} \subset \mathfrak{R}_d$ *an ideal with the following properties:*

$$V_{\mathbb{C}}(\mathfrak{a}) \ne \varnothing, \quad V_{\mathbb{C}}(\mathfrak{a}) \cap V_{\mathbb{C}}(f) = \varnothing,$$
$$\mathfrak{a} = \sqrt{\mathfrak{a}} = \{h \in \mathfrak{R}_d : h(c) = 0 \text{ for every } c \in V_{\mathbb{C}}(\mathfrak{a})\}. \tag{21.33}$$

If $\mathfrak{N}$ *is a Noetherian* $\mathfrak{R}_d$-*module with a prime filtration* $\mathfrak{N} = \mathfrak{N}_s \supset \cdots \supset \mathfrak{N}_0 = \{0\}$ *such that* $\mathfrak{N}_j/\mathfrak{N}_{j-1} \cong \mathfrak{R}_d/(f)$ *for* $j = 1, \ldots, s$, *then* $\mathfrak{a} \cdot \mathfrak{N} \cap \mathfrak{N}_j = \mathfrak{a} \cdot \mathfrak{N}_j$ *and*

$$\mathfrak{N}_j/(\mathfrak{N}_{j-1} + \mathfrak{a} \cdot \mathfrak{N}_j) \cong \mathfrak{N}_j/(\mathfrak{N}_{j-1} + \mathfrak{a} \cdot \mathfrak{N}) \tag{21.34}$$

for $j = 1, \ldots, s$.

PROOF. First we prove that $(f) \cap \mathfrak{a} = f \cdot \mathfrak{a}$. Indeed, if there exists an element $h \in ((f) \cap \mathfrak{a}) \smallsetminus f \cdot \mathfrak{a}$, then $h = gf$ for some $g \in \mathfrak{R}_d \smallsetminus \mathfrak{a}$. Since $\mathfrak{a} = \{h \in \mathfrak{R}_d : h(c) = 0$ for every $c \in V_{\mathbb{C}}(\mathfrak{a})\}$ we know that $V_{\mathbb{C}}(f) \cup V_{\mathbb{C}}(g) \supset V_{\mathbb{C}}(\mathfrak{a})$. However, $V_{\mathbb{C}}(f) \cap V_{\mathbb{C}}(\mathfrak{a}) = \varnothing$ by assumption, so that $V_{\mathbb{C}}(g) \supset V_{\mathbb{C}}(\mathfrak{a})$ and $g \in \mathfrak{a}$. This contradiction implies that $(f) \cap \mathfrak{a} = f \cdot \mathfrak{a}$, as claimed.

Now suppose that there exists a $j \in \{1, \ldots, s\}$ and an $a \in (\mathfrak{N}_j \cap \mathfrak{a} \cdot \mathfrak{N}) \smallsetminus \mathfrak{a} \cdot \mathfrak{N}_j$. We choose the smallest $k > j$ such that $a \in \mathfrak{a} \cdot \mathfrak{N}_k$ and note that there exist $b_1, \ldots, b_n$ in $\mathfrak{N}_k$ and $h_1, \ldots, h_n$ in $\mathfrak{a}$ such that $a = \sum_{i=1}^n h_i \cdot b_i$, and $\{b_1, \ldots, b_n\} \not\subset \mathfrak{N}_{k-1}$ due to the minimality of k. Choose $c \in \mathfrak{N}_k$ such that the map $h \mapsto h \cdot c + \mathfrak{N}_{k-1}$, $h \in \mathfrak{R}_d$, induces an isomorphism of $\mathfrak{R}_d/(f)$ with $\mathfrak{N}_k/\mathfrak{N}_{k-1}$. Then there exist $g_i \in \mathfrak{R}_d$ with $b_i' = b_i - g_i \cdot c \in \mathfrak{N}_{k-1}$ for $i = 1, \ldots, n$. Since $\{b_1, \ldots, b_n\} \not\subset \mathfrak{N}_{k-1}$ we know that $\{g_1, \ldots, g_n\} \not\subset (f)$, whereas

$$\sum_{i=1}^n h_i g_i \cdot c = \sum_{i=1}^n h_i \cdot (b_i - b_i') \in \mathfrak{N}_{k-1},$$

since $a = \sum_{i=1}^n h_i \cdot b_i \in \mathfrak{N}_j \subset \mathfrak{N}_{k-1}$. Hence

$$\sum_{i=1}^n h_i g_i \in \mathfrak{a} \cap (f) = \mathfrak{a} \cdot (f),$$

and there exist $p_1, \ldots, p_m$ in (f) and $q_1, \ldots, q_m$ in $\mathfrak{a}$ such that $\sum_{i=1}^n h_i g_i = \sum_{l=1}^m p_l q_l$ and

$$a = \sum_{l=1}^m p_l q_l \cdot c + \sum_{i=1}^n h_i \cdot b_i' \in \mathfrak{a} \cdot \mathfrak{N}_{k-1},$$

which contradicts the minimality of k. We conclude that $\mathfrak{N}_j \cap \mathfrak{a} \cdot \mathfrak{N} = \mathfrak{a} \cdot \mathfrak{N}_j$ for every $j = 1, \ldots, s$, which implies (21.34). $\square$

LEMMA 21.11. *Let $\mathfrak{p} \subset \mathfrak{R}_d$ be a non-zero prime ideal, and let $h \in \mathfrak{R}_d \smallsetminus \mathfrak{p}$. If $\mathfrak{M}, \mathfrak{N}$ are $\mathfrak{R}_d$-modules associated with $\mathfrak{p}$, and if $h \cdot \mathfrak{N} \subset \mathfrak{M} \subset \mathfrak{N}$, then*

$$h(\alpha^{\mathfrak{M}}) = h(\alpha^{\mathfrak{N}}),$$

$$\limsup_{\mathbf{p}} \frac{1}{|\mathbb{Z}^d / \Lambda_{\mathbf{p}}|} \log |Y_{\mathbf{p}}^{\mathfrak{M}}| = \limsup_{\mathbf{p}} \frac{1}{|\mathbb{Z}^d / \Lambda_{\mathbf{p}}|} \log |Y_{\mathbf{p}}^{\mathfrak{N}}|,$$

$$\limsup_{\substack{\langle \Lambda \rangle \to \infty \\ \Omega(\Lambda) \cap V_{\mathbb{C}}(\mathfrak{p}) = \varnothing}} \frac{1}{|\mathbb{Z}^d / \Lambda|} |\mathfrak{M}/\mathfrak{b}(\Lambda) \cdot \mathfrak{M}| = \limsup_{\substack{\langle \Lambda \rangle \to \infty \\ \Omega(\Lambda) \cap V_{\mathbb{C}}(\mathfrak{p}) = \varnothing}} \frac{1}{|\mathbb{Z}^d / \Lambda|} |\mathfrak{N}/\mathfrak{b}(\Lambda) \cdot \mathfrak{N}|, \qquad (21.35)$$

$$\liminf_{\substack{\langle \Lambda \rangle \to \infty \\ \Omega(\Lambda) \cap V_{\mathbb{C}}(\mathfrak{p}) = \varnothing}} \frac{1}{|\mathbb{Z}^d / \Lambda|} |\mathfrak{M}/\mathfrak{b}(\Lambda) \cdot \mathfrak{M}| = \liminf_{\substack{\langle \Lambda \rangle \to \infty \\ \Omega(\Lambda) \cap V_{\mathbb{C}}(\mathfrak{p}) = \varnothing}} \frac{1}{|\mathbb{Z}^d / \Lambda|} |\mathfrak{N}/\mathfrak{b}(\Lambda) \cdot \mathfrak{N}|,$$

where $\limsup_{\mathbf{p}}$ is interpreted as in (21.25). Note that the last two equations in (21.35) may be vacuous, since there need not exist any subgroups $\Lambda \subset \mathbb{Z}^d$ with finite index such that $\Omega(\Lambda) \cap V_{\mathbb{C}}(\mathfrak{p}) = \varnothing$.

PROOF. Lemma 21.7 implies the existence of an integer $M = M(\mathfrak{p}) \geq 1$ such that $|Y_{\mathbf{p}}^{\mathfrak{K}}| = |\mathfrak{K}/\mathfrak{j}_{\mathbf{p}} \cdot \mathfrak{K}| < \infty$ for every prime element $\mathbf{p} \in \mathbb{Z}^d$ satisfying (21.19), and every Noetherian $\mathfrak{R}_d$-module $\mathfrak{K}$ associated with $\mathfrak{p}$. Similarly, if $\Lambda \subset \mathbb{Z}^d$ is a subgroup with finite index such that $\Omega(\Lambda) \cap V_{\mathbb{C}}(\mathfrak{p}) = \varnothing$, then Theorem 6.5 (3) implies that $P_\Lambda(\alpha^{\mathfrak{K}}) = |\mathfrak{K}/\mathfrak{b}(\Lambda) \cdot \mathfrak{K}|$ for every $\mathfrak{R}_d$-module $\mathfrak{K}$ associated with $\mathfrak{p}$. Let $\mathfrak{K}, \mathfrak{L}$ be Noetherian $\mathfrak{R}_d$-modules associated with $\mathfrak{p}$ such that $h \cdot \mathfrak{K} \subset \mathfrak{L} \subset \mathfrak{K}$. For every prime element $\mathbf{p} \in \mathbb{Z}^d$ and every subgroup $\Lambda \subset \mathbb{Z}^d$ of finite index with the properties just described, we have that

$$|\mathfrak{K}/\mathfrak{j}_{\mathbf{p}} \cdot \mathfrak{K}| = |\mathfrak{K}/(\mathfrak{L} + \mathfrak{j}_{\mathbf{p}} \cdot \mathfrak{K})| \cdot |(\mathfrak{L} + \mathfrak{j}_{\mathbf{p}} \cdot \mathfrak{K})/\mathfrak{j}_{\mathbf{p}} \cdot \mathfrak{K}|$$
$$= |(\mathfrak{K}/\mathfrak{L})/\mathfrak{j}_{\mathbf{p}} \cdot (\mathfrak{K}/\mathfrak{L})| \cdot |\mathfrak{L}/(\mathfrak{L} \cap \mathfrak{j}_{\mathbf{p}} \cdot \mathfrak{K}|$$
$$\leq |(\mathfrak{K}/\mathfrak{L})/\mathfrak{j}_{\mathbf{p}} \cdot (\mathfrak{K}/\mathfrak{L})| \cdot |\mathfrak{L}/\mathfrak{j}_{\mathbf{p}} \cdot \mathfrak{L}|,$$
$$|P_\Lambda(\alpha^{\mathfrak{K}})| = |\mathfrak{K}/\mathfrak{b}(\Lambda) \cdot \mathfrak{K}| \qquad (21.36)$$
$$= |\mathfrak{K}/(\mathfrak{L} + \mathfrak{b}(\Lambda) \cdot \mathfrak{K})| \cdot |(\mathfrak{L} + \mathfrak{b}(\Lambda) \cdot \mathfrak{K})/\mathfrak{b}(\Lambda) \cdot \mathfrak{K}|$$
$$= |(\mathfrak{K}/\mathfrak{L})/\mathfrak{b}(\Lambda) \cdot (\mathfrak{K}/\mathfrak{L})| \cdot |\mathfrak{L}/(\mathfrak{L} \cap \mathfrak{b}(\Lambda) \cdot \mathfrak{K}|$$
$$\leq P_\Lambda(\alpha^{\mathfrak{K}/\mathfrak{L}}) P_\Lambda(\alpha^{\mathfrak{L}}).$$

In (21.36) we have used the fact that $\mathfrak{j}_{\mathbf{p}} \cdot \mathfrak{L} \subset \mathfrak{L} \cap \mathfrak{j}_{\mathbf{p}} \cdot \mathfrak{K}$ and $\mathfrak{b}(\Lambda) \cdot \mathfrak{L} \subset \mathfrak{L} \cap \mathfrak{b}(\Lambda) \cdot \mathfrak{K}$. Since every prime ideal associated with $\mathfrak{K}/\mathfrak{L}$ must contain $\mathfrak{p} + h\mathfrak{R}_d$, none of the prime ideals associated with $\mathfrak{K}/\mathfrak{L}$ is principal. By Corollary 18.5, Proposition 18.6, and Lemma 21.3,

$$\limsup_{\mathbf{p}} \frac{1}{|\mathbb{Z}^d/\Lambda_{\mathbf{p}}|} \log |(\mathfrak{K}/\mathfrak{L})/j_{\mathbf{p}} \cdot (\mathfrak{K}/\mathfrak{L})| \leq p^+(\alpha^{\mathfrak{K}/\mathfrak{L}}) \leq h(\alpha^{\mathfrak{K}/\mathfrak{L}}) = 0,$$

and (21.36) and the addition formula (14.1) yield that

$$\limsup_{\mathbf{p}} \frac{1}{|\mathbb{Z}^d/\Lambda_{\mathbf{p}}|} \log |\mathfrak{K}/j_{\mathbf{p}} \cdot \mathfrak{K}|$$
$$\leq \limsup_{\mathbf{p}} \frac{1}{|\mathbb{Z}^d/\Lambda_{\mathbf{p}}|} \log |\mathfrak{L}/j_{\mathbf{p}} \cdot \mathfrak{L}|,$$
$$\lim_{\substack{\langle\Lambda\rangle\to\infty \\ \Omega(\Lambda)\cap V_{\mathbb{C}}(\mathfrak{p})=\varnothing}} \frac{1}{|\mathbb{Z}^d/\Lambda|} \log |\mathfrak{K}/\mathfrak{b}(\Lambda) \cdot \mathfrak{K}| \tag{21.37}$$
$$\leq \lim_{\substack{\langle\Lambda\rangle\to\infty \\ \Omega(\Lambda)\cap V_{\mathbb{C}}(\mathfrak{p})=\varnothing}} \frac{1}{|\mathbb{Z}^d/\Lambda|} \log |\mathfrak{L}/\mathfrak{b}(\Lambda) \cdot \mathfrak{L}|,$$
$$h(\alpha^{\mathfrak{K}}) = h(\alpha^{\mathfrak{L}}).$$

Now assume that $\mathfrak{L} = f \cdot \mathfrak{K}$. The map $\mathfrak{K} \longmapsto \mathfrak{L}/j_{\mathbf{p}} \cdot \mathfrak{L}$ consisting of multiplication by f, followed by the quotient map, is surjective, and its kernel contains $j_{\mathbf{p}} \cdot \mathfrak{K}$. Hence $|\mathfrak{L}/j_{\mathbf{p}} \cdot \mathfrak{L}| \leq |\mathfrak{K}/j_{\mathbf{p}} \cdot \mathfrak{K}|$, and similarly we see that $|\mathfrak{L}/\mathfrak{b}(\Lambda) \cdot \mathfrak{L}| \leq |\mathfrak{K}/\mathfrak{b}(\Lambda) \cdot \mathfrak{K}|$. According to (21.36),

$$\limsup_{\mathbf{p}} \frac{1}{|\mathbb{Z}^d/\Lambda_{\mathbf{p}}|} (\log |\mathfrak{K}/j_{\mathbf{p}} \cdot \mathfrak{K}| - \log |\mathfrak{L}/j_{\mathbf{p}} \cdot \mathfrak{L}|) = 0,$$
$$\lim_{\substack{\langle\Lambda\rangle\to\infty \\ \Omega(\Lambda)\cap V_{\mathbb{C}}(\mathfrak{p})=\varnothing}} \frac{1}{|\mathbb{Z}^d/\Lambda|} (\log |\mathfrak{K}/\mathfrak{b}(\Lambda) \cdot \mathfrak{K}| - \log |\mathfrak{L}/\mathfrak{b}(\Lambda) \cdot \mathfrak{L}|) = 0,$$

so that

$$\limsup_{\mathbf{p}} \frac{1}{|\mathbb{Z}^d/\Lambda_{\mathbf{p}}|} \log |Y_{\mathbf{p}}^{\mathfrak{K}}| = \limsup_{\mathbf{p}} \frac{1}{|\mathbb{Z}^d/\Lambda_{\mathbf{p}}|} \log |Y_{\mathbf{p}}^{f \cdot \mathfrak{K}}|,$$
$$\lim_{\substack{\langle\Lambda\rangle\to\infty \\ \Omega(\Lambda)\cap V_{\mathbb{C}}(\mathfrak{p})=\varnothing}} \frac{1}{|\mathbb{Z}^d/\Lambda|} \log |\mathfrak{K}/\mathfrak{b}(\Lambda) \cdot \mathfrak{K}| \tag{21.38}$$
$$= \lim_{\substack{\langle\Lambda\rangle\to\infty \\ \Omega(\Lambda)\cap V_{\mathbb{C}}(\mathfrak{p})=\varnothing}} \frac{1}{|\mathbb{Z}^d/\Lambda|} \log |f \cdot \mathfrak{K}/\mathfrak{b}(\Lambda) \cdot f \cdot \mathfrak{K}|,$$
$$p^\pm(\alpha^{\mathfrak{K}}) = p^\pm(\alpha^{f \cdot \mathfrak{K}}).$$

We apply (21.36)–(21.38) first with $\mathfrak{L} = \mathfrak{N}$ and $\mathfrak{K} = \mathfrak{M}$, and then with $\mathfrak{L} = f \cdot \mathfrak{M}$ and $\mathfrak{K} = \mathfrak{N}$, and obtain (21.35). $\square$

LEMMA 21.12. *Let* $\mathfrak{p} \subset \mathfrak{R}_d$ *be a non-zero prime ideal, and let* $\mathfrak{M}$ *be a Noetherian* $\mathfrak{R}_d$-*module associated with* $\mathfrak{p}$. *Then*

$$\limsup_{\mathbf{p}} \frac{1}{|\mathbb{Z}^d/\Lambda_{\mathbf{p}}|} \log |Y_{\mathbf{p}}^{\mathfrak{M}}| = p^+(\alpha^{\mathfrak{M}}) = h(\alpha^{\mathfrak{M}}),$$

where $\limsup_{\mathbf{p}}$ *is interpreted as in* (21.25).

PROOF. According to Proposition 6.1 there exist integers $1 \leq t \leq s$ and submodules $\mathfrak{M} = \mathfrak{N}_s \supset \cdots \supset \mathfrak{N}_0 = \{0\}$ such that, for every $i = 1, \ldots, s$, $\mathfrak{N}_i/\mathfrak{N}_{i-1} \cong \mathfrak{R}_d/\mathfrak{q}_i$ for some prime ideal $\mathfrak{p} \subset \mathfrak{q}_i \subset \mathfrak{R}_d$, $\mathfrak{q}_i = \mathfrak{p}$ for $i = 1, \ldots, t$, and $\mathfrak{q}_i \supsetneq \mathfrak{p}$ for $i = t+1, \ldots, s$. For each $i = t+1, \ldots, s$, choose a $g_i \in \mathfrak{q}_i \smallsetminus \mathfrak{p}$, and set $h = g_{t+1} \cdot \ldots \cdot g_s$. Then $h \notin \mathfrak{p}$, and $h \cdot \mathfrak{M} \subset \mathfrak{N}_t \subset \mathfrak{M}$. Lemma 21.11 shows that $\limsup_{\mathbf{p}} \frac{1}{|\mathbb{Z}^d/\Lambda_{\mathbf{p}}|} \log |Y_{\mathbf{p}}^{\mathfrak{M}}| = \limsup_{\mathbf{p}} \frac{1}{|\mathbb{Z}^d/\Lambda_{\mathbf{p}}|} \log |Y_{\mathbf{p}}^{\mathfrak{N}_t}|$, $p^{\pm}(\alpha^{\mathfrak{M}}) = p^{\pm}(\alpha^{\mathfrak{N}_t})$, and $h(\alpha^{\mathfrak{M}}) = h(\alpha^{\mathfrak{N}_t})$, so that we may assume without loss in generality that $s = t$, and that $\mathfrak{q}_i = \mathfrak{p}$ for all $i = 1, \ldots, s$.

If $\mathfrak{p}$ is not principal, then Corollary 18.5 and Proposition 18.6 imply that $h(\alpha^{\mathfrak{M}}) = 0$, and the Lemmas 21.3 and 21.7 yield that $\limsup_{\mathbf{p}} \frac{1}{|\mathbb{Z}^d/\Lambda_{\mathbf{p}}|} \log |Y_{\mathbf{p}}^{\mathfrak{M}}| = p^+(\alpha^{\mathfrak{M}}) = 0$. If $\mathfrak{p} = (f)$ for a non-zero irreducible polynomial $f \in \mathfrak{R}_d$, we apply the Lemmas 21.6 and 21.10 to find a constant $M \geq 1$ such that, for every prime element $\mathbf{p} = (p_1, \ldots, p_d) \in \mathbb{Z}^d$ satisfying (21.19), and for every $j = 1, \ldots, s$, $\mathfrak{N}_j \cap \mathfrak{j}_{\mathbf{p}} \cdot \mathfrak{M} = \mathfrak{j}_{\mathbf{p}} \cdot \mathfrak{N}_j$ and hence $|\mathfrak{N}_j/(\mathfrak{N}_{j-1} + \mathfrak{j}_{\mathbf{p}} \cdot \mathfrak{M})| = |\mathfrak{N}_j/(\mathfrak{N}_{j-1} + \mathfrak{j}_{\mathbf{p}} \cdot \mathfrak{N}_j)|$. It follows that

$$\begin{aligned}
|\mathfrak{M}/\mathfrak{j}_{\mathbf{p}} \cdot \mathfrak{M}| &= \prod_{j=1}^{s} |\mathfrak{N}_j/(\mathfrak{N}_{j-1} + \mathfrak{j}_{\mathbf{p}} \cdot \mathfrak{M})| \\
&= \prod_{j=1}^{s} |\mathfrak{N}_j/(\mathfrak{N}_{j-1} + \mathfrak{j}_{\mathbf{p}} \cdot \mathfrak{N}_j)| = |\mathfrak{R}_d/(\mathfrak{p} + \mathfrak{j}_{\mathbf{p}})|^s.
\end{aligned} \tag{21.39}$$

From Proposition 18.6 we know that $h(\alpha) = s h(\alpha^{\mathfrak{R}_p/\mathfrak{p}})$, and (21.39) is equivalent to the statement that $|\mathfrak{M}/\mathfrak{j}_{\mathbf{p}} \cdot \mathfrak{M}| = |\mathfrak{R}_d/(\mathfrak{p} + \mathfrak{j}_{\mathbf{p}})|^s$ whenever the prime element $\mathbf{p} \in \mathbb{Z}^d$ satisfies (21.19). An application of Lemma 21.9 completes the proof. $\square$

LEMMA 21.13. *Let* $\mathfrak{M}$ *be a Noetherian torsion* $\mathfrak{R}_d$-*module. Then*

$$\limsup_{\mathbf{p}} \frac{1}{|\mathbb{Z}^d/\Lambda_{\mathbf{p}}|} \log |Y_{\mathbf{p}}^{\mathfrak{M}}| = p^+(\alpha^{\mathfrak{M}}) = h(\alpha^{\mathfrak{M}}).$$

PROOF. Let $\{\mathfrak{W}_1, \ldots, \mathfrak{W}_m\}$ be a reduced primary decomposition of $\mathfrak{M}$ with associated (non-zero) primes $\{\mathfrak{p}_1, \ldots, \mathfrak{p}_m\}$ (cf. (6.5)). Then the diagonal map $\theta \colon a \mapsto (a + \mathfrak{W}_1, \ldots, a + \mathfrak{W}_m)$ from $\mathfrak{M}$ to $\bigoplus_{i=1}^{m} \mathfrak{M}/\mathfrak{W}_i = \mathfrak{K}$ is injective, and the dual map $\hat{\theta} \colon V = X^{\mathfrak{K}} = \bigoplus_{i=1}^{m} \widehat{\mathfrak{M}/\mathfrak{W}_i} \longmapsto X$ is surjective. Let $Z = \ker(\hat{\theta}) \subset V$

be the kernel of $\hat{\theta}$. The addition formula (14.1) shows that

$$\sum_{i=1}^{m} h(\alpha^{\mathfrak{M}/\mathfrak{W}_i}) = h(\alpha^{\mathfrak{K}}) = h(\alpha) + h(\alpha^Z),$$

where α^Z is the restriction of $\alpha^{\mathfrak{K}}$ to $Z \subset V = X^{\mathfrak{K}}$.

We claim that $h(\alpha^Z) = 0$. If $\mathfrak{N}$ is an $\mathfrak{R}_d$-module, and if $\mathfrak{N}_1, \ldots, \mathfrak{N}_k$ are submodules of $\mathfrak{N}$ such that $\mathfrak{N} = \mathfrak{N}_1 + \cdots + \mathfrak{N}_k$, then $\mathfrak{N}$ is a homomorphic image of $\bigoplus_{i=1}^{k} \mathfrak{N}_i$, where the homomorphism is given by addition, and hence $\sum_{j=1}^{k} h(\alpha^{\mathfrak{N}_i}) = h(\alpha^{\bigoplus_{i=1}^{k} \mathfrak{N}_i}) \geq h(\alpha^{\mathfrak{N}})$. In order to apply this observation to α^Z we regard the i-th summand $\mathfrak{M}/\mathfrak{W}_i$ of $\mathfrak{K}$ as a subgroup of $\mathfrak{K}$ and note that $Z = \widehat{\mathfrak{K}/\theta(\mathfrak{M})}$ and $\mathfrak{K}/\theta(\mathfrak{M}) = (\mathfrak{M}/\mathfrak{W}_1)/\theta(\mathfrak{M}) + \cdots + (\mathfrak{M}/\mathfrak{W}_m)/\theta(\mathfrak{M})$. If we can prove that $h(\alpha^{(\mathfrak{M}/\mathfrak{W}_i)/\theta(\mathfrak{M})}) = 0$ for $i = 1, \ldots, m$, then $h(\alpha^{\mathfrak{K}/\theta(\mathfrak{M})}) = 0$ by the above argument. Fix $i \in \{1, \ldots, m\}$, and consider an element

$$(\mathfrak{W}_1, \ldots, \mathfrak{W}_{i-1}, a + \mathfrak{W}_i, \mathfrak{W}_{i+1}, \ldots, \mathfrak{W}_m) + \theta(\mathfrak{M})$$
$$= (-a + \mathfrak{W}_1, \ldots, -a + \mathfrak{W}_{i-1}, \mathfrak{W}_i, -a + \mathfrak{W}_{i+1}, \ldots, -a + \mathfrak{W}_m) + \theta(\mathfrak{M})$$

in $(\mathfrak{M}/\mathfrak{W}_i)/\theta(\mathfrak{M}) \subset \mathfrak{K}/\theta(\mathfrak{M})$ whose annihilator $\mathfrak{q}$ is a prime ideal. Then $\mathfrak{q} \supset \mathfrak{p}_i + \prod_{j \neq i} \mathfrak{p}_j$ and is therefore non-principal. In other words, every prime ideal associated with $(\mathfrak{M}/\mathfrak{W}_i)/\theta(\mathfrak{M})$ is non-principal, and Corollary 18.5 and Proposition 19.4 imply that $h(\alpha^{(\mathfrak{M}/\mathfrak{W}_i)/\theta(\mathfrak{M})}) = 0$. From the Lemmas 21.3 and 21.7 we know that

$$\limsup_{\mathbf{p}} \frac{1}{|\mathbb{Z}^d/\Lambda_{\mathbf{p}}|} \log |Y_{\mathbf{p}}^{(\mathfrak{M}/\mathfrak{W}_i)/\theta(\mathfrak{M})}| = p^+(\alpha^Z) = h(\alpha^Z) = 0,$$

and that

$$\sum_{i=1}^{m} h(\alpha^{\mathfrak{M}/\mathfrak{W}_i}) = h(\alpha^{\mathfrak{K}}) = h(\alpha). \tag{21.40}$$

From (21.36), applied to $\mathfrak{K}$ and $\mathfrak{L} = \theta(\mathfrak{M})$, we see that

$$\sum_{i=1}^{m} \limsup_{\mathbf{p}} \frac{1}{|\mathbb{Z}^d/\Lambda_{\mathbf{p}}|} \log |Y_{\mathbf{p}}^{(\mathfrak{M}/\mathfrak{W}_i)}| = \limsup_{\mathbf{p}} \frac{1}{|\mathbb{Z}^d/\Lambda_{\mathbf{p}}|} \log |Y_{\mathbf{p}}^{\mathfrak{K}}|$$
$$= \limsup_{\mathbf{p}} \frac{1}{|\mathbb{Z}^d/\Lambda_{\mathbf{p}}|} \log |Y_{\mathbf{p}}^{\mathfrak{L}}| = \limsup_{\mathbf{p}} \frac{1}{|\mathbb{Z}^d/\Lambda_{\mathbf{p}}|} \log |Y_{\mathbf{p}}^{\mathfrak{M}}|. \tag{21.41}$$

The proof is completed by combining (21.40)–(21.41) with Lemma 21.12. $\square$

PROOF OF THEOREM 21.1 (2). We write $\mathfrak{M} = \hat{X}$ for the $\mathbb{Z}^d$-module arising from Lemma 5.1. Since α satisfies the d.c.c, $\mathfrak{M}$ is Noetherian by Proposition 5.4, and the finiteness of the entropy of α implies that $\mathfrak{M}$ is a torsion module: otherwise the prime ideal $\{0\}$ is associated with $\mathfrak{M}$, which means that $\mathfrak{M}$ has a submodule $\mathfrak{N} \cong \mathfrak{R}_d$, and $\alpha^{\mathfrak{M}} \geq \alpha^{\mathfrak{N}} = \infty$ by Theorem 17.1. Now apply Lemma 21.13. $\square$

The last remaining assertion of Theorem 21.1 concerns the cases where $d = 1$, or where α is expansive.

LEMMA 21.14. *Let $d \geq 1$, and let $\mathfrak{p} \subset \mathfrak{R}_d$ be a non-zero prime ideal. If $d = 1$, or if $\alpha = \alpha^{\mathfrak{R}_d/\mathfrak{p}}$ is expansive, then*

$$h(\alpha) = p^+(\alpha) = p^-(\alpha) = \lim_{\langle \Lambda \rangle \to \infty} \frac{1}{|\mathbb{Z}^d/\Lambda|} \log |\text{Fix}_\Lambda(\alpha)|. \qquad (21.42)$$

PROOF. If $\mathfrak{p}$ is not principal, or if $\mathfrak{p} = (f)$ for some generalized cyclotomic polynomial $f \in \mathfrak{R}_d$, then Lemma 21.3, Corollary 18.5, and Theorem 19.5 show that $h(\alpha) = p^\pm(\alpha) = 0$.

If $\mathfrak{p} = (f)$ for some irreducible polynomial $0 \neq f \in \mathfrak{R}_d$ which is not generalized cyclotomic, then we set $\mathfrak{M} = \mathfrak{R}_d/(f)$ and note that $h(\alpha) = \log \text{M}(f) \geq p^+(\alpha) \geq p^-(\alpha)$ by Lemma 21.4 and Theorem 18.1.

If $d = 1$ then (21.42) is an immediate consequence of Lemma 21.8. Suppose therefore that $d > 1$ and α is expansive, and fix a subgroup $\Lambda \subset \mathbb{Z}^d$ of index $|\mathbb{Z}^d/\Lambda| < \infty$. From Theorem 6.5 (1), (3)–(4), and (21.13), we know that $f(\omega) \neq 0$ for every $\omega \in \Omega(\Lambda) = V_{\mathbb{C}}(\mathfrak{b}(\Lambda))$, so that $\text{Fix}_\Lambda(\alpha)$ is finite. As in the proof of Lemma 21.9 we set $\mathfrak{N} = \mathfrak{R}_d/\mathfrak{b}(\Lambda)$, where $\mathfrak{b}(\Lambda)$, and denote by β_f multiplication by f on $\mathfrak{N}$. Choose a map $c \colon \mathbb{Z}^d/\Lambda \longmapsto \mathbb{Z}^d$ with $c(\mathbf{z}) + \Lambda = \mathbf{z}$ for all $\mathbf{z} \in \mathbb{Z}^d/\Lambda$, and define a group isomorphism $\psi \colon \mathbb{Z}^{\mathbb{Z}^d/\Lambda} \longmapsto \mathfrak{N}$ by setting, for every $w \colon \mathbb{Z}^d/\Lambda \longmapsto \mathbb{Z}$ in $\mathbb{Z}^{\mathbb{Z}^d/\Lambda}$,

$$\psi(w) = \sum_{\mathbf{z} \in \mathbb{Z}^d/\Lambda} w(\mathbf{z}) u^{c(\mathbf{z})} + \mathfrak{b}(\Lambda).$$

Then $\beta'_f = \psi \cdot \beta_f \cdot \psi^{-1} \colon \mathbb{Z}^{\mathbb{Z}^d/\Lambda} \longmapsto \mathbb{Z}^{\mathbb{Z}^d/\Lambda}$ is a homomorphism, and

$$\begin{aligned} P_\Lambda(\alpha) &= |\mathfrak{M}/\mathfrak{b}(\Lambda) \cdot \mathfrak{M}| = |\mathfrak{R}_d/((f) + \mathfrak{b}(\Lambda))| \\ &= |\mathfrak{N}/f \cdot \mathfrak{N}| = |\mathbb{Z}^{\mathbb{Z}^d/\Lambda}/\beta'_f(\mathbb{Z}^{\mathbb{Z}^d/\Lambda})|. \end{aligned} \qquad (21.43)$$

In order to calculate the last term in (21.43) we embed $\mathbb{Z}^{\mathbb{Z}^d/\Lambda}$ linearly in $\mathbb{C}^{\mathbb{Z}^d/\Lambda}$ and consider the linear map $\beta''_f \colon \mathbb{C}^{\mathbb{Z}^d/\Lambda} \longmapsto \mathbb{C}^{\mathbb{Z}^d/\Lambda}$ defined by β'_f. For every $\omega \in \Omega(\Lambda))$, the indicator function $w_\omega = 1_{\{\omega\}} \in \mathbb{C}^{\mathbb{Z}^d/\Lambda}$ is an eigenvector of β''_f with non-zero eigenvalue $f(\omega)$, and $\{w_\omega : \omega \in \Omega(\Lambda))\}$ is a $\mathbb{C}$-basis of $\mathbb{C}^{\mathbb{Z}^d/\Lambda}$ consisting of eigenvectors of β''_f. Hence

$$P_\Lambda(\alpha) = |\mathbb{Z}^{\mathbb{Z}^d/\Lambda}/\beta'_f(\mathbb{Z}^{\mathbb{Z}^d/\Lambda})| = |\det \beta''_f| = \prod_{w \in \Omega(\Lambda))} |f(\omega)|,$$

so that

$$\frac{1}{|\mathbb{Z}^d/\Lambda|} \log P_\Lambda(\alpha) = \frac{1}{|\mathbb{Z}^d/\Lambda|} \sum_{w \in \Omega_\Lambda} \log |f(\omega)|. \qquad (21.44)$$

Since α is expansive, $f(\mathbf{s}) \neq 0$ for every $\mathbf{s} \in \mathbb{S}^d$ (Theorem 6.5 (4)), and $\log |f|$ is continuous and hence Riemann-integrable. The second term in (21.44) is a Riemann sum approximation to the integral of $\log |f|$ on $\mathbb{S}^d$, and Theorem 18.1 implies that

$$p^-(\alpha) = p^+(\alpha) = \lim_{\langle \Lambda \rangle \to \infty} \frac{1}{|\mathbb{Z}^d/\Lambda|} \log P_\Lambda(\alpha) = \lim_{\langle \Lambda \rangle \to \infty} \frac{1}{|\mathbb{Z}^d/\Lambda|} \log |\mathrm{Fix}_\Lambda(\alpha)|$$

$$= \lim_{\langle \Lambda \rangle \to \infty} \frac{1}{|\mathbb{Z}^d/\Lambda|} \sum_{\omega \in \Omega_\Lambda} \log |f(\omega)| \tag{21.45}$$

$$= \log \mathrm{M}(f) = h(\alpha). \quad \square$$

LEMMA 21.15. *Let $\mathfrak{p} \subset \mathfrak{R}_d$ be a non-zero prime ideal, and let $\mathfrak{M}$ be a Noetherian $\mathfrak{R}_d$-module associated with $\mathfrak{p}$. If $d = 1$, or if $\alpha^{\mathfrak{R}_d/\mathfrak{p}}$ is expansive, then $p^+(\alpha^{\mathfrak{M}}) = p^-(\alpha^{\mathfrak{M}}) = h(\alpha^{\mathfrak{M}})$.*

PROOF. If $\mathfrak{p}$ is non-principal, or if $\mathfrak{p} = (f)$ for a generalized cyclotomic polynomial f, then $p^\pm(\alpha^{\mathfrak{M}}) = h(\alpha^{\mathfrak{M}}) = 0$ by Lemma 21.3, Proposition 18.6, and Corollary 18.5. If $\mathfrak{p} = (f)$ for a polynomial $f \in \mathfrak{R}_d$ which is not generalized cyclotomic, the proof is identical to that of Lemma 21.12, except that we use $\mathfrak{b}(\Lambda)$ and Lemma 21.14 instead of $j_\mathfrak{p}$ and Lemma 21.9 (note that $\Omega(\Lambda) \cap V_{\mathbb{C}}(\mathfrak{p}) = \varnothing$ for every subgroup $\Lambda \subset \mathbb{Z}^d$ of finite index. $\quad \square$

PROOF OF THEOREM 21.1 (3). Proceed as in the proof of Theorem 21.1 (2), using Lemma 21.15 instead of Lemma 21.13. $\quad \square$

CONCLUDING REMARKS 21.16. (1) The exposition in this section follows [63] and [62]. The rôle of Gelfond's theorem in the proof of Lemma 21.8 is discussed in [60].

(2) The necessity of the d.c.c. for Theorem 21.1 is clear from Example 5.6 (1)–(3). The assumption that $h(\alpha) < \infty$ can also not be dropped: if α is the shift-action of $\mathbb{Z}^d$ on $\mathbb{T}^{\mathbb{Z}^d}$, then $\mathrm{Fix}_\Lambda(\alpha)$ is connected for every subgroup $\Lambda \subset \mathbb{Z}^d$ of finite index, and hence $0 = p^\pm(\alpha) < h(\alpha) = \infty$. In general, if $\mathfrak{M}$ is a Noetherian $\mathfrak{R}_d$-module, and if $\mathfrak{M}^{(t)} = \{a \in \mathfrak{M} : f \cdot a = 0 \text{ for some } f \in \mathfrak{R}_d\}$ is the torsion submodule of $\mathfrak{M}$, then one would conjecture that

$$p^+(\alpha^{\mathfrak{M}}) = h(\alpha^{\mathfrak{M}^{(t)}}). \tag{21.46}$$

The proof of (21.45) may not be completely straightforward, since there exist Noetherian $\mathfrak{R}_d$-modules $\mathfrak{M}$ with $\mathfrak{M}^{(t)} = \{0\}$, for which $\mathrm{Fix}_\Lambda(\alpha^{\mathfrak{M}})$ is disconnected for certain subgroups $\Lambda \subset \mathbb{Z}^d$ of finite index. A simple example of this phenomenon is obtained by setting $\mathfrak{M} = 2\mathfrak{R}_1 + (u_1 - 1) \subset \mathfrak{R}_1$. If $m \geq 1$, then the torsion submodule $\mathfrak{N}^{(t)}$ of the module $\mathfrak{N} = \mathfrak{M}/(u_1^m - 1) \cdot \mathfrak{M}$ has cardinality 2, so that $P_{m\mathbb{Z}}(\alpha^{\mathfrak{M}}) = 2$ for every $m \geq 1$. However, $\lim_{m \to \infty} \frac{1}{m} \log P_{m\mathbb{Z}}(\alpha^{\mathfrak{M}}) = 0$, and $h(\alpha^{\mathfrak{M}}) = \infty$.

(3) Theorem 21.1 (3) can be strengthened slightly. If $d = 2$, and if $\mathfrak{p} = (f)$ with $f = 1 + u_2 + u_2 \in \mathfrak{R}_2$, then $\alpha = \alpha^{\mathfrak{R}_2/(f)}$ is non-expansive, and $\mathrm{Fix}_{3\mathbb{Z}^2}(\alpha)$ is not discrete (Example 18.16 (1)). By taking a careful look at the proof of

Lemma 21.14 we see that $p^-(\alpha) = p^+(\alpha) = h(\alpha)$; more generally, if $d \geq 2$, and if $f \in \mathfrak{R}_d$ is an irreducible polynomial with

$$V_{\mathbb{C}}(f) \cap \mathbb{S}^d \subset \Omega = \{\omega = (\omega_1, \ldots, \omega_d) \in \mathbb{C}^d :$$
$$\omega_1^k = \cdots = \omega_d^k = 1 \text{ for some } k \geq 1\}, \tag{21.47}$$

then

$$p^-(\alpha^{\mathfrak{R}_d/(f)}) = p^+(\alpha^{\mathfrak{R}_d/(f)}) = h(\alpha^{\mathfrak{R}_d/(f)}). \tag{21.48}$$

The difficulty in proving (21.47) for *every* irreducible polynomial $f \in \mathfrak{R}_d$ arises from the fact that $V_{\mathbb{C}}(f)$ may intersect $\mathbb{S}^d$ in points whose coordinates are not all roots of unity, and that not enough is known about rational approximation to the logarithms of these points. A general proof of (21.47) would require a stronger version of Gelfond's result and may turn out to be quite difficult (cf. [102] for a quantitative form of Gelfond's result due to Feldman which does not, however, appear to be strong enough for our purposes). Nevertheless it seems likely that (21.47) holds for *all* non-zero polynomials f, and that

$$p^{\pm}(\alpha^{\mathfrak{M}}) = h(\alpha^{\mathfrak{M}^{(t)}}) \tag{21.49}$$

for every Noetherian $\mathfrak{R}_d$-module $\mathfrak{M}$ (cf. (21.45)).

22. The distribution of periodic points

Let α be a $\mathbb{Z}^d$-action by automorphisms of a compact, abelian group X. For every subgroup $\Lambda \subset \mathbb{Z}^d$ of finite index we define $\langle \Lambda \rangle$ and $\text{Fix}_\Lambda(\alpha)$ by (21.1) and Theorem 6.5 (3)), respectively, and we write

$$\mu_\Lambda^\alpha = \lambda_{\text{Fix}_\Lambda(\alpha)} \tag{22.1}$$

for the normalized Haar measure of the subgroup $\text{Fix}_\Lambda(\alpha) \subset X$, considered as a probability measure on $\mathfrak{B}_X$.

For any probability measure μ on $\mathfrak{B}_X$ we denote by $\hat{\mu} \colon \hat{X} \longmapsto \mathbb{C}$ the *Fourier transform* $\hat{\mu}(a) = \int \langle x, a \rangle \, d\mu(x)$, $a \in \hat{X}$, of μ. A sequence $(\mu_n, n \geq 1)$ of probability measures on $\mathfrak{B}_X$ converges weakly to μ if and only if $\lim_{n \to \infty} \hat{\mu}_n(a) = \hat{\mu}(a)$ for every $a \in \hat{X}$.

THEOREM 22.1. *Let $d \geq 1$, and let α be a $\mathbb{Z}^d$-action by automorphisms of a compact, abelian group X with completely positive entropy. If $d = 1$, or if α is expansive, then $\lim_{\langle \Lambda \rangle \to \infty} \mu_\Lambda^\alpha = \lambda_X$ in the topology of weak convergence.*

We begin the proof of Theorem 22.1 with two lemmas.

LEMMA 22.2. *Let $\mathfrak{p} \subset \mathfrak{R}_d$ be a positive prime ideal (Definition 19.3), and let $\mathfrak{M}$ be a Noetherian $\mathfrak{R}_d$-module associated with $\mathfrak{p}$. If $d = 1$, or if $\alpha^{\mathfrak{R}_d/\mathfrak{p}}$ is expansive, then there exists, for every non-zero element $a \in \mathfrak{M}$, an $L \geq 1$ such that $a \notin \mathfrak{b}(\Lambda) \cdot \mathfrak{M}$ for every subgroup of finite index $\Lambda \subset \mathbb{Z}^d$ with $\langle \Lambda \rangle > L$.*

PROOF. According to Proposition 6.1 we can find submodules $\mathfrak{M} = \mathfrak{N}_s \supset \cdots \supset \mathfrak{N}_0 = \{0\}$ and an integer $t \in \{2, \ldots, s\}$ such that, for $j = 1, \ldots, s$, $\mathfrak{N}_j/\mathfrak{N}_{j-1} \cong \mathfrak{R}_d/\mathfrak{q}_j$ for some prime ideal $\mathfrak{p}_j \subset \mathfrak{R}_d$, $\mathfrak{q}_j = \mathfrak{p}$ for $j = 1, \ldots, t$, and $\mathfrak{q}_j \supsetneq \mathfrak{p}$ for $j = t+1, \ldots, s$. Since $\mathfrak{p}$ is principal and $\Omega(\Lambda) \cap V_{\mathbb{C}}(\mathfrak{p}) = \varnothing$ (Theorem 6.5), Lemma 21.10 implies that $\mathfrak{p} \cap \mathfrak{b}(\Lambda) = \mathfrak{p} \cdot \mathfrak{b}(\Lambda)$ and $\mathfrak{N}_j \cap \mathfrak{b}(\Lambda) \cdot \mathfrak{N}_t = \mathfrak{b}(\Lambda) \cdot \mathfrak{N}_j$ for every subgroup $\Lambda \subset \mathbb{Z}^d$ of finite index. We choose and fix polynomials $g_j \in \mathfrak{q}_j \smallsetminus \mathfrak{p}$ for $j = t+1, \ldots, s$, set $g = g_{t+1} \cdots g_s$, and note that $0 \neq g \cdot a \in \mathfrak{N}_t$ for every non-zero element $a \in \mathfrak{M}$.

Suppose that $a \in \mathfrak{M}$ is a non-zero element with $a \in \mathfrak{b}(\Lambda_n) \cdot \mathfrak{M}$ for some sequence $(\Lambda_n, n \geq 1)$ of subgroups of finite index in $\mathbb{Z}^d$ with $\lim_{n \to \infty}\langle \Lambda_n \rangle = \infty$. Then $b = g \cdot a$ is a non-zero element in $\mathfrak{b}(\Lambda_n) \cdot \mathfrak{N}_t$ for every $n \geq 1$, and we choose $k \in \{1, \ldots, t\}$ with $b \in \mathfrak{N}_k \smallsetminus \mathfrak{N}_{k-1}$ and note that $b \in \mathfrak{N}_k \cap \mathfrak{b}(\Lambda_n) \cdot \mathfrak{N}_t = \mathfrak{b}(\Lambda_n) \cdot \mathfrak{N}_k$ for all sufficiently large $n \geq 1$. There exists an element $c \in \mathfrak{N}_k \smallsetminus \mathfrak{N}_{k-1}$ such that the map $f \mapsto f \cdot c + \mathfrak{N}_{k-1}$ induces an isomorphism of $\mathfrak{R}_d/\mathfrak{q}_k = \mathfrak{R}_d/\mathfrak{p}$ and $\mathfrak{N}_k/\mathfrak{N}_{k-1}$, and we choose $h \in \mathfrak{R}_d$ such that $h \cdot c \in b + \mathfrak{N}_{k-1}$. Our choice of k implies that $h \notin \mathfrak{p}$, and Lemma 21.14 yields that

$$0 < h(\alpha^{\mathfrak{R}_d/\mathfrak{p}}) = h(\alpha^{\mathfrak{N}_k/\mathfrak{N}_{k-1}}) = \lim_{n \to \infty} \frac{1}{|\mathbb{Z}^d/\Lambda_n|} \log P_{\Lambda_n}(\alpha^{\mathfrak{N}_k/\mathfrak{N}_{k-1}})$$

$$= \lim_{n \to \infty} \frac{1}{|\mathbb{Z}^d/\Lambda_n|} \log |\mathfrak{N}_k/(\mathfrak{N}_{k-1} + \mathfrak{b}(\Lambda_n) \cdot \mathfrak{N}_k)|$$

$$= \lim_{n \to \infty} \frac{1}{|\mathbb{Z}^d/\Lambda_n|} \log |\mathfrak{N}_k/(\mathfrak{N}_{k-1} + \mathfrak{b}(\Lambda_n) \cdot \mathfrak{N}_k + \mathfrak{R}_d \cdot b)| \quad (22.2)$$

$$= \lim_{n \to \infty} \frac{1}{|\mathbb{Z}^d/\Lambda_n|} \log |\mathfrak{R}_d/(\mathfrak{p} + h\mathfrak{R}_d + \mathfrak{b}(\Lambda_n))| = h(\alpha^{\mathfrak{R}_d/(\mathfrak{p}+(h))}),$$

since $\alpha^{\mathfrak{R}_d/(\mathfrak{p}+(h))}$ is expansive by Theorem 6.5 (4) whenever $\alpha^{\mathfrak{R}_d/\mathfrak{p}}$ is expansive. As $h \notin \mathfrak{p}$, none of the prime ideals associated with $\mathfrak{R}_d/(\mathfrak{p}+(h))$ can be principal, and Corollary 18.5 and Proposition 19.4 imply that $h(\alpha^{\mathfrak{R}_d/(\mathfrak{p}+(h))}) = 0$, which is impossible in view of (22.2). This contradiction shows that every non-zero $a \in \mathfrak{M}$ must satisfy that $a \notin \mathfrak{b}(\Lambda) \cdot \mathfrak{M}$ whenever $\langle \Lambda \rangle$ is sufficiently large. $\square$

LEMMA 22.3. *Let $\mathfrak{M}$ be a Noetherian $\mathfrak{R}_d$-module whose associated prime ideals $\mathfrak{p}_1, \ldots, \mathfrak{p}_m$ are all non-zero and positive. If $d = 1$, or if $\alpha^{\mathfrak{R}_d/\mathfrak{p}_j}$ is expansive for every $j = 1, \ldots, m$, then there exists, for every non-zero element $a \in \mathfrak{M}$, an integer $L \geq 1$ such that $a \notin \mathfrak{b}(\Lambda) \cdot \mathfrak{M}$ for every subgroup of finite index $\Lambda \subset \mathbb{Z}^d$ with $\langle \Lambda \rangle > L$.*

PROOF. We choose a reduced primary decomposition $\{\mathfrak{W}_1, \ldots, \mathfrak{W}_m\}$ of $\mathfrak{M}$ corresponding to the prime ideals $\mathfrak{p}_1, \ldots, \mathfrak{p}_m$ (cf. (6.5)) and consider the injective homomorphism $a \mapsto (a + \mathfrak{W}_1, \ldots, a + \mathfrak{W}_m)$ from $\mathfrak{M}$ to $\mathfrak{N} = \mathfrak{M}/\mathfrak{W}_1 \times \cdots \times \mathfrak{M}/\mathfrak{W}_m$. If there exists a non-zero element $a \in \mathfrak{M}$ and a sequence $(\Lambda_n, n \geq 1)$ of subgroups of finite index in $\mathbb{Z}^d$ such that $\lim_{n \to \infty}\langle \Lambda_n \rangle = \infty$ and $a \in \mathfrak{b}(\Lambda_n) \cdot \mathfrak{M}$ for every $n \geq 1$, then we can obviously also find a non-zero element $\bar{a} =$

$(a + \mathfrak{W}_1, \ldots, a + \mathfrak{W}_m) \in \mathfrak{N}$ such that $\bar{a} \in \mathfrak{b}(\Lambda_n) \cdot \mathfrak{N}$ for every $n \geq 1$. Alas, the latter is impossible by Lemma 22.2. $\square$

PROOF OF THEOREM 22.1. Lemma 5.1, Proposition 5.4, and (4.10) allow us to assume that $\alpha = \alpha^{\mathfrak{M}}$ and $X = X^{\mathfrak{M}} = \widehat{\mathfrak{M}}$ for some Noetherian $\mathfrak{R}_d$-module $\mathfrak{M}$. If α is expansive, then Theorem 6.5 (4) shows that $\alpha^{\mathfrak{R}_d/\mathfrak{p}}$ is expansive for every prime ideal $\mathfrak{p} \subset \mathfrak{R}_d$ associated with $\mathfrak{M}$. The Fourier transform $\widehat{\mu_\Lambda^\alpha}$ of μ_Λ^α is equal to the indicator function of the submodule $\mathfrak{b}(\Lambda) \cdot \mathfrak{M} \subset \mathfrak{M}$, and the sequence μ_Λ^α converges weakly to λ_X if and only if there exists, for every non-zero $a \in \mathfrak{M}$, an $L \geq 1$ with $a \notin \mathfrak{b}(\Lambda) \cdot \mathfrak{M}$ whenever $\langle \Lambda \rangle > L$. However, this is precisely what was proved in Lemma 22.3. $\square$

THEOREM 22.4. *Let α be an expansive action of $\mathbb{Z}^d$ by automorphisms of a compact, abelian group X, and let ν be a weak limit point of μ_Λ^α as $\langle \Lambda \rangle \to \infty$. Then $h_\nu(\alpha) = h(\alpha)$, i.e. ν is a measure of maximal entropy for α (cf. Proposition 13.5).*

PROOF. By Lemma 5.1, Proposition 5.4, and (4.10) we may assume that $\alpha = \alpha^{\mathfrak{M}}$ and $X = X^{\mathfrak{M}} = \widehat{\mathfrak{M}}$ for some Noetherian $\mathfrak{R}_d$-module $\mathfrak{M}$. Theorem 20.8 shows that there exists a unique maximal submodule $\mathfrak{N} \subset \mathfrak{M}$ such that every prime ideal associated with $\mathfrak{N}$ is null, and $\alpha^{\mathfrak{M}/\mathfrak{N}}$ has completely positive entropy. Every prime ideal $\mathfrak{q}$ associated with $\mathfrak{M}/\mathfrak{N}$ is positive by Theorem 20.7 and contains a prime ideal associated with $\mathfrak{M}$, and Theorem 6.5 (4) implies that $\alpha^{\mathfrak{R}_d/\mathfrak{q}}$ is expansive.

For every $\Lambda \subset \mathbb{Z}^d$ and $a \in \mathfrak{M}$, $\widehat{\mu_\Lambda^\alpha}(a) \in \{0, 1\}$. It follows that $\hat{\nu}(a) \in \{0, 1\}$ for every $a \in \mathfrak{M}$, so that ν is the normalized Haar measure of a closed, and necessarily α-invariant, subgroup $Z \subset X$, and that $\mathcal{S}(\hat{\nu}) = \{a \in \mathfrak{M} : \hat{\nu}(a) = 1\} = Z^\perp$ is a submodule of $\mathfrak{M}$.

If $a \in \mathcal{S}(\hat{\nu})$ then there exists a sequence $(\Lambda_n, n \geq 1)$ of subgroups of finite index in $\mathbb{Z}^d$ with $\lim_{n\to\infty} \langle \Lambda_n \rangle = \infty$ and $a \in \mathcal{S}(\widehat{\mu_{\Lambda_n}^\alpha}) = \mathfrak{b}(\Lambda_n) \cdot \mathfrak{M}$ for every $n \geq 1$, and Lemma 22.3, applied to $\mathfrak{M}/\mathfrak{N}$, implies that a must lie in $\mathfrak{N}$. Hence $\mathcal{S}(\hat{\nu}) = Z^\perp \subset \mathfrak{N}$, and $Z \supset Y = \mathfrak{N}^\perp$. By Proposition 13.5, Theorem 20.8, and by the addition formula (14.1), $h(\alpha) \geq h_\nu(\alpha) = h(\alpha^Z) = h(\alpha^{Z/Y}) + h(\alpha^Y) = h(\alpha^Y) = h(\alpha)$. $\square$

We conclude this section with two examples which show that in the absence of completely positive entropy the measures μ_Λ^α need not converge as $\langle \Lambda \rangle \to \infty$.

EXAMPLES 22.5. (1) Let $\mathfrak{a} = (u_1 - 2, u_2 - 3) \subset \mathfrak{R}_2$, $\mathfrak{M} = \mathfrak{R}_2/\mathfrak{a}$, and let $\alpha = \alpha^{\mathfrak{M}}$ and $X = X^{\mathfrak{M}}$ (cf. Example 5.2 (2)). Then $\mathfrak{M} \cong \mathbb{Z}[\frac{1}{6}]$, and $\hat{\alpha}_{(1,0)}$ and $\hat{\alpha}_{(0,1)}$ correspond to multiplication by 2 and 3 on $\mathbb{Z}[\frac{1}{6}]$. We identify $\mathfrak{M}$ with $\mathbb{Z}[\frac{1}{6}]$ and choose sequences $(j_n, n \geq 1)$ and $(k_n, n \geq 1)$ of positive integers such that $\lim_{n\to\infty} j_n = \lim_{n\to\infty} k_n = \infty$, and $2^{j_n} \equiv 1 \pmod{p_k}$, $3^{k_n} \equiv 1 \pmod{p_k}$ for every $n \geq 1$ and $k = 1, \ldots, n$, where p_k is the k-th prime exceeding 3:

$p_1 = 5$, $p_2 = 7$, etc. If $\Lambda_n = \{(rj_n, sk_n) : r, s \in \mathbb{Z}\} \subset \mathbb{Z}^2$, and if $a \in \mathfrak{b}(\Lambda_n) \cdot \mathfrak{M} = (2^{j_n} - 1)\mathbb{Z}[\frac{1}{6}] + (3^{j_n} - 1)\mathbb{Z}[\frac{1}{6}]$, then $a = \frac{t}{6^l}$ for some $l \geq 1$ and some $t \in \mathbb{Z}$ which is divisible by $p_1 \cdot \cdots \cdot p_n$, and we conclude that every non-zero element $a \in \mathfrak{M} = \mathbb{Z}[\frac{1}{6}]$ can only lie in finitely many $\mathfrak{b}(\Lambda_n) \cdot \mathfrak{M}$. As we have see in the proof of Theorem 22.1, this implies that $\lim_{n \to \infty} \mu_{\Lambda_n}^\alpha = \lambda_X$.

To see that other limits are possible, choose a sequence $(k_n', n \geq 1)$ such that $\lim_{n \to \infty} k_n' = \infty$ and $3^{k_n'} \equiv 6 \pmod{25}$ for every $n \geq 1$. Next we choose an increasing sequence $(j_n', n \geq 1)$ in $\mathbb{N}$ such that, for every $n \geq 1$, $2^{j_n'} \equiv 1 \pmod 5$, and $2^{j_n'} \equiv 2 \pmod p$ for every prime factor $p \neq 5$ of $3^{k_n'} - 1$. We define Λ_n', $n \geq 1$ as above with j_n' and k_n' replacing j_n and k_n, and note that $\mathfrak{b}(\Lambda_n') \cdot \mathfrak{M} = 5\mathbb{Z}[\frac{1}{6}]$ for every $n \geq 1$. In particular, the measures $\mu_{\Lambda_n'}^\alpha$, $n \geq 1$, are all equal, and distinct from λ_X.

(2) Let $\mathfrak{a} = (2, 1 + u_1 + u_2) \subset \mathfrak{R}_2$, $\mathfrak{M} = \mathfrak{R}_2/\mathfrak{a}$, $\alpha = \alpha^{\mathfrak{M}}$, and $X = X^{\mathfrak{M}}$ (cf. Examples 5.3 (3) and 6.18 (5)). By considering the realization of $X \subset \{0, \frac{1}{2}\}^{\mathbb{Z}^2} \subset \mathbb{T}^{\mathbb{Z}^2}$ of Example 5.3 (3) one can easily verify that the identity $\mathbf{0}_X$ is the only element in X which is fixed under $\alpha_{(2^k, 0)}$ for any $k \geq 0$ (cf. also Example 5.6 (2)). In particular, if $\Lambda_k = \{(r2^k, s2^k) : r, s \in \mathbb{Z}\}$, then $\mathrm{Fix}_{\Lambda_k}(\alpha) = \{\mathbf{0}_X\}$ for every $k \geq 0$. However, if $\Lambda_k' = \{(r(2^k - 1), s(2^k - 1)) : r, s \in \mathbb{Z}\}$, then $\mathrm{Fix}_{\Lambda_k'}(\alpha)$ consists of all points $x \in X \subset \{0, \frac{1}{2}\}^{\mathbb{Z}^2} \subset \mathbb{T}^{\mathbb{Z}^2}$ which have horizontal period $2^k - 1$, and from (5.7) it is clear that any non-zero $a \in \mathfrak{M}$ can lie in $\mathrm{Fix}_{\Lambda_k'}(\alpha)^\perp = \mathfrak{b}(\Lambda_k') \cdot \mathfrak{M}$ for only finitely many $k \geq 0$. As we have seen in the proof of Theorem 22.1, this guarantees that $\lim_{k \to \infty} \mu_{\Lambda_k'}^\alpha = \lambda_X$. $\boxdot$

CONCLUDING REMARKS 22.6. The material in this section is taken from [107]. For $d = 1$, Theorem 22.1 was proved in [60] (cf. [59] and [69]), and for expansive $\mathbb{Z}^d$-action it is due to [107]. Like Theorem 21.1 (3), Theorem 22.1 is a direct consequence of Lemma 21.14 (cf. (22.2)), and can be strengthened in exactly the same way (cf. Remark 21.16 (3)).

23. Bernoullicity

Let $d \geq 1$. A measure-preserving $\mathbb{Z}^d$-action T on a probability space $(X, \mathfrak{S}, \mu)$ is *Bernoulli* if there exists a probability space $(Y, \mathfrak{T}, \nu)$ such that T is measurably conjugate to the shift-action σ of $\mathbb{Z}^d$ on $(Y^{\mathbb{Z}^d}, \mathfrak{T}^{\mathbb{Z}^d}, \nu^{\mathbb{Z}^d})$, where $\mathfrak{T}^{\mathbb{Z}^d}$ is the product Borel field on $Y^{\mathbb{Z}^d}$, and where σ is defined as in (2.1) or (5.8). In particular T is Bernoulli if and only if there exists a countably generated sigma-algebra $\mathfrak{U} \subset \mathfrak{S}$ with the following properties:

(1) $\mathfrak{U}$ is *independent* under T, i.e.

$$\nu(B_0 \cap T_{-\mathbf{n}_1}(B_1) \cap \cdots \cap T_{-\mathbf{n}_k}(B_k)) = \nu(B_0) \cdot \ldots \cdot \nu(B_k)$$

whenever $k \geq 1$, $B_0, \ldots, B_k$ lie in $\mathfrak{U}$, and $\mathbf{0}, \mathbf{n}_1, \ldots, \mathbf{n}_k$ are distinct elements in $\mathbb{Z}^d$,

(2) $\Sigma(\bigcup_{\mathbf{n}\in\mathbb{Z}^d} T_{-\mathbf{n}}(\mathfrak{U})) = \mathfrak{S}$ (mod ν), where $\Sigma(\mathcal{C})$ is the sigma-algebra generated by a collection of sets $\mathcal{C} \subset \mathfrak{S}$.

If a (countably generated) sigma-algebra $\mathfrak{U} \subset \mathfrak{S}$ satisfies (1), but not necessarily (2), then $\mathfrak{V} = \Sigma(\bigcup_{\mathbf{n}\in\mathbb{Z}^d} T_{-\mathbf{n}}(\mathfrak{U}))$ is called a *Bernoulli factor* of T.

Since Bernoulli actions of $\mathbb{Z}^d$ are measurably conjugate if and only if they have the same entropy ([76]), Bernoullicity is an important property in the study of conjugacy of $\mathbb{Z}^d$-actions. In this section we prove the following theorem.

THEOREM 23.1. *Let $d \geq 1$, and let α be a $\mathbb{Z}^d$-action by automorphisms of a compact, abelian group X. Then α is Bernoulli on $(X, \mathfrak{B}_X, \lambda_X)$ if and only if it has completely positive entropy.*

Before discussing the proof of Theorem 23.1, let us consider the following examples, where Bernoullicity allows us to conclude the measurable conjugacy of topologically non-conjugate $\mathbb{Z}^d$-actions.

EXAMPLES 23.2. (1) Let $d = 2$, $f^{(i,j)} = 1 + u_1^i + u_2^j \in \mathfrak{R}_2$, $(i,j) \in \{1,-1\}^2$ and let $\mathfrak{M}^{(i,j)} = \mathfrak{R}_2/f^{(i,j)}\mathfrak{R}_2$, $\alpha^{(i,j)} = \alpha^{\mathfrak{M}^{(i,j)}}$ and $X^{(i,j)} = X^{\mathfrak{M}^{(i,j)}}$. Lemma 18.4 and its proof show that $\alpha^{(i,j)} = \alpha_A^{(1,1)}$ with $A = \left(\begin{smallmatrix} i & 0 \\ 0 & j \end{smallmatrix}\right)$, where $\alpha_A^{(1,1)}$ is the $\mathbb{Z}^2$-action $\mathbf{n} \mapsto \alpha_{A\mathbf{n}}^{(1,1)}$. By applying either Proposition 13.1 or an elementary change of variable in Theorem 18.1 we see that $h(\alpha^{(i,j)}) = h(\alpha^{(1,1)})$ for every $(i,j) \in \{1,-1\}^2$, and Proposition 19.7 gives the exact value of $h(\alpha^{(1,1)})$. According to Theorem 20.8, $\alpha^{(i,j)}$ has completely positive entropy, and Theorem 23.1 implies that the $\mathbb{Z}^2$-actions $\alpha^{(i,j)}$ are Bernoulli with equal entropy. By [76] all these actions are measurably conjugate; however, since the prime ideals associated with the various $\mathfrak{R}_2$-modules $\mathfrak{M}^{(i,j)} = \mathfrak{R}_2/(f^{(i,j)})$ are all distinct, Theorem 5.9 implies that the $\mathbb{Z}^2$-actions $\alpha^{(i,j)}$ and $\alpha^{(i',j')}$ are topologically non-conjugate whenever $(i,j) \neq (i',j')$.

(2) More generally, if α is a $\mathbb{Z}^d$-action by automorphisms of a compact, abelian group X with completely positive entropy, and if $A \in \mathrm{GL}(d, \mathbb{Z})$, then Proposition 13.1 implies that the $\mathbb{Z}^d$-action $\alpha_A \colon \mathbf{n} \mapsto \alpha_{A\mathbf{n}}$ has completely positive entropy, and that $h(\alpha) = h(\alpha_A)$. By Theorem , α and α_A are Bernoulli, and hence measurably conjugate by [76].

(3) If we consider, in the notation of Example (1), the prime ideals $\mathfrak{p}^{(i,j)} = (2, f^{(i,j)}) = 2\mathfrak{R}_2 + f^{(i,j)}\mathfrak{R}_2 \subset \mathfrak{R}_2$ and set $\mathfrak{N}^{(i,j)} = \mathfrak{R}_2/\mathfrak{p}^{(i,j)}$ for every $(i,j) \in \{1,-1\}^2$, then Theorem 25.15 will imply that the $\mathbb{Z}^2$-actions $\alpha^{\mathfrak{N}^{(i,j)}}$ and $\alpha^{\mathfrak{N}^{(i',j')}}$ are measurably non-isomorphic whenever $(i,j) \neq (i',j')$. This is an indication that in the absence of completely positive entropy (and hence of Bernoullicity) the measurable conjugacy problem for $\mathbb{Z}^d$-actions by automorphisms of compact, abelian groups becomes considerably more complicated (cf. Chapters 7 and 9). $\boxdot$

If T is a measure preserving $\mathbb{Z}^d$-action on a probability space $(X, \mathfrak{S}, \mu)$, then a direct proof of the Bernoullicity of T amounts to the explicit construction a sigma-algebra $\mathfrak{T} \subset \mathfrak{S}$ satisfying the conditions (1) and (2) at the beginning of this section, which is in general very difficult. Fortunately Bernoullicity can be shown to be equivalent to variety of apparently weaker and more easily verifiable properties (cf. Theorem 23.4 and [34]). In order to formulate some of these conditions we recall that, if $\mathfrak{M}$ is a Noetherian $\mathfrak{R}_d$-module, then Example 5.3 (4) shows that there exists an integer $k \geq 1$ such that the $\mathbb{Z}^d$-action $\alpha^{\mathfrak{M}}$ has a natural realization as the restriction of the shift-action σ of $\mathbb{Z}^d$ on $(\mathbb{T}^k)^{\mathbb{Z}^d}$ to some closed, shift-invariant subgroup $X = X^{\mathfrak{M}} \subset (\mathbb{T}^k)^{\mathbb{Z}^d}$. The Bernoullicity of $\alpha^{\mathfrak{M}}$ is equivalent to the statement that the measure-preserving $\mathbb{Z}^d$-action σ on $((\mathbb{T}^k)^{\mathbb{Z}^d}, \mathfrak{B}_{(\mathbb{T}^k)^{\mathbb{Z}^d}}, \lambda_X)$ is Bernoulli, where λ_X is viewed as a shift-invariant probability measure on $(\mathbb{T}^k)^{\mathbb{Z}^d}$.

For the following definitions we adopt a slightly more general point of view by considering shift-invariant probability measures on $Y^{\mathbb{Z}^d}$, where (Y, δ) is a compact, metric space with diameter $\mathrm{diam}(Y) = \max_{y,y' \in Y} \delta(y, y') = 1$, and where the shift-action σ of $\mathbb{Z}^d$ on $Y^{\mathbb{Z}^d}$ is defined as in (5.8). For every subset $F \subset \mathbb{Z}^d$ we denote by Y^F the compact, metrizable space of all maps $x \colon F \longmapsto Y$, write $\mathfrak{B}_{Y^F}$ for the Borel field of Y^F, and denote by $M_1(Y^F)$ the weak*-compact set of probability measures on $\mathfrak{B}_{Y^F}$. In the special case where $F = \mathbb{Z}^d$ we write $M_1(Y^{\mathbb{Z}^d})^\sigma \subset M_1(Y^{\mathbb{Z}^d})$ for the set of σ-invariant probability measures on $Y^{\mathbb{Z}^d}$, which is again compact in the weak*-topology. If $\varnothing \neq F' \subset F \subset \mathbb{Z}^d$ then the coordinate projection (or restriction) $\pi_{F'} \colon Y^F \longmapsto Y^{F'}$ induces a continuous, surjective map $\mu \mapsto \mu_{F'} = \mu \pi_{F'}^{-1}$ from $M_1(Y^F)$ to $M_1(Y^{F'})$, defined by

$$\mu_{F'}(B) = \mu \pi_{F'}^{-1}(B) = \mu(\pi_{F'}^{-1}(B)) \tag{23.1}$$

for every $\mu \in M_1(Y^F)$ and $B \in \mathfrak{B}_{Y^{F'}}$.

For every finite or countably infinite partition $\mathcal{A}$ of a set $F \subset \mathbb{Z}^d$ and every $\mu \in M_1(Y^F)$ we set

$$\mu^{\mathcal{A}} = \prod_{A \in \mathcal{A}} \mu_A \in M_1(Y^F), \tag{23.2}$$

where μ_A is described in (23.1), and where $\prod_{A \in \mathcal{A}} \mu_A$ is the product measure on $Y^F \cong \prod_{A \in \mathcal{A}} Y^A$ of the measures μ_A, $A \in \mathcal{A}$. If $\mathcal{A}'$ is a partition of F which refines $\mathcal{A}$, then

$$(\mu^{\mathcal{A}})^{\mathcal{A}'} = (\mu^{\mathcal{A}'})^{\mathcal{A}} = \mu^{\mathcal{A}'}.$$

For a *partial cover* $\mathcal{A}$ of F (i.e. for a collection of disjoint subsets of F) we set

$$[\mathcal{A}] = \bigcup_{A \in \mathcal{A}} A,$$

put $A_\infty = F \smallsetminus [\mathcal{A}]$, denote by $\mathcal{A}_\infty = \mathcal{A} \cup \{A_\infty\}$ the partition of F obtained by adding A_∞ to $\mathcal{A}$, and define $\mu^{\mathcal{A}} \in M_1(Y^F)$ by

$$\mu^{\mathcal{A}} = \mu^{\mathcal{A}_\infty} \qquad (23.3)$$

for every $\mu \in M_1(Y^F)$.

Let $F \subset \mathbb{Z}^d$ be non-empty, and let $\mu_1, \mu_2 \in M_1(Y^F)$. We define the set $C(\mu_1, \mu_2)$ of *couplings* of μ_1 and μ_2 by

$$C(\mu_1, \mu_2) = \{\nu \in M_1(Y^F \times Y^F) : \nu(B \times Y^F) = \mu_1(B), \ \nu(Y^F \times B) = \mu_2(B)$$
$$\text{for every } B \in \mathfrak{B}_{Y^F}\}.$$

The elements of $C(\mu_1, \mu_2)$ are precisely those measures in $M_1(Y^F \times Y^F)$ whose projections onto the two coordinates coincide with μ_1 and μ_2, respectively. For $F = \mathbb{Z}^d$ and $\mu_1, \mu_2 \in M_1(Y^{\mathbb{Z}^d})^\sigma$ we denote by

$$J(\mu_1, \mu_2) = C(\mu_1, \mu_2) \cap M_1(Y^{\mathbb{Z}^d} \times Y^{\mathbb{Z}^d})^{\sigma \times \sigma}$$

the set of $\sigma \times \sigma$-invariant couplings or *joinings* of $\mu_1 \times \mu_2$. Note that the sets $C(\mu_1, \mu_2) \subset M_1(Y^F \times Y^F)$ and $J(\mu_1, \mu_2) \subset M_1(Y^{\mathbb{Z}^d} \times Y^{\mathbb{Z}^d})$ are closed and hence compact.

In order to introduce the $\bar{d}$-metric we write a typical element $z \in Y^F \times Y^F \cong (Y \times Y)^F, F \subset \mathbb{Z}^d$, as $z = ((z_{\mathbf{n}}^{(1)}, z_{\mathbf{n}}^{(2)}), \mathbf{n} \in F)$ with $z_{\mathbf{n}}^{(i)} \in Y$ and set $\pi_{\{\mathbf{n}\}}^{(i)} : z \mapsto z_{\mathbf{n}}^{(i)}$ for every $\mathbf{n} \in F$, $i = 1, 2$. If F is finite and $\mu_1, \mu_2 \in M_1(Y^F)$, put

$$\bar{d}_F(\mu_1, \mu_2) = \inf_{\nu \in C(\mu_1, \mu_2)} \frac{1}{|F|} \sum_{\mathbf{n} \in F} \int \delta(\pi_{\{\mathbf{n}\}}^{(1)}, \pi_{\{\mathbf{n}\}}^{(2)}) \, d\nu. \qquad (23.4)$$

Since $C(\mu_1, \mu_2)$ is compact, the infimum in (23.4) is actually a minimum. If F_1, F_2 are subsets of $\mathbb{Z}^d$ and $F \subset F_1 \cap F_2$ is finite and non-empty, then $\bar{d}_F(\mu_1, \mu_2)$ is defined as

$$\bar{d}_F(\mu_1, \mu_2) = \bar{d}_F(\mu_1 \pi_F^{-1}, \mu_2 \pi_F^{-1})$$

whenever $\mu_i \in M_1(Y^{F_i})$, $i = 1, 2$. For $F = \mathbb{Z}^d$ and $\mu_1, \mu_2 \in M_1(Y^{\mathbb{Z}^d})^\sigma$ we put

$$\bar{d}(\mu_1, \mu_2) = \limsup_{M \to \infty} \bar{d}_{B_M}(\mu_1, \mu_2) = \min_{\nu \in J(\mu_1, \mu_2)} \int \delta(\pi_{\{\mathbf{0}\}}^{(1)}, \pi_{\{\mathbf{0}\}}^{(2)}) \, d\nu \quad (23.5)$$

with

$$B_n = \{-n, \dots, n\}^d \subset \mathbb{Z}^d \qquad (23.6)$$

for every $n \geq 0$, where the last identity in (23.5) follows from the ergodic theorem and the compactness of $J(\mu_1, \mu_2)$.

For every non-empty, finite subset $F \subset \mathbb{Z}^d$, the map $\bar{d}_F : M_1(Y^F) \times M_1(Y^F) \longmapsto \mathbb{R}$ has the following properties.

(1) $\bar{d}_F$ is a metric on $M_1(Y^F)$ which induces the weak*-topology;
(2) If $\varnothing \neq F' \subset F$ then

$$\frac{|F'|}{|F|}\bar{d}_{F'}(\mu_1, \mu_2) \leq \bar{d}_F(\mu_1, \mu_2)$$

$$\leq \frac{|F'|}{|F|}\bar{d}_{F'}(\mu_1, \mu_2) + \left(1 - \frac{|F'|}{|F|}\right) \qquad (23.7)$$

for all $\mu_1, \mu_2 \in M_1(Y^F)$;
(3) If Y is finite then

$$\bar{d}_F(\mu_1, \mu_2) = \frac{1}{2}\sum_{z \in Y^F}\big|\mu_1(\{z\}) - \mu_2(\{z\})\big|. \qquad (23.8)$$

The map $\bar{d} \colon M_1(Y^{\mathbb{Z}^d})^\sigma \times M_1(Y^{\mathbb{Z}^d})^\sigma \longmapsto \mathbb{R}$ is again a metric which no longer induces the weak*-topology, but which has the property that certain dynamically important subsets of $M_1(Y^{\mathbb{Z}^d})^\sigma$ are closed in its topology (cf. e.g. [90]). If we call a probability measure $\mu \in M_1(Y^{\mathbb{Z}^d})^\sigma$ Bernoulli if the $\mathbb{Z}^d$-action σ on $(Y^{\mathbb{Z}^d}, \mathfrak{B}_{Y^{\mathbb{Z}^d}}, \mu)$ is Bernoulli, then $\bar{d}$ has the following remarkable property.

LEMMA 23.3 ([76], [91]). *The set of Bernoulli measures in* $M_1(Y^{\mathbb{Z}^d})^\sigma$ *is* $\bar{d}$-*closed.*

We quote two further properties of Bernoulli actions of $\mathbb{Z}^d$ from [76].

LEMMA 23.4 ([76]). *Let T be a measure preserving $\mathbb{Z}^d$-action on a probability space $(X, \mathfrak{S}, \mu)$ which is Bernoulli. Then every non-trivial T-invariant sigma-algebra $\mathfrak{V} \subset \mathfrak{S}$ is a Bernoulli factor of T.*

LEMMA 23.5 ([76]). *Let T be a measure preserving $\mathbb{Z}^d$-action on a probability space $(X, \mathfrak{S}, \mu)$. If there exists an increasing sequence $(\mathfrak{V}_n, n \geq 1)$ of Bernoulli factors of T such that $\mathfrak{V}_n \nearrow \mathfrak{S}$ as $n \to \infty$ then T is Bernoulli on $(X, \mathfrak{S}, \mu)$.*

A non-empty, finite set $F \subset \mathbb{Z}^d$ is (k, ε)-invariant if

$$\big|(F + \mathbf{n}) \cap F\big| \geq (1 - \varepsilon)|F|$$

for every $\mathbf{n} \in B_k$. If F is (k, ε)-invariant then

$$|\{\mathbf{m} \in F : B_k + \mathbf{m} \subset F\}| \geq (1 - \varepsilon|B_k|)|F|,$$
$$|F + B_k| \leq (1 + \varepsilon|B_k|)|F|. \qquad (23.9)$$

A partial cover $\mathcal{A}$ of of an arbitrary subset $F \subset \mathbb{Z}^d$ is a (k, K, ε)-cover if each $A \in \mathcal{A}$ is (k, ε)-invariant and contained in a translate of B_K, and if

$$|[\mathcal{A}]| \geq (1 - \varepsilon)|F|.$$

For every $N \geq 0$ we denote by

$$\mathcal{B}_N = \{B_N + \mathbf{n} : \mathbf{n} \in (2N+1)\mathbb{Z}^d\} \qquad (23.10)$$

the tiling of $\mathbb{Z}^d$ by translates of B_N.

DEFINITION 23.6. Let $\mu \in M_1(Y^{\mathbb{Z}^d})^\sigma$.

(1) μ is *sporadically almost block independent* (*sporadically a.b.i.*) if there exist, for every $\varepsilon > 0$, an integer $K \geq 0$, an increasing sequence $(N_j, j \geq 1)$ in $\mathbb{N}$ and, for each $j \geq 1$, a partition $\mathcal{A}^{(j)}$ of B_{N_j}, such that each $A \in \mathcal{A}^{(j)}$ is contained in some translate of B_K and

$$\bar{d}_{B_{N_j}}(\mu, \mu^{\mathcal{A}^{(j)}}) < \varepsilon.$$

(2) μ is *universally almost block independent* (*universally a.b.i.*) if there exist, for every $\varepsilon > 0$, an integer $k \geq 0$ and an $\varepsilon' > 0$ with

$$\bar{d}_{B_N}(\mu, \mu^{\mathcal{A}}) < \varepsilon$$

for every $K, N \geq 1$, and for every (k, K, ε')-cover $\mathcal{A}$ of B_N.

(3) μ is *almost box independent* if

$$\lim_{N \to \infty} \limsup_{M \to \infty} \bar{d}_{B_M}(\mu, \mu^{\mathcal{B}_N}) = 0.$$

(4) μ is *summably Vershik* if there exists an increasing sequence $(N_j, j \geq 1)$ in $\mathbb{N}$ and, for each $j \geq 2$, a partial cover $\mathcal{A}^{(j)}$ of B_{N_j} by translates of $B_{N_{j-1}}$ with the following properties.

  (a) $\sum_{j \geq 2}(1 - \|[\mathcal{A}^{(j)}]\|/|B_{N_j}|) < \infty$;

  (b) $\sum_{j \geq 2} \bar{d}_{B_{N_j}}(\mu, \mu^{\mathcal{A}^{(j)}}) < \infty$.

For a discussion of these conditions (in particular of condition (4)) and a proof of the following theorem we refer to [91].

THEOREM 23.7 ([91]). *The following conditions are equivalent for every shift-invariant probability measure μ on $Y^{\mathbb{Z}^d}$.*

  (1) *μ is almost box independent;*
  (2) *μ is sporadically a.b.i.;*
  (3) *μ is universally a.b.i.;*
  (4) *μ is Bernoulli.*

Furthermore, if $\mu \in M_1(Y^{\mathbb{Z}^d})^\sigma$ is summably Vershik then it is Bernoulli.

The summable Vershik condition turns out to be particularly suited for proving the Bernoullicity of $\mathbb{Z}^d$-actions of the form $\alpha^{\mathfrak{R}_d/\mathfrak{p}}$, where $\mathfrak{p} \subset \mathfrak{R}_d$ is a positive prime ideal (Definition 19.3). We begin our exploration of the rôle played by the summable Vershik condition with two lemmas from [41]. Let

$(Y, \mathfrak{T}, \nu)$ be a probability space, and let $\varepsilon > 0$. Two finite partitions $\mathcal{P}, \mathcal{Q}$ in $\mathfrak{T}$ are ε-*independent* if

$$\sum_{P \in \mathcal{P}, Q \in \mathcal{Q}} |\nu(P \cap Q) - \nu(P)\nu(Q)| < \varepsilon. \tag{23.11}$$

LEMMA 23.8. *Let $(Y, \mathfrak{T}, \nu)$ be a probability space, $\mathcal{P}, \mathcal{Q}$ finite partitions in $\mathfrak{T}$, and $\varepsilon > 0$. Suppose that there exist a set $E \in \mathfrak{T}$ and non-negative, measurable maps $f_P, g_Q : Y \longmapsto \mathbb{R}$, $P \in \mathcal{P}$, $Q \in \mathcal{Q}$, such that the following conditions are satisfied.*

(1) $\nu(E) < \varepsilon^2$;
(2) $f_P(y) \geq 1$ *for every* $P \in \mathcal{P}$ *and* $y \in P \smallsetminus E$, *and* $g_Q(y) \geq 1$ *for every* $Q \in \mathcal{Q}$ *and* $y \in Q \smallsetminus E$;
(3) $\sum_{P \in \mathcal{P}} \int f_P \, d\nu < 1 + \varepsilon^2$ *and* $\sum_{Q \in \mathcal{Q}} \int g_Q \, d\nu < 1 + \varepsilon^2$;
(4) $\int f_P g_Q \, d\nu = \int f_P \, d\nu \int g_Q \, d\nu$ *for every* $P \in \mathcal{P}, Q \in \mathcal{Q}$.

Then $\mathcal{P}$ and $\mathcal{Q}$ are 30ε-independent.

PROOF. If $\varepsilon > \frac{1}{10}$ the assertion is trivial, so that we may assume that $\varepsilon \leq \frac{1}{10}$. Let

$$\mathcal{P}_0 = \left\{ P \in \mathcal{P} : \int_P f_P \, d\nu \leq (1 + \varepsilon)\nu(P) \right\},$$

$$\mathcal{P}_1 = \left\{ P \in \mathcal{P} : \int_P f_P \, d\nu \geq (1 - \varepsilon)\nu(P) \right\},$$

$$\mathcal{Q}_0 = \left\{ Q \in \mathcal{Q} : \int_Q g_Q \, d\nu \leq (1 + \varepsilon)\nu(Q) \right\},$$

$$\mathcal{Q}_1 = \left\{ Q \in \mathcal{Q} : \int_Q g_Q \, d\nu \geq (1 - \varepsilon)\nu(Q) \right\},$$

$$\mathcal{P}' = \mathcal{P}_0 \cap \mathcal{P}_1, \quad \mathcal{Q}' = \mathcal{Q}_0 \cap \mathcal{Q}_1.$$

Then

$$\sum_{P \notin \mathcal{P}_0} \int f_P \, d\nu \geq (1 + \varepsilon) \sum_{P \notin \mathcal{P}_0} \nu(P),$$

$$\sum_{P \in \mathcal{P}_0} \int f_P \, d\nu \geq \sum_{P \in \mathcal{P}_0} \nu(P \smallsetminus E) \geq \sum_{P \in \mathcal{P}_0} \nu(P) - \varepsilon^2.$$

According to assumption (2),

$$1 + \varepsilon^2 \geq \sum_{P \in \mathcal{P}} \int f_P \, d\nu \geq (1 + \varepsilon) \sum_{P \notin \mathcal{P}_0} \nu(P) + \sum_{P \in \mathcal{P}_0} \nu(P) - \varepsilon^2,$$

so that

$$\sum_{P \notin \mathcal{P}_0} \nu(P) \leq 2\varepsilon. \tag{23.12}$$

If $P \notin \mathcal{P}_1$, then $\nu(P \cap E) \geq \varepsilon\nu(A)$, $\varepsilon^2 > \nu(E) \geq \sum_{P \notin \mathcal{P}_1} \nu(P \cap E) \geq \varepsilon \sum_{P \notin \mathcal{P}_1} \nu(P)$, and hence

$$\sum_{P \notin \mathcal{P}_1} \nu(P) \leq \varepsilon. \tag{23.13}$$

According to (23.12)–(23.13),

$$\sum_{P \notin \mathcal{P}'} \nu(P) \leq 3\varepsilon, \quad \sum_{Q \notin \mathcal{Q}'} \nu(Q) \leq 3\varepsilon, \tag{23.14}$$

where the second inequality is proved in the same way as the first. Now consider the partition $\mathcal{V} = \mathcal{P} \vee \mathcal{Q}$, set $h_V = f_P g_Q$ for every (non-empty) element $V = P \cap Q \in \mathcal{V}$, and put

$$\mathcal{V}_0 = \left\{ V \in \mathcal{V} : \int h_V \, d\nu \leq (1+\varepsilon)\nu(V) \right\},$$

$$\mathcal{V}_1 = \left\{ V \in \mathcal{V} : \int h_V \, d\nu \geq (1-\varepsilon)\nu(V) \right\},$$

$$\mathcal{V}' = \mathcal{V}_0 \cap \mathcal{V}_1.$$

The assumptions (1)–(3) imply that

$$\sum_{V \in \mathcal{V}} \int h_V \, d\nu < (1+\varepsilon^2)^2 < 1 + 3\varepsilon^2,$$

$$h_V(y) \geq 1 \text{ for every } V \in \mathcal{V} \text{ and } y \in V \setminus E,$$

and we argue as in (23.14) that

$$\sum_{V \notin \mathcal{V}_0} \nu(V) \leq 4\varepsilon, \quad \sum_{V \notin \mathcal{V}_1} \nu(V) \leq \varepsilon, \quad \sum_{V \notin \mathcal{V}'} \nu(V) < 5\varepsilon. \tag{23.15}$$

If $P \in \mathcal{P}'$, $Q \in \mathcal{Q}'$ and $P \cap Q \in \mathcal{V}'$, then (4) shows that

$$\nu(P \cap Q) \leq (1-\varepsilon)^{-1} \int f_P g_Q \, d\nu = (1-\varepsilon)^{-1} \int f_P \, d\nu \int g_Q \, d\nu$$

$$\leq (1-\varepsilon)^{-1}(1+\varepsilon)^2 \nu(P)\nu(Q),$$

$$\nu(P \cap Q) \geq (1+\varepsilon)^{-1} \int f_P g_Q \, d\nu = (1+\varepsilon)^{-1} \int f_P \, d\nu \int g_Q \, d\nu$$

$$\geq (1+\varepsilon)^{-1}(1-\varepsilon)^2 \nu(P)\nu(Q),$$

and hence that

$$|\nu(P \cap Q) - \nu(P)\nu(Q)| \leq \frac{4\varepsilon}{(1-\varepsilon)^2} \cdot \nu(P \cap Q) < 5\varepsilon\nu(P \cap Q).$$

If $P \in \mathcal{P}'$, $Q \in \mathcal{Q}'$, but $P \cap Q \notin \mathcal{V}'$, then we only know that

$$\nu(P)\nu(Q) \leq \frac{1}{(1-\varepsilon)^2} \int f_P g_Q \, d\nu \leq \frac{5}{4} \int f_P g_Q \, d\nu.$$

According to (23.14)–(23.15),

$$\sum_{P \cap Q \in \mathcal{V}} |\nu(P \cap Q) - \nu(P)\nu(Q)|$$

$$\leq \sum_{P \in \mathcal{P}', Q \in \mathcal{Q}', P \cap Q \in \mathcal{V}'} |\nu(P \cap Q) - \nu(P)\nu(Q)|$$

$$+ \sum_{P \notin \mathcal{P}', Q \in \mathcal{Q}} \nu(P \cap Q) + \nu(P)\nu(Q)$$

$$+ \sum_{Q \notin \mathcal{Q}', P \in \mathcal{P}} \nu(P \cap Q) + \nu(P)\nu(Q)$$

$$+ \sum_{P \in \mathcal{P}', Q \in \mathcal{Q}', P \cap Q \notin \mathcal{V}'} \nu(P \cap Q) + \nu(P)\nu(Q)$$

$$\leq 5\varepsilon + 6\varepsilon + 6\varepsilon + 5\varepsilon + \frac{5}{4} \sum_{V \notin \mathcal{V}'} \int h_V \, d\nu$$

$$\leq 22\varepsilon + \frac{5}{4}\left(1 + 3\varepsilon^2 - \sum_{V \in \mathcal{V}'} \int h_V \, d\nu \right) < 30\varepsilon. \quad \square$$

For every $k \geq 2$ we write $\mathcal{P}^{(k)} \subset \mathfrak{B}_{\mathbb{T}}$ for the partition of $\mathbb{T}$ into k intervals of equal length: $\mathcal{P}^{(k)} = \{P_0^{(k)}, \ldots, P_{k-1}^{(k)}\}$ with $P_j^{(k)} = [\frac{j}{k}, \frac{j+1}{k}) + \mathbb{Z} \subset \mathbb{T}$ for $j = 0, \ldots, k-1$.

LEMMA 23.9. *Let $k \geq 2$ and $m \geq 1$. There exists a Borel set $E^{(k,m)} \subset \mathbb{T}$ and continuous, non-negative functions $h_j^{(k,m)} \colon \mathbb{T} \longmapsto \mathbb{R}$, $j = 0, \ldots, k-1$, with the following properties.*

(1) *$\lambda_{\mathbb{T}}(E^{(k,m)}) < \frac{1}{m^2}$, and $\sum_{j=0}^{k-1} h_j^{(k,m)} \leq 1 + km^2$ for every $t \in \mathbb{T}$;*

(2) *For every $j = 0, \ldots, k-1$, $t_1 \in P_j^{(k)} \smallsetminus E^{(k,m)}$, $t_2 \in \mathbb{T} \smallsetminus (P_j^{(k)} \cup E^{(k,m)})$,*

$$h_j^{(k,m)}(t_1) \geq 1, \quad h_j^{(k,m)}(t_2) \leq \frac{1}{m^2};$$

(3) *For every $j = 0, \ldots, k-1$,*

$$\int h_j^{(k,m)}(t) e^{-2\pi i n t} \, d\lambda_{\mathbb{T}}(t) = 0$$

whenever $|n| > 8k^3m^6$, i.e. $h_j^{(k,m)}$ is a trigonometric polynomial involving only trigonometric functions $e^{2\pi i n t}$ with $|n| \leq 8k^3m^6$.

PROOF. For every $n \geq 64$ and $t \in \mathbb{T}$ we set

$$a_n(t) = \frac{1}{\sqrt{n}} \cdot \sum_{l=0}^{n-1} e^{2\pi i l x} = \frac{1}{\sqrt{n}} \cdot \frac{e^{2\pi i n t} - 1}{e^{2\pi i t} - 1},$$

$$b_n(t) = |a_n(t)|^2 = \sum_{l=-n+1}^{n-1} \frac{n-|l|}{n} \cdot e^{2\pi i l t}.$$

Then $b_n \geq 0$, $\int b_n \, d\lambda_{\mathbb{T}} = 1$ and, for every $t \in \mathbb{T}$ with $\|t\| = \min\{|t - m| : m \in \mathbb{Z}\} \geq n^{-1/3}$,

$$|a_n(t)| \leq \frac{1}{\sqrt{n}} \cdot \frac{1}{|\sin \pi t|} \leq \frac{1}{4n^{1/6}}, \quad b_n(t) \leq \frac{1}{16n^{1/3}}. \qquad (23.16)$$

Put $F^{(n)} = \{t \in \mathbb{T} : |t - \frac{l}{n}| < \frac{1}{n^{1/3}}$ for some $l \in \mathbb{Z}\}$ and set, for every $j \in \{0, \ldots, k-1\}$, $g_j^{(n)} = 1_{P_j^{(k)}} * b_n + \frac{1}{16n^{1/3}}$, where $*$ denotes convolution. Since b_n is non-negative, $g_j^{(n)}$ is non-negative, $g_j^{(n)} \leq 1 + \frac{1}{16n^{1/3}}$,

$$\sum_{j=0}^{k-1} g_j^{(n)} = b_n * \left(\sum_{j=0}^{k-1} 1_{P_j^{(k)}} \right) + \frac{k}{16n^{1/3}} = 1 + \frac{k}{16n^{1/3}},$$

and the inequality (23.16) implies that $g_j^{(n)}(t) \geq 1$ for every $t \in P_j^{(k)} \smallsetminus F^{(n)}$. The proof is completed by setting $n = 8k^3m^6$, $E^{(k,m)} = F^{(n)}$ and $h_j^{(k,m)} = g_j^{(n)}$ for every $j = 0, \ldots, k-1$. $\quad\square$

Let $d \geq 1$, and let $f = \sum_{\mathbf{n} \in \mathbb{Z}^d} c_f(\mathbf{n}) u^{\mathbf{n}} \in \mathfrak{R}_d$ be a non-zero, irreducible Laurent polynomial which is not generalized cyclotomic. From the Theorems 19.5 and 20.8 we know that $\alpha^{\mathfrak{R}_d/(f)}$ has completely positive entropy. If f is a constant multiple of a monomial, i.e. if $f = pu^{\mathbf{n}}$ for some $\mathbf{n} \in \mathbb{Z}^d$ and some rational prime $p > 1$, then $\alpha^{\mathfrak{R}_d/(f)}$ is algebraically conjugate to the shift-action of $\mathbb{Z}^d$ on $\mathbb{Z}_{/p}^{\mathbb{Z}^d}$, and hence Bernoulli. If f is not of this simple form, the Bernoullicity of $\alpha^{\mathfrak{R}_d/(f)}$ is considerably more difficult to establish. The following observation helps to overcome a minor technical difficulty in proving that $\alpha^{\mathfrak{R}_d/(f)}$ is Bernoulli.

LEMMA 23.10. *Let $A \in \mathrm{GL}(d, \mathbb{Z})$, and let $f^A = \sum_{\mathbf{n} \in \mathbb{Z}^d} c_f(\mathbf{n}) u^{A\mathbf{n}} \in \mathfrak{R}_d$. Then $\alpha = \alpha^{\mathfrak{R}_d/(f)}$ is Bernoulli if and only if $\alpha^{\mathfrak{R}_d/(f^A)}$ is Bernoulli.*

PROOF. The proof of Lemma 18.4 shows that there exists a continuous group isomorphism $\psi_A \colon \mathbb{T}^{\mathbb{Z}^d} \longmapsto \mathbb{T}^{\mathbb{Z}^d}$ such that $\psi_A(X^{\mathfrak{R}_d/(f)}) = X^{\mathfrak{R}_d/(f^A)}$ and $\psi_A \cdot \alpha_{\mathbf{n}}^{\mathfrak{R}_d/(f)} = \alpha_{A\mathbf{n}}^{\mathfrak{R}_d/(f^A)} \cdot \psi_A$ for every $\mathbf{n} \in \mathbb{Z}^d$. In particular, $\alpha = \alpha^{\mathfrak{R}_d/(f)}$ is algebraically conjugate to the $\mathbb{Z}^d$-action $\alpha_A^{\mathfrak{R}_d/(f^A)} \colon \mathbf{n} \mapsto \alpha_{A\mathbf{n}}^{\mathfrak{R}_d/(f^A)}$, and the definition of Bernoullicity shows that $\alpha^{\mathfrak{R}_d/(f^A)}$ is Bernoulli if and only if $\alpha \cong \alpha_A^{\mathfrak{R}_d/(f^A)}$ is Bernoulli. $\quad\square$

Motivated by Lemma 23.10 we make the following *ad hoc* definition.

DEFINITION 23.11. An irreducible element $f = \sum_{\mathbf{n} \in \mathbb{Z}^d} c_f(\mathbf{n}) u^{\mathbf{n}} \in \mathfrak{R}_d$ is nice if $c_f(\mathbf{0}) \neq 0$, and if there exists an element $\mathbf{r}^* = (r_1^*, \ldots, r_d^*) \in \mathbb{Z}^d$ such that $r_i^* > 0$ for $i = 1, \ldots, d$, $c_f(\mathbf{r}^*) \neq 0$, and $c_f(\mathbf{n}) = 0$ for every $\mathbf{n} \in \mathbb{Z}^d \setminus Q_{\mathbf{r}^*}^*$, where

$$Q_{\mathbf{r}^*}^* = \{\mathbf{0}, \mathbf{r}^*\} \cup \{\mathbf{n} = (n_1, \ldots, n_d) \in \mathbb{Z}^d : 0 < n_i < r_i^* \text{ for all } i = 1, \ldots, d\}.$$

If $d = 1$ every non-constant, irreducible polynomial $f \in \mathfrak{R}_1$ with non-zero constant term is nice; if $d \geq 2$, and if $f \in \mathfrak{R}_d$ is irreducible and has at least two non-zero coefficients, then we can find $\mathbf{m} \in \mathbb{Z}^d$ and $A \in \mathrm{GL}(d, \mathbb{Z})$ such that $u^{\mathbf{m}} f^A$ is nice.

EXAMPLE 23.12. Let $d = 2$, and let $f = 1 + u_1 + u_2 \in \mathfrak{R}_2$. The support $\mathcal{S}(f) = \{\mathbf{m} \in \mathbb{Z}^2 : c_f(\mathbf{m}) \neq 0\}$ has the form $\mathcal{S}(f) = \{(0,0), (1,0), (0,1)\}$, i.e. f is not nice. However, if $A = \left(\begin{smallmatrix} 7 & 3 \\ 2 & 1 \end{smallmatrix}\right) \in \mathrm{SL}(2, \mathbb{Z})$, then $\mathcal{S}(f^A) = \{(0,0), (7,2), (3,1)\}$, and f^A is therefore nice. $\boxdot$

Until further notice we assume that $d \geq 1$, $f \in \mathfrak{R}_d$ is a nice, irreducible polynomial which is not generalized cyclotomic, and that $\alpha = \alpha^{\mathfrak{R}_d/(f)}$ is the shift-action of $\mathbb{Z}^d$ on $X = X^{\mathfrak{R}_d/(f)} \subset \mathbb{T}^{\mathbb{Z}^d}$ (cf. (5.9)–(5.10)). Let $k \geq 2$, define the partition $\mathcal{P}^{(k)} \subset \mathfrak{B}_{\mathbb{T}}$ as in the paragraph preceding Lemma 23.9, and put

$$\mathcal{P}_{\mathbf{n}}^{(k)} = \pi_{\{\mathbf{n}\}}^{-1}(\mathcal{P}^{(k)}) \subset \mathfrak{B}_X \tag{23.17}$$

for every $\mathbf{n} \in \mathbb{Z}^d$, where $\pi_{\{\mathbf{n}\}} : X \longmapsto \mathbb{T}$ is the projection onto the $\mathbf{n}$-th coordinate. For every $n \geq 1$ we define $B_n \subset \mathbb{Z}^d$ by (23.6) and consider the finite partition

$$\mathcal{Q}_n^{(k)} = \bigvee_{\mathbf{n} \in B_n} \mathcal{P}_{\mathbf{n}}^{(k)}$$

Our first task in verifying the summable Vershik condition for α consists of showing that, for $j = 1, \ldots, d$, the partitions

$$\begin{aligned}
\mathcal{A}(k, n, j)^- &= \alpha_{(2^n + n^2)\mathbf{e}^{(j)}}\left(\mathcal{Q}_{2^n}^{(k)}\right), \\
\mathcal{A}(k, n, j)^+ &= \alpha_{-(2^n + n^2)\mathbf{e}^{(j)}}\left(\mathcal{Q}_{2^n}^{(k)}\right),
\end{aligned} \tag{23.18}$$

become rapidly independent as $n \to \infty$, where $\mathbf{e}^{(j)}$ is the j-th unit vector in $\mathbb{Z}^d$. Set, for every $n \geq 0$ and $j = 1, \ldots, d$,

$$Q(2^n, j)^{\pm} = B_{2^n} \pm (2^n + n^2)\mathbf{e}^{(j)}, \quad E(n, j)^{\pm} = \bigcup_{\mathbf{r} \in Q(2^n, j)^{\pm}} \alpha_{-\mathbf{r}}(E^{(k, |\mathbf{r}|^{2d})}),$$

where $|\mathbf{r}| = \max\{|r_1|, \ldots, |r_d|\}$ for every $\mathbf{r} \in \mathbb{Z}^d$. For all atoms

$$A^- = \bigcap_{\mathbf{r} \in Q(2^n, j)^-} \alpha_{-\mathbf{r}}(P_{j_{\mathbf{r}}}^{(k)}) \in \mathcal{A}(k, n, j)^-,$$

$$A^+ = \bigcap_{\mathbf{r} \in Q(2^n, j)^+} \alpha_{-\mathbf{r}}(P_{j\mathbf{r}}^{(k)}) \in \mathcal{A}(k, n, j)^+$$

with $j_{\mathbf{r}} \in \{0, \dots, k-1\}$ for every $\mathbf{r}$ we let

$$g_{A^-} = \prod_{\mathbf{r} \in Q(2^n, j)^-} h_{j_{\mathbf{r}}}^{(k, |\mathbf{r}|^{2d})} \cdot \pi_{\{0\}} \cdot \alpha_{\mathbf{r}} = \prod_{\mathbf{r} \in Q(2^n, j)^-} h_{j_{\mathbf{r}}}^{(k, |\mathbf{r}|^{2d})} \cdot \pi_{\{\mathbf{r}\}},$$

$$g_{A^+} = \prod_{\mathbf{r} \in Q(2^n, j)^+} h_{j_{\mathbf{r}}}^{(k, |\mathbf{r}|^{2d})} \cdot \pi_{\{0\}} \cdot \alpha_{\mathbf{r}} = \prod_{\mathbf{r} \in Q(2^n, j)^+} h_{j_{\mathbf{r}}}^{(k, |\mathbf{r}|^{2d})} \cdot \pi_{\{\mathbf{r}\}},$$

and apply Lemma 23.9 to obtain the following for every $n \geq 1$ and $j = 1, \dots, d$.

(a) The sets $E(n, j)^{\pm} \subset X$ have the property that

$$\lambda_X(E(n, j)^{\pm}) \leq \sum_{\mathbf{r} \in Q(2^n, j)^{\pm}} \frac{1}{|\mathbf{r}|^{4d}}$$

$$< (2d - 1) \sum_{r=n^2}^{\infty} (2r + 1)^{d-1} r^{-4d} < 2^d n^{-6d};$$

(b) For every $A^{\pm} \in \mathcal{A}(k, n, j)^{\pm}$, $g_{A^{\pm}} \geq 0$;

(c) The continuous maps $g_{A^{\pm}} : X \longmapsto \mathbb{R}$, $A^{\pm} \in \mathcal{A}(k, n, j)^{\pm}$, satisfy that

$$\sum_{A^+ \in \mathcal{A}(k,n,j)^+} g_{A^+} \leq \prod_{\mathbf{r} \in Q(2^n, j)^+} \left(1 + \frac{k}{|\mathbf{r}|^{4d}}\right)$$

$$< \prod_{r=n^2}^{\infty} \left(1 + \frac{k}{r^{4d}}\right)^{(2d-1)(2r+1)^{d-1}},$$

$$\sum_{A^- \in \mathcal{A}(k,n,j)^-} g_{A^-} \leq \prod_{\mathbf{r} \in Q(2^n, j)^-} \left(1 + \frac{k}{|\mathbf{r}|^{4d}}\right)$$

$$< \prod_{r=n^2}^{\infty} \left(1 + \frac{k}{r^{4d}}\right)^{(2d-1)(2r+1)^{d-1}};$$

(d) If $A^{\pm} \in \mathcal{A}(k, n, j)^{\pm}$, and if

$$g_{A^{\pm}}(x) = \sum_{a \in \Re_d/(f)} \hat{g}_{A^{\pm}}(a) \cdot \langle x, a \rangle$$

is the Fourier series of $g_{A^{\pm}}$, then every $a \in \Re_d/(f)$ with $\hat{g}_{A^-}(a) \neq 0$ for some $A^- \in \mathcal{A}(k, n, j)^-$ is of the form $a = \phi + (f)$, where

$$\phi = \sum_{\mathbf{r} \in Q(2^n, j)^-} c_{\phi}(\mathbf{r}) u^{\mathbf{r}} \in \Re_d$$

with

$$|c_{\phi}(\mathbf{r})| \leq 8k^3 |\mathbf{r}|^{12d}$$

for every $\mathbf{r} \in Q(2^n, j)^-$; similarly, if $\hat{g}_{A^+}(a) \neq 0$ for some $A^+ \in \mathcal{A}(k, n, j)^+$ and $a \in \mathfrak{R}_d/(f)$, then $a = \psi + (f)$, where

$$\psi = \sum_{\mathbf{r} \in Q(2^n, j)^+} c_\phi(\mathbf{r}) u^\mathbf{r} \in \mathfrak{R}_d$$

with

$$|c_\phi(\mathbf{r})| \leq 8k^3 |\mathbf{r}|^{12d}$$

for every $\mathbf{r} \in Q(2^n, j)^+$.

The integer r_j^* in the next lemma is the degree of f in the variable u_j (cf. Definition 23.11).

LEMMA 23.13. *For $j = 1, \ldots, d$, and for every integer $n \geq 2kd$ with*

$$e^{-2n^2 h(\alpha)} \cdot (k^3 2^{14dn})^{r_j^*} < 1, \tag{23.19}$$

the partitions $\mathcal{A}(k, n, j)^-$ and $\mathcal{A}(k, n, j)^+$ in (23.18) are $30k2^{d/2}n^{-3d}$-independent.

PROOF. All the definitions and properties involved are invariant under permutations of coordinates; it will thus suffice to prove the lemma for $j = 1$. As f is nice (Definition 23.11), we can write it in the form $f = \sum_{j=0}^{r_1^*} g_j u_1^j$ with $g_j \in \mathbb{Z}[u_2, \ldots, u_d]$ for every $j = 0, \ldots, r_1^*$, where $g_0 = c_f(0) \neq 0$ and $g_{r_1^*} = c_{\mathbf{r}^*} u_2^{r_2^*} \cdot \ldots \cdot u_d^{r_d^*} \neq 0$.

Assume for the moment that we have proved the following for every integer n satisfying (23.19): if

$$\psi^- = \sum_{\mathbf{r} \in Q(2^n, 1)^-} c_{\psi^-}(\mathbf{r}) u^\mathbf{r}, \quad \psi^+ = \sum_{\mathbf{r} \in Q(2^n, 1)^+} c_{\psi^+}(\mathbf{r}) u^\mathbf{r} \tag{23.20}$$

are elements of $\mathfrak{R}_d$ with

$$|c_{\psi^-}(\mathbf{r})| \leq 8k^3 |\mathbf{r}|^{12d}, \quad |c_{\psi^+}(\mathbf{r}')| \leq 8k^3 |\mathbf{r}'|^{12d} \tag{23.21}$$

for every $\mathbf{r} \in Q(2^n, 1)^-$, $\mathbf{r}' \in Q(2^n, 1)^+$, and if

$$\psi^- + (f) = \psi^+ + (f), \tag{23.22}$$

then

$$\psi^- \in (f), \quad \psi^+ \in (f). \tag{23.23}$$

If $A^- \in \mathcal{A}(k, n, 1)^-$, $A^+ \in \mathcal{A}(k, n, 1)^+$, then condition (d) preceding the statement of this lemma and (23.20)–(23.23) together show that no non-trivial character $a \in \hat{X} = \mathfrak{R}_d/(f)$ has the property that a occurs in the Fourier series of g_{A^-} and $-a$ in the Fourier series of g_{A^+}. Hence $\int g_{A^-} g_{A^+} \, d\lambda_X =$

$\int g_{A^-} \, d\lambda_X \cdot \int g_{A^-} \, d\lambda_X$ for every $A^- \in \mathcal{A}(k, n, 1)^-$, $A^+ \in \mathcal{A}(k, n, 1)^+$. From the conditions (a)–(c) above we obtain that $\lambda_X(E(n, 1)^{\pm}) < 2^d n^{-6d}$, and that

$$\log\left(\sum_{A^{\pm} \in \mathcal{A}(k,n,1)^{\pm}} g_{A^{\pm}}(x) \right) < \sum_{r \geq n^2} (2d - 1)(2r + 1)^{d-1} \log\left(1 + \frac{k}{r^{4d}} \right)$$

$$< (2d - 1) \cdot \sum_{r \geq n^2} (2r + 1)^{d-1} \frac{k}{r^{4d}}$$

$$< 2^d (2d - 1) \cdot \sum_{r \geq n^2} \frac{k}{r^{3d+1}} < 2^d \frac{k}{n^{6d}}$$

for every $x \in X$. Since $\log t \geq \frac{1}{2}(t - 1) \geq \frac{1}{k}(t - 1)$ whenever $1 \leq t \leq 2$ we see that

$$\sum_{A^{\pm} \in \mathcal{A}(k,n,1)^{\pm}} g_{A^{\pm}}(x) \leq 2^d \frac{k^2}{n^{6d}}$$

whenever $n \geq 2kd$, and Lemma 23.8 yields that $\mathcal{A}(k, n, 1)^-$ and $\mathcal{A}(k, n, 1)^+$ are $30k2^{d/2}n^{-3d}$-independent.

In order to prove that (23.20)–(23.22) imply (23.23) whenever $n \geq 2kd$ satisfies (23.19) we assume that $p_2, \ldots, p_d$ are distinct rational primes with $p_l > r_l^*$ for every $l = 2, \ldots, d$, choose primitive p_l-th unit roots ω_l, $l = 2, \ldots, d$, and set

$$\mathbf{p} = (p_2, \ldots, p_d), \quad [\mathbf{p}] = \prod_{l=2}^{d}(p_l - 1), \quad \boldsymbol{\omega} = (\omega_2, \ldots, \omega_d).$$

Let $\mathbb{L} = \mathbb{Q}[\omega_2, \ldots, \omega_d]$, put $h(u_1) = f(u_1, \omega_2, \ldots, \omega_d)/c_0(f) \in \mathbb{L}[u_1]$, and consider the decomposition $h = h_1 \cdot \ldots \cdot h_q$ of h into irreducible elements of $\mathbb{L}[u_1]$ with constant terms 1. By applying the Galois group $\mathrm{Gal}[\mathbb{L} : \mathbb{Q}]$ of $\mathbb{L}$ over $\mathbb{Q}$ to $h = h_1 \cdot \ldots \cdot h_q$ we obtain, for every $\kappa \in \mathrm{Gal}[\mathbb{L} : \mathbb{Q}]$, a corresponding decomposition $h^{\kappa} = h_1^{\kappa} \cdot \ldots \cdot h_q^{\kappa}$, where h^{κ} and h_i^{κ} are the images of h and h_i under κ. Put, for every $i = 1, \ldots, q$,

$$H_i = \prod_{\kappa \in \mathrm{Gal}[\mathbb{L}:\mathbb{Q}]} h_i^{\kappa} \in \mathbb{Q}[u_1],$$

$$H(u_1) = \prod_{i=1}^{q} H_i(u_1) = c_f(0)^{-[\mathbf{p}]} \cdot \prod_{j_2=1}^{p_2-1} \cdots \prod_{j_d=1}^{p_d-1} f(u_1, \omega_2^{j_2}, \ldots, \omega_d^{j_d}).$$

Elementary Galois theory shows that there exists, for every $i \in \{1, \ldots, q\}$, an integer $t_i \geq 1$ such that

$$H_i = G_i^{t_i} \tag{23.24}$$

for some irreducible polynomial $G_i \in \mathbb{Q}[u_1]$ with constant term 1. Let $\zeta(i) = \zeta(\mathbf{p}, \boldsymbol{\omega}, i)$ be a root of h_i, put $\mathbb{K}_i = \mathbb{Q}[\zeta(i)]$, $\mathbb{L}_i = \mathbb{L}[\zeta(i)]$, and recall that, if v is a place of $\mathbb{K}_i$, then

$$|a|_v^{m_i} = \prod_{\{w \in P^{\mathbb{L}_i} : w \text{ lies above } v\}} |a|_w \tag{23.25}$$

for every $a \in \mathbb{K}_i$, where $m_i = [\mathbb{L}_i : \mathbb{K}_i]$ (cf. Section 7, the second proof of Proposition 17.2 and Example 17.3 (2)).

Let us now return to the assertion of the lemma and assume that $n \geq 2kd$ obeys (23.19), and that $\psi^-, \psi^+ \in \mathfrak{R}_d$ satisfy (23.20)–(23.22). As the ideal $(f) = f\mathfrak{R}_d \subset \mathfrak{R}_d$ is prime, (23.22) is equivalent to the condition that

$$
\begin{aligned}
\psi^-(\xi) &= \sum_{\mathbf{r} \in Q(2^n,1)^-} c_{\psi^-}(\mathbf{r})\xi^{\mathbf{r}} \\
&= \sum_{\mathbf{r} \in Q(2^n,1)^+} c_{\psi^+}(\mathbf{r})\xi^{\mathbf{r}} = \psi^+(\xi) \\
&= \psi(\xi), \quad \text{say,}
\end{aligned}
\tag{23.26}
$$

for every $\xi = (\xi_1, \ldots, \xi_d)$ in the variety $V(f) = \{c \in (\overline{\mathbb{Q}}^\times)^d : f(c) = 0\}$, where $\xi^{\mathbf{r}} = \xi_1^{r_1} \cdot \ldots \cdot \xi_d^{r_d}$ for every $\mathbf{r} = (r_1, \ldots, r_d) \in \mathbb{Z}^d$, and (23.23) is equivalent to claiming that

$$\psi(\xi) = 0 \tag{23.27}$$

for every $\xi \in V_{\mathbb{C}}(f)$. In particular, if $\zeta(i) = \zeta(\mathbf{p}, \boldsymbol{\omega}, i)$ is the root of h_i chosen above, then $f(\zeta(i), \omega_2, \ldots, \omega_d) = 0$, i.e. $\xi(i) = \xi(\mathbf{p}, \boldsymbol{\omega}, i) = (\zeta(i), \omega_2, \ldots, \omega_d) \in V(f)$, and we have to show that $\psi(\xi(i)) = 0$.

Note that

$$
|\psi(\xi(i))|_w = \left| \sum_{\mathbf{r} \in Q(2^n,1)^-} c_{\psi^-}(\mathbf{r})\zeta(i)^{\mathbf{r}} \right|_w
= \left| \sum_{\mathbf{r} \in Q(2^n,1)^+} c_{\psi^+}(\mathbf{r})\zeta(i)^{\mathbf{r}} \right|_w
\tag{23.28}
$$

for every $w \in P^{\mathbb{L}_i}$. Put $|a|_w^- = \min\{|a|_w, |a|_w^{-1}\}$ for every non-zero element $a \in \mathbb{L}$ and conclude from (23.21) and (23.28) that

$$
|\psi(\xi(i))|_w \leq
\begin{cases}
\left(|\zeta(i)|_w^-\right)^{n^2} & \text{if } w \text{ is finite,} \\
\left(|\zeta(i)|_w^-\right)^{n^2} \cdot \left|\sum_{\mathbf{r} \in Q(2^n,1)^+} 8k^3 |\mathbf{r}|^{12d}\right|_w & \text{if } w \text{ is infinite,}
\end{cases}
$$

since $|\omega_j|_w = 1$ for all $j = 2, \ldots, d$ and $w \in P^{\mathbb{L}_i}$. An elementary calculation shows that

$$\sum_{\mathbf{r} \in Q(2^n,1)^+} 8k^3 |\mathbf{r}|^{12d} < \sum_{j=n^2}^{2^n+n^2} 8k^3 (2d-1)(2j+1)^{d-1} j^{12d}$$

$$< 8k^3(2d-1)2^{d-1} \cdot \sum_{j=n^2}^{2^n+n^2} (j+1)^{13d-1}$$

$$< 2^d k^3 (2^n + n^2 + 1)^{13d} < k^3 2^{14dn},$$

so that

$$|\psi(\xi(i))|_w \leq \begin{cases} \left(|\zeta(i)|_w^-\right)^{n^2} & \text{if } w \text{ is finite,} \\ \left(|\zeta(i)|_w^-\right)^{n^2} \cdot |k^3 2^{14dn}|_w & \text{if } w \text{ is infinite,} \end{cases}$$

for every $w \in P^{\mathbb{L}_i}$. According to (23.25),

$$\prod_{\{w \in P^{\mathbb{L}_i} : w \text{ lies above } v\}} |\psi(\xi(i))|_w \leq \begin{cases} \left(|\zeta(i)|_v^-\right)^{m_i n^2} & \text{if } v \text{ is finite,} \\ \left(|\zeta(i)|_v^-\right)^{m_i n^2} \cdot |k^3 2^{14dn}|_v^{m_i} & \text{if } v \text{ is infinite,} \end{cases}$$

for every $v \in P^{\mathbb{K}_i}$, and

$$\prod_{w \in P^{\mathbb{L}_i}} |\psi(\xi(i))|_w \leq \left(\prod_{v \in P^{\mathbb{K}_i}} |\zeta(i)|_v^- \right)^{m_i n^2} \cdot (k^3 2^{14dn})^{m_i [\mathbb{K}_i : \mathbb{Q}]}.$$

By varying $i \in \{1, \ldots, q\}$ and using (23.24) we obtain that

$$\prod_{i=1}^{q} \left(\prod_{w \in P^{\mathbb{L}_i}} |\psi(\xi(i))|_w \right)^{t_i/m_i[\mathbf{p}]}$$

$$\leq \prod_{i=1}^{q} \left(\prod_{v \in P^{\mathbb{K}_i}} |\zeta(i)|_v^- \right)^{t_i n^2/[\mathbf{p}]} \cdot \prod_{i=1}^{q} (k^3 2^{14dn})^{t_i [\mathbb{K}_i : \mathbb{Q}]/[\mathbf{p}]}. \tag{23.29}$$

According to (17.16),

$$\mathbb{M}(G_i)^{-2} = \left(\prod_{v \in P^{\mathbb{K}_i}} \exp(\log^+ |\zeta(i)|_v) \right)^{-2} = \prod_{v \in P^{\mathbb{K}_i}} |\zeta(i)|_v^-,$$

and by taking the product over $i = 1, \ldots, q$ and using (23.25) we see that

$$\prod_{i=1}^{q} \left(\prod_{v \in P^{\mathbb{K}_i}} |\zeta(i)|_v^- \right)^{t_i n^2/[\mathbf{p}]} = \mathbb{M}(F^{(\mathbf{P})})^{-2n^2/[\mathbf{p}]}, \tag{23.30}$$

where

$$F^{(\mathbf{P})}(u_1) = \prod_{j_2=1}^{p_2-1} \cdots \prod_{j_d=1}^{p_d-1} f(u_1, \omega_2^{j_2}, \ldots, \omega_d^{j_d})$$

$$= H(u_1) \cdot c_f(0)^{[\mathbf{p}]} = \prod_{i=1}^{q} G_i(u_1)^{t_i} \cdot c_f(0)^{[\mathbf{p}]}.$$

As in (21.32) we calculate that

$$\log \mathrm{M}\,(F^{(\mathbf{p})})^{-2n^2/[\mathbf{p}]} = \frac{1}{[\mathbf{p}]} \sum_{j_2=1}^{p_2-1} \cdots \sum_{j_d=1}^{p_d-1} \int_{\mathbb{S}} \log |f(s, \omega_2^{j_2}, \ldots, \omega_d^{j_d})|\, d\lambda_{\mathbb{S}}(s)$$

$$\to -2n^2 \cdot \int_{\mathbb{S}^d} \log |f(\mathbf{s})|\, d\lambda_{\mathbb{S}^d}(\mathbf{s}) = -2n^2 h(\alpha) \qquad (23.31)$$

as the distinct primes $(p_2, \ldots, p_d)$ tend to ∞, and by counting degrees we see that

$$\prod_{i=1}^{q} (k^3 2^{14dn})^{t_i [\mathbb{K}_i : \mathbb{Q}]/[\mathbf{p}]} = (k^3 2^{14dn})^{r_1^*} \qquad (23.32)$$

for every $\mathbf{p} = (p_2, \ldots, p_d)$. From (23.29)–(23.32) we conclude that

$$\prod_{i=1}^{q} \left(\prod_{w \in P^{\mathbb{L}_i}} |\psi(\xi(i))|_w \right)^{t_i/m_i[\mathbf{p}]} \leq \mathrm{M}\,(F(\mathbf{p}))^{-2n^2/[\mathbf{p}]} \cdot (k^3 2^{14dn})^{r_1^*}$$

$$\to e^{-2n^2 h(\alpha)} \cdot (k^3 2^{14dn})^{r_1^*}$$

as the primes p_i tend to ∞. In particular, since n satisfies (23.19),

$$\prod_{i=1}^{q} \left(\prod_{w \in P^{\mathbb{L}_i}} |\psi(\xi(i))|_w \right)^{t_i/m_i[\mathbf{p}]} < 1$$

whenever the distinct primes $p_2, \ldots, p_d$ are sufficiently large. The product formula (17.15) shows that

$$\prod_{i=1}^{q} \left(\prod_{w \in P^{\mathbb{L}_i}} |\psi(\xi(i))|_w \right)^{t_i/m_i[\mathbf{p}]} = \begin{cases} 0 & \text{if } \prod_{i=1}^{q} \psi(\xi(i)) = 0, \\ 1 & \text{otherwise,} \end{cases}$$

and hence that

$$\prod_{i=1}^{q} \psi(\xi(\mathbf{p}, \boldsymbol{\omega}, i)) = 0 \qquad (23.33)$$

for every vector $\mathbf{p} = (p_2, \ldots, p_d)$ consisting of distinct and sufficiently large rational primes, and for all corresponding primitive unit roots $\boldsymbol{\omega} = (\omega_2, \ldots, \omega_d)$.

For every $\mathbf{s} = (s_2, \ldots, s_d) \in \mathbb{S}^{d-1}$ we set $f(u_1, \mathbf{s}) = f(u_1, s_2, \ldots, s_d)$ and denote by $\zeta(\mathbf{s}, 1), \ldots, \zeta(\mathbf{s}, r_1^*)$ the roots of $f(u_1, \mathbf{s})$. By (23.26), (23.33) and continuity,

$$\prod_{i=1}^{r_1^*} \psi^-(\zeta(\mathbf{s}, i), s_2, \ldots, s_d) = \prod_{i=1}^{r_1^*} \psi^+(\zeta(\mathbf{s}, i), s_2, \ldots, s_d) = 0$$

for every $\mathbf{s} \in \mathbb{S}^{d-1}$, and we conclude that $\psi^-, \psi^+ \in (f) = f\Re_d$, since the complex varieties $V_{\mathbb{C}}(\psi^-)$ and $V_{\mathbb{C}}(\psi^+)$ of ψ^- and ψ^+ contain generic points of $V_{\mathbb{C}}(f)$ (cf. [83]). This shows that (23.20)–(23.22) implies (23.23) and completes the proof of the lemma. $\square$

Define a Borel map $\phi^{(k)} \colon X \longrightarrow \mathbb{Z}_{/k}^{\mathbb{Z}^d}$ by setting

$$(\phi^{(k)}(x))_{\mathbf{n}} = \sum_{j=0}^{k-1} k \cdot 1_{P_j(k)}(x_{\mathbf{n}})$$

for every $x = (x_{\mathbf{n}}) \in X \subset \mathbb{T}^{\mathbb{Z}^d}$ and $\mathbf{n} \in \mathbb{Z}^d$, where $\mathcal{P}^{(k)} = \{P_0^{(k)}, \ldots, P_{k-1}^{(k)}\}$ is the partition of $\mathbb{T}$ defined before the statement of Lemma 23.9, denote by $T^{(k)}$ the shift-action (5.8) of $\mathbb{Z}^d$ on $\mathbb{Z}_{/k}^{\mathbb{Z}^d}$, and set $\mu^{(k)} = \lambda_X(\phi^{(k)})^{-1}$. From the definition of $\phi^{(k)}$ it is clear that

$$\phi^{(k)} \cdot \alpha_{\mathbf{n}} = T_{\mathbf{n}}^{(k)} \cdot \phi^{(k)}$$

for every $\mathbf{n} \in \mathbb{Z}^d$, and that $\mu^{(k)} \in M_1(\mathbb{Z}_{/k}^{\mathbb{Z}^d})^{T^{(k)}}$.

LEMMA 23.14. *The measure* $\mu = \mu^{(k)} \in M_1(\mathbb{Z}_{/k}^{\mathbb{Z}^d})^{T^{(k)}}$ *is summably Vershik and hence Bernoulli.*

PROOF. For every $n \geq 1$ we put $m_n = \sum_{l=1}^{n}(l+d+1)^2$, $N_n = 2^n + 2m_n$, and define a partial cover

$$\mathcal{A}^{(n)} = \left\{ B_{N_{n-1}} + \sum_{l=1}^{d} i_l((n+d+1)^2 + 2^{n-1} + m_{n-1})e^{(l)} \right.$$
$$\left. : (i_1, \ldots, i_d) \in \{1, -1\}^d \right\}$$

of B_{N_n}, $n \geq 2$, by translates of $B_{N_{n-1}}$ with

$$\sum_{n \geq 2} \left(1 - \frac{|[\mathcal{A}^{(n)}]|}{|B_{N_n}|} \right) < \infty. \tag{23.34}$$

From (23.7) we know that

$$\bar{d}_{B_{N_n}}(\mu, \mu^{\mathcal{A}^{(n)}}) \leq \bar{d}_{[\mathcal{A}^{(n)}]}(\mu, \mu^{\mathcal{A}^{(n)}}) + \left(1 - \frac{|[\mathcal{A}^{(n)}]|}{|B_{N_n}|} \right)$$

for every $n \geq 2$, and we claim that

$$\bar{d}_{[\mathcal{A}^{(n)}]}(\mu, \mu^{\mathcal{A}^{(n)}}) < 30dk2^{d/2}n^{-3d}$$

whenever $n \geq 2kd$ satisfies (23.19). Indeed,

$$[\mathcal{A}^{(n)}] \subset Q(2^{n+1}, j)^+ \cup Q(2^{n+1}, j)^-$$

for $j = 1, \ldots, d$, and Lemma 23.13 shows that the partitions

$$\bigvee_{\mathbf{n} \in [\mathcal{A}^{(n)}] \cap Q(2^{n+1}, j)+} \mathcal{P}_{\mathbf{n}}^{(k)}, \qquad \bigvee_{\mathbf{n} \in [\mathcal{A}^{(n)}] \cap Q(2^{n+1}, j)-} \mathcal{P}_{\mathbf{n}}^{(k)}$$

are $30k2^{d/2}n^{-3d}$-independent whenever n is sufficiently large. We define partitions $\mathcal{B}^{(n,j)}$, $j = 0, \ldots, d$, of $[\mathcal{A}^{(n)}]$ by setting $\mathcal{B}^{(n,0)} = \{[\mathcal{A}^{(n)}]\}$ and

$$\mathcal{B}^{(n,j)} = \{[\mathcal{A}^{(n)}] \cap Q(2^{n+1}, 1)^{\pm} \cap \cdots \cap Q(2^{n+1}, j)^{\pm}\}$$

for every $j = 1, \ldots, d$. From (23.8) it is clear that

$$\bar{d}_{[\mathcal{A}^{(n)}]}(\mu^{\mathcal{B}^{(n,j-1)}}, \mu^{\mathcal{B}^{(n,j)}}) \leq 30k2^{d/2}n^{-3d}$$

for $j = 1, \ldots, d$. Since $\mathcal{B}^{(n,0)} = \{[\mathcal{A}^{(n)}]\}$ and $\mathcal{B}^{(n,d)} = \mathcal{A}^{(n)}$ we conclude that

$$\bar{d}_{[\mathcal{A}^{(n)}]}(\mu, \mu^{\mathcal{A}^{(n)}}) \leq \sum_{j=1}^{d} \bar{d}_{[\mathcal{A}^{(n)}]}(\mu^{\mathcal{B}^{(n,j-1)}}, \mu^{\mathcal{B}^{(n,j)}}) \leq 30dk2^{d/2}n^{-3d}$$

and, by (23.7), that

$$\bar{d}_{B_{N_n}}(\mu, \mu^{\mathcal{A}^{(n)}}) \leq 30dk2^{d/2}n^{-3d} + \left(1 - \frac{|[\mathcal{A}^{(n)}]|}{|B_{N_n}|}\right)$$

for every sufficiently large $n \geq 2$. In conjunction with (23.34) this proves that $\mu = \mu^{(k)}$ is summably Vershik, and the Bernoullicity of $\mu^{(k)}$ follows from Theorem 23.7. $\square$

PROPOSITION 23.15. *If $\mathfrak{p} \subset \mathfrak{R}_d$ is a prime ideal with $h(\alpha^{\mathfrak{R}_d/\mathfrak{p}}) > 0$ then $\alpha^{\mathfrak{R}_d/(f)}$ is Bernoulli.*

PROOF. Let $\mathfrak{p} \subset \mathfrak{R}_d$ be a prime ideal, and let $\alpha = \alpha^{\mathfrak{R}_d/\mathfrak{p}}$. If $h(\alpha) > 0$, then $\mathfrak{p}$ is principal by Corollary 18.5. The cases where $\mathfrak{p} = \{0\}$ or $\mathfrak{p} = (p)$ for some rational prime $p > 1$ were dealt with in the discussion preceding Lemma 23.10. In order to complete the proof of the proposition we assume that $\mathfrak{p} = (f) = f\mathfrak{R}_d$ for some non-zero, irreducible polynomial $f = \sum_{\mathbf{n} \in \mathbb{Z}^d} c_f(\mathbf{n})u^{\mathbf{n}} \in \mathfrak{R}_d$ with at least two non-zero coefficients and apply Lemma 23.10 to assume in addition that f is nice in the sense of Definition 23.11. We realize $\alpha = \alpha^{\mathfrak{R}_d/(f)}$ as the shift-action of $\mathbb{Z}^d$ on the closed, shift-invariant subgroup $X = X^{\mathfrak{R}_d/(f)} \subset \mathbb{T}^{\mathbb{Z}^d}$ in (5.9)–(5.10) and observe that the Bernoullicity of α is equivalent to the assertion that the measure $\lambda_X \in M_1(\mathbb{T}^{\mathbb{Z}^d})^{\sigma}$ is Bernoulli, where σ is the shift-action (5.8) of $\mathbb{Z}^d$ on $\mathbb{T}^{\mathbb{Z}^d}$. If we identify each partition $\mathcal{P}_{\mathbf{n}}^{(k)}$ in (23.17) with the corresponding partition of $\mathbb{T}^{\mathbb{Z}^d} \supset X$, then Lemma 23.14 amounts to saying that the σ-invariant sigma-algebra $\mathfrak{V}^{(k)} = \bigvee_{\mathbf{n} \in \mathbb{Z}^d} \sigma_{-\mathbf{n}}(\mathcal{P}_0^{(k)}) = \bigvee_{\mathbf{n} \in \mathbb{Z}^d} \mathcal{P}_{\mathbf{n}}^{(k)} \subset \mathfrak{B}_{\mathbb{T}^{\mathbb{Z}^d}}$ is a Bernoulli factor of σ on $(\mathbb{T}^{\mathbb{Z}^d}, \mathfrak{B}_{\mathbb{T}^{\mathbb{Z}^d}}, \lambda_X)$ for every $k \geq 2$. Since $\mathfrak{V}^{(k)} \nearrow \mathfrak{B}_{\mathbb{T}^{\mathbb{Z}^d}}$ as $k \to \infty$, the Bernoullicity of λ_X follows from Lemma 23.5. $\square$

In order to derive Theorem 23.1 from Proposition 23.15 we require a further proposition.

PROPOSITION 23.16. *Let* $\mathfrak{p} \subset \mathfrak{R}_d$ *be a prime ideal with* $h(\alpha^{\mathfrak{R}_d/\mathfrak{p}}) > 0$, *and let* $\mathfrak{N}$ *be a Noetherian* $\mathfrak{R}_d$-*module with submodules*

$$\mathfrak{N} = \mathfrak{N}_s \supset \cdots \supset \mathfrak{N}_0 = \{0\} \tag{23.35}$$

such that

$$\mathfrak{N}_j/\mathfrak{N}_{j-1} \cong \mathfrak{R}_d/\mathfrak{p} \tag{23.36}$$

for $j = 1, \ldots, s$, *where* $s \geq 1$. *Then* $\alpha^{\mathfrak{N}}$ *is Bernoulli.*

For the proof of Proposition 23.16 we require a relative version of one of the characterizations of Bernoullicity in Definition 23.6 and Theorem 23.7. Suppose that (Y, δ) and (Z, δ') are compact, metric spaces with diameter 1, and let $\sigma^{(Y)}$, $\sigma^{(Z)}$ and $T = \sigma^{(Y)} \times \sigma^{(Z)}$ be the shift-actions (5.8) of $\mathbb{Z}^d$ on $Y^{\mathbb{Z}^d}$, $Z^{\mathbb{Z}^d}$ and $(Y \times Z)^{\mathbb{Z}^d}$, respectively. We write a typical element in $(Y \times Z)^{\mathbb{Z}^d} \cong Y^{\mathbb{Z}^d} \times Z^{\mathbb{Z}^d}$ as (y, z) with $y = (y_{\mathbf{n}}) \in Y^{\mathbb{Z}^d}$ and $z = (z_{\mathbf{n}}) \in Z^{\mathbb{Z}^d}$ and denote by $\pi^{(Y)}(y, z) = y$ and $\pi^{(Z)}(y, z) = z$ the projections of (y, z) onto its coordinates in $Y^{\mathbb{Z}^d}$ and $Z^{\mathbb{Z}^d}$. Let $\mu \in M_1((Y \times Z)^{\mathbb{Z}^d})^T$, set $\mu^{(Z)} = \mu(\pi^{(Z)})^{-1} \in M_1(Z^{\mathbb{Z}^d})^{\sigma^{(Z)}}$, and apply standard decomposition theory to obtain a Borel map $z \mapsto \mu_z$ from $Z^{\mathbb{Z}^d}$ to $M_1(Y^{\mathbb{Z}^d})$ such that

$$\int h \, d\mu = \int_{Z^{\mathbb{Z}^d}} \int_{Y^{\mathbb{Z}^d}} h(y, z) \, d\mu_z(y) \, d\mu^{(Z)}(z) \tag{23.37}$$

for every continuous map $h \colon Y^{\mathbb{Z}^d} \times Z^{\mathbb{Z}^d} \longmapsto \mathbb{R}$, and

$$\mu_{\sigma_{\mathbf{n}}^{(Y)}} = \mu_z \sigma_{-\mathbf{n}}^{(Y)} \tag{23.38}$$

for every $z \in Z^{\mathbb{Z}^d}$ and $\mathbf{n} \in \mathbb{Z}^d$.

DEFINITION 23.17. *The measure* $\mu \in M_1((Y \times Z)^{\mathbb{Z}^d})^T$ *is relatively almost box independent with respect to* $Z^{\mathbb{Z}^d}$ *if, for* $\mu^{(Z)}$-*a.e.* $z \in Z^{\mathbb{Z}^d}$,

$$\lim_{N \to \infty} \limsup_{M \to \infty} \bar{d}_{BM}(\mu_z, \mu_z^{\mathcal{B}_N}) = 0.$$

PROPOSITION 23.18 ([91]). *Suppose that* $\mu \in M_1((Y \times Z)^{\mathbb{Z}^d})^T$ *satisfies the following conditions.*

(1) $\mu^{(Z)} \in M_1(Z^{\mathbb{Z}^d})^{\sigma^{(Z)}}$ *is Bernoulli;*

(2) μ *is relatively almost box independent with respect to* $Z^{\mathbb{Z}^d}$.

Then μ *is Bernoulli.*

We turn to the proof of Proposition 23.16 and denote, for every $l \geq 1$, by $\sigma^{(l)}$ the shift-action (5.8) of $\mathbb{Z}^d$ on $V^l = (\mathbb{T}^{\mathbb{Z}^d})^l \cong (\mathbb{T}^l)^{\mathbb{Z}^d}$. If $l = 1$ we write σ instead of $\sigma^{(1)}$. For the next two lemmas we assume that $f = \sum_{\mathbf{n} \in \mathbb{Z}^d} c_f(\mathbf{n}) u^{\mathbf{n}} \in \mathfrak{R}_d$ is a non-zero, irreducible polynomial with $h(\alpha^{\mathfrak{R}_d/(f)}) > 0$, which is nice in the sense of Definition 23.11. We regard $Y = X^{\mathfrak{R}_d/(f)}$ as the closed, shift-invariant subgroup (5.9) of V, identify $\alpha^{\mathfrak{R}_d/(f)}$ with the restriction of σ to Y (cf. (5.8) and (5.10)), and view λ_Y as an element of $M_1(V)^\sigma$. For every $v \in V$ we define a probability measure $\lambda^{(v)} \in M_1(V)$ by setting

$$\lambda^{(v)}(B) = \lambda_Y(B + v) \tag{23.39}$$

for every $B \in \mathfrak{B}_V$.

LEMMA 23.19. *The measures $\lambda^{(v)}$, $v \in V$, are uniformly almost box independent in the sense that*

$$\lim_{N \to \infty} \limsup_{M \to \infty} \left(\sup_{v \in V} \bar{d}_{B_M}(\lambda^{(v)}, (\lambda^{(v)})^{\mathcal{B}_N}) \right) = 0 \tag{23.40}$$

PROOF. Proposition 23.15 and Theorem 23.7 together imply that $\lambda_Y \in M_1(V)^\sigma$ is Bernoulli and thus almost box independent. Hence there exists, for every $M, N \geq 1$, a probability measure $\nu^{(N)} \in C(\lambda_Y, \lambda_Y^{\mathcal{B}_N}) \subset M_1(V^2)$ which is invariant under the $\mathbb{Z}^d$-action $\mathbf{n} \mapsto \sigma^{(2)}_{(2N+1)\mathbf{n}}$ on V^2, and which satisfies that

$$\lim_{M \to \infty} \bar{d}_{B_M}(\lambda_Y, \lambda_Y^{\mathcal{B}_N}) = \lim_{M \to \infty} \frac{1}{|B_M|} \sum_{\mathbf{n} \in B_M} \int_{V^2} \delta(\pi^{(1)}_{\{\mathbf{n}\}}, \pi^{(2)}_{\{\mathbf{n}\}}) \, d\nu^{(N)},$$

where $\pi^{(i)}_{\{\mathbf{n}\}}(v^{(1)}, v^{(2)}) = v_{\mathbf{n}}^{(i)}$ for every $(v^{(1)}, v^{(2)}) \in V^2$, $i = 1, 2$ and $\mathbf{n} \in \mathbb{Z}^d$. For every $v \in V$ we define a homeomorphism $R_v \colon V^2 \longmapsto V^2$ by $R_v(v^{(1)}, v^{(2)}) = (v^{(1)} + v, v^{(2)} + v)$ for every $(v^{(1)}, v^{(2)}) \in V^2$. The measure $\nu^{(N)} R_v \in M_1(V^2)$ satisfies that

$$\bar{d}_{B_M}(\lambda^{(v)}, (\lambda^{(v)})^{\mathcal{B}_N}) = \frac{1}{|B_M|} \sum_{\mathbf{n} \in B_M} \int_{V^2} \delta(\pi^{(1)}_{\{\mathbf{n}\}}, \pi^{(2)}_{\{\mathbf{n}\}}) \, d\nu^{(N)} R_v$$

$$= \bar{d}_{B_M}(\lambda_Y, \lambda_Y^{\mathcal{B}_N})$$

for every $v \in V$ and $M, N \geq 0$, and by letting first M and then N tend to infinity we obtain (23.40). $\square$

LEMMA 23.20. *Let $\mathfrak{N}$ be an $\mathfrak{R}_d$-module which satisfies (23.35)–(23.36) for some $s \geq 1$, and for $\mathfrak{p} = (f) = f\mathfrak{R}_d$. Then $\alpha^{\mathfrak{N}}$ is Bernoulli.*

PROOF. We prove the Bernoullicity of $\alpha^{\mathfrak{N}}$ by induction on the integer s in (23.35). If $s = 1$ then $\mathfrak{N} = \mathfrak{R}_d/(f)$, and $\alpha^{\mathfrak{N}}$ is Bernoulli by Proposition 23.15. Assume therefore that $s > 1$, and that we have proved the Bernoullicity of $\alpha^{\mathfrak{N}'}$ for every $\mathfrak{R}_d$-module $\mathfrak{N}'$ of the form $\mathfrak{N}' = \mathfrak{N}'_{s-1} \supset \cdots \supset \mathfrak{N}'_0 = \{0\}$ with $\mathfrak{N}'_j/\mathfrak{N}'_{j-1} \cong \mathfrak{R}_d/(f)$ for every $j = 1, \dots, s-1$.

Let $\mathfrak{N}$ be an $\mathfrak{R}_d$-module with submodules $\mathfrak{N} = \mathfrak{N}_s \supset \cdots \supset \mathfrak{N}_0 = \{0\}$ such that $\mathfrak{N}_j/\mathfrak{N}_{j-1} \cong \mathfrak{R}_d/(f)$ for every $j = 1, \ldots, s$. Choose elements $a_1, \ldots, a_s$ in $\mathfrak{N}$ such that $\mathfrak{N}_j = \mathfrak{R}_d \cdot a_j + \mathfrak{N}_{j-1}$ for $j = 1, \ldots, s$, and consider the corresponding surjective homomorphism $\hat{\psi} \colon \mathfrak{R}_d^s \longmapsto \mathfrak{N}$ with $\hat{\psi}(f_1, \ldots, f_s) = \sum_{i=1}^s f_i \cdot a_i$ for every $(f_1, \ldots, f_s) \in \mathfrak{R}_d^s$. The injective dual homomorphism $\psi \colon X^{\mathfrak{N}} \longmapsto V^s = \widehat{\mathfrak{R}_d^s}$ satisfies that $\sigma_{\mathbf{n}}^{(s)} \cdot \psi(x) = \psi \cdot \alpha_{\mathbf{n}}^{\mathfrak{N}}(x)$ for every $x \in X$ and $\mathbf{n} \in \mathbb{Z}^d$, and allows us to regard $X = X^{\mathfrak{N}}$ as a closed, shift-invariant subgroup of V^s (cf. Example 5.2 (4)). Furthermore, if $X_j = \mathfrak{N}_j^{\perp} \subset X \subset V^s$, then $X_0 = X$ and

$$X_j = \{x \in X : \pi^{(1)}(x) = \cdots = \pi^{(j)}(x) = 0\},$$
$$X_{j-1}/X_j \cong X^{\mathfrak{R}_d/(f)}, \qquad (23.41)$$
$$X^{\mathfrak{N}_j} = X/X_j \cong \eta^{(j)}(X) \subset V^j$$

for every $j = 1, \ldots, s$.

We set $W = \eta^{(s-1)}(X) \subset V^{s-1}$ and note that $\pi^{(s)}(X_{s-1}) = Y = X^{\mathfrak{R}_d/(f)} \subset V$. According to (23.41) our induction hypothesis implies that the restriction of $\sigma^{(s-1)}$ to $W \cong X^{\mathfrak{N}_{s-1}}$ is Bernoulli, and Proposition 23.15 guarantees that $\sigma = \sigma^{(1)}$ is Bernoulli on Y. As in (23.37)–(23.38) we obtain a family $\{\mu_w : w \in V^{s-1}\} \in M_1(V)$ with

$$\int h \, d\lambda_X = \int_{V^{s-1}} \int_V h(v^{(1)}, \ldots, v^{(s)}) \, d\mu_{(v^{(1)}, \ldots, v^{(s-1)})}(v^{(s)})$$
$$d\lambda_W(v^{(1)}, \ldots, v^{(s-1)}),$$

and (23.41) implies that there exists, for λ_W-a.e. $w \in V^{s-1}$, an element $v(w) \in V$ with

$$\mu_w = \lambda^{(v(w))}.$$

According to Lemma 23.19 and Definition 23.17, λ_X is relatively almost box independent with respect to V^{s-1}, and Proposition 23.18 shows that $\lambda_X = \lambda_{x^{\mathfrak{N}}} \in M_1(V^s)^{\sigma^{(s)}}$ is Bernoulli. $\square$

PROOF OF PROPOSITION 23.16. Let $\mathfrak{p} \subset \mathfrak{R}_d$ be a prime ideal with $h(\alpha^{\mathfrak{R}_d/\mathfrak{p}}) > 0$, and let $\mathfrak{N}$ be an $\mathfrak{R}_d$-module satisfying (23.35)–(23.36). Then $\mathfrak{p}$ is principal by Corollary 18.5, and the description at the beginning of the proof of Lemma 23.19 shows that there exists an $s \geq 1$ and elements $a_1, \ldots, a_s$ in $\mathfrak{N}$ such that $\mathfrak{N} = \sum_{j=1}^s \mathfrak{R}_d \cdot a_j$ and $\{h \in \mathfrak{R}_d : h \cdot a_j \in \mathfrak{N}_{j-1}\} = \mathfrak{p}$ for every $j = 1, \ldots, s$, where $\mathfrak{N}_0 = \{0\}$ and $\mathfrak{N}_j = \sum_{i=1}^j \mathfrak{R}_d \cdot a_i$ for $j = 1, \ldots, s$.

In particular, if $\mathfrak{p} = \{0\}$, then $\{a_1, \ldots, a_s\}$ is linearly independent over $\mathfrak{R}_d$, $\mathfrak{N} \cong \mathfrak{R}_d^s$, and $\alpha^{\mathfrak{N}}$ is conjugate to the shift-action (5.8) of $\mathbb{Z}^d$ on $(\mathbb{T}^s)^{\mathbb{Z}^d}$ and hence Bernoulli.

If $\mathfrak{p} = (p)$ for some rational prime $p > 1$ we set $X_j = \mathfrak{N}_j^{\perp} \subset X = X^{\mathfrak{N}} = \widehat{\mathfrak{N}}$, $j = 0, \ldots, s$, and observe that $X_s = \{0\}$, $X_{s-1} \cong X^{\mathfrak{R}_d/\mathfrak{p}}$, and use Remark 6.19 (4) to identify the $\mathbb{Z}^d$-action $\alpha^{X_{s-1}}$ induced by $\alpha = \alpha^{\mathfrak{N}}$ on X_{s-1} with the shift-action σ of $\mathbb{Z}^d$ on $\mathbb{Z}_{/p}^{\mathbb{Z}^d}$. We claim that there exists a Haar measure preserving Borel isomorphism $\phi\colon X \longmapsto X/X_{s-1} \times (\mathbb{Z}/p\mathbb{Z})^{\mathbb{Z}^d}$ which carries α to the cartesian product $\alpha^{X/X_{s-1}} \times \sigma$, where $\alpha^{X/X_{s-1}}$ is the $\mathbb{Z}^d$-action induced by α on X/X_{s-1}.

In order to construct ϕ we set $W = X/X_{s-1}$ and choose a Borel map $\zeta\colon W \longmapsto X$ with $\zeta(x + X_{s-1}) + X_{s-1} = x + X_{s-1}$ for every $x \in X$ (cf. Lemma 1.5.1 in [78]). Define a Borel isomorphism $\psi\colon X \longmapsto W \times X_{s-1}$ by setting $\psi(x) = (x + X_{s-1}, x - \zeta(x + X_{s-1}))$ for every $x \in X$, and use the identification of X_{s-1} with $V = (\mathbb{Z}/p\mathbb{Z})^{\mathbb{Z}^d}$ to regard ψ as a Borel isomorphism $\psi\colon X \longmapsto W \times V$. The $\mathbb{Z}^d$-action α' on $W \times V$ defined by $\alpha'_{\mathbf{n}} = \psi \cdot \alpha_{\mathbf{n}} \cdot \psi^{-1}$ is of the form

$$\alpha'_{\mathbf{n}}(w, v) = (\alpha_{\mathbf{n}}^W(w), \sigma_{\mathbf{n}}(v) + c(\mathbf{n}, w)) \tag{23.42}$$

for every $\mathbf{n} \in \mathbb{Z}^d$, where $c\colon \mathbb{Z}^d \times W \longmapsto V$ is a Borel map with

$$\sigma_{\mathbf{m}}(c(\mathbf{n}, w)) + c(\mathbf{m}, \alpha_{\mathbf{n}}^W(w)) = c(\mathbf{m} + \mathbf{n}, w) \tag{23.43}$$

for all $\mathbf{m}, \mathbf{n} \in \mathbb{Z}^d$ and $w \in W$. If c is of the form

$$c(\mathbf{n}, \cdot) = \sigma_{\mathbf{n}} \cdot \gamma - \gamma \cdot \alpha_{\mathbf{n}}^W \tag{23.44}$$

for every $\mathbf{n} \in \mathbb{Z}^d$, where $\gamma\colon W \longmapsto V$ is Borel, then the map $\phi(w, v) = (w, v + \gamma(w))$ from $W \times V$ to $W \times V$ carries α' to the product action $\alpha^W \times \sigma$ of $\mathbb{Z}^d$ on $W \times V$. In order to find a solution γ of (23.44) we write $\pi_{\{\mathbf{n}\}}\colon V \longmapsto \mathbb{Z}_{/p}$ for the $\mathbf{n}$-th coordinate projection and set $c_{\mathbf{n}}(\mathbf{m}, w) = \pi_{\{\mathbf{n}\}}(c(\mathbf{m}, w))$ for every $\mathbf{n} \in \mathbb{Z}^d$ and $w \in W$. Then (23.44) is equivalent to the solution of the equations

$$c_{\mathbf{m}}(\mathbf{n}, w) = \gamma_{\mathbf{m}+\mathbf{n}}(w) - \gamma_{\mathbf{m}}(\alpha_{\mathbf{n}}^W(w)) \tag{23.45}$$

for every $\mathbf{m}, \mathbf{n} \in \mathbb{Z}^d$ and $w \in W$ in terms of Borel maps $\gamma_{\mathbf{m}}\colon W \longmapsto \mathbb{Z}_{/p}$, $\mathbf{m} \in \mathbb{Z}^d$; if all these equations can be solved, then $\gamma\colon W \longmapsto V$ is obtained by setting $\pi_{\{\mathbf{m}\}} \cdot \gamma = \gamma_{\mathbf{m}}$ for every $\mathbf{m} \in \mathbb{Z}^d$. In order to solve (23.45) we set, for every $w \in W$, $\gamma_{\mathbf{0}}(w) = 0$ and use (23.43) to solve (23.45) inductively for $\mathbf{m} = \mathbf{0}$ and for every $\mathbf{n} \in \mathbb{Z}^d$.

This shows that α is indeed conjugate to $\alpha^{X/X_{s-1}} \times \sigma$ on $X/X^{s-1} \times V$, and by replacing $\mathfrak{N} = \mathfrak{N}_s$ with $\mathfrak{N}_{s-1}$, X with $\widehat{\mathfrak{N}_{s-1}} = X/X_{s-1}$, and α with $\alpha^{\mathfrak{N}_{s-1}}$ we see that α is conjugate to $\alpha^{X/X_{s-1}} \times \sigma \times \sigma$ on $X/X_{s-2} \times V^2$. By using induction we obtain after s steps that α is conjugate to $\sigma \times \cdots \times \sigma$ on $V \times \cdots \times V = V^s$, and hence Bernoulli.

Finally we have to deal with the case where $\mathfrak{p} = (f)$ for some irreducible element $f = \sum_{\mathbf{n} \in \mathbb{Z}^d} c_f(\mathbf{n}) u^{\mathbf{n}} \in \mathfrak{R}_d$ which has at least two non-zero coefficients. The same consideration as in Lemma 23.10 allows us to assume that f is nice (Definition 23.11), in which case the Bernoullicity of $\alpha = \alpha^{\mathfrak{N}}$ is proved in Lemma 23.20. $\square$

PROOF OF THEOREM 23.1. Let α be a $\mathbb{Z}^d$-action by automorphisms of a compact, abelian group X. If α is Bernoulli, then it is clear that α has completely positive entropy ([76]).

Conversely, if α has completely positive entropy, and if $\mathfrak{M} = \hat{X}$ is the $\mathfrak{R}_d$-module defined by Lemma 5.1, then Theorem 20.8 shows that every prime ideal $\mathfrak{p} \subset \mathfrak{R}_d$ associated with $\mathfrak{M}$ is positive, i.e. satisfies that $h(\alpha^{\mathfrak{R}_d/\mathfrak{p}}) > 0$. If $\mathfrak{M}$ is not Noetherian there exists an increasing sequence of finitely generated—and hence Noetherian—submodules $(\mathfrak{M}_k, \, k \geq 1)$ of $\mathfrak{M}$ with $\mathfrak{M} = \bigcup_{k \geq 1} \mathfrak{M}_k$. Since every prime ideal associated with any of the $\mathfrak{M}_k$ is also associated with $\mathfrak{M}$, and since every prime ideal associated with $\mathfrak{M}$ is associated with $\mathfrak{M}_k$ for some $k \geq 1$, Theorem 20.8 guarantees that $\alpha^{\mathfrak{M}}$ has completely positive entropy if and only if $\alpha^{\mathfrak{M}_k}$ has completely positive entropy for every $k \geq 1$. For every $k \geq 1$, $\alpha^{\mathfrak{M}_k}$ is the $\mathbb{Z}^d$-action induced by α on the quotient group $X/\mathfrak{M}_k^\perp$. As the α-invariant subgroups $\mathfrak{M}_k^\perp \subset X$ decrease to $\{0\}$ as $k \to \infty$, Lemma 23.5 shows that α is Bernoulli if and only if $\alpha^{\mathfrak{M}_k}$ is Bernoulli for every $k \geq 1$. This implies that it is enough to prove Theorem 23.1 under the additional assumption that the $\mathfrak{R}_d$-module $\mathfrak{M} = \hat{X}$ is Noetherian.

We denote by $\{\mathfrak{p}_1, \ldots, \mathfrak{p}_m\}$ the set of prime ideals associated with the module $\mathfrak{M}$ (which is now assumed to be Noetherian) and consider the Noetherian $\mathfrak{R}_d$-module $\mathfrak{N} = \mathfrak{N}^{(1)} \oplus \cdots \oplus \mathfrak{N}^{(m)} \supset \mathfrak{M}$ constructed in Corollary 6.3. Since each $\mathfrak{p}_i$ is positive, and since each of the modules $\mathfrak{N}^{(i)}$ satisfies (23.35)–(23.36) for some $s \geq 1$, and for $\mathfrak{p} = \mathfrak{p}_i$, Proposition 23.16 shows that $\alpha^{\mathfrak{N}^{(i)}}$ is Bernoulli for $i = 1, \ldots, m$. Hence $\alpha^{\mathfrak{N}^{(1)} \oplus \cdots \oplus \mathfrak{N}^{(m)}} = \alpha^{\mathfrak{N}^{(1)}} \times \cdots \times \alpha^{\mathfrak{N}^{(m)}}$ is Bernoulli, and α is Bernoulli by (6.7) and Lemma 23.4. $\square$

CONCLUDING REMARK 23.21. The exposition in this section follows [91], except that the proofs of some of the purely measure theoretic results presented there are omitted. Example 23.2 (1) is taken from [106]. For $d = 1$, Theorem 23.1 was established in [3], [4], [41], [58] and [72] (note that, for $d = 1$, ergodicity and completely positive entropy are equivalent by Corollary 20.6 and Theorem 20.14). The idea to use valuations in the proof of Bernoullicity of certain automorphisms of solenoids already appears in some unpublished notes of Katznelson. A proof of Theorem 23.1 for $d = 1$, based on a suggestion due to Lind to explore the connection between the product formula (17.15) for global fields and a certain asymptotic independence property ensuring Bernoullicity, can be found in [61]. For $d = 2$, the Bernoullicity of $\alpha^{\mathfrak{R}_2/(1+u_1+u_2)}$ was appears in [106], and of expansive $\mathbb{Z}^2$-actions with completely positive entropy in [108].

CHAPTER VII

Zero entropy

24. Entropy and dimension

One of the most interesting phenomena which arises in the transition from $\mathbb{Z}$-actions to $\mathbb{Z}^d$-actions by automorphisms of compact groups is the existence of non-trivial (e.g. mixing) actions with zero entropy. Theorem 19.5 characterizes the principal prime ideals $\mathfrak{p} \subset \mathfrak{R}_d$ which are null, and from Corollary 18.5 we know that every non-principal prime ideal $\mathfrak{p} \subset \mathfrak{R}_d$ is null. Theorem 6.5 (2) and Proposition 19.4 yield an abundance of mixing $\mathbb{Z}^d$-actions with zero entropy. One class of such actions on compact, connected, abelian groups is introduced in Section 7, where we investigate actions of the form $\alpha^{\mathfrak{R}_d/\mathfrak{a}}$ arising from prime ideals $\mathfrak{a} \subset \mathfrak{R}_d$ for which $V_{\mathbb{C}}(\mathfrak{a})$ is finite, and other zero entropy $\mathbb{Z}^d$-actions are considered in the Examples 4.16 (1), 5.3 (5), 6.18 (5), and 8.5 (1). Before introducing further examples of mixing $\mathbb{Z}^d$ with zero entropy we shall discuss briefly the restrictions $\alpha^{(\Gamma)}$ of a $\mathbb{Z}^d$-action α by automorphisms of a compact, abelian group X to various subgroups $\Gamma \subset \mathbb{Z}^d$. If $h(\alpha) = 0$, then $h(\alpha^{(\Gamma)})$ may be positive (even infinite) for certain subgroups $\Gamma \subset \mathbb{Z}^d$ of rank $r < d$. For a $\mathbb{Z}^d$-action of the form $\alpha = \alpha^{\mathfrak{R}_d/\mathfrak{p}}$, where $\mathfrak{p} \subset \mathfrak{R}_d$ is a prime ideal, this dependence of entropy on the rank of Γ involves the number $r(\mathfrak{p})$ introduced in the Propositions 8.2–8.3.

We begin our investigation with prime ideals $\mathfrak{p} \subset \mathfrak{R}_d$ with $p(\mathfrak{p}) > 0$, where $p(\mathfrak{p})$ is the characteristic of $\mathfrak{R}_d/\mathfrak{p}$ defined as in (6.2). For notation we refer to Section 13; in particular,

$$Q(m) = \{-m, \ldots, m-1\}^d \subset \mathbb{Z}^d \text{ for every } m \geq 1. \tag{24.1}$$

PROPOSITION 24.1. *Suppose that* $\mathfrak{p} \subset \mathfrak{R}_d$ *is a prime ideal such that* $p = p(\mathfrak{p}) > 0$ *and* $\alpha = \alpha^{\mathfrak{R}_d/\mathfrak{p}}$ *is ergodic, and let* $r = r(\mathfrak{p}) \in \{1, \ldots, d\}$ *be the integer defined in* Proposition 8.2. *If s is a non-negative real number, then*

$$\limsup_{m \to \infty} \frac{1}{\mathcal{U}} \frac{1}{(2m)^s} \log N \left(\bigvee_{\mathbf{m} \in Q(m)} \alpha_{-\mathbf{m}}(\mathcal{U}) \right)$$

$$= \limsup_{m \to \infty} \frac{1}{\mathcal{P}} \frac{1}{(2m)^s} \log h_{\lambda_X} \left(\bigvee_{\mathbf{m} \in Q(m)} \alpha_{-\mathbf{m}}(\mathcal{P}) \right) \quad (24.2)$$

$$= \begin{cases} \infty & \text{if and only if } s < r(\mathfrak{p}), \\ 0 & \text{if and only if } s > r(\mathfrak{p}), \end{cases}$$

where the suprema are taken over all open covers $\mathcal{U}$ and all finite, measurable partitions $\mathcal{P}$ of $X = X^{\mathfrak{R}_d/\mathfrak{p}}$, and where λ_X is the normalized Haar measure of X. In particular, the integer $r(\mathfrak{q})$ does not depend on the choice of the primitive subgroup $\Gamma \subset \mathbb{Z}^d$ in Proposition 8.2, and is a measurable conjugacy invariant.

PROOF. We realize the totally disconnected group $X = X^{\mathfrak{R}_d/\mathfrak{p}}$ as the closed, shift-invariant subgroup (6.19) of $\mathbb{F}_p^{\mathbb{Z}^d}$ and assume that $\alpha = \alpha^{\mathfrak{R}_d/\mathfrak{p}}$ is the shift-action of $\mathbb{Z}^d$ on X. As in (24.3) we denote by

$$\mathcal{P}_0 = \{[0]_0, \ldots, [p-1]_0\} \quad (24.3)$$

the state partition of X, where $[i]_{\mathbf{m}} = \{x = (x_{\mathbf{n}}) \in X : x_{\mathbf{m}} = i\}$ for $i = 0, \ldots, p-1$ and $\mathbf{m} \in \mathbb{Z}^d$. Then $\mathcal{P}_0$ generates the topology of X under α, and we set, for every $m \geq 0$,

$$\mathcal{P}_0(m) = \bigvee_{\mathbf{m} \in Q(m)} \alpha_{-\mathbf{m}}(\mathcal{P}_0).$$

Let $\Gamma, Q \subset \mathbb{Z}^d$ be the primitive subgroup and the finite set appearing in Proposition 8.2, and let $\bar{\Gamma} = \Gamma + Q$. As explained in the proof of that proposition, we may assume without loss in generality that $\Gamma = \{\mathbf{n} = (n_1, \ldots, n_d) \in \mathbb{Z}^d : n_{r+1} = \cdots = n_d = 0\}$. In this case we may also take it that $n_1 = \cdots = n_r = 0$ for every $\mathbf{n} = (n_1, \ldots, n_d) \in Q$; these additional assumptions will obviously not affect our claim. For every $m \geq 0$ we put $Q(m)' = Q(m) \cap \bar{\Gamma}$ and denote by $S(m)' \supset Q(m)'$ the unique maximal subset of $\mathbb{Z}^d$ with the property that $\pi_{S(m)'}(x) = \pi_{S(m)'}(x')$ for all $x, x' \in X$ with $\pi_{Q(m)'}(x) = \pi_{Q(m)'}(x')$, where π_S, $S \subset \mathbb{Z}^d$, denotes, as usual, the coordinate projection. If $S(m) = S(m') \cap Q(m)$, then it is clear from the proof of Proposition 8.2 that there exists a positive constants c such that $Q(cm) \subset S(m) \subset Q(m)$ for all $m \geq 1$. Since $\mathcal{P}_0$ generates to topology and hence the Borel field $\mathfrak{B}_X$ of X, we know that

$$\limsup_{m \to \infty} \frac{1}{\mathcal{U}} \frac{1}{(2m)^s} \log N \left(\bigvee_{\mathbf{m} \in Q(m)} \alpha_{-\mathbf{m}}(\mathcal{U}) \right) = \lim_{m \to \infty} \frac{1}{(2m)^s} \log N(\mathcal{P}_0(m))$$

and

$$\limsup_{m \to \infty} \frac{1}{(2m)^s} \log h_{\lambda_X} \left(\bigvee_{\mathbf{m} \in Q(m)} \alpha_{-\mathbf{m}}(\mathcal{P}) \right) = \lim_{m \to \infty} \frac{1}{(2m)^s} \log h_{\lambda_X}(\mathcal{P}_0(m)).$$

Furthermore,

$$\lim_{m \to \infty} \frac{1}{(2m)^s} \log N(\mathcal{P}_0(cm)) = \lim_{m \to \infty} \frac{1}{(2m)^s} \log h_{\lambda_X}(\mathcal{P}_0(cm))$$

$$\leq \lim_{m \to \infty} \frac{1}{(2m)^s} \log h_{\lambda_X} \left(\bigvee_{\mathbf{m} \in S(m)} \alpha_{-\mathbf{m}}(\mathcal{P}_0) \right)$$

$$= \lim_{m \to \infty} \frac{1}{(2m)^s} \log h_{\lambda_X} \left(\bigvee_{\mathbf{m} \in Q(m)'} \alpha_{-\mathbf{m}}(\mathcal{P}_0) \right)$$

$$= \lim_{m \to \infty} \frac{1}{(2m)^s} p^{|Q|(2m)^r}$$

$$\leq \lim_{m \to \infty} \frac{1}{(2m)^s} \log N(\mathcal{P}_0(cm))$$

$$= \lim_{m \to \infty} \frac{1}{(2m)^s} \log h_{\lambda_X}(\mathcal{P}_0(cm))$$

for every sufficiently large $m \geq 1$, which proves (24.2). $\square$

If the prime ideal $\mathfrak{p} \subset \mathfrak{R}_d$ satisfies that $p(\mathfrak{p}) = 0$, then the following examples show that we cannot expect a precise analogue of Proposition 24.1.

EXAMPLES 24.2. (1) Let ϕ_L the L-th cyclotomic polynomial for some $L \geq 1$, and let $f(u_1, \ldots, u_d) = \phi_L(u_d) \in \mathfrak{R}_d$. We use (5.9) to realize $\alpha = \alpha^{\mathfrak{R}_d/(f)}$ as the shift-action of $\mathbb{Z}^d$ on $X = X^{\mathfrak{R}_d/(f)} \subset \mathbb{T}^{\mathbb{Z}^d}$. Since $\phi_L(u_d)$ divides $u_d^L - 1$, every $x \in X$ satisfies that $\alpha_{L\mathbf{e}^{(1)}}(x) = x$, where $\mathbf{e}^{(j)} \in \mathbb{Z}^d$ is the j-th unit vector. We define $Q(m)$, $m \geq 1$, as in (24.1) and obtain that

$$\sup_{\mathcal{U}} \lim_{m \to \infty} \frac{1}{(2m)^s} \log N \left(\bigvee_{\mathbf{m} \in Q(m)} \alpha_{-\mathbf{m}}(\mathcal{U}) \right)$$

$$= \sup_{\mathcal{U}} \lim_{m \to \infty} \frac{1}{(2m)^s} \log N \left(\bigvee_{\mathbf{m} \in Q(m)} \alpha_{-\mathbf{m}}(\mathcal{U}') \right) \quad (24.4)$$

$$= \begin{cases} \infty & \text{if } s \leq r(\mathfrak{p}) = d - 1, \\ 0 & \text{if } s > r(\mathfrak{p}), \end{cases}$$

where the supremum is taken over all finite, open covers of $\mathcal{U}$ of X, and where, for each such cover $\mathcal{U}$, $\mathcal{U}' = \bigvee_{k=0}^{L-1} \alpha_{-k\mathbf{e}^{(1)}}(\mathcal{U})$ is a finite, open cover of X which is invariant under $\alpha_{\mathbf{e}^{(1)}}$.

(2) Let $0 \neq f \in \mathfrak{R}_d$ be an irreducible Laurent polynomial with $\mathbb{M}(f) > 1$, and let $\alpha = \alpha^{\mathfrak{R}_d/(f)}$ be the shift-action of $\mathbb{Z}^d$ on the subgroup $X = X^{\mathfrak{R}_d/(f)} \subset$

$\mathbb{T}^{\mathbb{Z}^d}$ described in (5.9). Then $h(\alpha) = \log \mathrm{M}(f) > 0$. If $\mathcal{U}$ is an open cover of X, then (24.4) changes to

$$\sup_{\mathcal{U}} \lim_{m \to \infty} \frac{1}{(2m)^s} \log N\left(\bigvee_{\mathbf{m} \in Q(m)} \alpha_{-\mathbf{m}}(\mathcal{U}) \right)$$

$$= \begin{cases} \infty & \text{if } s < r(\mathfrak{p}) + 1 = d, \\ h(\alpha) & \text{if } s = r(\mathfrak{p}) + 1, \\ 0 & \text{if } s > r(\mathfrak{p}) + 1. \end{cases} \qquad (24.5)$$

In Example 24.2 (1)–(2), $s = r(\mathfrak{p}) + 1$ is the smallest integer such that the expressions in (24.4) and (24.5) are finite. The next proposition shows that this property characterizes $r(\mathfrak{p})$ for every prime ideal $\mathfrak{p} \subset \mathfrak{R}_d$ such that $p(\mathfrak{p}) = 0$ and $\alpha^{\mathfrak{R}_d/\mathfrak{p}}$ is ergodic.

PROPOSITION 24.3. *Let* $\mathfrak{p} \subset \mathfrak{R}_d$ *be a prime ideal such that* $p(\mathfrak{p}) = 0$ *and* $\alpha = \alpha^{\mathfrak{R}_d/\mathfrak{p}}$ *is ergodic. If* $r = r(\mathfrak{p})$ *is defined as in Proposition 8.3, then* $s = r(\mathfrak{p}) + 1$ *is the smallest integer for which*

$$\sup_{\mathcal{U}} \lim_{m \to \infty} \frac{1}{(2m)^s} \log N\left(\bigvee_{\mathbf{m} \in Q(m)} \alpha_{-\mathbf{m}}(\mathcal{U}) \right)$$

$$= \sup_{\mathcal{P}} \lim_{m \to \infty} \frac{1}{(2m)^s} \log h_{\lambda_X}\left(\bigvee_{\mathbf{m} \in Q(m)} \alpha_{-\mathbf{m}}(\mathcal{P}) \right) < \infty. \qquad (24.6)$$

Here $Q(m)$, $m \geq 1$, *is defined in* (24.1), *the suprema in* (24.6) *are taken over all open covers* $\mathcal{U}$ *and all finite measurable partitions* $\mathcal{P}$ *of* $X = X^{\mathfrak{R}_d/\mathfrak{p}}$, *and* λ_X *is the normalized Haar measure of* X. *In particular, the integer* $r(\mathfrak{q})$ *does not depend on the choice of the primitive subgroup* $\Gamma \subset \mathbb{Z}^d$ *in Proposition 8.3, and is a measurable conjugacy invariant.*

PROOF. First we deal with the case where $\mathfrak{p} = \mathfrak{j}_c$ for some $c \in (\overline{\mathbb{Q}}^{\times})^d$, and where $r(\mathfrak{p}) = 0$ by Remark 8.4 (1). Then

$$\sup_{\mathcal{U}} \lim_{m \to \infty} \frac{1}{(2m)^s} \log N\left(\bigvee_{\mathbf{m} \in Q(m)} \alpha_{-\mathbf{m}}(\mathcal{U}) \right)$$

$$= \sup_{\mathcal{P}} \lim_{m \to \infty} \frac{1}{(2m)^s} \log h_{\lambda_X}\left(\bigvee_{\mathbf{m} \in Q(m)} \alpha_{-\mathbf{m}}(\mathcal{P}) \right)$$

$$= \begin{cases} h(\alpha) < \infty & \text{if } s = r(\mathfrak{p}) + 1 = 1, \\ 0 & \text{if } s > r(\mathfrak{p}) + 1. \end{cases}$$

This proves the proposition for prime ideals of the form $\mathfrak{p} = \mathfrak{j}_c$.

If $\mathfrak{p}$ is not of this form, then Proposition 8.3 shows that $r(\mathfrak{p}) \geq 1$. If $\mathfrak{p} = \{0\}$, the assertion of the Proposition is obvious. If $\mathfrak{p} \neq \{0\}$, let $\Gamma, Q \subset \mathbb{Z}^d$

be the primitive subgroup and the finite set appearing in Proposition 8.3, put $\bar{\Gamma} = \Gamma + Q$, and assume as in the proof of Proposition 24.1 that $\Gamma = \{ \mathbf{n} = (n_1, \ldots, n_d) \in \mathbb{Z}^d : n_{r+1} = \cdots = n_d = 0 \}$, and that $n_1 = \cdots = n_r = 0$ for every $\mathbf{n} = (n_1, \ldots, n_d) \in Q$. If $f_{r+1}, \ldots, f_d$ are the Laurent polynomials appearing in the proof of Proposition 8.3, then our choice of Γ implies that each f_j is a function of the variables $u_1, \ldots, u_j$. We write the Laurent polynomials f_j in the form (5.2) and define a linear subspace $S \subset \mathbb{R}^{\mathbb{Z}^d}$ by setting

$$S = \left\{ z = (z_\mathbf{m}) \subset \mathbb{R}^{\mathbb{Z}^d} : \sum_{\mathbf{n} \in \mathbb{Z}^d} c_{f_j}(\mathbf{n}) z_{\mathbf{m}+\mathbf{n}} = 0 \quad \text{for all } \mathbf{m} \in \mathbb{Z}^d \text{ and } j = r+1, \ldots, d \right\}$$

(cf. (6.9)). For every $m \geq 1$ we set $Q(m)' = Q(m) \cap \bar{\Gamma}$ and denote by $S(m)' \supset Q(m)'$ the largest subset of $\mathbb{Z}^d$ such that $\pi_{S(m)'}(z) = \pi_{S(m)'}(z')$ for all $z, z' \in S$ with $\pi_{Q(m)'}(z) = \pi_{Q(m)'}(z')$, where π_F is, as usual, the coordinate projection which restricts each $z \colon \mathbb{Z}^d \longmapsto \mathbb{R}$ in S to a set $F \subset \mathbb{Z}^d$. If $S(m) = S(m)' \cap Q(m)$, then it is clear that there exists a positive constant c such that $Q(cm) \subset S(m) \subset Q(cm)$ for all $m \geq 1$.

We realize $X = X^{\mathfrak{R}_d/\mathfrak{p}} \subset \mathbb{T}^{\mathbb{Z}^d}$ and $\alpha = \alpha^{\mathfrak{R}_d/\mathfrak{p}}$ as in (5.9) and denote by $\vartheta(s, t) = \|s - t\|$ the metric (17.7) on $\mathbb{T}$. The Propositions 13.1–13.2 and 13.7 guarantee that

$$\sup_{\mathcal{U}} \lim_{m \to \infty} \frac{1}{(2m)^s} \log N \left(\bigvee_{\mathbf{m} \in Q(m)} \alpha_{-\mathbf{m}}(\mathcal{U}) \right)$$

$$= \sup_{\mathcal{P}} \lim_{m \to \infty} \frac{1}{(2m)^s} \log h_{\lambda_X} \left(\bigvee_{\mathbf{m} \in Q(m)} \alpha_{-\mathbf{m}}(\mathcal{P}) \right)$$

$$= \lim_{\varepsilon \to 0} \lim_{m \to \infty} \frac{1}{(2m)^s} \log \lambda_X (B'_\vartheta(Q(m), \varepsilon)),$$

where $B'_\vartheta(\varepsilon) = \{ x \in X : \|x_0\| < \varepsilon \}$ and $B'_\vartheta(Q(m), \varepsilon) = \bigcap_{\mathbf{m} \in Q(m)} (\alpha_{-\mathbf{m}}(B'_\vartheta(\varepsilon)).$ From (5.9) and the choice of $f_{r+1}, \ldots, f_d$ in the proof of Proposition 8.3 it is clear that we can find a constant $C > 0$ such that, for all sufficiently small $\varepsilon > 0$ and all sufficiently large m,

$$\lambda_X \left(\bigcap_{\mathbf{m} \in S(m)} \alpha_{-\mathbf{m}}(B'_\vartheta(\varepsilon)) \right) \geq C^m \lambda_X \left(\bigcap_{\mathbf{m} \in Q(m)'} \alpha_{-\mathbf{m}}(B'_\vartheta(\varepsilon)) \right)$$

$$= C^m \varepsilon^{|Q(m)'|} = C^m \varepsilon^{|Q|(2m)^r}.$$

Hence

$$\lim_{\varepsilon \to 0} \lim_{m \to \infty} \frac{1}{(2m)^s} \log \lambda_X (B'_\vartheta(Q(m), \varepsilon))$$

$$\leq \lim_{\varepsilon \to 0} \lim_{m \to \infty} \frac{1}{(2m)^s} \log \lambda_X \left(\bigcap_{\mathbf{m} \in S(m/c)} \alpha_{-\mathbf{m}}(B'_\vartheta(\varepsilon)) \right)$$

$$\leq \lim_{\varepsilon \to 0} \lim_{m \to \infty} \frac{1}{(2m)^s} \log(C^{m/c} \varepsilon^{|Q|(2m/c+1)^r}),$$

so that

$$\lim_{\varepsilon \to 0} \lim_{m \to \infty} \frac{1}{(2m)^s} \log \lambda_X(B'_\vartheta(Q(m), \varepsilon)) < \infty$$

if $s \geq r(\mathfrak{p}) + 1$. On the other hand, if $s \leq r(\mathfrak{p})$, then

$$\lim_{\varepsilon \to 0} \lim_{m \to \infty} \frac{1}{(2m)^s} \log \lambda_X(B'_\vartheta(Q(m), \varepsilon))$$

$$\geq \lim_{\varepsilon \to 0} \lim_{m \to \infty} \frac{1}{(2m)^s} \log \lambda_X \left(\bigcap_{\mathbf{m} \in Q(m)'} \alpha_{-\mathbf{m}}(B'_\vartheta(\varepsilon)) \right)$$

$$= \lim_{\varepsilon \to 0} \lim_{m \to \infty} \frac{1}{(2m)^s} \log \varepsilon^{|Q|(2m)^r(\mathfrak{p})} = \infty,$$

which completes the proof of (24.6). The remaining assertions of this proposition are immediate consequences of (24.6). $\square$

REMARK 24.4. The Propositions 24.1 and 24.3 introduce a notion of 'entropy dimension' for an ergodic $\mathbb{Z}^d$-action α by automorphisms of a compact, abelian group X: put

$$\dim_h(\alpha) = s(\alpha), \tag{24.7}$$

where $s(\alpha)$ is the unique positive real number such that

$$\sup_{\mathcal{U}} \lim_{m \to \infty} \frac{1}{(2m)^s} \log N \left(\bigvee_{\mathbf{m} \in Q(m)} \alpha_{-\mathbf{m}}(\mathcal{U}) \right)$$

$$= \sup_{\mathcal{P}} \lim_{m \to \infty} \frac{1}{(2m)^s} \log h_{\lambda_X} \left(\bigvee_{\mathbf{m} \in Q(m)} \alpha_{-\mathbf{m}}(\mathcal{P}) \right) \tag{24.8}$$

$$= \begin{cases} \infty & \text{if } s < s(\alpha), \\ 0 & \text{if } s > s(\alpha). \end{cases}$$

If $\alpha = \alpha^{\mathfrak{R}_d/\mathfrak{p}}$ for a prime ideal $\mathfrak{p} \subset \mathfrak{R}_d$, then (24.2) and (24.4)–(24.6) imply that $\dim_h(\alpha) = r(\mathfrak{p})$ if $p(\mathfrak{p}) > 0$, and that $\dim_h(\alpha) \in (r(\mathfrak{p}), r(\mathfrak{p}) + 1]$ if $p(\mathfrak{p}) = 0$. If α is an arbitrary, ergodic $\mathbb{Z}^d$-action by automorphisms of a compact group X satisfying the d.c.c., then $\dim_h(\alpha)$ is the maximum of the values $\dim_h(\alpha^{\mathfrak{R}_d/\mathfrak{p}})$ as $\mathfrak{p}$ runs through the set of prime ideals associated with the Noetherian module $\mathfrak{M} = \hat{X}$ (cf. Lemma 5.1 and Proposition 5.4). Note that $\dim_h(\alpha)$ is also related to the dimensions of those subgroups $\Gamma \subset \mathbb{Z}^d$ such that the restrictions $\alpha^{(\Gamma)}$

of α to Γ have positive entropy (cf. Example 24.5 below). Other dynamical properties of the actions $\alpha^{(\Gamma)}$ were discussed in Remark 8.4.

We end this section with a few more examples.

EXAMPLES 24.5. (1) In Example 24.2 (1), $h(\alpha) = 0$, and $\dim_h(\alpha) = d-1$. If $k \in \{1, \ldots, d-1\}$, then every subgroup $\Gamma \subset \mathbb{Z}^d$ of rank k with $\Gamma \cap \{k\mathbf{e}^{(1)} : k \in \mathbb{Z}\} = \{\mathbf{0}\}$ satisfies that $h(\alpha^{(\Gamma)}) = \infty$. However, if $\Gamma \subset \mathbb{Z}^d$ is a subgroup of rank k with $\Gamma \cap \{k\mathbf{e}^{(1)} : k \in \mathbb{Z}\} \supsetneqq \{\mathbf{0}\}$, then $h(\alpha^{(\Gamma)}) = 0$.

(2) In Example 24.2 (2), $0 < h(\alpha) < \infty$, and $\dim_h(\alpha) = d$. If $k \in \{1, \ldots, d-1\}$, then every subgroup $\Gamma \subset \mathbb{Z}^d$ of rank k satisfies that $h(\alpha^{(\Gamma)}) = \infty$.

(3) Let $d = 3$, $\mathfrak{p} = (1 - 2u_2, 1 - 3u_3) \subset \mathfrak{R}_3$, and let $\alpha = \alpha^{\mathfrak{R}_3/\mathfrak{p}}$. Then $h(\alpha) = 0$ by Proposition 17.5, and $\dim_h(\alpha) = 2$. If $\Gamma \subset \mathbb{Z}^3$ is subgroup of rank 2, then $0 \le h(\alpha^{(\Gamma)}) < \infty$: for example,

$$h(\alpha^{(\Gamma)}) = \begin{cases} \log 2 & \text{if } \Gamma = \{(n_1, n_2, 0) : (n_1, n_2) \in \mathbb{Z}^2\}, \\ \log 3 & \text{if } \Gamma = \{(n_1, 0, n_2) : (n_1, n_2) \in \mathbb{Z}^2\}, \\ 0 & \text{if } \Gamma = \{(0, n_1, n_2) : (n_1, n_2) \in \mathbb{Z}^2\}. \end{cases}$$

If $\Gamma \subset \mathbb{Z}^3$ has rank 1, then $h(\alpha^{(\Gamma)}) = \infty$.

(4) Let $d = 3$, $\mathfrak{p} = (1 - u_2, 1 - 3u_3) \subset \mathfrak{R}_3$, and let $\alpha = \alpha^{\mathfrak{R}_3/\mathfrak{p}}$. Then $h(\alpha) = 0$, $\dim_h(\alpha) = 2$, and

$$h(\alpha^{(\Gamma)}) = \begin{cases} 0 & \text{if } \Gamma = \{(n_1, n_2, 0) : (n_1, n_2) \in \mathbb{Z}^2\}, \\ \log 3 & \text{if } \Gamma = \{(n_1, 0, n_2) : (n_1, n_2) \in \mathbb{Z}^2\}, \\ 0 & \text{if } \Gamma = \{(0, n_1, n_2) : (n_1, n_2) \in \mathbb{Z}^2\}. \end{cases}$$

If $\Gamma' = \{(0, n, 0) : n \in \mathbb{Z}\} \subset \mathbb{Z}^3$, then $h(\alpha^{(\Gamma)}) = 0$. However, $h(\alpha^{(\Gamma)}) = \infty$ for all subgroups $\Gamma \subset \mathbb{Z}^3$ of rank 1 which are not contained in Γ'.

(5) Let $\mathfrak{p} = j_c$ for some $c \in (\overline{\mathbb{Q}}^\times)^d$ such that $\alpha = \alpha^{\mathfrak{R}_d/j_c}$ is ergodic (Proposition 7.2). Then $\dim_h(\alpha) = 1$, and $h(\alpha^{(\Gamma)}) < \infty$ for every subgroup $\Gamma \subset \mathbb{Z}^d$ of rank 1. $\square$

CONCLUDING REMARK 24.6. If $\mathfrak{p} \subset \mathfrak{R}_d$ is a prime ideal, and if $\alpha = \alpha^{\mathfrak{R}_d/\mathfrak{p}}$ is mixing, then the 'entropy dimension' $\dim_h(\alpha)$ in Remark 24.4 is related to the *Krull dimension* $\dim(\mathfrak{p})$ of the ideal $\mathfrak{p}$, which is defined as the maximum of the lengths n of all possible chains of prime ideal $\mathfrak{p} = \mathfrak{p}_0 \supsetneqq \mathfrak{p}_1 \supsetneqq \cdots \supsetneqq \mathfrak{p}_n = \{0\}$. For example, if $\mathfrak{p} = j_c$ for some $c \in (\overline{\mathbb{Q}}^\times)^d$, then $\dim(\mathfrak{p}) = \dim_h(\alpha) = 1$. In Example (4), $\dim(\mathfrak{p}) = \dim_h(\alpha) = 3$. In Example (2), $\dim(\mathfrak{p}) = \dim_h(\alpha) = d$. In Example (1), $d = \dim(\mathfrak{p}) > \dim_h(\alpha) = d - 1$, but α is obviously non-mixing (cf. Theorem 6.5 (2)).

25. Shift-invariant subgroups of $(\mathbb{Z}/p\mathbb{Z})^{\mathbb{Z}^2}$

Let $p > 1$ be a rational prime, $d > 1$, and let $\mathbb{F}_{p^k}$ be the field with p^k elements for every $k \geq 1$. From Remark 6.19 (4) we know that every closed, shift-invariant subgroup $X \subset \mathbb{F}_p^{\mathbb{Z}^d}$ is isomorphic to $X^{\mathfrak{R}_d/\mathfrak{a}} \cong X^{\mathfrak{R}_d^{(p)}/\mathfrak{b}}$ for some ideals $(p) \subset \mathfrak{a} \subset \mathfrak{R}_d$ and $\mathfrak{b} \subset \mathfrak{R}_d^{(p)}$ (cf. (6.1)), and that this isomorphism carries the shift-action σ of $\mathbb{Z}^d$ on X to the actions $\alpha^{\mathfrak{R}_d/\mathfrak{a}}$ and $\alpha^{\mathfrak{R}_d^{(p)}/\mathfrak{b}}$, respectively. If the ideal $\mathfrak{a}$ is principal, then $\mathfrak{b} = \{0\}$, $X = \mathbb{F}_p^{\mathbb{Z}^d}$, and $h(\sigma) = \log p$. If $\mathfrak{a}$ is non-principal (or, equivalently, if $\mathfrak{b} \neq \{0\}$), then every prime ideal associated with $\mathfrak{R}_d/\mathfrak{a}$ contains $\mathfrak{a}$ and is therefore non-principal and null (Definition 19.3 and Corollary 18.5), and Proposition 19.4 implies that $h(\sigma) = 0$. Conversely, if $\mathfrak{b} \subset \mathfrak{R}_d^{(p)}$ is an ideal, then we may regard $X^{\mathfrak{R}_d^{(p)}/\mathfrak{b}}$ as a closed, shift-invariant subgroup of $\mathbb{F}_p^{\mathbb{Z}^d}$ and $\alpha^{\mathfrak{R}_d^{(p)}/\mathfrak{b}}$ as the shift-action σ of $\mathbb{Z}^d$ on $X^{\mathfrak{R}_d^{(p)}/\mathfrak{b}}$ (cf. (6.19)–(6.20)).

If $p = 2$, (6.19) assumes a particularly simple form: we write $\mathcal{S}(f)$ for the support of a Laurent polynomial $f \in \mathfrak{R}_d^{(2)}$, assume that $\mathfrak{a} = (f_1, \ldots, f_k) = \sum_{i=1}^{k} f_i \mathfrak{R}_d^{(2)}$, and set $F_i = \mathcal{S}(f_i)$. Then $X^{\mathfrak{R}_d^{(2)}/\mathfrak{a}}$ can be viewed as the closed, shift-invariant subgroup of $\mathbb{F}_2^{\mathbb{Z}^d} \cong \widehat{\mathfrak{R}_d^{(2)}}$ consisting of all points $x = (x_{\mathbf{n}}) \in \mathbb{Z}_{/2}^{\mathbb{Z}^d}$ whose coordinates add up to 0 (mod 2) in every translate of every $F_i \subset \mathbb{Z}^d$, $i = 1, \ldots, k$, and $\alpha^{\mathfrak{R}_d^{(2)}/(f)}$ as the shift-action of $\mathbb{Z}^d$ on X. If the ideal $\mathfrak{a} \subset \mathfrak{R}_d^{(2)}$ is principal, $\mathfrak{a} = (f)$, say, then $X^{\mathfrak{R}_d^{(2)}/(f)}$ is of the form X_F for the finite subset $F = \mathcal{S}(f) \subset \mathbb{Z}^d$, where

$$X_F = \left\{ x = (x_{\mathbf{n}}) \in \mathbb{Z}_{/2}^{\mathbb{Z}^d} : \sum_{\mathbf{m} \in F} x_{\mathbf{m}+\mathbf{n}} = 0 \ (\text{mod } 2) \atop \text{for every } \mathbf{n} \in \mathbb{Z}^d \right\}, \tag{25.1}$$

and σ^F is the shift-action of $\mathbb{Z}^2$ on X_F. We have proved the following proposition.

PROPOSITION 25.1. *Let $X \subset \mathbb{F}_2^{\mathbb{Z}^d}$ be a closed, shift-invariant subgroup, and let σ be the shift-action of $\mathbb{Z}^d$ on X. Then there exist non-empty, finite subsets $F_1, \ldots, F_k$, $k \geq 1$, in $\mathbb{Z}^d$ such that*

$$X = \bigcap_{i=1}^{k} X_{F_i}, \tag{25.2}$$

where each X_{F_i} is defined by (25.1).

EXAMPLE 25.2. If $X \subset \mathbb{F}_2^{\mathbb{Z}^d}$ is a closed, shift-invariant subgroup, then the sets F_i, $i = 1, \ldots, k$, in Proposition 25.1 are obviously determined only up to translation by an element of $\mathbb{Z}^d$. In the following examples we assume that $d = 2$, and consider subgroups $X \subset \mathbb{F}_2^{\mathbb{Z}^2}$ of the form $X = X_F$ for certain finite sets $F \subset \mathbb{Z}^2$; since the location of these sets in $\mathbb{Z}^2$ is irrelevant, we represent them graphically in what should be a self-explanatory form. Let

$$
\begin{aligned}
F_1 &= \{(0,0),(1,0),(0,1)\} = \;\raisebox{-0.3ex}{$\vdots$}\,\cdot\,, \\
F_2 &= \{(0,0),(1,0),(1,1)\} = \;\cdot\,\raisebox{-0.3ex}{$\vdots$}\,, \\
F_3 &= \{(0,0),(1,0),(0,1),(1,1)\} = \;\raisebox{-0.3ex}{$\vdots$}\,\raisebox{-0.3ex}{$\vdots$}\,, \\
F_4 &= \{(0,0),(1,0),(0,1),(-1,0),(0,-1)\} = \;\cdot\,\raisebox{-0.3ex}{$\vdots$}\,\cdot\,, \\
F_5 &= \{(1,0),(0,1),(-1,0),(0,-1)\} = \;\cdot\,\raisebox{-0.3ex}{$\circ$}\,\cdot\,, \\
F_6 &= \{(0,0),(1,0),(0,1),(0,2)\} = \;\raisebox{-0.3ex}{$\vdots$}\,\cdot\,, \\
F_7 &= \{(0,0),(1,0),(0,2)\} = \;\raisebox{-0.3ex}{$\circ$}\,\cdot\,.
\end{aligned}
\tag{25.3}
$$

All these examples have zero entropy and are ergodic, and $\dim_h(\sigma^{F_i}) = 1$ for $i = 1, \ldots, 7$ (cf. Theorem 6.5 (1), Remark 6.19 (4), Proposition 24.1, and Remark 24.4). However, not all of them are isomorphic. We know that σ^{F_1} is mixing (Example 5.3 (5) and Example 6.18 (5)), and σ^{F_4} was proved to be mixing in [56]; more generally, Theorem 6.2 (2) implies that σ^{F_i} is mixing for every $i \in \{1, \ldots, 7\}$ with the exception of σ^{F_3} and σ^{F_5} ($\sigma^{F_3} \cong \alpha^{\mathfrak{R}_2/(2,(1-u_1)(1-u_2))}$, and $\sigma^{F_5} \cong \alpha^{\mathfrak{R}_2/(2,(1-u_1 u_2)(1-u_1^{-1} u_2))}$). $\quad\square$

The purpose of this and the following section is to discuss further the closed, shift-invariant subgroups of $\mathbb{F}_p^{\mathbb{Z}^d}$, and to derive conjugacy invariants related to entropy, which will allow us to distinguish between them dynamically. In this section we deal with the special case where $d = 2$, which is very intuitive and simple, but which nevertheless holds many mysteries. In Section 26 we shall extend these invariants to $d \geq 2$.

The *highest common factor* $\gcd(\mathfrak{a})$ of an ideal $\mathfrak{a} \subset \mathfrak{R}_d^{(p)}$ is the highest common factor of all elements of $\mathfrak{a}$. Note that $\gcd(\mathfrak{a})$ is determined uniquely up to multiplication by a monomial.

LEMMA 25.3. *Let $f_1, f_2 \in \mathfrak{R}_2^{(p)}$, and let $(f_1, f_2) = f_1 \mathfrak{R}_2^{(p)} + f_2 \mathfrak{R}_2^{(p)}$. Then $\mathfrak{R}_2^{(p)}/(f_1, f_2)$ is finite if and only if $\gcd(f_1, f_2) = \gcd\{f_1, f_2\} = 1$.*

PROOF. If f_1 and f_2 have a non-trivial common factor h, then $(f_1, f_2) \subset (h)$ and $\mathfrak{R}_2^{(p)}/(h)$ is infinite, so that $\mathfrak{R}_2^{(p)}/(f_1, f_2)$ must be infinite. Conversely, assume that $\gcd\{f_1, f_2\} = 1$. For every $\mathbf{m} \in \mathbb{Z}^2$ and $A \in \mathrm{GL}(2, \mathbb{Z})$, the cardinalities $|\mathfrak{R}_2^{(p)}/(f_1, f_2)|$ and $|\mathfrak{R}_2^{(p)}/(u^{\mathbf{m}} f_1^A, f_2^A)|$ are equal (cf. Lemma 18.4). We choose $\mathbf{m} \in \mathbb{Z}^2$ and $A \in \mathrm{GL}(2, \mathbb{Z})$ so that $g_1 = u^{\mathbf{m}} f_1^A = a + \phi_1 u_2 + \cdots + \phi_{D-1} u_2^{D-1} + b u_1^{D'} u_2^D$ for some $D \geq 1$, $D' \in \mathbb{Z}$, $a, b \in \mathbb{F}_p$, and $\phi_1, \ldots, \phi_{D-1}$ in $\mathfrak{R}_1^{(p)} = \mathbb{F}_p[u_1^{\pm 1}]$. The $\mathfrak{R}_2^{(p)}$-module $\mathfrak{N} = \mathfrak{R}_2^{(p)}/(g_1)$ is also an $\mathfrak{R}_1^{(p)}$-module, and we write $\mathfrak{N}'$ instead of $\mathfrak{N}$ if $\mathfrak{N}' = \mathfrak{N}$ is to be regarded as an $\mathfrak{R}_1^{(p)}$-module. From our choice of g_1 it is clear that $\mathfrak{N}'$ is isomorphic to $(\mathfrak{R}_1^{(p)})^D$, and hence Noetherian. Since $\gcd\{f_1, f_2\} = \gcd\{g_1, g_2\} = 1$, where $g_2 = f_2^A$, the map

$(g_2)_{\mathfrak{N}'}: \mathfrak{N}' \longmapsto \mathfrak{N}'$ consisting of multiplication by g_2 is injective. The isomorphism $\mathfrak{N}' \cong (\mathfrak{R}_1^{(p)})^D$ carries $(g_2)_{\mathfrak{N}'}$ to a non-singular matrix $A \in M_D(\mathfrak{R}_1^{(p)})$, and we write $B = \mathrm{adj}(A)$ for the adjoint matrix in $M_D(\mathfrak{R}_1^{(p)})$ satisfying that $AB = BA = \det(A) \cdot I$, where I is the $D \times D$ identity matrix. Since $\mathfrak{R}_1^{(p)}/h\mathfrak{R}_1^{(p)}$ is finite for every non-zero Laurent polynomial $h \in \mathfrak{R}_1^{(p)}$, $\mathfrak{R}_1^{(p)}/\det(A)\mathfrak{R}_1^{(p)}$ is finite, and

$$
\begin{aligned}
|\mathfrak{R}_2^{(p)}/(f_1, f_2)| &= |\mathfrak{R}_2^{(p)}/(g_1, g_2)| = |\mathfrak{N}'/f_{\mathfrak{N}'} \cdot \mathfrak{N}'| = |(\mathfrak{R}_1^{(p)})^D/A(\mathfrak{R}_1^{(p)})^D| \\
&\leq |(\mathfrak{R}_1^{(p)})^D/A(\mathfrak{R}_1^{(p)})^D| \cdot |(\mathfrak{R}_1^{(p)})^D/B(\mathfrak{R}_1^{(p)})^D| \\
&= |(\mathfrak{R}_1^{(p)})^D/\det(A) \cdot (\mathfrak{R}_1^{(p)})^D| = |\mathfrak{R}_1^{(p)}/\det(A)\mathfrak{R}_1^{(p)}|^D < \infty. \quad \square
\end{aligned}
$$

LEMMA 25.4. *If $\mathfrak{a} \subset \mathfrak{R}_2^{(p)}$ is an ideal and $f = \gcd(\mathfrak{a})$, then $(f)/\mathfrak{a}$ is finite.*

PROOF. We use induction on the number of generators of $\mathfrak{a}$. If $\mathfrak{a} = (f_1, f_2)$, $\gcd(\mathfrak{a}) = f$, and $f_i = fg_i$, $i = 1, 2$, then $(f_1, f_2) \subset (f)$, $\mathfrak{R}_2^{(p)}/(g_1, g_2)$ is finite by Lemma 25.3, and the sequence

$$
0 \longrightarrow (g_1, g_2) \longrightarrow \mathfrak{R}_2^{(p)} \xrightarrow{\psi} \mathfrak{R}_2^{(p)}/(f_1, f_2) \xrightarrow{\eta} \mathfrak{R}_2^{(p)}/(f) \longrightarrow 0
$$

is exact, where ψ is multiplication by f, followed by the quotient map $\mathfrak{R}_2^{(p)} \longmapsto \mathfrak{R}_2^{(p)}/(f_1, f_2)$, and where $\eta: \mathfrak{R}_2^{(p)}/(f_1, f_2) \longmapsto \mathfrak{R}_2^{(p)}/(f)$ is the quotient map. In particular, $\ker(\eta) \cong \mathfrak{R}_2^{(p)}/(g_1, g_2)$ is finite by Lemma 25.3, which implies that $(f)/(f_1, f_2)$ is finite.

Suppose that the lemma has been proved for every ideal in $\mathfrak{R}_2^{(p)}$ with n generators, and assume that $\mathfrak{a} = (f_1, \ldots, f_{n+1}) \subset \mathfrak{R}_2^{(p)}$. Let $f' = \gcd\{f_1, \ldots, f_n\}$ and $f = \gcd\{f', f_{n+1}\} = \gcd(\mathfrak{a})$. By the induction hypothesis, $(f')/(f_1, \ldots, f_n)$ and $(f)/(f', f_{n+1})$ are finite. Since

$$
\begin{aligned}
|(f', f_{n+1})/\mathfrak{a}| &= |((f') + (f_n + 1))/((f_1, \ldots, f_n) + (f_n + 1))| \\
&\leq |(f')/(f_1, \ldots, f_n)| < \infty
\end{aligned}
$$

and $|(f)/(f', f_{n+1})| < \infty$, we obtain that

$$
|(f)/\mathfrak{a}| = |(f)/(f', f_{n+1})| \cdot |(f', f_{n+1})/\mathfrak{a}| < \infty,
$$

as claimed. $\square$

PROPOSITION 25.5. *Let $X \subset \mathbb{F}_p^{\mathbb{Z}^2}$ be a closed, shift-invariant subgroup, and let σ be the shift-action of $\mathbb{Z}^2$ on X. If $\mathfrak{a} \subset \mathfrak{R}_2^{(p)}$ is the ideal (6.20), then σ is ergodic if and only if $\mathfrak{a}$ is principal. Furthermore, if $\mathfrak{a} = (f)$ is principal, and if $\mathbf{n} \in \mathbb{Z}^2$ is primitive, then $\sigma_{\mathbf{n}}$ is ergodic on X if and only if f is not divisible by any polynomial of the form $g(u^{\mathbf{n}})$ with $g \in \mathfrak{R}_1^{(p)}$.*

PROOF. If $\mathfrak{a}$ is principal, then $V(\mathfrak{p})$ is infinite for every prime ideal $\mathfrak{p} \subset \mathfrak{R}_2^{(p)}$ associated with $\mathfrak{R}_2^{(p)}/\mathfrak{a}$, since every such prime ideal is principal (cf. Theorem 6.5 (1) and Remark 6.19 (4)). If $\mathfrak{a}$ is non-principal, and if $f = \gcd(\mathfrak{a})$, then Lemma 25.4 implies that $\widehat{\mathfrak{R}_2^{(p)}/(f)}$ is a closed, shift-invariant subgroup of finite index in $\widehat{\mathfrak{R}_2^{(p)}/\mathfrak{a}}$, which shows that $\mathfrak{a}$ cannot be ergodic.

Now assume that $\mathfrak{a} = (f)$ is principal. The prime ideals associated with $\mathfrak{R}_2^{(p)}/(f)$ are the principal ideals arising from the prime divisors of f in $\mathfrak{R}_2^{(p)}$, and the last assertion follows from Proposition 6.6 and Remark 6.19 (4). □

For every non-empty subset $S \subset \mathbb{R}^d$ and every primitive element $\mathbf{n} \in \mathbb{Z}^d$ we denote by $\pi^{(\mathbf{n})}(S)$ the orthogonal projection of S onto the one-dimensional subspace $\{t\mathbf{n} : t \in \mathbb{R}\} \subset \mathbb{R}^d$. If $F \subset \mathbb{Z}^d \subset \mathbb{R}^d$ is a non-empty, finite set, we write $\mathcal{C}(F) \subset \mathbb{R}^d$ for the convex hull of F and denote by

$$w(\mathbf{n}, F) = |\pi_{\mathbf{n}}(\mathbb{Z}^d) \cap \pi_{\mathbf{n}}(\mathcal{C}(F))| - 1 \tag{25.4}$$

the *width* of F in the direction $\mathbf{n}$. The *face* of $\mathcal{C}(F)$ in the direction $\mathbf{n} = (n_1, \ldots, n_d)$ is defined as

$$\Phi_{\mathbf{n}}(F) = \left\{ \mathbf{t} \in \mathcal{C}(F) : \langle \mathbf{t}, \mathbf{n} \rangle = \max_{\mathbf{s} \in \mathcal{C}(F)} \langle \mathbf{s}, \mathbf{n} \rangle \right\}, \tag{25.5}$$

where $\langle \mathbf{s}, \mathbf{n} \rangle = s_1 n_1 + \cdots + s_d n_d$ for every $\mathbf{s} = (s_1, \ldots, s_d) \in \mathbb{R}^d$. If $d = 2$ we set

$$\omega(\mathbf{n}, F) = |\mathbb{Z}^2 \cap \Phi_{\mathbf{n}}(F)| - 1. \tag{25.6}$$

LEMMA 25.6. *Let $F \subset \mathbb{Z}^d$ be a non-empty, finite set, $\mathbf{n} \in \mathbb{Z}^d$ a primitive element, $A \in \mathrm{GL}(d, \mathbb{Z})$, and let $A^\top$ be the transpose of A. Then*

$$w(\mathbf{n}, AF) = w(A^\top \mathbf{n}, F). \tag{25.7}$$

If $d = 2$, then

$$\omega(\mathbf{n}, AF) = \omega(A^\top \mathbf{n}, F). \tag{25.8}$$

PROOF. For every $\mathbf{m} \in \mathbb{Z}^d$, $\langle A^\top \mathbf{n}, A^{-1}\mathbf{m} \rangle = \langle \mathbf{n}, \mathbf{m} \rangle$, so that

$$w(\mathbf{n}, F) = \left| \left\{ \langle \mathbf{n}, \mathbf{m} \rangle : \mathbf{m} \in \mathbb{Z}^d \text{ and } \min_{\mathbf{k} \in F} \langle \mathbf{n}, \mathbf{k} \rangle \le \langle \mathbf{n}, \mathbf{m} \rangle \le \max_{\mathbf{k} \in F} \langle \mathbf{n}, \mathbf{k} \rangle \right\} \right|$$

$$= \Big| \Big\{ \langle A^\top \mathbf{n}, A^{-1}\mathbf{m} \rangle : \mathbf{m} \in \mathbb{Z}^d \text{ and }$$

$$\min_{\mathbf{k} \in F} \langle A^\top \mathbf{n}, A^{-1}\mathbf{k} \rangle \le \langle A^\top \mathbf{n}, A^{-1}\mathbf{m} \rangle \le \max_{\mathbf{k} \in F} \langle A^\top \mathbf{n}, A^{-1}\mathbf{k} \rangle \Big\} \Big|$$

$$= \Big| \Big\{ \langle A^\top \mathbf{n}, \mathbf{m} \rangle : \mathbf{m} \in \mathbb{Z}^d$$

$$\text{and } \min_{\mathbf{k} \in A^{-1}F} \langle An, \mathbf{k} \rangle \le \langle A^\top \mathbf{n}, \mathbf{m} \rangle \le \max_{\mathbf{k} \in A^{-1}F} \langle A^\top \mathbf{n}, \mathbf{k} \rangle \Big\} \Big|$$

$$= w(A^\top \mathbf{n}, A^{-1}F).$$

By replacing F with $A^{-1}F$ we obtain (25.7), and (25.8) is proved similarly. □

PROPOSITION 25.7. *Let $X \subset \mathbb{F}_p^{\mathbb{Z}^2}$ be a closed, shift-invariant subgroup such that the shift-action σ of $\mathbb{Z}^2$ on X is ergodic, and let $f \in \mathfrak{R}_2^{(p)}$ be a polynomial with $(f) = \mathfrak{a}$, where $\mathfrak{a} \subset \mathfrak{R}_d^{(p)}$ the principal ideal (6.20) (cf. Proposition 25.5). If $\mathbf{n} = (n_1, n_2) \in \mathbb{Z}^2$ is primitive we denote by $\mathbf{n}^\perp = (n_1', n_2') \in \mathbb{Z}^2$ the unique primitive element with $\langle \mathbf{n}, \mathbf{n}^\perp \rangle = 0$ and $n_1 n_2' - n_2 n_1' > 0$. Then*

$$h(\sigma_\mathbf{n}) = w(\mathbf{n}^\perp, \mathcal{S}(f)) \cdot \log p. \tag{25.9}$$

The automorphism $\sigma_\mathbf{n}$ is expansive on X if and only if

$$\omega(\mathbf{n}^\perp, \mathcal{S}(f)) = \omega(-\mathbf{n}^\perp, \mathcal{S}(f)) = 0; \tag{25.10}$$

in this case $\sigma_\mathbf{n}$ is algebraically conjugate to the shift on $(\mathbb{F}_p^{w(\mathbf{n}^\perp, \mathcal{S}(f))})^\mathbb{Z}$.

PROOF. Assume for the moment that $\mathbf{n} = \mathbf{e}^{(1)}$, where $\mathbf{e}^{(j)}$ is the j-th unit vector in $\mathbb{Z}^2$, and set $S(-m, n) = \mathbb{Z} \times \{-m, \dots, n\} \subset \mathbb{Z}^2$ and $X(-m, n) = \pi_{S(-m,n)}(X) \subset \mathbb{F}_p^{S(-m,n)}$ for all $m, n \geq 0$. If $\omega(\pm \mathbf{e}^{(2)}, \mathcal{S}(f)) = 0$ and $m + n + 1 = w(\mathbf{e}^{(2)}, \mathcal{S}(f))$, then every $x \in X$ is completely determined by its projection onto the strip $S(m, n)$. It follows that $\sigma_{\mathbf{e}^{(1)}}$ is expansive, and that it is conjugate to the shift on $(\mathbb{F}_p^{w(\mathbf{e}^{(2)}, \mathcal{S}(f))})^\mathbb{Z}$.

If $\omega(\mathbf{e}^{(2)}, \mathcal{S}(f)) + \omega((0, -1), \mathcal{S}(f)) > 0$ and $m + n + 1 = w(\mathbf{e}^{(2)}, \mathcal{S}(f))$, then the obvious projections from $X(-m, n+1)$ and $X(-m-1, n)$ onto $X(-m, n)$ have kernels of size $p^{\omega(\mathbf{e}^{(2)}, \mathcal{S}(f))}$ and $p^{\omega(-\mathbf{e}^{(2)}, \mathcal{S}(f))}$, respectively. Hence $h(\sigma_{\mathbf{e}^{(1)}}) = w(\mathbf{e}^{(2)}, \mathcal{S}(f)) \cdot \log p$, and $\sigma_{\mathbf{e}^{(1)}}$ is non-expansive whenever

$$\omega(\mathbf{e}^{(2)}, \mathcal{S}(f)) + \omega(-\mathbf{e}^{(2)}, \mathcal{S}(f)) > 0$$

(it helps to draw a picture and to study the Examples 25.2).

In order to prove (25.9) for an arbitrary primitive element $\mathbf{n} \in \mathbb{Z}^2$ we use Lemma 11.3 to find a matrix $A \in \mathrm{GL}(2, \mathbb{Z})$ with $A^\top \mathbf{n} = \mathbf{e}^{(1)}$. Assume that f is of the form (5.2), set $f^{A^\top} = \sum_{\mathbf{m} \in \mathbb{Z}^2} c_f(\mathbf{m}) u^{A^\top \mathbf{m}}$, and consider the shift-action $\sigma' = \alpha^{\mathfrak{R}_2^{(p)}/(f^{A^\top})}$ of $\mathbb{Z}^2$ on the closed, shift-invariant subgroup $X' = X^{\mathfrak{R}_2^{(p)}/(f^{A^\top})} \subset \mathbb{F}_p^{\mathbb{Z}^2}$. As we saw in Lemma 18.4, there exists a continuous group isomorphism $\psi_{A^\top} \colon X \longmapsto X'$ with $\psi_{A^\top} \cdot \sigma_\mathbf{m} = \sigma'_{A^\top \mathbf{m}} \cdot \psi_{A^\top}$ for every $\mathbf{m} \in \mathbb{Z}^2$. The first part of this proof and (25.7) together imply that $h(\sigma_\mathbf{n}) = h(\sigma'_{\mathbf{e}^{(1)}}) = w(\mathbf{e}^{(2)}, \mathcal{S}(f^{A^\top})) \cdot \log p = w(\mathbf{e}^{(2)}, A^\top \mathcal{S}(f)) \cdot \log p = w(A\mathbf{e}^{(2)}, \mathcal{S}(f)) \cdot \log p = w(\mathbf{n}^\perp, \mathcal{S}(f)) \cdot \log p$ (since $0 = \langle \mathbf{n} \perp, \mathbf{n} \rangle = \langle \mathbf{e}^{(2)}, \mathbf{e}^{(1)} \rangle = \langle A\mathbf{e}^{(2)}, (A^\top)^{-1} \mathbf{e}^{(1)} \rangle = \langle A\mathbf{e}^{(2)}, \mathbf{n} \rangle$, we know that $\mathbf{n}^\perp = \pm A\mathbf{e}^{(2)}$). An analogous argument shows that $\sigma_\mathbf{n}$ is expansive if and only if (25.10) is satisfied. $\square$

Proposition 25.7 shows that, for every non-zero $f \in \mathfrak{R}_2^{(p)}$, the widths $w(\mathbf{n}, \mathcal{S}(f))$ are measurable conjugacy invariants of the $\mathbb{Z}^2$-action $\alpha = \alpha^{\mathfrak{R}_d/(f)}$ on $X = X^{\mathfrak{R}_2^{(p)}/(f)}$, as $\mathbf{n}$ varies over the primitive elements of $\mathbb{Z}^2$.

EXAMPLES 25.8. Let $F_i \subset \mathbb{Z}^2$, $i = 1, \ldots, 7$, be the sets in Example 25.2, and let σ^{F_i} be the shift-action of $\mathbb{Z}^2$ on the subgroup X_{F_i} defined by (25.1).

(1) If $F = F_1 = \{(0,0), (1,0), (0,1)\}$, then $w(\mathbf{e}^{(1)}, F) = w(\mathbf{e}^{(2)}, F) = 1$, $w((1,1), F) = 2$, and $w((-1,1), F) = 1$. Hence $h(\sigma^F_{\mathbf{e}^{(2)}}) = h(\sigma^F_{\mathbf{e}^{(1)}}) = h(\sigma^F_{(-1,1)}) = \log 2$, and $h(\sigma^F_{(1,1)}) = \log 4$. Furthermore,

$$w((-1,0), F) = w((0,-1), F) = w((1,1), F) = 1,$$

so that $\sigma^F_{\mathbf{e}^{(2)}}$, $\sigma^F_{\mathbf{e}^{(1)}}$, and $\sigma^F_{(1,-1)}$ are non-expansive, but $\sigma^F_{\mathbf{n}}$ is expansive for every primitive $\mathbf{n} \notin \{\pm(1,0), \pm(0,1), \pm(1,-1)\}$.

(2) Let $F = F_2 = \{(0,0), (1,0), (1,1)\}$. Since $w((1,1), F_2) = 1$ we know from Proposition 25.7 and Example (1) that $\log 2 = h(\sigma^{F_2}_{(1,1)}) \neq h(\sigma^{F_1}_{(1,1)}) = \log 4$, so that the $\mathbb{Z}^2$-actions σ^{F_1} and σ^{F_2} cannot be measurably conjugate.

(3) Since σ^{F_3} is not mixing, it cannot be measurably conjugate to σ^{F_i} for $i = 1, 2$. Another way of distinguishing σ^{F_3} from σ^{F_i}, $i = 1, 2$, is by observing that $w((1,1), F_3) = w((-1,1), F_3) = 2$, which is obviously untrue for F_1 and F_2.

(4) By using the widths $w(\mathbf{n}, F)$ (or, equivalently, the directional entropies $h(\sigma^F_{\mathbf{n}})$) we can distinguish the $\mathbb{Z}^2$-actions σ^{F_i}, $i = 4, 5$, from all the other σ^{F_j} in Example 25.2, but not from each other. However, σ^{F_4} is mixing, whereas σ^{F_5} is non-mixing.

(5) Similarly we see that σ^{F_i} cannot be measurably conjugate to σ^{F_j} whenever $i \in \{6,7\}$ and $j \in \{1,2,3,4,5\}$, but that $h(\sigma^{F_6}_{\mathbf{n}}) = h(\sigma^{F_7}_{\mathbf{n}})$ for all $\mathbf{n} \in \mathbb{Z}^2$. Since both σ^{F_6} and σ^{F_7} are mixing, we are unable to distinguish these actions with the tools developed so far. $\boxdot$

Although the widths $w(\mathbf{n}, \mathcal{S}(f))$ provide a certain amount of information about the convex hull $\mathcal{C}(f)$ of the support of a Laurent polynomial $f \in \mathfrak{R}^{(p)}_d$, they do not—in general—allow us to distinguish between Laurent polynomials whose supports have quite dissimilar convex hulls.

EXAMPLES 25.9. (1) Let $p = d = 2$, and let $f_1 = 1 + u_1^2 + u_2^2$, $f_2 = u_1 + u_2 + u_1^2 + u_2^2 + u_1^2 u_2 + u_1 u_2^2$. In the representation of Example 25.2 we have that

$$F_1 = \mathcal{C}(f_1) \cap \mathbb{Z}^2 = \begin{matrix} \bullet \\ \bullet \bullet \bullet \end{matrix}, \qquad F_2 = \mathcal{C}(f_2) \cap \mathbb{Z}^2 = \begin{matrix} \bullet \bullet \\ \bullet \bullet \end{matrix},$$

and $w(\mathbf{n}, \mathcal{S}(f_1)) = w(\mathbf{n}, F_1) = w(\mathbf{n}, F_2) = w(\mathbf{n}, \mathcal{S}(f_2))$ for every primitive element $\mathbf{n} \in \mathbb{Z}^2$.

(2) Let $p = d = 2$, $0 \neq f \in \mathfrak{R}^{(2)}_2$, and let $f_1 = f^2$, $f_2 = f f^*$, where $f^* = \sum_{\mathbf{m} \in \mathbb{Z}^2} c_f(\mathbf{m}) u^{-\mathbf{m}}$ (cf. (5.2)). Then $w(\mathbf{n}, \mathcal{S}(f_1)) = w(\mathbf{n}, \mathcal{S}(f_2))$ for every primitive element $\mathbf{n} \in \mathbb{Z}^2$. Example (1) arises by setting $f = 1 + u_1 + u_2$. $\boxdot$

By using the notion of relative entropy for $\mathbb{Z}^d$-actions developed in [33], one can obtain further measurable conjugacy invariants for the $\mathbb{Z}^d$-actions $\alpha^{\mathfrak{R}_d^{(p)}/(f)}$, $f \in \mathfrak{R}_d^{(p)}$, which will allow us to prove that the convex hull $\mathcal{C}(f)$ is— up to translation—a measurable conjugacy invariant of $\alpha^{\mathfrak{R}_d^{(p)}/(f)}$. In the spirit of this section we restrict ourselves to the case where $d = 2$ before presenting a more general picture in Section 26. Those proofs which are not significantly simplified by assuming that $d = 2$ will be postponed until Section 26 in order to avoid excessive duplication.

If $p > 1$ is a rational prime, $X \subset \mathbb{F}_p^{\mathbb{Z}^2}$ a closed, shift-invariant subgroup, and σ the shift-action of $\mathbb{Z}^2$ on X, then the state partition (24.3) generates $\mathfrak{B}_X$ under the σ. More generally, suppose that $\mathcal{Q} \subset \mathfrak{B}_X$ is a finite partition. Put

$$\mathfrak{A}_X(\mathcal{Q})_k = \bigvee_{\{\mathbf{m}=(m_1,m_2)\in\mathbb{Z}^2:m_2\leq k\}} \sigma_{-\mathbf{m}}(\mathcal{Q}), \quad k \in \mathbb{Z},$$

$$\mathfrak{A}_X(\mathcal{Q})_{-\infty} = \bigcap_{k\in\mathbb{Z}} \mathfrak{A}_X(\mathcal{Q})_k. \tag{25.11}$$

If the space X is understood we suppress the subscript X in these definitions.

For the following lemmas we assume that $X \subsetneqq \mathbb{F}_p^{\mathbb{Z}^2}$ is a closed, shift-invariant subgroup such that the shift-action σ of $\mathbb{Z}^2$ is ergodic on X, and choose a (non-zero) polynomial $f \in \mathfrak{R}_2^{(p)}$ with $(f) = \mathfrak{a}$, where $\mathfrak{a} \subset \mathfrak{R}_d^{(p)}$ is the principal ideal (6.20) (cf. Proposition 25.5). The state partition (24.3) of X is denoted by $\mathcal{P}_0$.

LEMMA 25.10. *For every finite partition $\mathcal{Q} \subset \mathfrak{B}_X$, and for every $k \in \mathbb{Z}$,*

$$\mathfrak{A}(\mathcal{Q})_k = \sigma_{-k\mathbf{e}^{(2)}}(\mathfrak{A}(\mathcal{Q})_0), \quad \sigma_{\mathbf{e}^{(1)}}(\mathfrak{A}(\mathcal{Q})_k) = \mathfrak{A}(\mathcal{Q})_k,$$
$$\text{and } \mathfrak{A}(\mathcal{Q})_{-1} \subset \mathfrak{A}(\mathcal{Q})_0. \tag{25.12}$$

If $\mathcal{Q}$ is a generator, then

$$\lim_{k\to\infty} \mathfrak{A}(\mathcal{Q})_k = \mathfrak{B}_X, \tag{25.13}$$

and

$$\mathfrak{A}(\mathcal{Q})_{-\infty} = \mathfrak{A}(\mathcal{Q})_0 = \mathfrak{A}(\mathcal{P}_0)_{-\infty} = \mathfrak{A}(\mathcal{P}_0)_0 = \mathfrak{B}_X \tag{25.14}$$

whenever $\omega(\mathbf{e}^{(2)}, \mathcal{S}(f)) = 0$.

PROOF. The relations (25.12)–(25.13) are obvious from the definition of $\mathfrak{A}(\mathcal{Q})_0$. In order to prove (25.14) we assume that $\omega(\mathbf{e}^{(2)}, \mathcal{S}(f)) = 0$. Fix $\varepsilon > 0$ and use (25.13) to find an integer $M \geq 0$ such that $\sigma_{M\mathbf{e}^{(2)}}(\mathcal{P}_0) \underset{\varepsilon}{\subset} \mathfrak{A}(\mathcal{Q})_0$ (this means that there exists, for every $P \in \sigma_{(0,M)}(\mathcal{P}_0)$, a set P' in the sigma-algebra $\mathfrak{A}(\mathcal{Q})_0$ with $\lambda_X(P\triangle P') < \varepsilon$, where λ_X is the normalized Haar measure on X). According to (25.12), $\sigma_{(m_1,m_2)}(\mathcal{P}_0) \underset{\varepsilon}{\subset} \mathfrak{A}(\mathcal{Q})_0$ for all $m_2 \geq M$ and $m_1 \in \mathbb{Z}$. For every $\mathbf{k} = (k_1, k_2) \in \mathbb{Z}^2$ we can choose an integer $j \geq 1$ and an element $\mathbf{m} \in \mathbb{Z}^2$

such that $\mathbf{k} \in \mathcal{S}(u^{\mathbf{m}} f^{p^j}) = \mathbf{m} + p^j \mathcal{S}(f)$, and $\mathcal{S}(u^{\mathbf{m}} f^{p^j}) \smallsetminus \{\mathbf{k}\} \subset \{\mathbf{n} = (n_1, n_2) \in \mathbb{Z}^2 : n_2 \le -M\}$. Since $u^{\mathbf{m}} f^{p^j} \in \mathfrak{a}$, $x_{\mathbf{k}} = -\sum_{\mathbf{n} \in \mathcal{S}(f) \smallsetminus \{\mathbf{k}\}} c_f(\mathbf{n}) x_{\mathbf{m} + p^j \mathbf{n}}$ for every $x = (x_{\mathbf{n}}) \in X$, so that $x_{\mathbf{k}}$ is completely determined by the coordinates $x_{\mathbf{n}}$ with $\mathbf{n} \in (\mathbf{m} + p^j \mathcal{S}(f)) \smallsetminus \{\mathbf{k}\}$. In particular we conclude that $\sigma_{-\mathbf{k}}(\mathcal{P}_0) \subset_{s\varepsilon} \mathfrak{A}(\mathcal{Q})_0$ for every $\mathbf{k} \in \mathbb{Z}^2$, where $s = |\mathcal{S}(f)|$. As ε was arbitrary, we obtain that $\sigma_{-\mathbf{k}}(\mathcal{P}_0) \subset \mathfrak{A}(\mathcal{Q})_0$ for every $\mathbf{k} \in \mathbb{Z}^2$, so that $\mathfrak{A}(\mathcal{Q})_0 = \mathfrak{B}_X$ for every finite generator $\mathcal{Q}$ of $\mathfrak{B}_X$. It follows that $\mathfrak{A}(\mathcal{Q})_k = \sigma_{-k\mathbf{e}^{(2)}}(\mathfrak{A}(\mathcal{Q})_0) = \mathfrak{B}_X$ for every $k \in \mathbb{Z}$, and that $\mathfrak{A}(\mathcal{Q})_{-\infty} = \mathfrak{B}_X$. In particular, if $\mathcal{Q} = \mathcal{P}_0$, then $\mathfrak{A}(\mathcal{P}_0)_{-\infty} = \mathfrak{B}_X$, which completes the proof of (25.14). $\square$

LEMMA 25.11. *If $\omega(\mathbf{e}^{(2)}, \mathcal{S}(f)) > 0$, then $\mathfrak{A}(\mathcal{P}_0)_{-\infty} \ne \mathfrak{B}_X$ and $H_{\lambda_X}(\mathfrak{B}_X | \mathfrak{A}(\mathcal{P}_0)_{-\infty}) = \infty$. If f is in addition irreducible, then $\mathfrak{A}(\mathcal{P}_0)_{-\infty} = \{\varnothing, X\}$ (mod λ_X).*

PROOF. For every $k \in \mathbb{Z}$ we set

$$\Omega(X)_k = \{x = (x_{\mathbf{n}}) \in X : x_{\mathbf{n}} = 0 \tag{25.15}$$
$$\text{for every } \mathbf{n} = (n_1, n_2) \in \mathbb{Z}^2 \text{ with } n_2 \le k\},$$

and write

$$\Omega(X)_{-\infty} = \overline{\bigcup_{k \in \mathbb{Z}} \Omega(X)_k} \tag{25.16}$$

for the closure of $\bigcup_{k \in \mathbb{Z}} \Omega(X)_k$. Then $\Omega(X)_k$ and $\Omega(X)_{-\infty}$ are closed subgroups of X, $\Omega(X)_k$ is invariant under $\sigma_{\mathbf{e}^{(1)}}$, $\Omega(X)_{-\infty}$ is shift-invariant, and

$$\mathfrak{A}(\mathcal{P}_0)_k = \mathfrak{B}_{X/\Omega(X)_k} \text{ for every } k \in \mathbb{Z} \cup \{-\infty\}. \tag{25.17}$$

Let

$$H = \{\mathbf{n} = (n_1, n_2) \in \mathbb{Z}^2 : n_2 \le 0\}, \quad H' = \{\mathbf{n} = (n_1, n_2) \in \mathbb{Z}^2 : n_2 < 0\}.$$

If $x \in X$ is a point with $\pi_{H'}(x) = \{\mathbf{0}_{\mathbb{F}_p^{H}}\}$, then the coordinates $x_{(m,0)}$, $m \in \mathbb{Z}$, are completely determined by the coordinates $x_{(k,0)}$ with $0 \le k < \omega(\mathbf{e}^{(2)}, \mathcal{S}(f))$; conversely, if we choose $i_k \in \mathbb{F}_p$ arbitrarily for $k = 0, \ldots, \omega(\mathbf{e}^{(2)}, \mathcal{S}(f)) - 1$, then there exists a point $y \in X$ with $y_{\mathbf{n}} = 0$ for $\mathbf{n} \in H'$, and $y_{(k,0)} = i_k$ for $k = 0, \ldots, \omega(\mathbf{e}^{(2)}, \mathcal{S}(f)) - 1$. Hence

$$|\Omega(X)_{-1}/\Omega(X)_0| = |\pi_H(\Omega(X)_{-1})| = p^{\omega(\mathbf{e}^{(2)}, \mathcal{S}(f))}. \tag{25.18}$$

If $\omega(\mathbf{e}^{(2)}, \mathcal{S}(f)) > 0$ then $|\Omega(X)_{-\infty}| \ge |\Omega(X)_{-m}/\Omega(X)_0| = p^{m\omega(\mathbf{e}^{(2)}, \mathcal{S}(f))}$ for every $m \ge 1$, so that $\Omega(X)_{-\infty}$ is infinite, $\mathfrak{A}(\mathcal{P}_0)_{-\infty} = \mathfrak{B}_{X/\Omega(X)_{-\infty}} \ne \mathfrak{B}_X$, and

$$H_{\lambda_X}(\mathfrak{B}_X | \mathfrak{A}(\mathcal{P}_0)_{-\infty}) = \log |\Omega(X)_{-\infty}| = \infty. \tag{25.19}$$

The annihilators of the subgroups $\{\mathbf{0}_{\mathbb{F}_p^{\mathbb{Z}^2}}\} \subset \Omega(X)_{-\infty} \subset X \subset \mathbb{F}_p^{\mathbb{Z}^2}$ satisfy that $\mathfrak{R}_2^{(p)} \supset (\Omega(X)_{-\infty})^{\perp} \supset (f) \supset \{0\}$, and the shift-invariance of $\Omega(X)_{-\infty}$

shows that $\mathfrak{b} = (\Omega(X)_{-\infty})^{\perp}$ is an ideal containing (f). If f is irreducible, then Lemma 25.3 implies that either $\mathfrak{b} = (f)$ and $\Omega(X)_{-\infty} = X$, or that $|\mathfrak{R}_2^{(p)}/\mathfrak{b}| = |\Omega(X)_{-\infty}| < \infty$. In the first case the lemma is proved, and the second case is impossible by (25.19). $\square$

LEMMA 25.12. *For every finite generator $\mathcal{Q} \subset \mathfrak{B}_X$,*

$$\mathfrak{A}(\mathcal{Q})_{-\infty} \supset \mathfrak{A}(\mathcal{P_0})_{-\infty} \pmod{\lambda_X}. \tag{25.20}$$

Furthermore, if $\mathcal{Q} \subset \mathfrak{B}_X$ is a finite partition satisfying (25.20), then

$$H_{\lambda_X}(\mathfrak{A}(\mathcal{Q})_0|\mathfrak{A}(\mathcal{Q})_{-1}) \leq H_{\lambda_X}(\mathfrak{A}(\mathcal{P_0})_0|\mathfrak{A}(\mathcal{P_0})_{-1})$$
$$= \omega(\mathbf{e}^{(2)}, \mathcal{S}(f)) \log p \leq h_{\lambda_X}(\sigma_{\mathbf{e}^{(2)}}). \tag{25.21}$$

PROOF. Lemma 25.11 implies that (25.20) is trivially satisfied if f is irreducible; a general proof of (25.20) will be postponed until Lemma 26.6. For every finite partition $\mathcal{Q} \subset \mathfrak{B}_X$ satisfying (25.20) we set $\mathcal{Q}^{(m)} = \bigvee_{|k| \leq m} \sigma_{-k\mathbf{e}^{(1)}}(\mathcal{Q})$ and $\mathfrak{A}^{(m)}(\mathcal{Q}) = \bigvee_{k \leq 0} \sigma_{-k\mathbf{e}^{(2)}}(\mathcal{Q}^{(m)})$. Then

$$H_{\lambda_X}(\mathcal{Q}^{(m)}|\mathfrak{A}(\mathcal{Q})_{-1}) \leq H_{\lambda_X}(\mathcal{Q}^{(m)}|\sigma_{\mathbf{e}^{(2)}}(\mathfrak{A}^{(m)}(\mathcal{Q}))) \leq h(\sigma_{\mathbf{e}^{(2)}}),$$

so that $H_{\lambda_X}(\mathfrak{A}(\mathcal{Q})_0|\mathfrak{A}(\mathcal{Q})_{-1}) \leq h_{\lambda_X}(\sigma_{\mathbf{e}^{(2)}}) < \infty$.

The decreasing martingale theorem, combined with (25.20), allows us to find, for every $\varepsilon > 0$, an $N \geq 1$ such that

$$H_{\lambda_X}(\mathfrak{A}(\mathcal{Q})_0|\mathfrak{A}(\mathcal{Q})_{-1}) \geq H_{\lambda_X}(\mathfrak{A}(\mathcal{Q})_0|\mathfrak{A}(\mathcal{Q})_{-1} \vee \mathfrak{A}(\mathcal{P_0})_{-n})$$
$$\geq H_{\lambda_X}(\mathfrak{A}(\mathcal{Q})_0|\mathfrak{A}(\mathcal{Q})_{-1}) - \varepsilon$$

for every $n \geq N$. It follows that, for every $t \geq 1$ and $n \geq N$,

$$tH_{\lambda}(\mathfrak{A}(\mathcal{Q})_0|\mathfrak{A}(\mathcal{Q})_{-1}) = H_{\lambda_X}(\mathfrak{A}(\mathcal{Q})_0|\mathfrak{A}(\mathcal{Q})_{-t})$$
$$\geq H_{\lambda_X}(\mathfrak{A}(\mathcal{Q})_0|\mathfrak{A}(\mathcal{Q})_{-t} \vee \mathfrak{A}(\mathcal{P_0})_{-n-t}) \tag{25.22}$$
$$\geq tH_{\lambda}(\mathfrak{A}(\mathcal{Q})_0|\mathfrak{A}(\mathcal{Q})_{-1}) - t\varepsilon.$$

Since $\mathfrak{A}(\mathcal{P_0})_m$ increases to $\mathfrak{B}_X$ as $m \to \infty$, there exists, for every $\varepsilon > 0$, an $M \geq 1$ with $H_{\lambda_X}(\mathfrak{A}(\mathcal{Q})_0|\mathfrak{A}(\mathcal{Q})_{-1} \vee \mathfrak{A}(\mathcal{P_0})_m) < \varepsilon$ for every $m \geq M$. Hence

$$H_{\lambda_X}(\mathfrak{A}(\mathcal{Q})_0|\mathfrak{A}(\mathcal{Q})_{-t} \vee \mathfrak{A}(\mathcal{P_0})_m) < t\varepsilon \tag{25.23}$$

for every $t \geq 1$ and $m \geq M$. Finally,

$$tH_{\lambda_X}(\mathfrak{A}(\mathcal{Q})_0|\mathfrak{A}(\mathcal{Q})_{-1}) = H_{\lambda_X}(\mathfrak{A}(\mathcal{Q})_0|\mathfrak{A}(\mathcal{Q})_{-t})$$
$$\leq H_{\lambda_X}(\mathfrak{A}(\mathcal{Q})_0|\mathfrak{A}(\mathcal{Q})_{-t} \vee \mathfrak{A}(\mathcal{P_0})_{-n-t}) + t\varepsilon \qquad \text{by (25.22)}$$
$$\leq H_{\lambda_X}(\mathfrak{A}(\mathcal{Q})_0 \vee \mathfrak{A}(\mathcal{P_0})_m|\mathfrak{A}(\mathcal{Q})_{-t} \vee \mathfrak{A}(\mathcal{P_0})_{-n-t}) + t\varepsilon$$
$$= H_{\lambda_X}(\mathfrak{A}(\mathcal{Q})_0|\mathfrak{A}(\mathcal{P_0})_m \vee \mathfrak{A}(\mathcal{Q})_{-t})$$
$$\quad + H_{\lambda_X}(\mathfrak{A}(\mathcal{P_0})_m|\mathfrak{A}(\mathcal{Q})_{-t} \vee \mathfrak{A}(\mathcal{P_0})_{-n-t}) + t\varepsilon$$
$$\leq H_{\lambda_X}(\mathfrak{A}(\mathcal{P_0})_m|\mathfrak{A}(\mathcal{P_0})_{-n-t}) + 2t\varepsilon \qquad \text{by (25.23)}$$

$$= (m + n + t)H_{\lambda_X}(\mathfrak{A}(\mathcal{P}_0)_0|\mathfrak{A}(\mathcal{P}_0)_{-1}) + 2t\varepsilon.$$

If we divide by t and let $t \to \infty$ we obtain that $H_{\lambda_X}(\mathfrak{A}(\mathcal{Q})_0|\mathfrak{A}(\mathcal{Q})_{-1}) \leq H_{\lambda_X}(\mathfrak{A}(\mathcal{P}_0)_0|\mathfrak{A}(\mathcal{P}_0)_{-1})$.

The identity

$$H_{\lambda_X}(\mathfrak{A}(\mathcal{P}_0)_0|\mathfrak{A}(\mathcal{P}_0)_{-1}) = \omega(\mathbf{e}^{(2)}, \mathcal{S}(f)) \log p$$

follows from (25.15)–(25.18), and from the fact that the conditional information function $I_{\lambda_X}(\mathfrak{A}(\mathcal{P}_0)_0|\mathfrak{A}(\mathcal{P}_0)_{-1})$ is constant and equal to $\log |\Omega(X)_{-1}/\Omega(X)_0| = \log |\pi_H(\Omega(X)_{-1})| = \log p^{\omega(\mathbf{e}^{(2)}, \mathcal{S}(f))}$. $\qquad\square$

EXAMPLES 25.13. In the following examples we choose a non-zero polynomial $f \in \mathfrak{R}_2^{(p)}$ and write $\sigma = \alpha^{\mathfrak{R}_2^{(p)}/(f)}$ for the shift-action of $\mathbb{Z}^2$ on the closed, shift-invariant subgroup $X = X^{\mathfrak{R}_2^{(p)}/(f)} \subset \mathbb{F}_p^{\mathbb{Z}^2}$.

(1) Let $p = 2$, $f = 1 + u_1 + u_1^2 + u_2 + u_1 u_2 \in \mathfrak{R}_2^{(2)}$, and put

$$F = \mathcal{S}(f) = \begin{smallmatrix} \bullet & \bullet \\ \bullet & \bullet \\ & \bullet \end{smallmatrix}$$

(cf. Example 25.2). Then $X = X^{\mathfrak{R}_2^{(2)}/(f)} = X_F \subset \mathbb{F}_2^{\mathbb{Z}^2}$ (cf. (25.1)), $\omega(\mathbf{e}^{(2)}, F) = \omega(-\mathbf{e}^{(1)}, F) = \omega((1,1), F) = 1$, $\omega(-\mathbf{e}^{(2)}, F) = 2$, and $\omega(\mathbf{n}, F) = 0$ for all other primitive elements $\mathbf{n} \in \mathbb{Z}^2$. If a point $x \in X$ satisfies that $x_{\mathbf{n}} = 0$ for all $\mathbf{n} = (n_1, n_2) \in \mathbb{Z}^2$ with $n_2 < 0$, then $x_{(m,0)} + x_{(m+1,0)} = 0$ for all $m \in \mathbb{Z}$, so that the choice of $x_{(0,0)}$ will determine $x_{(m,0)}$ for all $m \in \mathbb{Z}$. In particular, $|\pi_H(\Omega(X)_{-1})| = |\Omega(X)_{-1}/\Omega(X)_0| = 2 = 2^{\omega(\mathbf{e}^{(2)}, F)}$, and $H_{\lambda_X}(\mathfrak{A}(\mathcal{P}_0)_0|\mathfrak{A}(\mathcal{P}_0)_{-1}) = \int I_{\lambda_X}(\mathfrak{A}(\mathcal{P}_0)_0|\mathfrak{A}(\mathcal{P}_0)_{-1})d\lambda_X = \log |\Omega(X)_{-1}/\Omega(X)_0| = \log |\pi_H(\Omega(X)_{-1})| = \log 2^{\omega(\mathbf{e}^{(2)}, \mathcal{S}(f))} = \log 2$. Since f is irreducible and $\omega(\mathbf{e}^{(2)}, \mathcal{S}(f)) > 0$, $\mathfrak{A}(\mathcal{P}_0)_{-\infty} = \{\varnothing, X\}$ by Lemma 25.11.

(2) If $p = 2$, $f = 1 + u_1 + u_2 + u_1 u_2 + u_1^2 u_2 \in \mathfrak{R}_2^{(2)}$, and

$$F = \mathcal{S}(f) = \begin{smallmatrix} \bullet & \bullet & \bullet \\ \bullet & \bullet & \end{smallmatrix},$$

then $X = X_F \subset \mathbb{F}_2^{\mathbb{Z}^2}$, $\omega(\mathbf{e}^{(2)}, \mathcal{S}(f)) = 2$, and $|\pi_H(\Omega(X)_{-1})| = |\Omega(X)_{-1}/\Omega(X)_0| = 4 = 2^{\omega(\mathbf{e}^{(2)}, \mathcal{S}(f))}$. The irreducibility of f again implies that $\mathfrak{A}(\mathcal{P}_0)_{-\infty} = \{\varnothing, X\}$.

(3) If p is arbitrary and $f = 1 + u_2 + u_1 u_2$, then $\omega(\mathbf{e}^{(2)}, \mathcal{S}(f)) = 1 > 0$, and Lemma 25.11 implies that $\mathfrak{A}(\mathcal{P}_0)_{-\infty} = \{\varnothing, X\}$. The automorphism $\sigma_{\mathbf{e}^{(1)}}$ is ergodic (by Proposition 25.5) and has finite entropy (by Proposition 25.7), and there exists a finite partition $\mathcal{Q} \subset \mathfrak{B}_X$ which is a generator for $\sigma_{\mathbf{e}^{(1)}}$ (cf. [48]). Then $\bigvee_{k \in \mathbb{Z}} \sigma_{-k\mathbf{e}^{(1)}}(\mathcal{Q}) = \mathfrak{A}(\mathcal{Q})_0 = \mathfrak{A}(\mathcal{Q})_{-\infty} = \mathfrak{B}_X$, and $H_{\lambda_X}(\mathfrak{A}(\mathcal{Q})_0|\mathfrak{A}(\mathcal{Q})_{-1}) = 0$. Moreover, the partition $\mathcal{Q}' = \mathcal{Q} \vee \mathcal{P}_0$ is a finite generator for σ which refines $\mathcal{P}_0$, but $\mathfrak{A}(\mathcal{Q}')_{-\infty} = \mathfrak{A}(\mathcal{Q}')_0 = \mathfrak{B}_X$ and $H_{\lambda_X}(\mathfrak{A}(\mathcal{Q}')_0|\mathfrak{A}(\mathcal{Q}')_{-1}) = 0$. This shows that the inclusion in (25.20) may be strict, and that the conditional entropy $H_{\lambda_X}(\mathfrak{A}(\mathcal{Q})_0|\mathfrak{A}(\mathcal{Q})_{-1})$ is not a monotonic function of $\mathcal{Q}$. $\qquad\boxdot$

THEOREM 25.14. *Let $p_i, i = 1, 2$, be rational primes, and let $f_i \in \mathfrak{R}_2^{(p_i)}$ be non-zero Laurent polynomials. If the $\mathbb{Z}^2$-actions $\alpha^{(1)} = \alpha^{\mathfrak{R}_2^{(p)}/(f_1)}$ and $\alpha^{(2)} = \alpha^{\mathfrak{R}_2^{(p)}/(f_2)}$ are measurably conjugate, then $p_1 = p_2$ and $\mathcal{C}(f_1) = \mathcal{C}(f_2) + \mathbf{m}$ for some $\mathbf{m} \in \mathbb{Z}^2$. Furthermore, if $X^{(i)} = X^{\mathfrak{R}_2^{(p)}/(f_i)}$, and if $\phi \colon X^{(1)} \longmapsto X^{(2)}$ is a measure preserving isomorphism with $\phi \cdot \alpha_{\mathbf{n}}^{(1)} = \alpha_{\mathbf{n}}^{(2)} \cdot \phi$ for every $\mathbf{n} \in \mathbb{Z}^2$, then*

$$\phi(\mathfrak{A}_{X^{(1)}}(\mathcal{P}_0^{(1)})_{-\infty}) = \mathfrak{A}_{X^{(2)}}(\mathcal{P}_0^{(2)})_{-\infty} \pmod{\lambda_{X^{(2)}}}, \qquad (25.24)$$

where $\mathcal{P}_0^{(i)}$ is the state partition (24.3) in $X^{(i)}$.

PROOF. Lemma 25.12 implies that, for every non-zero $f \in \mathfrak{R}_2^{(p)}$, the numbers $\omega(\mathbf{e}^{(2)}, \mathcal{S}(f))$ and p are measurable conjugacy invariants of the $\mathbb{Z}^2$-action $\alpha = \alpha^{\mathfrak{R}_2^{(p)}/(f)}$. In particular, if $f_1 \in \mathfrak{R}_2^{(p_1)}$ and $f_2 \in \mathfrak{R}_2^{(p_2)}$, and if $\alpha^{(1)}$ and $\alpha^{(2)}$ are measurably conjugate, then $p_1 = p_2 = p$, say, and we may assume that $X^{(i)} \subset \mathbb{F}_p^{\mathbb{Z}^2}$, and that $\sigma^{(i)} = \alpha^{(i)}$ is the restriction to $X^{(i)}$ of the shift-action σ of $\mathbb{Z}^2$ on $\mathbb{F}_p^{\mathbb{Z}^2}$. If $\sigma^{(1)}$ is measurably conjugate to $\sigma^{(2)}$, then the $\mathbb{Z}^2$-action $\sigma_{A^{-1}}^{(1)} \colon \mathbf{n} \mapsto \sigma_{A^{-1}\mathbf{n}}^{(1)}$ is conjugate to $\sigma_{A^{-1}}^{(2)}$ for every $A \in \mathrm{GL}(2, \mathbb{Z})$. According to the proof of Lemma 18.4, $\sigma_{A^{-1}}^{(i)}$ is (algebraically conjugate to) $\alpha^{\mathfrak{R}_2^{(p)}/(f_i^A)}$, and by applying Lemma 25.6 and Proposition 25.7 to f_i^A instead of f_i we see that

$$\omega(A^\top \mathbf{e}^{(2)}, \mathcal{S}(f_1)) = \omega(\mathbf{e}^{(2)}, A\mathcal{S}(f_1)) = \omega(\mathbf{e}^{(2)}, \mathcal{S}(f_1^A))$$
$$= \omega(\mathbf{e}^{(2)}, \mathcal{S}(f_2^A)) = \omega(A^\top \mathbf{e}^{(2)}, \mathcal{S}(f_2))$$

for every $A \in \mathrm{GL}(2, \mathbb{Z})$. From Lemma 11.3 we conclude that

$$\omega(\mathbf{n}, \mathcal{S}(f_1)) = \omega(\mathbf{n}, \mathcal{S}(f_2)) \qquad (25.25)$$

for every primitive element $\mathbf{n} \in \mathbb{Z}^2$, which implies that $\mathcal{C}(f_1)$ must be a translate of $\mathcal{C}(f_2)$. The equation (25.24) is a direct consequence of Lemma 25.12. $\square$

THEOREM 25.15. *Let p be a rational prime, and let f_1, f_2 be non-zero Laurent polynomials in $\mathfrak{R}_2^{(p)}$ such that f_1 is irreducible and $\alpha^{(1)} = \alpha^{\mathfrak{R}_2^{(p)}/(f_1)}$ is mixing on $X^{(1)} = X^{\mathfrak{R}_2^{(p)}/(f_1)}$. Let $\alpha^{(2)} = \alpha^{\mathfrak{R}_2^{(p)}/(f_2)}$, $X^{(2)} = X^{\mathfrak{R}_2^{(p)}/(f_2)}$, and assume that that there exists a measure preserving, surjective Borel map $\phi \colon X^{(1)} \longmapsto X^{(2)}$ such that $\phi \cdot \alpha_{\mathbf{n}}^{(1)} = \alpha_{\mathbf{n}}^{(2)} \cdot \phi \; \lambda_{X^{(1)}}$-a.e., for every $\mathbf{n} \in \mathbb{Z}^2$ (i.e. that $\alpha^{(2)}$ is a measurable factor of $\alpha^{(1)}$). Then $\omega(\mathbf{n}, \mathcal{S}(f_1)) \geq \omega(\mathbf{n}, \mathcal{S}(f_2))$ for every primitive element $\mathbf{n} \in \mathbb{Z}^2$. Furthermore there exists a finite set $F \subset \mathbb{Z}^2$ such that $\mathcal{C}(f_2) + \mathcal{C}(F) = \mathcal{C}(f_1)$.*

PROOF. Assume that $X^{(1)}$ and $X^{(2)}$ have been realized as closed, shift-invariant subgroups of $\mathbb{F}_p^{\mathbb{Z}^2}$ (cf. (6.19)–(6.20)), and that $\alpha^{(1)} = \sigma^{(1)}$ and $\alpha^{(2)} = \sigma^{(2)}$ are the shift-actions of $\mathbb{Z}^2$ on $X^{(1)}$ and $X^{(2)}$. We denote by $\mathcal{P}_0^{(1)}$ and $\mathcal{P}_0^{(2)}$

the state partitions (24.3) in $X^{(1)}$ and $X^{(2)}$, respectively. According to the Lemmas 25.10–25.11,

$$\mathfrak{A}_{X^{(1)}}(\mathcal{P}_0^{(1)})_{-\infty} = \begin{cases} \{\varnothing, X^{(1)}\} & \text{if } \omega(\mathbf{e}^{(2)}, \mathcal{S}(f_1)) > 0, \\ \mathfrak{B}_{X^{(1)}} & \text{if } \omega(\mathbf{e}^{(2)}, \mathcal{S}(f_1)) = 0. \end{cases} \qquad (25.26)$$

If $\omega(\mathbf{e}^{(2)}, \mathcal{S}(f_1)) = 0$, but $\omega(\mathbf{e}^{(2)}, \mathcal{S}(f_2)) > 0$, then $\mathfrak{A}_{X^{(2)}}(\mathcal{P}_0^{(2)})_{-\infty} \neq \mathfrak{B}_{X^{(2)}}$ by Lemma 25.11, so that $\{\mathbf{0}_{X^{(2)}}\} \neq \Omega(X^{(2)})_{-\infty} \subset X^{(2)}$ by (25.17). Hence $Y = \phi^{-1}(\Omega(X^{(2)})_{-\infty})$ is a non-trivial, closed, shift-invariant subgroup of $X^{(1)}$. The annihilators of the subgroups $\{\mathbf{0}_{\mathbb{F}_p^{\mathbb{Z}^2}}\} \subset Y \subset X^{(1)} \subset \mathbb{F}_p^{\mathbb{Z}^2}$ satisfy that $\mathfrak{R}_2^{(p)} \supset Y^\perp \supset (f_1) \supset \{0\}$, and the shift-invariance of Y implies that $\mathfrak{b} = Y^\perp$ is an ideal containing (f_1). Since f_1 is irreducible, we either have that $\mathfrak{b} = (f_1)$ and $Y = X^{(1)}$, or that $|\mathfrak{R}_2^{(2)}/\mathfrak{b}| = |Y| < \infty$. In the first case $\phi \colon X^{(1)} \longmapsto X^{(2)}$ must be an isomorphism, and Theorem 25.14 shows that $\omega(\mathbf{e}^{(2)}, \mathcal{S}(f_1)) = \omega(\mathbf{e}^{(2)}, \mathcal{S}(f_2)) > 0$. In the second case $\Omega(X^{(2)})_{-\infty}$ is finite, in violation of (25.19). These contradictions show that $\omega(\mathbf{e}^{(2)}, \mathcal{S}(f_2)) = 0$ whenever $\omega(\mathbf{e}^{(2)}, \mathcal{S}(f_1)) = 0$.

If $\omega(\mathbf{e}^{(2)}, \mathcal{S}(f_1)) > 0$ we put $\mathcal{Q} = \phi^{-1}(\mathcal{P}_0^{(2)})$ and apply the Lemmas 25.11–25.12 to obtain that $\{\varnothing, X^{(1)}\} = \mathfrak{A}_{X^{(1)}}(\mathcal{P}_0)_{-\infty} \subset \mathfrak{A}_{X^{(1)}}(\mathcal{Q})_{-\infty}$ and

$$0 = \omega(\mathbf{e}^{(2)}, \mathcal{S}(f_1)) \log p = H_{\lambda_{X^{(1)}}}(\mathfrak{A}_{X^{(1)}}(\mathcal{P}_0^{(1)})_0 | \mathfrak{A}_{X^{(1)}}(\mathcal{P}_0^{(1)})_{-1})$$

$$\geq H_{\lambda_{X^{(1)}}}(\mathfrak{A}_{X^{(1)}}(\mathcal{Q})_0 | \mathfrak{A}_{X^{(1)}}(\mathcal{Q})_{-1}) = H_{\lambda_{X^{(2)}}}(\mathfrak{A}_{X^{(2)}}(\mathcal{P}_0^{(2)})_0 | \mathfrak{A}_{X^{(2)}}(\mathcal{P}_0^{(2)})_{-1})$$

$$= \omega(\mathbf{e}^{(2)}, \mathcal{S}(f_2)) \log p,$$

so that

$$\omega(\mathbf{e}^{(2)}, \mathcal{S}(f_1)) \geq \omega(\mathbf{e}^{(2)}, \mathcal{S}(f_2)). \qquad (25.27)$$

For every $A \in \mathrm{GL}(2, \mathbb{Z})$, the map $\phi \colon X^{(1)} \longmapsto X^{(2)}$ satisfies that $\phi \cdot \sigma_{A^{-1}\mathbf{n}} = \sigma_{A^{-1}\mathbf{n}}^{(2)} \cdot \phi$ for every $\mathbf{n} \in \mathbb{Z}^2$. It follows that the $\mathbb{Z}^2$-action $\alpha^{\mathfrak{R}_2^{(p)}/(f_2^A)}$ is a factor of $\alpha^{\mathfrak{R}_2^{(p)}/(f_1^A)}$, where we are using the same notation as in the proofs of Lemma 18.4 and Theorem 25.14. By replacing f_i with f_i^A in (25.27) we obtain that $\omega(A^\top \mathbf{e}^{(2)}, \mathcal{S}(f_1)) = \omega(\mathbf{e}^{(2)}, \mathcal{S}(f_1^A)) \geq \omega(\mathbf{e}^{(2)}, \mathcal{S}(f_2^A)) = \omega(A^\top \mathbf{e}^{(2)}, \mathcal{S}(f_2))$ for every $A \in \mathrm{GL}(2, \mathbb{Z})$, and Lemma 11.3 yields that

$$\omega(\mathbf{n}, \mathcal{S}(f_1)) \geq \omega(\mathbf{n}, \mathcal{S}(f_2)) \qquad (25.28)$$

for every primitive element $\mathbf{n} \in \mathbb{Z}^2$.

We set $F_1 = \mathcal{S}(f)$ and $F_2 = \mathcal{S}(f^{(2)})$. For every finite set $E \subset \mathbb{Z}^2$, the boundary of the convex hull $\mathcal{C}(E)$ is a closed polygon in $\mathbb{R}^2$ with edges $\{v_{\mathbf{n}} = \omega(\mathbf{n}, E)\mathbf{n}^\perp : \mathbf{n} \in \mathbb{Z}^2 \text{ and } \omega(\mathbf{n}, E) > 0\}$, and we conclude that $\sum v_{\mathbf{n}} = 0$, where the sum is taken over all primitive elements in $\mathbb{Z}^2$. From (25.28) it is clear that $\omega(\mathbf{n}, F_1) \geq \omega(\mathbf{n}, F_2)$ for every primitive $\mathbf{n} \in \mathbb{Z}^2$. As one traverses the edges $v_{\mathbf{n}}$ of the polygon formed by the boundary of $\mathcal{C}(F_1)$, the vectors

$\mathbf{w_n} = (\omega(\mathbf{n}, F_1) - \omega(\mathbf{n}, F_2))\mathbf{n}^\perp$ traverse another polygon, which is the boundary of a convex set $F' \subset \mathbb{R}^2$ satisfying that $F' + \mathcal{C}(F_2) = \mathcal{C}(F_1)$. The proof is completed by setting $F = F' \cap \mathbb{Z}^2$. $\quad\square$

EXAMPLES 25.16. (1) Theorem 25.14 enables us to distinguish between all the $\mathbb{Z}^2$-actions σ^{F_i} in Example 25.2 with the exception of σ^{F_6} and σ^{F_7}: for all i, j with $1 \leq i < j \leq 7$ and $(i, j) \neq (6, 7)$, $\mathcal{C}(F_i) \neq \mathcal{C}(F_j)$, so that the $\mathbb{Z}^2$-actions σ^{F_i} and σ^{F_j} cannot be measurably conjugate. Theorem 25.14 does *not* enable us to prove that σ^{F_6} and σ^{F_7} are not measurably isomorphic. As we shall see in Example 28.10 (3), these actions can be distinguished by their higher order mixing behaviour.

(2) The sets $F_i \subset \mathbb{Z}^2$, $i = 1, 2$, in Example 25.9 (1) satisfy that $w(\mathbf{n}, F_1) = w(\mathbf{n}, F_2)$ for every primitive element $\mathbf{n} \in \mathbb{Z}^2$, so that, in the notation of (25.1), $h(\sigma_\mathbf{n}^{F_1}) = h(\sigma_\mathbf{n}^{F_2})$ for all $\mathbf{n} \in \mathbb{Z}^2$. However, since $0 = \omega(\mathbf{e}^{(2)}, F_1) \neq \omega(\mathbf{e}^{(2)}, F_2) = 1$, the $\mathbb{Z}^2$-actions σ^{F_1} and σ^{F_2} cannot be measurably conjugate by Theorem 25.14. However, since $\sigma_\mathbf{n}^{F_i}$ is ergodic and hence Bernoulli for every non-zero $\mathbf{n} \in \mathbb{Z}^2$ ([72]), $\sigma_\mathbf{n}^{F_1}$ is measurably conjugate to $\sigma_\mathbf{n}^{F_2}$ for every $\mathbf{n} \in \mathbb{Z}^2$.

(3) Equation (25.24) can be used to prove that certain $\mathbb{Z}^2$-actions of the form $\alpha^{\mathfrak{R}_2^{(p)}/(f_i)}$, $i = 1, 2$, are non-isomorphic, even if $\mathcal{C}(f_1) = \mathcal{C}(f_2)$. Let $d = 2$, and consider the polynomials $f_1 = 1 + u_1 + u_1^2 + u_1^3 u_2 + u_2^2 + u_1 u_2^2$, $f_2 = 1 + u_1 + u_1^2 + u_2 + u_1^3 u_2 + u_2^2 + u_1 u_2^2$ in $\mathfrak{R}_2^{(2)}$. In the representation of Example 25.2,

$$F_1 = \mathbb{S}(f_1) = \begin{smallmatrix} \bullet & \bullet \\ \circ & \circ & \circ & \bullet \\ \bullet & \bullet & \bullet \end{smallmatrix}, \quad F_2 = \mathbb{S}(f_2) = \begin{smallmatrix} \bullet & \bullet \\ \bullet & \circ & \circ & \bullet \\ \bullet & \bullet & \bullet \end{smallmatrix},$$

$X^{(1)} = X^{\mathfrak{R}_2^{(2)}/(f_1)} = X_{F_1} \subset \mathbb{F}_2^{\mathbb{Z}^2}$, $X^{(2)} = X^{\mathfrak{R}_2^{(2)}/(f_2)} = X_{F_2} \subset \mathbb{F}_2^{\mathbb{Z}^2}$, and $\sigma^{(1)} = \alpha^{\mathfrak{R}_2^{(2)}/(f_1)} = \sigma^{F_1}$, $\sigma^{(2)} = \alpha^{\mathfrak{R}_2^{(2)}/(f_2)} = \sigma^{F_2}$ are the restrictions of the shift-action of $\mathbb{Z}^2$ on $\mathbb{F}_2^{\mathbb{Z}^2}$ to the closed, shift-invariant subgroups $X^{(1)}$ and $X^{(2)}$ in $\mathbb{F}_2^{\mathbb{Z}^2}$. We write $\mathcal{P}_0^{(i)}$ for the state partition (24.3) of $X^{(i)}$. Since f_2 is irreducible and $\omega(\mathbf{e}^{(2)}, \mathbb{S}(f_2)) = \omega(\mathbf{e}^{(2)}, F_2) > 0$, Lemma 25.11 shows that $\mathfrak{A}_X(\mathcal{P}_0^{(2)})_{-\infty} = \{\varnothing, X'\}$. The polynomial f_1 is, however, reducible: $f_1 = g_1 g_2$ with $g_1 = 1 + u_1 + u_1^2 + u_2$, $g_2 = 1 + u_2 + u_1 u_2$. Furthermore, since $\{0\} \subsetneq (f_1) \subsetneq (g_2) \subsetneq \mathfrak{R}_2^{(2)}$, the annihilators of these groups satisfy that $\mathbb{F}_2^{\mathbb{Z}^2} \supsetneq X \supsetneq Y = (g_2)^\perp \supsetneq \{0_X\}$. We write $\mathcal{P}_0$ for the state partition of the closed, shift-invariant subgroup $Y = X^{\mathfrak{R}_2^{(2)}/(g_2)} \subset \mathbb{F}_2^{\mathbb{Z}^2}$ and apply Lemma 25.11 to the irreducible polynomial g_2 to obtain that $\mathfrak{A}_Y(\mathcal{P}_0)_{-\infty} = \{\varnothing, Y\}$. According to (25.16)–(25.17) this means that $\Omega(Y)_{-\infty} = Y$, and hence that $\{0_X\} \neq Y \subset \Omega(X)_{-\infty}$. It follows that $\{\varnothing, X\} \subsetneq \mathfrak{A}_X(\mathcal{P}_0^{(1)})_{-\infty} \subsetneq \mathfrak{B}_{X^{(1)}}$ (cf. Lemma 25.11), and (25.24) implies that $\sigma^{(1)} = \sigma^{F_1}$ and $\sigma^{(2)} = \sigma^{F_2}$ cannot be measurably conjugate, although $\mathcal{C}(F_1) = \mathcal{C}(f_1) = \mathcal{C}(f_2) = \mathcal{C}(F_2)$.

(4) If $p > 1$, and if $X \subset \mathbb{F}_p^{\mathbb{Z}^2}$ is a closed, shift-invariant subgroup such that the shift-action σ of $\mathbb{Z}^2$ on X is ergodic, then Lemma 25.12 reveals a remarkable

(potential) asymmetry between the automorphisms $\sigma_{\mathbf{e}^{(2)}}$ and $\sigma_{-\mathbf{e}^{(2)}} = \sigma_{\mathbf{e}^{(2)}}^{-1}$. If $\mathcal{P}_0$ is the state partition (24.3) of X, then $\mathfrak{A}(\mathcal{P}_0)_{-\infty}$ measures the extent to which every point $x \in X$ is determined by its coordinates $x_{(m_1, -m_2)}$ with m_2 very large. Similarly we see that the sigma-algebras

$$\mathfrak{A}(\mathcal{P}_0)_k^* = \bigvee_{\{\mathbf{m}=(m_1, m_2) \in \mathbb{Z}^2 : m_2 \leq k\}} \sigma_{\mathbf{m}}(\mathcal{P}_0), \quad k \in \mathbb{Z}, \quad \mathfrak{A}(\mathcal{P}_0)_{-\infty}^* = \bigcap_{k \in \mathbb{Z}} \mathfrak{A}(\mathcal{P}_0)_k^*,$$

measure the dependence of x on its coordinates $x_{(m_1, m_2)}$ with m_2 very large. If $p = 2$, $F = F_1 = \{(0,0), (1,0), (0,1)\}$ in Example 25.2, and $X = X_F \subset \mathbb{F}_2^{\mathbb{Z}^2}$, then

$$\mathfrak{A}(\mathcal{P}_0)_{-\infty} = \mathfrak{B}_X \quad \text{and} \quad H_{\lambda_X}(\mathfrak{A}(\mathcal{P}_0)_0 | \mathfrak{A}(\mathcal{P}_0)_{-1}) = 0,$$

but

$$\mathfrak{A}(\mathcal{P}_0)_{-\infty}^* = \{\varnothing, X\} \quad \text{and} \quad H_{\lambda_X}(\mathfrak{A}(\mathcal{P}_0)_0^* | \mathfrak{A}(\mathcal{P}_0)_{-1}^*) = \log 2.$$

In Example 25.13 (1) we had $p = 2$, $f = 1 + u_1 + u_1^2 + u_2 + u_1 u_2 \in \mathfrak{R}_2^{(2)}$, $F = \mathcal{S}(f)$ and $X = X_F \subset \mathbb{F}_2^{\mathbb{Z}^2}$, and by comparing the Examples 25.13 (1) and (2) one sees that

$$\mathfrak{A}(\mathcal{P}_0)_{-\infty} = \mathfrak{A}(\mathcal{P}_0)_{-\infty}^* = \{\varnothing, X\},$$

but

$$\log 2 = H_{\lambda_X}(\mathfrak{A}(\mathcal{P}_0)_0 | \mathfrak{A}(\mathcal{P}_0)_{-1}) \neq H_{\lambda_X}(\mathfrak{A}(\mathcal{P}_0)_0^* | \mathfrak{A}(\mathcal{P}_0)_{-1}^*) = \log 4. \quad \boxdot$$

CONCLUDING REMARKS 25.17. (1) The material in this section is taken from [46] and [47]. The Examples 25.9 were communicated to me by Gruber.

(2) Let $p > 1$, and let X, X' be closed, shift-invariant subgroups of $\mathbb{F}_p^{\mathbb{Z}^2}$. We write σ and σ' for the shift-actions of $\mathbb{Z}^2$ on X and X', respectively. If σ and σ' are measurably conjugate, then Theorem 25.14 implies that $\{\mathbf{n} \in \mathbb{Z}^2 : \sigma_{\mathbf{n}} \text{ is expansive}\} = \{\mathbf{n} \in \mathbb{Z}^2 : \sigma_{\mathbf{n}}' \text{ is expansive}\}$. If σ' is a measurable factor of σ, then Theorem 25.15 shows that $\{\mathbf{n} \in \mathbb{Z}^2 : \sigma_{\mathbf{n}} \text{ is expansive}\} \subset \{\mathbf{n} \in \mathbb{Z}^2 : \sigma_{\mathbf{n}}' \text{ is expansive}\}$.

26. Relative entropies and residual sigma-algebras

The proof of Theorem 25.14 depends on the fact that, for every closed, shift-invariant subgroup $X \subset \mathbb{F}_p^{\mathbb{Z}^2}$, the conditional entropy $H_{\lambda_X}(\mathfrak{A}(\mathcal{Q})_0 | \mathfrak{A}(\mathcal{Q})_{-1})$ in Lemma 25.12 is finite for every finite, measurable partition $\mathcal{Q}$ of X. However, if $d > 2$, then the analogous conditional entropy will—in general—no longer be finite, and we have to use a more sophisticated approach in order to extract the necessary information from the sigma-algebras $\mathfrak{A}(\mathcal{Q})_k$. The key tool for this is the notion of relative entropy in [33]. We begin with some general definitions.

Let T be a measure preserving $\mathbb{Z}^d$-action on a probability space $(Y, \mathfrak{T}, \mu)$, and let $\mathcal{Q} \subset \mathfrak{T}$ be a countable partition with $H_\mu(\mathcal{Q}) < \infty$. For every primitive element $\mathbf{n} \in \mathbb{Z}^d$ and every $t \in \mathbb{R}$ we set

$$\mathfrak{A}_T(\mathcal{Q}, \mathbf{n})_t = \bigvee_{\{\mathbf{m} \in \mathbb{Z}^d : \langle \mathbf{m}, \mathbf{n} \rangle \leq t\}} T_{-\mathbf{m}}(\mathcal{Q}),$$

$$\mathfrak{A}_T(\mathcal{Q}, \mathbf{n})_t^- = \bigvee_{\{\mathbf{m} \in \mathbb{Z}^d : \langle \mathbf{m}, \mathbf{n} \rangle < t\}} T_{-\mathbf{m}}(\mathcal{Q}), \qquad (26.1)$$

$$\mathfrak{A}_T(\mathcal{Q}, \mathbf{n})_{-\infty} = \bigcap_{t \in \mathbb{R}} \mathfrak{A}_T(\mathcal{Q}, \mathbf{n})_t,$$

where $\langle \cdot, \cdot \rangle$ is defined in (16.3). The T-invariant sigma-algebra $\mathfrak{A}_T(\mathcal{Q}, \mathbf{n})_{-\infty}$ is the *residual sigma-algebra* of $\mathcal{Q}$ in the direction $\mathbf{n}$.

Let $\mathbf{n} \in \mathbb{Z}^d$ be a primitive element, and let

$$\Xi_\mathbf{n} = \{\mathbf{m} = (m_1, \ldots, m_d) : \langle \mathbf{m}, \mathbf{n} \rangle = 0\} \cong \mathbb{Z}^{d-1} \subset \mathbb{Z}^d. \qquad (26.2)$$

For every subgroup $\Gamma \subset \Xi_\mathbf{n}$ and every countable partition $\mathcal{Q} \subset \mathfrak{T}$ with $h_\mu(\mathcal{Q}) < \infty$ we set

$$h_\mu^{(\mathbf{n})}(T^{(\Gamma)}, \mathcal{Q}) = h_\mu^{\mathfrak{A}_T(\mathcal{Q}, \mathbf{n})_0}(T^{(\Gamma)} | \mathfrak{A}_T(\mathcal{Q}, \mathbf{n})_0^-)$$

$$= \lim_{\langle Q \rangle \to \infty} \frac{1}{|Q|} H_\mu\left(\bigvee_{\mathbf{m} \in Q} T_{-\mathbf{m}}(\mathcal{Q}) \middle| \mathfrak{A}_T(\mathcal{Q}, \mathbf{n})_0^- \right), \qquad (26.3)$$

where the limit is taken over all sequences of rectangles $Q \subset \Gamma$ with $\langle \Gamma \rangle \to \infty$ (cf. (13.7) and Remark 13.4). By

$$h_\mu^{(\mathbf{n})}(T^{(\Gamma)}) = \sup_\mathcal{Q} h_\mu^{\mathfrak{A}_T(\mathcal{Q}, \mathbf{n})_0}(T^{(\Gamma)} | \mathfrak{A}_T(\mathcal{Q}, \mathbf{n})_0^-) \qquad (26.4)$$

we denote the *relative entropy* (*with respect to* $\mathbf{n}$) of the restriction $T^{(\Gamma)} : \mathbf{m} \mapsto T_\mathbf{m}$, $\mathbf{m} \in \Gamma$, of T to Γ. The supremum in (26.4) is taken over all countable partitions $\mathcal{Q} \subset \mathfrak{T}$ with finite entropy.

When exploring the properties of $h_\mu^{(\mathbf{n})}(T^{(\Gamma)})$ we may argue as in the proofs of the Theorems 25.14–25.15 and assume without loss in generality that $\mathbf{n} = \mathbf{e}^{(d)} = (0, \ldots, 0, 1) \in \mathbb{Z}^d$. In order to simplify notation in this case we set

$$\Xi = \{\mathbf{n} = (n_1, \ldots, n_d) \in \mathbb{Z}^d : n_d = 0\},$$
$$\mathfrak{A}_T(\mathcal{Q})_t = \mathfrak{A}_T(\mathcal{Q}, \mathbf{e}^{(d)})_t, \ t \in \mathbb{R}, \ \text{and} \ \mathfrak{A}_T(\mathcal{Q})_{-\infty} = \bigcap_{t \in \mathbb{R}} \mathfrak{A}_T(\mathcal{Q})_t. \qquad (26.5)$$

Then

$$\mathfrak{A}_T(\mathcal{Q})_{-1} = \mathfrak{A}_T(\mathcal{Q}, \mathbf{e}^{(d)})_0^-, \quad \mathfrak{A}_T(\mathcal{Q})_k \subset \mathfrak{A}_T(\mathcal{Q})_{k+1},$$
$$\mathfrak{A}_T(\mathcal{Q}, \mathbf{e}^{(d)})_{-\infty} = \mathfrak{A}_T(\mathcal{Q})_{-\infty}, \quad T_{-\mathbf{n}}(\mathfrak{A}_T(\mathcal{Q})_k) = \mathfrak{A}_T(\mathcal{Q})_k$$

for every $k \in \mathbb{Z}$ and $\mathbf{n} \in \Xi$.

EXAMPLE 26.1. Let $\mathfrak{a} \subset \mathfrak{R}_d^{(p)}$ be an ideal, $X = X^{\mathfrak{R}_d^{(p)}/\mathfrak{a}} \subset \mathbb{Z}_{/p}^{\mathbb{Z}^d}$, and let $\alpha = \alpha^{\mathfrak{R}_d^{(p)}/\mathfrak{a}}$ be the shift-action of $\mathbb{Z}^d$ on X (cf. (6.19)–(6.20)). We write $\lambda = \lambda_X$ for the normalized Haar measure on X and define the state partition $\mathcal{P}_0$ of X by (24.3). If $\mathbf{n} \in \mathbb{Z}^d$ is a primitive element, and if, for every $t \in \mathbb{R}$,

$$\Omega(X, \mathbf{n})_t = \{x = (x_\mathbf{m}) \in X : x_\mathbf{m} = 0 \text{ for all } \mathbf{m} \text{ with } \langle \mathbf{m}, \mathbf{n} \rangle \leq t\},$$

$$\Omega(X, \mathbf{n})_t^- = \{x = (x_\mathbf{m}) \in X : x_\mathbf{m} = 0 \text{ for all } \mathbf{m} \text{ with } \langle \mathbf{m}, \mathbf{n} \rangle < t\},$$

$$\Omega(X, \mathbf{n})_{-\infty} = \overline{\bigcup_{t \in \mathbb{R}} \Omega(X, \mathbf{n})_t}, \tag{26.6}$$

then we see as in (25.17) that

$$\mathfrak{A}_\alpha(\mathcal{P}_0, \mathbf{n})_t = \mathfrak{B}_{X/\Omega(X,\mathbf{n})_t}, \quad \mathfrak{A}_\alpha(\mathcal{P}_0)_t^- = \mathfrak{B}_{X/\Omega(X,\mathbf{n})_t^-} \text{ for every } t \in \mathbb{R} \cup \{-\infty\},$$

$$h_{\lambda_X}^{\mathfrak{A}_\alpha(\mathcal{P}_0,\mathbf{n})_0}(\alpha^{(\Gamma)}|\mathfrak{A}_\alpha(\mathcal{P}_0,\mathbf{n})_0^-) = h((\alpha^{(\Gamma)})^{\Omega(X,\mathbf{n})_0^- / \Omega(X,\mathbf{n})_0}) \tag{26.7}$$

for every subgroup $\Gamma \subset \Xi_\mathbf{n}$. □

LEMMA 26.2. *Let $\Gamma \subset \Xi$ be a subgroup, and let $\mathcal{Q} \subset \mathfrak{T}$ be a countable partition with finite entropy. Then*

$$h_\mu^{\mathfrak{A}_T(\mathcal{Q})_0}(T^{(\Gamma)}|\mathfrak{A}_T(\mathcal{Q})_{-n}) = n h_\mu^{\mathfrak{A}_T(\mathcal{Q})_0}(T^{(\Gamma)}|\mathfrak{A}_T(\mathcal{Q})_{-1})$$

for every $n \geq 1$, where $T^{(\Gamma)}$ is the Γ-action $\mathbf{n} \mapsto T_\mathbf{n}, \mathbf{n} \in \Gamma$.

PROOF. We shall prove the lemma for $n = 2$; the general case differs only in notation. The assertion is trivial if $h_\mu^{\mathfrak{A}_T(\mathcal{Q})_0}(T^{(\Gamma)}|\mathfrak{A}_T(\mathcal{Q})_{-1}) = \infty$. If $h_\mu^{\mathfrak{A}_T(\mathcal{Q})_0}(T^{(\Gamma)}|\mathfrak{A}_T(\mathcal{Q})_{-1}) < \infty$ we define, for every $m \geq 0$,

$$\begin{aligned} N(m) &= \{\mathbf{n} = (n_1, \ldots, n_d) \in \mathbb{Z}^d : \\ &\quad (0, \ldots, 0) \preceq (n_1, \ldots, n_{d-1}) \text{ and } |n_d| \leq m\}, \\ N(m)^- &= \{\mathbf{n} = (n_1, \ldots, n_d) \in \mathbb{Z}^d : \\ &\quad (0, \ldots, 0) \prec (n_1, \ldots, n_{d-1}) \text{ and } |n_d| \leq m\}, \end{aligned} \tag{26.8}$$

where $\prec$ and $\preceq$ denote the lexicographic order on $\mathbb{Z}^{d-1}$, and set

$$Q_m = \{0\} \times \{-m, \ldots, m\}^{d-1} \subset \Xi, \quad \mathcal{Q}^{(m)} = \bigvee_{\mathbf{m} \in Q_m} T_{-\mathbf{m}}(\mathcal{Q}),$$

$$\overline{\mathcal{Q}}(m) = \bigvee_{\mathbf{m} \in N(m)} T_{-\mathbf{m}}(\mathcal{Q}^{(m)}), \quad \overline{\mathcal{Q}}(m)^- = \bigvee_{\mathbf{m} \in N(m)^-} T_{-\mathbf{m}}(\mathcal{Q}^{(m)}). \tag{26.9}$$

Fix $\varepsilon > 0$ and note that there exists an integer $M \geq 0$ such that, for all $m \geq M$,

$$\left| h_\mu^{\mathfrak{A}_T(\mathcal{Q})_0}(T^{(\Gamma)}|\mathfrak{A}_T(\mathcal{Q})_{-1}) \right.$$
$$\left. - \frac{1}{|Q|} H_\mu \left(\bigvee_{\mathbf{n} \in Q} T_{-\mathbf{n}}(\mathcal{Q}^{(m)}) \,\middle|\, \mathfrak{A}_T(\mathcal{Q})_{-1} \right) \right| < \varepsilon \tag{26.10}$$

and

$$
\left| h_\mu^{\mathfrak{A}_T(\mathcal{Q})_0}(T^{(\Gamma)}|\mathfrak{A}_T(\mathcal{Q})_{-2}) \right.
$$
$$
\left. - \frac{1}{|Q|} H_\mu \left(\bigvee_{\mathbf{n} \in Q} T_{-\mathbf{n}}(\mathcal{Q}^{(m)} \vee T_{\mathbf{e}(d)}(\mathcal{Q}^{(m)})) \,\Big|\, \mathfrak{A}_T(\mathcal{Q})_{-2} \right) \right| < \varepsilon \qquad (26.11)
$$

for all rectangles $Q \subset \Gamma$ for which $\langle Q \rangle$ is sufficiently large (since $\Gamma \cong \mathbb{Z}^{d'}$ for some $d' < d$ we can still speak of rectangles in Γ, although the notion of a rectangle depends on the choice of coordinates in Γ—cf. Remark 13.4). There exists an integer $M' \geq 1$ such that

$$
|H_\mu(\mathcal{Q}|\mathfrak{A}_T(\mathcal{Q})_{-1}) - H_\mu(\mathcal{Q}|T_{\mathbf{e}(d)}\mathcal{Q}^{(m')} \vee \mathfrak{A}_T(\mathcal{Q})_{-2})| < \varepsilon \qquad (26.12)
$$

for all $m' \geq M'$. Since

$$
H_\mu \left(\bigvee_{\mathbf{n} \in Q} T_{-\mathbf{n}}(\mathcal{Q}^{(m)} \vee T_{\mathbf{e}(d)}(\mathcal{Q}^{(m+m')})) \,\Big|\, \mathfrak{A}_T(\mathcal{Q})_{-2} \right)
$$
$$
= H_\mu \left(\bigvee_{\mathbf{n} \in Q} T_{-\mathbf{n}}(\mathcal{Q}^{(m)}) \,\Big|\, \bigvee_{\mathbf{n} \in Q} T_{-\mathbf{n}+\mathbf{e}(d)}(\mathcal{Q}^{(m+m')}) \vee \mathfrak{A}_T(\mathcal{Q})_{-2} \right)
$$
$$
+ H_\mu \left(\bigvee_{\mathbf{n} \in Q} T_{-\mathbf{n}}(\mathcal{Q}^{(m+m')}) \,\Big|\, \mathfrak{A}_T(\mathcal{Q})_{-1} \right),
$$

(26.10)–(26.12) imply that, as $\langle Q \rangle \to \infty$,

$$
|h_\mu^{\mathfrak{A}_T(\mathcal{Q})_0}(T^{(\Gamma)}|\mathfrak{A}_T(\mathcal{Q})_{-2}) - 2h_\mu^{\mathfrak{A}_T(\mathcal{Q})_0}(T^{(\Gamma)}|\mathfrak{A}_T(\mathcal{Q})_{-1})| < 4\varepsilon.
$$

Since ε is arbitrary, the lemma is proved. □

The next two lemmas generalize Lemma 25.12 to $\mathbb{Z}^d$-actions and give a general proof of the inclusion (25.20).

LEMMA 26.3. *Let* $\Gamma \subset \Xi$ *be a subgroup, and let* $\mathcal{P}, \mathcal{Q} \subset \mathfrak{T}$ *be countable partitions with finite entropies. If* $\mathcal{P}$ *generates* $\mathfrak{T}$ *under* T *and* $\mathfrak{A}_T(\mathcal{P})_{-\infty} \subset \mathfrak{A}_T(\mathcal{Q})_{-\infty}$ (mod μ), *then*

$$
h_\mu^{\mathfrak{A}_T(\mathcal{Q})_0}(T^{(\Gamma)}|\mathfrak{A}_T(\mathcal{Q})_{-1}) \leq h_\mu^{\mathfrak{A}_T(\mathcal{P})_0}(T^{(\Gamma)}|\mathfrak{A}_T(\mathcal{P})_{-1}). \qquad (26.13)
$$

PROOF. The decreasing martingale theorem implies that there exists, for any given $\varepsilon' > 0$, an $N' \geq 0$ such that

$$
\int |E_\mu(Q|\mathfrak{A}_T(\mathcal{P})_{-n}) - E_\mu(Q|\mathfrak{A}_T(\mathcal{P})_{-\infty})| \, d\mu < \varepsilon'
$$

for all $n \geq N'$ and $Q \in \mathcal{Q}$. Furthermore, since conditional expectation is a contraction on $L^1(Y, \mathfrak{T}, \mu)$,

$$
\int |E_\mu(Q|\mathfrak{A}_T(\mathcal{P})_{-n} \vee \mathfrak{S}) - E(Q|\mathfrak{A}_T(\mathcal{P})_{-\infty} \vee \mathfrak{S})| \, d\mu
$$

$$\leq \int |E_\mu(Q|\mathfrak{A}_T(\mathcal{P})_{-n}) - E_\mu(Q|\mathfrak{A}_T(\mathcal{P})_{-\infty})|\, d\mu$$

for all $n \geq 1$, $Q \in \mathcal{Q}$, and for every sigma-algebra $\mathfrak{S} \subset \mathfrak{T}$. We write $I_\mu(\mathcal{Q}|\mathfrak{S})$ for the conditional information function of the partition $\mathcal{Q}$, given $\mathfrak{S}$, and note that the family of functions $\{I_\mu(\mathcal{Q}|\mathfrak{S}) : \mathfrak{S} \subset \mathfrak{T}\}$ is uniformly integrable. Hence there exists an $N \geq 1$ such that $|H_\mu(\mathcal{Q}|\mathfrak{A}_T(\mathcal{P})_{-n} \vee \mathfrak{S}) - H_\mu(\mathcal{Q}|\mathfrak{A}_T(\mathcal{P})_{-\infty} \vee \mathfrak{S})| < \varepsilon'$ for every sigma-algebra $\mathfrak{S} \subset \mathfrak{T}$ and every $n \geq N$. In particular,

$$|H_\mu(\mathcal{Q}|\mathfrak{A}_T(\mathcal{P})_{-n} \vee \mathfrak{A}_T(\mathcal{Q})_{-m}) - H_\mu(\mathcal{Q}|\mathfrak{A}_T(\mathcal{Q})_{-m})| < \varepsilon' \qquad (26.14)$$

for all $m \in \mathbb{Z}$ and $n \geq N$, since $\mathfrak{A}_T(\mathcal{P})_{-\infty} \subset \mathfrak{A}_T(\mathcal{Q})_{-m} \pmod{\mu}$.

Since $\mathcal{P}$ is a generator for T, the increasing martingale theorem implies the existence of an integer $m \geq 0$ such that

$$H_\mu(\mathcal{Q}|T_{-m\mathbf{e}^{(d)}}(\mathcal{P}^{(m)}) \vee \cdots \vee T_{m\mathbf{e}^{(d)}}(\mathcal{P}^{(m)})) < \varepsilon', \qquad (26.15)$$

where $\mathcal{P}^{(m)}$ is defined as in (26.9). We fix $\varepsilon > 0$ and apply Lemma 26.2 to obtain that, for every $t \geq 1$,

$$t h_\mu^{\mathfrak{A}_T(\mathcal{Q})_0}(T^{(\Gamma)}|\mathfrak{A}_T(\mathcal{Q})_{-1}) = h_\mu^{\mathfrak{A}_T(\mathcal{Q})_0}(T^{(\Gamma)}|\mathfrak{A}_T(\mathcal{Q})_{-t}).$$

If $h_\mu^{\mathfrak{A}_T(\mathcal{Q})_0}(T^{(\Gamma)}|\mathfrak{A}_T(\mathcal{Q})_{-1}) = \infty$, then there exists, for every $K \geq 1$, an $M \geq 1$ such that, for all $k \geq M$,

$$\lim_{\langle Q \rangle \to \infty} \frac{1}{|Q|} H_\mu\left(\bigvee_{\mathbf{n} \in Q} T_{-\mathbf{n}}(\mathcal{Q}^{(k)}) \,\Big|\, \mathfrak{A}_T(\mathcal{Q})_{-1} \right) > K,$$

and hence

$$\lim_{\langle Q \rangle \to \infty} \frac{1}{t|Q|} H_\mu\left(\bigvee_{\mathbf{n} \in Q} T_{-\mathbf{n}}(\mathcal{Q}^{(k)}) \vee \cdots \right.$$
$$\left. \cdots \vee T_{(t-1)\mathbf{e}^{(d)}}(\mathcal{Q}^{(k)})) \,\Big|\, \mathfrak{A}_T(\mathcal{Q})_{-t} \right) > K \qquad (26.16)$$

for all $t \geq 1$.

If $h_\mu^{\mathfrak{A}_T(\mathcal{Q})_0}(T^{(\Gamma)}|\mathfrak{A}_T(\mathcal{Q})_{-1}) < \infty$ and $\varepsilon > 0$, then we can find an integer $M \geq 1$ such that, for every $k \geq M$,

$$\left| h_\mu^{\mathfrak{A}_T(\mathcal{Q})_0}(\alpha^{(\Gamma)}|\mathfrak{A}_T(\mathcal{Q})_{-1}) - \lim_{\langle Q \rangle \to \infty} \frac{1}{|Q|} H_\mu\left(\bigvee_{\mathbf{n} \in Q} \alpha_{-\mathbf{n}}(\mathcal{Q}^{(k)}) \,\Big|\, \mathfrak{A}_T(\mathcal{Q})_{-1} \right) \right| < \varepsilon,$$

$$|H_\mu(\mathcal{Q}|\mathfrak{A}_T(\mathcal{Q})_{-1}) - H(\mathcal{Q}|\alpha_{\mathbf{e}_d}(\mathcal{Q}^{(k)}) \vee \mathfrak{A}_T(\mathcal{Q})_{-2})| < \varepsilon.$$

Then

$$h_\mu^{\mathfrak{A}_T(\mathcal{Q})_0}(T^{(\Gamma)}|\mathfrak{A}_T(\mathcal{Q})_{-1}) = \frac{1}{t} h_\mu^{\mathfrak{A}_T(\mathcal{Q})_0}(T^{(\Gamma)}|\mathfrak{A}_T(\mathcal{Q})_{-t})$$

$$\leq \lim_{\langle Q \rangle \to \infty} \frac{1}{t|Q|} H_\mu\left(\bigvee_{\mathbf{n} \in Q} T_{-\mathbf{n}}(\mathcal{Q}^{(k_0)}) \vee \cdots \right. \qquad (26.17)$$
$$\left. \cdots \vee T_{(t-1)\mathbf{e}^{(d)}}(\mathcal{Q}^{(k_{t-1}))}) \,\Big|\, \mathfrak{A}_T(\mathcal{Q})_{-t} \right) + \varepsilon$$

for all $t \geq 1$ and $k_0, \ldots, k_{t-1} \geq M$. Having chosen $M \geq 1$ as in (26.16) or (26.17), we fix $k \geq M$ and apply (26.14), with $\varepsilon' = \varepsilon(2tk + 1)^{1-d}$, to find an integer $n \geq 0$ with

$$\lim_{\langle Q \rangle \to \infty} \frac{1}{t|Q|} H_\mu \left(\bigvee_{\mathbf{n} \in Q} T_{-\mathbf{n}}(\mathcal{Q}^{(k)} \vee \cdots \vee T_{(t-1)\mathbf{e}^{(d)}}(\mathcal{Q}^{(tk)})) \,\Big|\, \mathfrak{A}_T(\mathcal{Q})_{-t} \right)$$

$$\leq \lim_{\langle Q \rangle \to \infty} \frac{1}{t|Q|} H_\mu \left(\bigvee_{\mathbf{n} \in Q} T_{-\mathbf{n}}(\mathcal{Q}^{(k)} \vee \cdots \right.$$

$$\left. \cdots \vee T_{(t-1)\mathbf{e}^{(d)}}(\mathcal{Q}^{(tk)})) \,\Big|\, \mathfrak{A}_T(\mathcal{Q})_{-t} \vee \mathfrak{A}_T(\mathcal{P})_{-n-t} \right) + \varepsilon \qquad (26.18)$$

for every $t \geq 1$. By setting $\varepsilon' = \varepsilon(2tk + 1)^{1-d}$ in (26.14) we see that there exists an $m \geq 0$ such that, for the integer n in (26.18), and for every $t \geq 1$,

$$\lim_{\langle Q \rangle \to \infty} \frac{1}{t|Q|} H_\mu \left(\bigvee_{\mathbf{n} \in Q} T_{-\mathbf{n}}(\mathcal{Q}^{(k)} \vee \cdots \right.$$

$$\left. \cdots \vee T_{(t-1)\mathbf{e}^{(d)}}(\mathcal{Q}^{(tk)})) \,\Big|\, \mathfrak{A}_T(\mathcal{Q})_{-t} \vee \mathfrak{A}_T(\mathcal{P})_{-n-t} \right) - \varepsilon$$

$$\leq \lim_{Q \nearrow \Gamma} \frac{1}{t|Q|} H_\mu \left(\bigvee_{\mathbf{n} \in Q} T_{-\mathbf{n}}(T_{-m\mathbf{e}^{(d)}}(\mathcal{P}^{(m+k)}) \vee \cdots \right.$$

$$\left. \cdots \vee T_{(t+m-1)\mathbf{e}^{(d)}}(\mathcal{P}^{(m+(t+2m)k)})) \,\Big|\, \mathfrak{A}_T(\mathcal{P})_{-n-t} \right) \qquad (26.19)$$

$$\leq \lim_{Q \nearrow \Gamma} \frac{1}{t|Q|} H_\mu \left(\bigvee_{\mathbf{n} \in Q} T_{-\mathbf{n}}(T_{-m\mathbf{e}^{(d)}}(\mathcal{P}^{(m+k)}) \vee \cdots \right.$$

$$\left. \cdots \vee T_{(t+m-1)\mathbf{e}^{(d)}}(\mathcal{P}^{(m+(t+2m)k)})) \,\Big|\, \mathfrak{A}_T(\mathcal{P})_{-n-m-t} \right)$$

$$\leq \frac{1}{t} h_\mu^{\mathfrak{A}_T(\mathcal{P})_0}(T^{(\Gamma)} | \mathfrak{A}_T(\mathcal{P})_{-n-m-t}) + \varepsilon$$

$$= \frac{t+m+n}{t} h_\mu^{\mathfrak{A}_T(\mathcal{P})_0}(T^{(\Gamma)} | \mathfrak{A}_T(\mathcal{P})_{-1}).$$

From (26.16)–(26.19) and the arbitrariness of K it is clear that $h_\mu^{\mathfrak{A}_T(\mathcal{P})_0}(T^{(\Gamma)} | \mathfrak{A}_T(\mathcal{P})_{-1}) = \infty$ whenever $h_\mu^{\mathfrak{A}_T(\mathcal{Q})_0}(T^{(\Gamma)} | \mathfrak{A}_T(\mathcal{Q})_{-1}) = \infty$. If $h_\mu^{\mathfrak{A}_T(\mathcal{Q})_0}(T^{(\Gamma)} | \mathfrak{A}_T(\mathcal{Q})_{-1}) < \infty$, we let $t \to \infty$ in (26.17)–(26.19); since ε was arbitrary, we obtain that

$$h_\mu^{\mathfrak{A}_T(\mathcal{Q})_0}(T^{(\Gamma)} | \mathfrak{A}_T(\mathcal{Q})_{-1}) \leq h_\mu^{\mathfrak{A}_T(\mathcal{P})_0}(T^{(\Gamma)} | \mathfrak{A}_T(\mathcal{P})_{-1}). \qquad \square$$

Lemma 25.12 corresponds to the special case of Lemma 26.3 where $\Gamma = \{0_{\mathbb{Z}^d}\}$. The relative entropies $h_\mu^{(\mathbf{n})}(T^{(\Gamma)})$ in (26.4) are—in general—very difficult to compute. One of the reasons for this difficulty is that $h_\mu^{\mathfrak{A}_T(\mathcal{Q})_0}(T^{(\Gamma)}, \mathfrak{A}_T(\mathcal{Q})_{-1})$ is not monotonic in $\mathcal{Q}$, as we saw in Example 25.13 (3). However, for certain $\mathbb{Z}^d$-actions T by automorphisms of compact, abelian groups X, they can be calculated explicitly, since there exist distinguished, finite generators $\mathcal{P}$ of X for which $\mathfrak{A}_T(\mathcal{P})_{-\infty}$ is minimal, and $h_\mu^{(\mathbf{n})}(T^{(\Gamma)}) = h_\mu^{\mathfrak{A}_T(\mathcal{P})_0}(T^{(\Gamma)}, \mathfrak{A}_T(\mathcal{P})_{-1})$. Before we can proceed with these calculations we need two definitions.

DEFINITION 26.4. (1) For every non-zero element $f \in \mathfrak{R}_d^{(p)}$ and every primitive element $\mathbf{n} \in \mathbb{Z}^d$ we define Laurent polynomials $f^{\mathbf{n},\blacktriangle}, f^{\mathbf{n},\blacktriangledown} \in \mathfrak{R}_d^{(p)}$ as follows. If $f = h_1 \cdot \ldots \cdot h_r$ is a decomposition of f into irreducible elements of $\mathfrak{R}_d^{(p)}$ such that $|\mathcal{S}(h_j)| > 1$ for every $j = 1, \ldots, k$ (i.e. none of the h_j is a unit in $\mathfrak{R}_d^{(p)}$), then we set

$$f^{\mathbf{n},\blacktriangle} = \prod_{\{j : j = 1, \ldots, k \text{ and } |\Phi_{\mathbf{n}}(\mathcal{S}(h_j))| = 1\}} h_j,$$

$$f^{\mathbf{n},\blacktriangledown} = \prod_{\{j : j = 1, \ldots, k \text{ and } |\Phi_{\mathbf{n}}(\mathcal{S}(h_j))| > 1\}} h_j.$$

The Laurent polynomials $f^{\mathbf{n},\blacktriangle}$ and $f^{\mathbf{n},\blacktriangledown}$ are determined up to multiplication by a unit in $\mathfrak{R}_d^{(p)}$, and $f = f^{\mathbf{n},\blacktriangle} f^{\mathbf{n},\blacktriangledown}$. For simplicity of notation we put

$$f^{\blacktriangle} = f^{\mathbf{e}^{(d)},\blacktriangle}, \quad f^{\blacktriangledown} = f^{\mathbf{e}^{(d)},\blacktriangledown}.$$

If $\mathfrak{a} \subset \mathfrak{R}_d^{(p)}$ is a non-zero ideal we write

$$\mathfrak{a}^{\mathbf{n},\blacktriangle} = (f^{\mathbf{n},\blacktriangle}, f \in \mathfrak{a}), \quad \mathfrak{a}^{\mathbf{n},\blacktriangledown} = (f^{\mathbf{n},\blacktriangledown}, f \in \mathfrak{a})$$

for the ideals in $\mathfrak{R}_d^{(p)}$ generated by $\{f^{\mathbf{n},\blacktriangle} : f \in \mathfrak{a}\}$ and $\{f^{\mathbf{n},\blacktriangledown} : f \in \mathfrak{a}\}$, and set $\mathfrak{a}^{\blacktriangle} = \mathfrak{a}^{\mathbf{e}^{(d)},\blacktriangle}$ and $\mathfrak{a}^{\blacktriangledown} = \mathfrak{a}^{\mathbf{e}^{(d)},\blacktriangledown}$.

(2) For every non-zero element $f \in \mathfrak{R}_d^{(p)}$ we define Laurent polynomials $f^{\triangle}, f^{\triangledown} \in \mathfrak{R}_{d-1}^{(p)}$ by writing f in the form $f = \sum_{j=k^-}^{k^+} u_d^j$ with $g_j \in \mathfrak{R}_{d-1}^{(p)}$ for $j = k^-, \ldots, k^+$ and $g_{k^-} \cdot g_{k^+} \neq 0$, and by setting

$$f^{\triangle} = g_{k^-}, \quad f^{\triangledown} = g_{k^+}.$$

If $\mathfrak{a} \subset \mathfrak{R}_d^{(p)}$ is a non-zero ideal we write

$$\mathfrak{a}^{\triangle} = (f^{\triangle}, f \in \mathfrak{a}), \quad \mathfrak{a}^{\triangledown} = (f^{\triangledown}, f \in \mathfrak{a})$$

for the ideals in $\mathfrak{R}_{d-1}^{(p)}$ generated by $\{f^{\triangle} : f \in \mathfrak{a}\}$ and $\{f^{\triangledown} : f \in \mathfrak{a}\}$.

EXAMPLES 26.5. (1) The polynomial $f = 1 + u_1 + u_1^2 + u_1^3 u_2 + u_2^2 + u_1 u_2^2$ in Example 25.16 (3) satisfies that $f = g_1 g_2$ with $g_1 = 1 + u_1 + u_1^2 + u_2$ and $g_2 = 1 + u_2 + u_1 u_2$, and $f^{\blacktriangle} = g_1, f^{\blacktriangledown} = g_2$. Furthermore, $f^{\triangle} = 1 + u_1 + u_1^2, f^{\triangledown} = 1 + u_1$.

(2) Let $f = 1 + u_1 + u_2 + u_3 \in \mathfrak{R}_3^{(p)}$. Then $f^{\blacktriangle} = f, f^{\blacktriangledown} = 1, f^{\triangle} = 1 + u_1 + u_2$, and $f^{\triangledown} = 1$.

(3) If $\mathfrak{a} \subset \mathfrak{R}_d^{(p)}$ is a non-zero ideal of the form $\mathfrak{a} = (f_1, \ldots, f_k) = f_1 \mathfrak{R}_d^{(p)} + \cdots + f_k \mathfrak{R}_d^{(k)}$ with non-zero generators $f_1, \ldots, f_k$ in $\mathfrak{R}_d^{(p)}$, then

$$\mathfrak{a}^{\mathbf{n},\blacktriangle} = (f_1^{\mathbf{n},\blacktriangle}, \ldots, f_k^{\mathbf{n},\blacktriangle}), \quad \mathfrak{a}^{\mathbf{n},\blacktriangledown} = (f_1^{\mathbf{n},\blacktriangledown}, \ldots, f_k^{\mathbf{n},\blacktriangledown}),$$

$$\mathfrak{a}^{\blacktriangle} = (f_1^{\blacktriangle}, \ldots, f_k^{\blacktriangle}), \quad \mathfrak{a}^{\blacktriangledown} = (f_1^{\blacktriangledown}, \ldots, f_k^{\blacktriangledown}),$$

$$\mathfrak{a}^\triangle = (f_1^\triangle, \ldots, f_k^\triangle), \quad \mathfrak{a}^\triangledown = (f_1^\triangledown, \ldots, f_k^\triangledown). \quad \square$$

LEMMA 26.6. *Let* $d \geq 2$, p *a rational prime*, $\mathfrak{a} \subset \mathfrak{R}_d^{(p)}$ *an ideal, and let* $\alpha = \alpha^{\mathfrak{R}_d^{(p)}/\mathfrak{a}}$ *be the shift-action of* $\mathbb{Z}^d$ *on* $X = X^{\mathfrak{R}_d^{(p)}/\mathfrak{a}} \subset \mathbb{Z}_{/p}^{\mathbb{Z}^d}$ *and* $\mathcal{P}_0$ *the state partition of* X (*cf.* (6.19)–(6.20) *and* (24.3)). *Then* $\mathfrak{A}_\alpha(\mathcal{P}_0)_{-\infty} \subset \mathfrak{A}_\alpha(\mathcal{Q})_{-\infty}$ (mod λ_X) *for every countable generator* $\mathcal{Q} \subset \mathfrak{B}_X$ *for* α *with finite entropy.*

PROOF. As in (25.15)–(25.17) we set, for every $m \in \mathbb{Z}$,

$$\Omega(X)_m = \{x = (x_{\mathbf{n}}) \in X : x_{\mathbf{n}} = 0 \text{ for all}$$
$$\mathbf{n} = (n_1, \ldots, n_d) \in \mathbb{Z}^d \text{ with } n_d \leq m\}, \qquad (26.20)$$
$$\Omega(X)_{-\infty} = \overline{\bigcup_{m \in \mathbb{Z}} \Omega(X)_m}.$$

Then $\Omega(X)_m$ is a closed, Ξ-invariant subgroup of X, $\Omega(X)_{-\infty}$ is α-invariant, and

$$\mathfrak{A}_\alpha(\mathcal{P}_0)_m = \mathfrak{B}_{X/\Omega(X)_m} \text{ for every } m \in \mathbb{Z} \cup \{-\infty\} \qquad (26.21)$$

Our first task is to determine the group $\Omega(X)_{-\infty}$.

If $\mathfrak{a} = 0$, then α is the shift-action of $\mathbb{Z}^d$ on $X = \mathbb{F}_p^{\mathbb{Z}^d}$, $\Omega(X)_{-\infty} = X$, and $\mathfrak{A}_\alpha(\mathcal{P}_0)_{-\infty} = \{\varnothing, X\} \subset \mathfrak{A}_\alpha(\mathcal{Q})_{-\infty}$ for every countable partition $\mathcal{Q} \subset \mathfrak{B}_X$.

Now assume that $\{0\} \neq \mathfrak{a} = (f) = f\mathfrak{R}_d^{(p)}$ for some non-zero elements $f \in \mathfrak{R}_d^{(p)}$. We write $\mathcal{R}^\pm \subset \mathfrak{R}_d^{(p)}$ for the subrings consisting of all Laurent polynomials involving only positive, resp. negative, powers of u_d. After multiplying f by a monomial we may take it that $f \in \mathcal{R}^-$, and that f has constant term 1. For all $m \geq 0$, the $\mathcal{R}^-$-module $\mathcal{N}_m = (u_d^{-m}\mathcal{R}^- + f\mathfrak{R}_d^{(p)})/f\mathfrak{R}_d^{(p)}$ is Noetherian, and its associated prime ideals are all principal and of the form $h_i\mathcal{R}^-$, where $f = h_1 \cdot \ldots \cdot h_r$ is a decomposition of f into irreducible elements of $\mathcal{R}^-$.

Put $\mathcal{J} = u_d^{-1}\mathcal{R}^-$, $\mathcal{L}_n = \bigcap_{k \geq 0} \mathcal{J}^k \cdot \mathcal{N}_m = \bigcap_{k \geq 0} u_d^{-k} \cdot \mathcal{N}_m$, and note that $\mathcal{J} \cdot \mathcal{L}_m = \mathcal{L}_m$. By Corollary 2.5 in [5] there exists an element $\psi \in \mathcal{R}^-$ such that $(1 + u_d^{-1}\psi) \cdot \mathcal{L}_m = \{0\}$. If $\mathcal{L}_m \neq \{0\}$ this means that $1 + u_d^{-1}\psi$ lies in at least one of the prime ideals associated with the Noetherian $\mathcal{R}^-$-module $\mathcal{N}_m \cong \mathcal{R}/f\mathcal{R}^-$, so that $1 + u_d^{-1}\psi$ and f have a non-trivial common factor in $\mathcal{R}^-$.

We assume without loss in generality that each h_i has constant term 1, and that $h_i \in 1 + u_d^{-1}\mathcal{R}^-$ for $i = 1, \ldots, s-1$, but $h_j \notin 1 + u_d^{-1}\mathcal{R}^-$ for $j = s, \ldots, r$. Then

$$f^\blacktriangle = \prod_{j=1}^{s-1} h_j, \quad f^\blacktriangledown = \prod_{j=s}^{r} h_j \qquad (26.22)$$

(cf. Definition 26.4), and the argument in the preceding paragraph shows that $f^\blacktriangle \cdot \mathcal{L}_m = \{0\}$ for all $m \geq 0$, since $1 + u_d^{-1}\psi$ cannot be divisible by any h_i with

$i \in \{s, \ldots, r\}$. We put $H_m = \{\mathbf{n} = (n_1, \ldots, n_d) \in \mathbb{Z}^2 : n_d \leq m\}$ and note that, for every $m \geq 0$ and $k \leq 0$,

$$\widehat{\pi_{H_m}(X)} = \mathcal{N}_m, \quad (\pi_{H_m}(\Omega(X)_k))^{\perp} = u_d^{k-m} \mathcal{N}_m \subset \mathcal{N}_m,$$

$$\text{and } (\pi_{H_m}(\Omega(X)_{-\infty}))^{\perp} = \mathcal{L}_m \subset \mathcal{N}_m. \tag{26.23}$$

For every $\psi \in \mathfrak{R}_d^{(p)}$ we denote by $\alpha_\psi \colon X \longmapsto X$ the homomorphism dual to multiplication by $\psi \in \mathcal{R}^- \subset \mathfrak{R}_d^{(p)}$ on $\hat{X} = \mathfrak{R}_d^{(p)}/(f)$. Since $f^{\blacktriangle} \cdot \mathcal{L}_m = \{0\}$, the last equation in (26.23) shows that $\pi_{H_m}(\alpha_{f^{\blacktriangle}}(x)) \in \pi_{H_m}(\Omega(X)_{-\infty})$ for every $m \geq 1$ and $x \in X$, and we conclude that $\alpha_{f^{\blacktriangle}}(X) \subset \Omega(X)_{-\infty}$.

The annihilators of the α-invariant subgroups $\{0_X\} \subset \alpha_{f^{\blacktriangle}}(X) \subset X \subset \mathbb{F}_p^{\mathbb{Z}^d}$ satisfy that $\mathfrak{R}_d^{(p)} \supset \alpha_{f^{\blacktriangle}}(X)^{\perp} \supset (f) \supset \{0\}$. Furthermore, if $\psi \in \mathfrak{R}_d^{(p)}$ lies in the annihilator of $\alpha_{f^{\blacktriangle}}(X)$, then $\langle \alpha_{f^{\blacktriangle}}(x), \psi \rangle = \langle x, f^{\blacktriangle}\psi \rangle = 1$ for all $x \in X$, so that $f^{\blacktriangle}\psi \in (f)$; as $f^{\blacktriangle}$ and $f^{\blacktriangledown}$ have no common factors and $f = f^{\blacktriangle} f^{\blacktriangledown}$ this implies that $\alpha_{f^{\blacktriangle}}(X)^{\perp} \subset (f^{\blacktriangledown})$. Conversely, every $\psi \in (f^{\blacktriangledown})$ annihilates $\alpha_{f^{\blacktriangle}}(X)$, since $\langle \alpha_{f^{\blacktriangle}}(x), f^{\blacktriangledown} \rangle = \langle x, f^{\blacktriangle} f^{\blacktriangledown} \rangle = \langle x, f \rangle = 1$ for every $x \in X$. We have proved that $\alpha_{f^{\blacktriangle}}(X)^{\perp} = (f^{\blacktriangledown})$, so that $(f^{\blacktriangledown})^{\perp} = \alpha_{f^{\blacktriangle}}(X) \subset \Omega(X)_{-\infty}$.

Since $f^{\blacktriangle} \in 1 + u_d^{-1}\mathcal{R}^-$, we see exactly as in the proof of Lemma 25.10 that $\Omega(X)_{-\infty} \cap \ker(\alpha_{f^{\blacktriangle}}) = \Omega(Y)_{-\infty} = \{0\}$, where $Y = \ker(\alpha_{f^{\blacktriangle}}) = \widehat{\mathfrak{R}_d^{(p)}/(f^{\blacktriangle})}$. If $\Omega(X)_{-\infty} \supsetneq (f^{\blacktriangledown})^{\perp}$, and if $x \in \Omega(X)_{-\infty} \setminus (f^{\blacktriangledown})^{\perp}$, then $0_X \neq \alpha_h(x) \in \Omega(X)_{-\infty} \cap \ker(\alpha_{f^{\blacktriangle}}) = \{0_X\}$, which is impossible (note that $\alpha_{f^{\blacktriangle}} \cdot \alpha_{f^{\blacktriangledown}}(x) = \alpha_f(x) = 0_X$ for every $x \in X$). This contradiction shows that

$$\Omega(X)_{-\infty} = \alpha_{f^{\blacktriangle}}(X) = (f^{\blacktriangledown})^{\perp} = \widehat{\mathfrak{R}_d^{(p)}/(f^{\blacktriangledown})}, \tag{26.24}$$

where $f^{\blacktriangle}$ and $f^{\blacktriangledown}$ are chosen as in (26.22).

If $\mathfrak{a}$ is non-principal we choose non-zero elements $f_1, \ldots, f_k$ in $\mathfrak{R}_d^{(p)}$ such that $\mathfrak{a} = (f_1, \ldots, f_k)$, assume without loss in generality that $f_j \in \mathcal{R}^-$ and $f(0, \ldots, 0) = 1$ for $j = 1, \ldots, k$, and that $f_j^{\blacktriangle}$ and $f_j^{\blacktriangledown}$ are chosen as in (26.22). Then (26.24) shows that

$$X = \bigcap_{j=1}^{k} \widehat{\mathfrak{R}_d^{(p)}/(f_i)} = \bigcap_{j=1}^{k} (f_j)^{\perp} \subset \mathbb{F}_p^{\mathbb{Z}^d},$$

$$\Omega(X)_{-\infty} = \bigcap_{i=1}^{k} \widehat{\mathfrak{R}_d^{(p)}/(f_i^{\blacktriangledown})} = (\mathfrak{a}^{\blacktriangledown})^{\perp} \subset \mathbb{F}_p^{\mathbb{Z}^d}, \tag{26.25}$$

where the notation is as in Definition 26.4.

Having identified $\Omega(X)_{-\infty}$ for $X = X^{\mathfrak{R}_d^{(p)}/\mathfrak{a}}$, where $\mathfrak{a} = (f_1, \ldots, f_k)$ is an arbitrary, non-zero ideal in $\mathfrak{R}_d^{(p)}$, we turn to the proof of the inclusion $\mathfrak{B}_{X/\Omega(X)_{-\infty}} = \mathfrak{A}_\alpha(\mathcal{P}_0)_{-\infty} \subset \mathfrak{A}_\alpha(\mathcal{Q})_{-\infty}$ for every countable partition $\mathcal{Q} \subset \mathfrak{B}_X$ with finite entropy. Without loss in generality we assume that $f_j \in \mathcal{R}^-$ and $f(0, \ldots, 0) = 1$, and set $f_j = f_j^{\blacktriangledown} f_j^{\blacktriangle}$ for $j = 1, \ldots, k$ (cf. (26.22)). For

$j = 1, \ldots, k$, $f_j^{\blacktriangle} = 1 + u_d^{-1} g_j$ for some $g_j \in \mathcal{R}^-$. We fix $j \in \{1, \ldots, k\}$ for the moment. For every $n \geq 1$, the Laurent polynomial $\psi_j^{(n)} = f_j^{\blacktriangle p^n} f_j^{\blacktriangledown} \in f_j \mathcal{R}_d^{(p)} \subset \mathfrak{a}$ is of the form $f_j^{\blacktriangledown} + g_j^{p^n} f_j^{\blacktriangledown}$, and $g_j^{p^n} f_j^{\blacktriangledown} \in u_d^{-p^n} \mathcal{R}^-$. We denote by $\chi_\psi = \langle \cdot, \psi \rangle$ the character on $\mathbb{F}_p^{\mathbb{Z}^2}$ (and hence on X) corresponding to an element $\psi \in \mathcal{R}_d^{(p)}$ and denote by $\| \cdot \|$ the L^2-norm on $L^2(X, \mathfrak{B}_X, \lambda_X)$. If $\mathcal{Q} \subset \mathfrak{B}_X$ is a finite generator, then the increasing martingale theorem shows that $\lim_{k \to \infty} \| \chi_{f_j^{\blacktriangledown}} - E_{\lambda_X}(\chi_{f_j^{\blacktriangledown}} | \mathfrak{A}(\mathcal{Q})_k) \| = 0$. It follows that, for every $\mathbf{m} = (m_1, \ldots, m_d) \in \mathbb{Z}^2$ with $m_d < 0$, $\lim_{k \to \infty} \| \chi_{u^{km} f_j^{\blacktriangledown}} - E_{\lambda_X}(\chi_{u^{km} f_j^{\blacktriangledown}} | \mathfrak{A}(\mathcal{Q})_m) \| = 0$ for every $m \in \mathbb{Z}$, so that

$$\lim_{n \to \infty} \| \chi_{g_j^{p^n} f_j^{\blacktriangledown}} - E_{\lambda_X}(\chi_{g_j^{p^n} f_j^{\blacktriangledown}} | \mathfrak{A}(\mathcal{Q})_m) \| = \| \chi_{f_j^{\blacktriangledown}} - E_{\lambda_X}(\chi_{f_j^{\blacktriangledown}} | \mathfrak{A}(\mathcal{Q})_m) \| = 0$$

for every $m \in \mathbb{Z}$ (since $\psi_j^{(n)} \in \mathfrak{a}$, $\chi_{\psi_j^{(n)}}(x) = \chi_{f_j^{\blacktriangledown}}(x) \chi_{g_j^{p^n} f_j^{\blacktriangledown}}(x) = 1$ for every $x \in X$ and $n \geq 1$). This shows that $\chi_{f_j^{\blacktriangledown}}$ is $\mathfrak{A}(\mathcal{Q})_m$-measurable for every $m \in \mathbb{Z}$, and hence $\mathfrak{A}(\mathcal{Q})_{-\infty}$-measurable. By (26.21), $\mathfrak{A}(\mathcal{P}_0)_{-\infty} = \mathfrak{B}_{X/\Omega(X)_{-\infty}}$ is the smallest α-invariant sigma-algebra with respect to which $\chi_{f_j^{\blacktriangledown}}$ is measurable for every $j = 1, \ldots, k$, so that $\mathfrak{A}(\mathcal{Q})_{-\infty} \supset \mathfrak{A}(\mathcal{P}_0)_{-\infty} \pmod{\mu}$. $\square$

LEMMA 26.7. *Let $d \geq 2$, p a rational prime, $0 \neq f \in \mathcal{R}_d^{(p)}$, and let λ_X be the normalized Haar measure, $\mathcal{P}_0$ the state partition, and $\alpha = \alpha^{\mathcal{R}_d^{(p)}/(f)}$ the shift-action of $\mathbb{Z}^d$ on $X = X^{\mathcal{R}_d^{(p)}/(f)} \subset \mathbb{F}_p^{\mathbb{Z}^d}$ (cf. (6.19)–(6.20) and (24.3)). Then*

$$h(\alpha^{(\Xi)}) = w(\mathbf{e}^{(d)}, \mathcal{S}(f)) \cdot \log p. \tag{26.26}$$

Furthermore, if $\Omega(X, \pm \mathbf{e}^{(d)})_k \subset X$ are the subgroups and $f^{\triangledown}, f^{\triangle} \in \mathcal{R}_{d-1}^{(p)}$ the Laurent polynomials in Definition 26.4, then the action $(\alpha^{(\Xi)})^{\Omega(X)_{-1}/\Omega(X)_0}$ of $\Xi \cong \mathbb{Z}^{d-1}$ induced by $\alpha^{(\Xi)}$ on the quotient group $\Omega(X)_{-1}/\Omega(X)_0$ is algebraically conjugate to the $\mathbb{Z}^{d-1}$-action $\alpha^{\mathcal{R}_{d-1}^{(p)}/(f^{\triangledown})}$ on the group $X^{\mathcal{R}_{d-1}^{(p)}/(f^{\triangledown})}$, and $(\alpha^{(\Xi)})^{\Omega(X, -\mathbf{e}^{(d)})_{-1}/\Omega(X, -\mathbf{e}^{(d)})_0}$ is conjugate to $\alpha^{\mathcal{R}_{d-1}^{(p)}/(f^{\triangle})}$ on $X^{\mathcal{R}_{d-1}^{(p)}/(f^{\triangle})}$. In particular,

$$\begin{aligned} h_{\lambda_X}^{\mathfrak{A}_\alpha(\mathcal{P}_0, \mathbf{e}^{(d)})_0}(\alpha^{(\Gamma)} | \mathfrak{A}_\alpha(\mathcal{P}_0, \mathbf{e}^{(d)})_{-1}) &= h(\alpha^{\mathcal{R}_{d-1}^{(p)}/(f^{\triangledown})}), \\ h_{\lambda_X}^{\mathfrak{A}_\alpha(\mathcal{P}_0, -\mathbf{e}^{(d)})_0}(\alpha^{(\Gamma)} | \mathfrak{A}_\alpha(\mathcal{P}_0, -\mathbf{e}^{(d)})_{-1}) &= h(\alpha^{\mathcal{R}_{d-1}^{(p)}/(f^{\triangle})}), \end{aligned} \tag{26.27}$$

for every subgroup $\Gamma \subset \Xi$. The Ξ-action $\alpha^{(\Xi)}$ is expansive on X if and only if the faces $\Phi_{\pm \mathbf{e}^{(d)}}(\mathcal{S}(f))$ consist of single points. If $\alpha^{(\Xi)}$ is expansive, then it is algebraically conjugate to the shift-action of Ξ on $(\mathbb{F}_p^{w(\mathbf{e}^{(d)}, \mathcal{S}(f))})^\Xi$.

PROOF. From (6.19) it is clear that the Ξ-actions

$$(\alpha^{(\Xi)})^{\Omega(X, \pm \mathbf{e}^{(d)})_{-1}/\Omega(X, \pm \mathbf{e}^{(d)})_0}$$

are algebraically conjugate to the $\mathbb{Z}^{d-1}$-action $\alpha^{\mathfrak{R}^{(p)}_{d-1}/(f^{\triangledown})}$ and $\alpha^{\mathfrak{R}^{(p)}_{d-1}/(f^{\triangle})}$, respectively, where we identify $\Xi = \{\mathbf{m} = (m_1, \ldots, m_d) \in \mathbb{Z}^d : m_d = 0\} \subset \mathbb{Z}^d$ with $\mathbb{Z}^{d-1}$. This proves (26.27).

Now suppose that the faces $\Phi_{\pm \mathbf{e}^{(d)}}(\mathcal{S}(f))$ consist of single points. Then $\Omega(X)_0 = \{0_X\}$, and every point $x = (x_{\mathbf{m}}) \in X$ is completely determined by its coordinates in the half-space $H = \{(m_1, \ldots, m_d) \in \mathbb{Z}^d : m_d \leq 0\}$. A glance at (6.19) reveals that the coordinates $x_{(1,k_2,\ldots,k_d)}$, $(k_2, \ldots, k_d) \in \mathbb{Z}^{d-1}$, are already determined by the coordinates of x in the strip $S = \{(m_1, \ldots, m_d) \in \mathbb{Z}^d : -w(\mathbf{e}^{(d)}, \mathcal{S}(f)) + 1 \leq m_d \leq 0\}$. By iterating this argument we see that $\pi_S(x)$ determines $x_{\mathbf{m}}$ for every $\mathbf{m} = (m_1, \ldots, m_d) \in \mathbb{Z}^d$ with $m_d \geq 0$. The same argument, applied to $\Phi_{-\mathbf{e}^{(d)}}(\mathcal{S}(f))$ and $\Omega(X, -\mathbf{e}^{(d)})_{w(\mathbf{e}^{(d)}, \mathcal{S}(f))-1}$, shows that the point x is determined completely by $\pi_S(x)$, i.e. that the projection $\pi_S \colon X \longmapsto \mathbb{Z}^S_{/p}$ is is a group isomorphism. In particular we see that $\alpha^{(\Xi)}$ is expansive, and conjugate to the shift-action of Ξ on $(\mathbb{F}^{w(\mathbf{e}^{(d)}, \mathcal{S}(f))}_p)^\Xi$.

Conversely, if $\alpha^{(\Xi)}$ is expansive, there must exist an integer $M \geq 0$ such that $\pi_{S(-M,0)} \colon X \longmapsto \mathbb{Z}^{S(-M,0)}_{/p}$ is injective, where $S(-k,l) = \{\mathbf{m} = (m_1, \ldots, m_d) \in \mathbb{Z}^d : -k \leq m_d \leq l\}$ for all $k, l \geq 0$, and this is easily seen to imply that $\Omega(X, \pm\mathbf{e}^{(d)})_{-\infty} = \Omega(X, \pm\mathbf{e}^{(d)}) = \{0_X\}$, and that the faces $\Phi_{\pm \mathbf{e}^{(d)}}(\mathcal{S}(f))$ consist of single points.

In general the faces $\Phi_{\pm \mathbf{e}^{(d)}}(\mathcal{S}(f))$ may consist of more than one point, in which case the projection $\pi_S \colon X \longmapsto \mathbb{F}^S_p$ is still surjective, but no longer injective. As in the proof of Proposition 25.7 we consider, for $m + n + 1 \geq w(\mathbf{e}^{(d)}, \mathcal{S}(f))$, the group $X(-m, n) = \pi_{S(-m,n)}(X) \subset \mathbb{F}^{S(-m,n)}_p$. The (obvious) projections of $X(-m, n+1)$ and $X(-m-1, n)$ onto $X(-m, n)$ have kernels isomorphic to $\Omega(X, \mathbf{e}^{(d)})_{-1}/\Omega(X, \mathbf{e}^{(d)})_0$ and to $\Omega(X, -\mathbf{e}^{(d)})_{-1}/\Omega(X, -\mathbf{e}^{(d)})_0$, respectively. Since the sets $S(-m, n) \subset \mathbb{Z}^d$ are invariant under translation by Ξ, the shift-action $\alpha^{(\Xi)}$ of Ξ on X induces a Ξ-action $(\alpha^{(\Xi)})^{X(-m,n)}$, and the addition formula (14.1) implies that

$$h((\alpha^{(\Xi)})^{X(-m-1,n+1)}) = h((\alpha^{(\Xi)})^{X(-m-1,n+1)/X(-m-1,n)})$$

$$+ h((\alpha^{(\Xi)})^{X(-m-1,n)/X(-m,n)}) + h((\alpha^{(\Xi)})^{X(-m,n)})$$

$$= h(\alpha^{\mathfrak{R}^{(p)}_{d-1}/(f^{\triangledown})}) + h(\alpha^{\mathfrak{R}^{(p)}_{d-1}/(f^{\triangle})}) + h((\alpha^{(\Xi)})^{X(-m,n)}).$$

Since the Laurent polynomials $f^{\triangledown}, f^{\triangle}$ are non-zero, the discussion at the beginning of Section 25 shows that $h(\alpha^{\mathfrak{R}^{(p)}_{d-1}/(f^t op)}) = h(\alpha^{\mathfrak{R}^{(p)}_{d-1}/(f^{\triangle})}) = 0$, and by using Lemma 13.6 we obtain that

$$h(\alpha^{(\Xi)}) = \lim_{m,n\to\infty} h((\alpha^{(\Xi)})^{X(-m,n)}) = h((\alpha^{(\Xi)})^{X(-w(\mathbf{e}^{(d)}, \mathcal{S}(f))-1,0)})$$

$$= w(\mathbf{e}^{(d)}, \mathcal{S}(f)) \cdot \log p,$$

as claimed in (26.26). $\square$

In Lemma 26.7 we dealt with principal ideals in $\mathfrak{R}_d^{(p)}$, but (26.27) has a straightforward analogue for arbitrary ideals.

LEMMA 26.8. *Let* $d \geq 2$, p *a rational prime,* $\mathfrak{a} \subset \mathfrak{R}_d^{(p)}$ *an ideal, and let* λ_X *be the normalized Haar measure,* $\mathcal{P}_0$ *the state partition,* $\Omega(X, \pm \mathbf{e}^{(d)})_k \subset X$ *the subgroups, and* $\alpha = \alpha^{\mathfrak{R}_d^{(p)}/\mathfrak{a}}$ *the shift-action of* $\mathbb{Z}^d$ *on* $X = X^{\mathfrak{R}_d^{(p)}/\mathfrak{a}} \subset \mathbb{F}_p^{\mathbb{Z}^d}$ *defined in (6.19)–(6.20), (24.3)), and (26.6). If* $\mathfrak{a} = \{0\}$, *then the actions*

$$(\alpha^{(\Xi)})^{\Omega(X, \pm \mathbf{e}^{(d)})_{-1}/\Omega(X, \pm \mathbf{e}^{(d)})_0}$$

of Ξ *induced by* $\alpha^{(\Xi)}$ *on the quotients* $\Omega(X, \pm \mathbf{e}^{(d)})_{-1}/\Omega(X, \pm \mathbf{e}^{(d)})_0$ *are algebraically conjugate to the shift-action of* Ξ *on* $\mathbb{F}_p^\Xi$, *and*

$$h_{\lambda_X}^{\mathfrak{A}(\mathcal{P}_0, \pm \mathbf{e}^{(d)})_0}(\alpha^{(\Xi)}|\mathfrak{A}_\alpha(\mathcal{P}_0, \pm \mathbf{e}^{(d)})_{-1}) = h((\alpha^{(\Xi)})^{\Omega(X, \pm \mathbf{e}^{(d)})_{-1}/\Omega(X, \pm \mathbf{e}^{(d)})_0})$$
$$= \log p. \tag{26.28}$$

If the ideal $\mathfrak{a}$ *is non-zero, then the actions* $(\alpha^{(\Xi)})^{\Omega(X, \pm \mathbf{e}^{(d)})_{-1}/\Omega(X, \pm \mathbf{e}^{(d)})_0}$ *of* $\Xi \cong \mathbb{Z}^{d-1}$ *induced by* $\alpha^{(\Xi)}$ *on* $\Omega(X, \pm \mathbf{e}^{(d)})_{-1}/\Omega(X, \pm \mathbf{e}^{(d)})_0$ *are algebraically conjugate to the* $\mathbb{Z}^{d-1}$-*actions* $\alpha^{\mathfrak{R}_{d-1}^{(p)}/\mathfrak{a}^\nabla}$ *on* $X^{\mathfrak{R}_{d-1}^{(p)}/\mathfrak{a}^\nabla}$ *and* $\alpha^{\mathfrak{R}_{d-1}^{(p)}/\mathfrak{a}^\triangle}$ *on* $X^{\mathfrak{R}_{d-1}^{(p)}/\mathfrak{a}^\triangle}$, *where* $\mathfrak{a}^\nabla$ *and* $\mathfrak{a}^\triangle$ *are the ideals appearing in Definition 26.4. In particular,*

$$h_{\lambda_X}^{\mathfrak{A}_\alpha(\mathcal{P}_0, \mathbf{e}^{(d)})_0}(\alpha^{(\Gamma)}|\mathfrak{A}_\alpha(\mathcal{P}_0, \mathbf{e}^{(d)})_{-1}) = h((\alpha^{\mathfrak{R}_{d-1}^{(p)}/\mathfrak{a}^\nabla})^{(\Gamma)}),$$
$$h_{\lambda_X}^{\mathfrak{A}_\alpha(\mathcal{P}_0, -\mathbf{e}^{(d)})_0}(\alpha^{(\Gamma)}|\mathfrak{A}_\alpha(\mathcal{P}_0, -\mathbf{e}^{(d)})_{-1}) = h((\alpha^{\mathfrak{R}_{d-1}^{(p)}/\mathfrak{a}^\triangle})^{(\Gamma)}), \tag{26.29}$$

for every subgroup $\Gamma \subset \Xi$.

PROOF. If $\mathfrak{a} = \{0\}$, then $X = \mathbb{F}_p^{\mathbb{Z}^d}$, $\Omega(X, \pm \mathbf{e}^{(d)})_{-1}/\Omega(X, \pm \mathbf{e}^{(d)})_0 \cong \mathbb{F}_p^{\mathbb{Z}^{d-1}}$, and the actions

$$(\alpha^{(\Xi)})^{\Omega(X, \pm \mathbf{e}^{(d)})_{-1}/\Omega(X, \pm \mathbf{e}^{(d)})_0}$$

are conjugate to the shift-action of Ξ on $\mathbb{F}_p^\Xi$. The equation (26.28) now follows from (26.7). If $\mathfrak{a}$ is non-zero, (26.29) is proved in exactly the same way as (26.27). $\square$

For the notation used in the following theorem we refer to (26.2)–(26.5), and to Definition 26.4.

THEOREM 26.9. *Let* $p > 1$ *be a rational prime,* $d \geq 2$, $\mathfrak{a} \subset \mathfrak{R}_d^{(p)}$ *an ideal,* $X = X^{\mathfrak{R}_d^{(p)}/\mathfrak{a}} \subset \mathbb{Z}_{/p}^{\mathbb{Z}^d}$, $\alpha = \alpha^{\mathfrak{R}_d^{(p)}/\mathfrak{a}}$, *and let* $\mathcal{P}_0$ *be the state partition of* X *(cf.* (6.19)–(6.20) *and* (24.3)). *For every* $A \in \mathrm{GL}(d, \mathbb{Z})$ *we set* $\mathfrak{a}^A = \{f^A : f \in \mathfrak{a}\}$ *(cf. Lemma 18.4). Then the following is true for every primitive element* $\mathbf{n} \in \mathbb{Z}^d$.

(1) *For every countable partition $\mathcal{Q} \subset \mathfrak{B}_X$ with finite entropy, and for every subgroup $\Gamma \subset \Xi_{\mathbf{n}}$,*

$$\mathfrak{A}_\alpha(\mathcal{P}_0, \mathbf{n})_{-\infty} = \mathfrak{B}_{X/\Omega(X,\mathbf{n})_{-\infty}}$$
$$\subset \mathfrak{A}_\alpha(\mathcal{Q}, \mathbf{n})_{-\infty} \; (\mathrm{mod} \; \lambda_X), \tag{26.30}$$

and

$$h_{\lambda_X}^{(\mathbf{n})}(\alpha^{(\Gamma)}) = h_{\lambda_X}^{\mathfrak{A}_\alpha(\mathcal{P}_0, \mathbf{n})_0}(\alpha^{(\Gamma)}|\mathfrak{A}_\alpha(\mathcal{P}_0, \mathbf{n})_0^-)$$
$$= h((\alpha^{(\Gamma)})^{\Omega(X,\mathbf{n})_0^-/\Omega(X,\mathbf{n})_0}); \tag{26.31}$$

(2) *If $\mathfrak{a} = \{0\}$, then*

$$\Omega(X, \mathbf{n})_{-\infty} = \{\varnothing, X\}, \quad h_{\lambda_X}^{(\mathbf{n})}(\alpha^{(\Xi_{\mathbf{n}})}) = \log p; \tag{26.32}$$

(3) *If $\mathfrak{a} \neq \{0\}$, and if $A^\top \mathbf{e}^{(d)} = \mathbf{n}$ for some $A \in \mathrm{GL}(d, \mathbb{Z})$, then*

$$\Omega(X, \mathbf{n})_{-\infty} = (\mathfrak{a}^{\mathbf{n}, \blacktriangledown})^\perp = \widehat{\mathfrak{R}_d^{(p)}/\mathfrak{a}^{\mathbf{n}, \blacktriangledown}} \subset \mathbb{F}_p^{\mathbb{Z}^d}, \tag{26.33}$$

and

$$h_{\lambda_X}^{(\mathbf{n})}(\alpha^{(\Gamma)}) = h((\alpha^{\mathfrak{R}_{d-1}^{(p)}/(\mathfrak{a}^A)^\triangledown})^{(A\Gamma^*)}) \tag{26.34}$$

for every subgroup $\Gamma \subset \Xi_{\mathbf{n}}$, where $\mathfrak{a}^A = \{f^A : f \in \mathfrak{a}\}$, and where $A\Gamma^* \subset \mathbb{Z}^{d-1}$ is the subgroup corresponding to $A\Gamma \subset \Xi$ under the obvious isomorphism of $\mathbb{Z}^{d-1}$ with $\Xi = \{\mathbf{m} = (m_1, \ldots, m - d) \in \mathbb{Z}^d : m_d = 0\}$.

PROOF. As in the proof of Lemma 18.4 we define, for every $A \in \mathrm{GL}(d, \mathbb{Z})$, a continuous group isomorphism $\psi_A \colon X \longmapsto X^A = X^{\mathfrak{R}_d^{(p)}/\mathfrak{a}^A}$ by $(\psi_A(x))_{\mathbf{n}} = x_{A\mathbf{n}}$ for every $x = (x_{\mathbf{n}}) \in X$. Then $\psi_A \cdot \alpha_{A^{-1}\mathbf{n}} = \alpha_{\mathbf{n}}^{\mathfrak{R}_d^{(p)}/\mathfrak{a}^A} \cdot \psi_A$ for every $\mathbf{n} \in \mathbb{Z}^d$, i.e. ψ_A is a conjugacy of the $\mathbb{Z}^d$-actions $\alpha_{A^{-1}} \colon \mathbf{n} \mapsto \alpha_{A^{-1}\mathbf{n}}$ and $\alpha^{\mathfrak{R}_d^{(p)}/\mathfrak{a}^A}$. Since $\psi_A(\Omega(X)_0) = \Omega(X^A, A^\top \mathbf{e}^{(d)})_0$, $\psi_A(\Omega(\bar{X})_{-1}) = \Omega(\bar{X}^A, A^\top \mathbf{e}^{(d)})_0^-$, and

$$\psi_A(\Omega(\bar{X})_{-\infty}) = \Omega(\bar{X}^A, A^\top \mathbf{e}^{(d)})_{-\infty}, \tag{26.35}$$

we obtain that

$$\psi_A \cdot (\alpha^{(\Gamma)})^{\Omega(X)_{-1}/\Omega(X)_0} \cdot \psi_A = (\alpha^{(A^{-1}\Gamma)})^{\Omega(\bar{X}^A, A^\top \mathbf{e}^{(d)})_0^-/\Omega(X^A, A^\top \mathbf{e}^{(d)})_0} \cdot \psi_A$$

for every subgroup $\Gamma \subset \Xi$, and (26.7) and the Lemmas 26.3–26.7 imply that

$$h_{\lambda_{X^A}}^{(A^\top \mathbf{e}^{(d)})}(\alpha^{(A^{-1}\Xi)}) = h((\alpha^{(\Xi)})^{\Omega(X)_{-1}/\Omega(X)_0})$$
$$= \log p \text{ if } \mathfrak{a} = \{0\},$$
$$h_{\lambda_{X^A}}^{(A^\top \mathbf{e}^{(d)})}(\alpha^{(A^{-1}\Gamma)}) = h((\alpha^{(\Gamma)})^{\Omega(X)_{-1}/\Omega(X)_0}) \tag{26.36}$$
$$= h(\alpha^{\mathfrak{R}_{d-1}^{(p)}/\mathfrak{a}^\triangledown}) \text{ if } \mathfrak{a} \neq \{0\}.$$

If $A^\top = \mathbf{n}$, and if $\Gamma \subset \Xi$ is replaced by $A^{-1}\Gamma \subset \Xi_\mathbf{n}$, then (26.35) becomes (26.33), and (26.36) becomes (26.32) and (26.34). $\square$

Before turning to the corollaries of Theorem 26.9 we need a general lemma.

LEMMA 26.10. *Let $p > 1$ be a rational prime, and let α be an expansive $\mathbb{Z}^d$-action by automorphisms of an infinite compact group X with the property that $px = 0$ for all $x \in X$. Then there exists an integer $t \geq 0$ such that $h(\alpha) = \log p^t$. If $h(\alpha) = 0$ there exist integers $r \in \{1, \ldots, d-1\}$ and $t \geq 1$, and a subgroup $\Gamma \cong \mathbb{Z}^r$ in $\mathbb{Z}^d$ such that $h(\alpha^{(\Gamma)}) = p^t$.*

PROOF. Since $px = 0$ for every $x \in X$, the Noetherian $\mathfrak{R}_d$-module $\mathfrak{M} = \hat{X}$ obtained from Lemma 5.1 and Proposition 5.4 is a well-defined $\mathfrak{R}_d^{(p)}$-module (cf. Remark 6.19 (4)). Every prime ideal $\mathfrak{p}$ associated with $\mathfrak{M}$ contains the constant (p), and is therefore non-principal unless $\mathfrak{p} = (p)$. In particular, if $\{0\} = \mathfrak{N}_0 \subset \cdots \subset \mathfrak{N}_s = \mathfrak{M}$ is a prime filtration of $\mathfrak{M}$, then every prime ideal $\mathfrak{q}_j \subset \mathfrak{R}_d$ with $\mathfrak{N}_j/\mathfrak{N}_{j-1} \cong \mathfrak{R}_d/\mathfrak{q}_j$ contains (p), hence $h(\alpha^{\mathfrak{R}_d/\mathfrak{q}_j}) \in \{0, \log p\}$ by Corollary 18.5, and Proposition 18.6 implies that $h(\alpha^{(1)}) = \log p^t$ for some $t \geq 0$.

More generally, if $r \geq 1$, and if $\mathfrak{M}$ is an $\mathfrak{R}_r^{(p_1)}$-module such that the $\mathbb{Z}^r$-action $\alpha^{\mathfrak{M}}$ has finite, positive entropy, then

$$h(\alpha^{\mathfrak{M}}) = \log p^s \tag{26.37}$$

for some integer $s \geq 1$; if $\mathfrak{M}$ is Noetherian this has just been established; if $\mathfrak{M}$ is not Noetherian, (26.37) follows from Proposition 18.6.

We complete the proof of the lemma by showing that there exists an integer $r \geq 1$ and a (primitive) subgroup $\Gamma \cong \mathbb{Z}^r$ in $\mathbb{Z}^d$ such that $0 < h(\alpha^{(\Gamma)}) < \infty$. If this assertion has been proved we may assume without loss in generality that $\Gamma = \{(n_1, \ldots, n_d) \in \mathbb{Z}^d : n_{r+1} = \cdots = n_d = 0\}$ and consider $\mathfrak{M}$ as an $\mathfrak{R}_r$-module (cf. Lemma 5.1). According to (26.37), $h(\alpha^{(\Gamma)}) = \log p^t$ for some integer $t \geq 1$, which is exactly what was claimed.

In order to prove the existence of the required subgroup Γ we return the prime filtration $\{0\} = \mathfrak{N}_0 \subset \cdots \subset \mathfrak{N}_s = \mathfrak{M}$ and note that there exist prime ideals $\mathfrak{q}_j \subset \mathfrak{R}_d^{(p)}$ such that $\mathfrak{N}_j/\mathfrak{N}_{j-1} \cong \mathfrak{R}_d^{(p)}/\mathfrak{q}_j$ for every $j = 1, \ldots, s$. If $\alpha^{\mathfrak{R}_d^{(p)}/\mathfrak{q}_j}$ is non-ergodic for some $j \in \{1, \ldots, t\}$, then Theorem 8.1 shows that $\mathfrak{q}_j$ is maximal and $|X^{\mathfrak{R}_d^{(p)}/\mathfrak{q}_j}| = |\mathfrak{R}_d^{(p)}/\mathfrak{q}_j| < \infty$, and we set $r(\mathfrak{q}_j) = 0$. If $\alpha^{\mathfrak{R}_d^{(p)}/\mathfrak{q}_j}$ is ergodic we define $r(\mathfrak{q}_j)$ by Proposition 8.2. Put $r = \max\{r(\mathfrak{q}_1), \ldots, r(\mathfrak{q}_t)\}$ and $S = \{j : 1 \leq j \leq t \text{ and } r(\mathfrak{q}_j) = r\}$, and note that $r > 0$, since X is infinite. Fix $j_0 \in S$, and apply Proposition 8.2 to find, for $\mathfrak{p} = \mathfrak{q}_{j_0}$, a primitive subgroup $\Gamma \cong \mathbb{Z}^r$ in $\mathbb{Z}^d$ with the properties described there. According to Proposition 8.2 (3), the Γ-action $(\alpha^{\mathfrak{R}_d^{(p)}/\mathfrak{q}_{j_0}})^{(\Gamma)} = (\alpha^{\mathfrak{R}_d/\mathfrak{p}_{j_0}})^{(\Gamma)}$ is algebraically conjugate to the shift-action of Γ on $(\mathbb{F}_p^Q)^\Gamma$ for some finite set Q, so that $0 < h((\alpha^{\mathfrak{R}_d^{(p)}/\mathfrak{q}_{j_0}})^{(\Gamma)}) = \log p^{|Q|} < \infty$. From Proposition 24.1 it is clear that $h((\alpha^{\mathfrak{R}_d^{(p)}/\mathfrak{q}_j})^{(\Gamma)}) < \infty$ for every $j \in S$, and that $h((\alpha^{\mathfrak{R}_d^{(p)}/\mathfrak{q}_j})^{(\Gamma)}) = 0$ whenever $j \in \{1, \ldots, s\} \smallsetminus S$. The addition formula (14.1) yields that $0 < h(\alpha^{(\Gamma)}) < \infty$ and completes the proof of the lemma. $\square$

For the following corollary of Theorem 26.9 we use the notation of Theorem 26.9.

COROLLARY 26.11. *Let $p_1, p_2 > 1$ be a rational primes, $d \geq 2$, and let, for $i = 1, 2$, $\mathfrak{a}_i \subset \mathfrak{R}_d^{(p_i)}$ be an ideal, $X^{(i)} = X^{\mathfrak{R}_d^{(p_i)}/\mathfrak{a}_i} \subset (\mathbb{Z}_{/p_i})^{\mathbb{Z}^d}$, and $\alpha^{(i)} = \alpha^{\mathfrak{R}_d^{(p_i)}/\mathfrak{a}_i}$ (cf. (6.19)–(6.20)). If $\alpha^{(1)}$ is ergodic and measurably conjugate to $\alpha^{(2)}$, then the following is true.*

(1) $p_1 = p_2 = p$, *say;*
(2) $\mathfrak{a}_1 = \{0\}$ *if and only if $\mathfrak{a}_2 = \{0\}$;*
(3) *If $\mathfrak{a}_1$ and $\mathfrak{a}_2$ are non-zero, then*

$$h((\alpha^{\mathfrak{R}_d^{(p)}/\mathfrak{a}_1^{\mathbf{n},\blacktriangledown}})^{(\Gamma)}) = h((\alpha^{\mathfrak{R}_d^{(p)}/\mathfrak{a}_2^{\mathbf{n},\blacktriangledown}})^{(\Gamma)}) \qquad (26.38)$$

for every primitive element $\mathbf{n} \in \mathbb{Z}^d$ and every subgroup $\Gamma \subset \mathbb{Z}^d$ with $h((\alpha^{(1)})^{(\Gamma)}) = h((\alpha^{(2)})^{(\Gamma)}) < \infty$, and

$$h((\alpha^{\mathfrak{R}_{d-1}^{(p)}/(\mathfrak{a}_1^A)^{\triangledown}})^{(\Gamma)}) = h((\alpha^{\mathfrak{R}_{d-1}^{(p)}/(\mathfrak{a}_2^A)^{\triangledown}})^{(\Gamma)}) \qquad (26.39)$$

for every $A \in \mathrm{GL}(d, \mathbb{Z})$ and every subgroup $\Gamma \subset \mathbb{Z}^{d-1}$.

PROOF. By Lemma 26.10 there exists an infinite subgroup $\Gamma \subset \mathbb{Z}^d$ such that the restrictions of the $\mathbb{Z}^d$-actions $\alpha^{(i)}$ to Γ satisfy that $\log p_1^{t_1} = h((\alpha^{(1)})^{(\Gamma)}) = h((\alpha^{(2)})^{(\Gamma)}) = \log p_2^{t_2}$ for some positive integers t_1, t_2, which proves (1). The assertion (2) is obvious, since $h(\alpha^{(i)}) = \log p > 0$ if and only if $\mathfrak{a} = \{0\}$ (cf. the discussion at the beginning of Section 25). In order to prove (3) we note that any measure preserving isomorphism $\phi: X^{(1)} \longmapsto X^{(2)}$ with $\phi \cdot \alpha_{\mathbf{m}}^{(1)} = \alpha_{\mathbf{m}}^{(2)} \cdot \phi$ for all $\mathbf{m} \in \mathbb{Z}^d$ must satisfy that $\phi(\mathfrak{B}_{X^{(1)}/\Omega(X^{(1)},\mathbf{n})_{-\infty}}) = \mathfrak{B}_{X^{(2)}/\Omega(X^{(2)},\mathbf{n})_{-\infty}} \pmod{\lambda_{X^{(2)}}}$ for every primitive $\mathbf{n} \in \mathbb{Z}^d$, by (26.30). Hence $h((\alpha^{(1)})^{(\Gamma)}|\mathfrak{B}_{X^{(1)}/\Omega(X^{(1)},\mathbf{n})_{-\infty}}) = h((\alpha^{(2)})^{(\Gamma)}|\mathfrak{B}_{X^{(2)}/\Omega(X^{(2)},\mathbf{n})_{-\infty}})$ for every subgroup $\Gamma \subset \mathbb{Z}^d$, and the addition formula (14.1) shows that

$$
\begin{aligned}
h(((\alpha^{(1)})^{(\Gamma)})&^{\Omega(X^{(1)},\mathbf{n})_{-\infty}}) \\
&= h((\alpha^{(1)})^{(\Gamma)}) - h((\alpha^{(1)})^{(\Gamma)}|\mathfrak{B}_{X^{(1)}/\Omega(X^{(1)},\mathbf{n})_{-\infty}}) \\
&= h((\alpha^{(2)})^{(\Gamma)}) - h((\alpha^{(2)})^{(\Gamma)}|\mathfrak{B}_{X^{(2)}/\Omega(X^{(2)},\mathbf{n})_{-\infty}}) \\
&= h(((\alpha^{(2)})^{(\Gamma)})^{\Omega(X^{(2)},\mathbf{n})_{-\infty}}),
\end{aligned} \qquad (26.40)
$$

since all the terms occurring in the differences are finite. Hence (26.33) implies (26.38), and (26.39) follows from (26.34). $\square$

COROLLARY 26.12. *Let p_1, p_2 be rational primes, $d \geq 2$, and let $f_i \in \mathfrak{R}_d^{(p_i)}$ be non-zero Laurent polynomials such that the $\mathbb{Z}^d$-actions $\alpha^{(i)} = \alpha^{\mathfrak{R}_d^{(p_i)}/(f_i)}$, $i = 1, 2$, are measurably conjugate. Then $p_1 = p_2$, and $\mathcal{C}(f_1) = \mathcal{C}(f_2) + \mathbf{m}$ for some $\mathbf{m} \in \mathbb{Z}^d$.*

PROOF. From Corollary 26.11 we know that $p_1 = p_2 = p$, say, and that

$$h((\alpha^{\mathfrak{R}_{d-1}^{(p)}/((f_1^A)^{\triangledown})})^{(\Gamma)}) = h((\alpha^{\mathfrak{R}_{d-1}^{(p)}/((f_2^A)^{\triangledown})})^{(\Gamma)})$$

for every $A \in \mathrm{GL}(d, \mathbb{Z})$, and for every subgroup $\Gamma \in \mathbb{Z}^{d-1}$. In particular, if $\Xi' = \{(n_1, \ldots, n_{d-1}) \in \mathbb{Z}^{d-1} : n_{d-1} = 0\}$, and if $\mathbf{e}'^{(j)}$ is the j-th unit vector in $\mathbb{Z}^{d-1}$, then (26.26) shows that $h((\alpha^{\mathfrak{R}_{d-1}^{(p)}/((f_i^A)^{\triangledown})})^{(\Xi')}) = w(\mathbf{e}'^{(d-1)}, \mathcal{S}((f_i^A)^{\triangledown})) \cdot \log p$, and hence that

$$w(\mathbf{e}'^{(d-1)}, \mathcal{S}((f_1^A)^{\triangledown})) = w(\mathbf{e}'^{(d-1)}, \mathcal{S}((f_2^A)^{\triangledown})) \tag{26.41}$$

for every $A \in \mathrm{GL}(d, \mathbb{Z})$.

For every finite set $E \subset \mathbb{Z}^d$ we consider the face $\Phi_{\mathbf{e}(d)}(E)$, and choose $m \in \mathbb{Z}$ such that $\Phi_{\mathbf{e}(d)}(E) - m\mathbf{e}^{(d)} \subset \{(t_1, \ldots, t_d) \in \mathbb{R}^d : t_d = 0\}$. By dropping the last coordinate we regard $\Phi'_{\mathbf{e}(d)}(E) = \Phi_{\mathbf{e}(d)}(E) - m\mathbf{e}^{(d)}$ as a subset of $\mathbb{R}^{d-1}$ and denote by $w(\mathbf{e}'^{(d-1)}, \Phi'_{\mathbf{e}(d)}(E))$ the width (25.4) of $\Phi'_{\mathbf{e}(d)}(E)$ in the direction $\mathbf{e}'^{(d-1)}$. If $F \subset \mathbb{Z}^d$ is a finite set, then an easy geometric argument shows that the convex hull $\mathcal{C}(F)$ is determined up to translation by the numbers $w(\mathbf{e}'^{(d-1)}, \Phi'_{\mathbf{e}(d)}(AF))$, $A \in \mathrm{GL}(d, \mathbb{Z})$. Since, for every $A \in \mathrm{GL}(d, \mathbb{Z})$, $w(\mathbf{e}'^{(d-1)}, \mathcal{S}((f_i^A)^{\triangledown})) = w(\mathbf{e}'^{(d-1)}, \Phi'_{\mathbf{e}(d)}(A\mathcal{S}(f_i)))$, (26.40) shows that $\mathcal{C}(f_1)$ and $\mathcal{C}(f_2)$ coincide up to translation. $\square$

COROLLARY 26.13. *Let p be a rational prime, $\mathfrak{a} \subset \mathfrak{R}_d^{(p)}$ an ideal, and let $\alpha = \alpha^{\mathfrak{R}_d^{(p)}/\mathfrak{a}}$ and $X = X^{\mathfrak{R}_d^{(p)}/\mathfrak{a}} \subset \mathbb{F}_p^{\mathbb{Z}^d}$ (cf. (6.19)–(6.20)). The following conditions are equivalent for every primitive element $\mathbf{n} \in \mathbb{Z}^d$.*

(1) *$\alpha^{(\Xi_{\mathbf{n}})}$ is expansive;*
(2) *$\Omega(X, \mathbf{n})_0 = \Omega(X, -\mathbf{n})_0 = X$;*
(3) *$h_{\lambda_X}^{(\mathbf{n})}(\alpha^{(\Gamma)}) = h_{\lambda_X}^{(-\mathbf{n})}(\alpha^{(\Gamma)}) = 0$ for every subgroup $\Gamma \subset \Xi_{\mathbf{n}}$, where λ_X is the normalized Haar measure on X.*

In particular, if $\mathfrak{a}'$ is a second ideal in $\mathfrak{R}_d^{(p)}$ such that the $\mathbb{Z}^d$-action $\alpha' = \alpha^{\mathfrak{R}_d^{(p)}/\mathfrak{a}'}$ is measurably conjugate to α, then the set $\{\mathbf{n} \in \mathbb{Z}^d : \alpha^{(\Xi_{\mathbf{n}})}$ is expansive$\} = \{\mathbf{n} \in \mathbb{Z}^d : \alpha'^{(\Xi_{\mathbf{n}})}$ is expansive$\}$ is a measurable conjugacy invariant of α.

PROOF. The equivalence of (1) and (2) is proved as in Proposition 25.7 and Lemma 26.7, and the other assertions follow from Theorem 26.9. $\square$

EXAMPLES 26.14. (1) Consider the polynomial $f = 1 + u_1 + u_1^2 + u_1^3 u_2 + u_2^2 + u_1 u_2^2 \in \mathfrak{R}_2^{(2)}$ in Example 25.16 (3), and set $\alpha = \alpha^{\mathfrak{R}_2^{(2)}/(f)}$ and $X =$

$X^{\mathfrak{R}_2^{(2)}/(f)}$. Then $f = g_1 g_2$ with $g_1 = 1 + u_1 + u_1^2 + u_2$ and $g_2 = 1 + u_1 + u_1 u_2$, and we put, as in Example 25.2,

$$F = \mathcal{S}(f) = \quad, \qquad F_1 = \mathcal{S}(g_1) = \quad, \qquad F_2 = \mathcal{S}(g_2) = \quad.$$

According to Definition 26.4 (1),

$$f^{\mathbf{n},\mathbf{\nabla}} = \begin{cases} g_1 & \text{if } \mathbf{n} \in \{(2,1),(0,-1)\}, \\ g_2 & \text{if } \mathbf{n} \in \{(0,1),(1,-1)\}, \\ f & \text{if } \mathbf{n} = (-1,0), \\ 1 & \text{in all other cases,} \end{cases}$$

and Theorem 26.9 allow us to calculate the groups $\Omega(X,\mathbf{n})_{-\infty}$ for every $\mathbf{n} \in \mathbb{Z}^2$:

$$\Omega(X,\mathbf{n})_{-\infty} = \begin{cases} (g_1)^{\perp} = X^{\mathfrak{R}_2^{(2)}/(g_1)} & \text{if } \mathbf{n} \in \{(2,1),(0,-1)\}, \\ (g_2)^{\perp} = X^{\mathfrak{R}_2^{(2)}/(g_2)} & \text{if } \mathbf{n} \in \{(0,1),(1,-1)\}, \\ X & \text{if } \mathbf{n} = (-1,0), \\ \{\mathbf{0}_X\} & \text{in all other cases.} \end{cases}$$

In general, if $f \in \mathfrak{R}_d^{(p)}$ is a Laurent polynomial whose irreducible factors all have supports with dissimilar convex hulls, then Theorem 26.9 allows us to recognize this fact, and Corollary 26.12 shows that these convex hulls are measurable conjugacy invariants.

(2) Let $f = 1 + u_1 + u_1^2 + u_1^3 + u_2^3 \in \mathfrak{R}_2^{(2)}$, and let $\alpha = \alpha^{\mathfrak{R}_2^{(2)}/(f)}$ and $X = X^{\mathfrak{R}_2^{(2)}/(f)}$. Then $f = g_1 g_2$ with $g_1 = 1 + u_1 + u_2$ and $g_2 = 1 + u_1^2 + u_2 + u_1 u_2 + u_2^2$, and the sets

$$F = \mathcal{S}(f) = \quad, \qquad F_1 = \mathcal{S}(g_1) = \quad, \qquad F_2 = \mathcal{S}(g_2) = \quad,$$

have similar convex hulls (cf. Example 25.2). In particular,

$$f^{\mathbf{n},\mathbf{\nabla}} = \begin{cases} f & \text{if } \mathbf{n} \in \{(0,1),(1,0),(1,1)\}, \\ 1 & \text{in all other cases,} \end{cases}$$

and

$$\Omega(X,\mathbf{n})_{-\infty} = \begin{cases} X & \text{if } \mathbf{n} \in \{(1,0),(0,1),(1,1)\}, \\ \{\mathbf{0}_X\} & \text{in all other cases.} \end{cases}$$

In this example the groups $\Omega(X,\mathbf{n})_{-\infty}$ give no indication of the fact that f is reducible, and the measurable conjugacy invariants in Theorem 26.9 and its corollaries will not distinguish α from the $\mathbb{Z}^2$-action $\alpha' = \alpha^{\mathfrak{R}_2^{(2)}/(f')}$, where $f' = 1 + u_1 + u_1^2 + u_1^3 + u_2 + u_2^3 \in \mathfrak{R}_2^{(2)}$ and

$$\mathcal{S}(f) = \quad. \qquad \square$$

In our discussion of relative entropies and residual sigma-algebras we have up to now only considered examples of $\mathbb{Z}^d$-actions of the form $\alpha^{\mathfrak{R}_d^{(p)}/\mathfrak{a}}$, where $d \geq 2$, $p > 1$ is a rational prime, and $\mathfrak{a} \subset \mathfrak{R}_d^{(p)}$ is an ideal. However, relative entropies can also be calculated explicitly for certain $\mathbb{Z}^d$-actions on connected groups, as the following example shows.

EXAMPLE 26.15. Let $d = 2$, $\mathfrak{p} = (2 - u_1, 3 - u_2) \subset \mathfrak{R}_2$, and let $\alpha = \alpha^{\mathfrak{R}_2/\mathfrak{p}}$ be the shift-action of $\mathbb{Z}^2$ on $X = X^{\mathfrak{R}_2/\mathfrak{p}} \subset \mathbb{T}^{\mathbb{Z}^2}$ (cf. Examples 5.3 (4) and 7.7 (1)). The partition $\mathcal{P} = \{A_0, \ldots, A_5\}$ with

$$A_j = \left\{ x = (x_{\mathbf{n}}) \in X : \frac{j}{6} \leq x_{\mathbf{0}} < \frac{j+1}{6} \ (\mathrm{mod}\ 1) \right\}$$

for $j = 0, \ldots, 5$ generates $\mathfrak{B}_X$ under α, and we define $\mathfrak{A}_\alpha(\mathcal{P}, \mathbf{n})_t, \mathfrak{A}_\alpha(\mathcal{P}, \mathbf{n})_t^-$ and $\mathfrak{A}_\alpha(\mathcal{P}, \mathbf{n})_{-\infty}$ by (26.1) for every $\mathbf{n} \in \mathbb{Z}^2$ and $t \in \mathbb{R}$. In order to determine these sigma-algebras explicitly we write X as in Example 7.7 (1) in the form $X = (\mathbb{R} \times \mathbb{Z}_2 \times \mathbb{Z}_3)/i(\mathbb{Z})$ with $i(\mathbb{Z}) = \{(n, n, n) \in \mathbb{R} \times \mathbb{Z}_2 \times \mathbb{Z}_3 : n \in \mathbb{Z}\}$, and regard $\alpha_{(n_1, n_2)}$ as multiplication by $2^{n_1} 3^{n_2}$ on X. The character of X defined by an element $a \in \hat{X} = \mathbb{Z}[\frac{1}{6}]$ is given by

$$\langle a, x \rangle = e^{2\pi i (\mathrm{Int}(ax_1) + \mathrm{Frac}(ax_2) + \mathrm{Frac}(ax_3))}$$

for every $x = (x_1, x_2, x_3) + i(\mathbb{Z}) \in X$ with $x_1 \in \mathbb{R}$, $x_2 \in \mathbb{Z}_2$ and $x_3 \in \mathbb{Z}_3$ (cf. Example 7.7 (1)).

Write $\mathbf{e}^{(j)}$, $j = 1, 2$, for the j-th unit vector in $\mathbb{Z}^2$, and consider the closed subgroups

$$Y^{(1)} = (\{\mathbf{0}_{\mathbb{R}}\} \times \mathbb{Z}_2 \times \{\mathbf{0}_{\mathbb{Z}_3}\})/i(\mathbb{Z}), \quad Y^{(2)} = (\{\mathbf{0}_{\mathbb{R}}\} \times \{\mathbf{0}_{\mathbb{Z}_2}\} \times \mathbb{Z}_3)/i(\mathbb{Z})$$

of X. Then $\langle 1, y \rangle = 1$ for every $y \in Y^{(j)}$, $j = 1, 2$, and each of the sets

$$A_k = \left\{ x \in X : \langle 1, x \rangle = e^{2\pi i s} \text{ for some } s \in \left[\frac{k}{6}, \frac{k+1}{6} \right) \right\}, \ k = 0, \ldots, 5,$$

is invariant under translation by $Y^{(1)}$ and $Y^{(2)}$. Furthermore, since

$$\alpha_{(n_1, n_2)}(Y^{(j)}) \subset Y^{(j)}$$

whenever $j = 1, 2$, $(n_1, n_2) \in \mathbb{Z}^2$, and $n_j \geq 0$, every set in the sigma-algebra

$$\mathfrak{A}_\alpha(\mathcal{P}, -\mathbf{e}^{(j)})_0 = \bigvee_{\{(n_1, n_2) \in \mathbb{Z}^2 : n_j \geq 0\}} \alpha_{-(n_1, n_2)}(\mathcal{P})$$

is invariant under translation by $Y^{(j)}$, and

$$\mathfrak{A}_\alpha(\mathcal{P}, -\mathbf{e}^{(j)})_m = \mathfrak{B}_{X/\alpha_{m\mathbf{e}^{(j)}}(Y^{(j)})},$$
$$\mathfrak{A}_\alpha(\mathcal{P}, -\mathbf{e}^{(j)})_0^- = \mathfrak{A}_\alpha(\mathcal{P}, \mathbf{e}^{(j)})_{-1} = \mathfrak{B}_{X/\alpha_{-\mathbf{e}^{(j)}}(Y^{(j)})}$$

for every $j = 1, 2$ and $m \in \mathbb{Z}$. In particular,

$$H_{\lambda_X}(\mathfrak{A}_\alpha(\mathcal{P}, -\mathbf{e}^{(1)})_0|\mathfrak{A}_\alpha(\mathcal{P}, -\mathbf{e}^{(1)})_0^-) = |Y^{(1)}/\alpha_{\mathbf{e}^{(1)}}(Y^{(1)})| = \log 2,$$
$$H_{\lambda_X}(\mathfrak{A}_\alpha(\mathcal{P}, -\mathbf{e}^{(2)})_0|\mathfrak{A}_\alpha(\mathcal{P}, -\mathbf{e}^{(2)})_0^-) = |Y^{(2)}/\alpha_{\mathbf{e}^{(2)}}(Y^{(2)})| = \log 3.$$

Since the groups $\bigcup_{m \in \mathbb{Z}} \alpha_{(m,0)}(Y^{(1)})$ and $\bigcup_{m \in \mathbb{Z}} \alpha_{(m,0)}(Y^{(2)})$ are dense in X,

$$\mathfrak{A}_\alpha(\mathcal{P}, -\mathbf{e}^{(j)})_{-\infty} = \bigcap_{m \in \mathbb{Z}} \mathfrak{A}_\alpha(\mathcal{P}, -\mathbf{e}^{(j)})_m = \{\varnothing, X\} \pmod{\mu},$$

and Lemma 26.3 implies that

$$H_{\lambda_X}(\mathfrak{A}_\alpha(\mathcal{Q}, -\mathbf{e}^{(j)})_0|\mathfrak{A}_\alpha(\mathcal{Q}, -\mathbf{e}^{(j)})_0^-) \leq H_{\lambda_X}(\mathfrak{A}_\alpha(\mathcal{P}, -\mathbf{e}^{(j)})_0|\mathfrak{A}_\alpha(\mathcal{P}, -\mathbf{e}^{(j)})_0^-)$$
$$= \log(j + 1)$$

for $j = 1, 2$, and for every countable partition $\mathcal{Q} \subset \mathfrak{B}_X$ with finite entropy which generates $\mathfrak{B}_X$ under α.

Finally, if $\mathbf{n} \in \mathbb{Z}^2 \setminus \{\mathbf{0}, -\mathbf{e}^{(1)}, -\mathbf{e}^{(2)}\}$, then it is easy to see that $\mathfrak{A}_\alpha(\mathcal{P}, \mathbf{n})_m = \mathfrak{B}_X$ for every $m \in \mathbb{Z}$. $\square$

CONCLUDING REMARKS 26.16. (1) Except for Example 26.15, the exposition in this section follows [47].

(2) If T is a measure preserving $\mathbb{Z}^d$-action on a probability space $(Y, \mathfrak{T}, \mu)$ with $h_\mu(T) > 0$, and if $\mathcal{Q} \subset \mathfrak{T}$ is a generator for T with finite entropy, then

$$h_\mu^{\mathfrak{A}_T(\mathcal{Q}, \mathbf{n})}(T^{(\Xi_\mathbf{n})}|\mathfrak{A}_T(\mathcal{Q}, \mathbf{n})_0^-) = h_\mu(T)$$

for every primitive element $\mathbf{n} \in \mathbb{Z}^d$; this is essentially the situation discussed in [33]. In this case $h_\mu^{\mathfrak{A}_T(\mathcal{Q}, \mathbf{n})}(T^{(\Gamma)}|\mathfrak{A}_T(\mathcal{Q}, \mathbf{n})_0^-) = \infty$ for all subgroups $\Gamma \subset \Xi_\mathbf{n}$ with infinite index (cf. Proposition 13.2).

CHAPTER VIII

Mixing

27. Multiple mixing and additive relations in fields

Let $d \geq 1$, and let T be a measure preserving $\mathbb{Z}^d$-action on a probability space $(Y, \mathfrak{T}, \mu)$. A non-empty subset $F \subset \mathbb{Z}^d$ is *mixing* for T if, for all collections of sets $\{B_\mathbf{n} : \mathbf{n} \in F\} \subset \mathfrak{T}$,

$$\lim_{k \to \infty} \mu\left(\bigcap_{\mathbf{n} \in F} T_{-k\mathbf{n}}(B_\mathbf{n})\right) = \prod_{\mathbf{n} \in F} \mu(B_\mathbf{n}), \tag{27.1}$$

and *non-mixing* otherwise. If T is r-mixing in the sense of (20.9), then every set $F \subset \mathbb{Z}^d$ of cardinality r is mixing, but the reverse implication is far from clear. If $F \subset \mathbb{Z}^d$ is a non-empty set then we can translate F and assume that $\mathbf{0} \in F$ without affecting its mixing behaviour. Furthermore, if $F \subset \mathbb{Z}^d$ is non-mixing, then every set $F' \supset F$ is non-mixing. Finally, if $b \cdot F = \{b\mathbf{m} : \mathbf{m} \in F\} \subset \mathbb{Z}^d$ is non-mixing for some positive $b \in \mathbb{Q}$, then F is non-mixing.

It is not known whether every measure preserving, mixing $\mathbb{Z}$-action T on a probability space $(Y, \mathfrak{T}, \mu)$ is r-mixing for every $r \geq 2$, or whether every non-empty subset $F \subset \mathbb{Z}$ is mixing for T. Perhaps surprisingly, if $d \geq 2$, and if T is a measure preserving $\mathbb{Z}^d$-action on a probability space, then mixing no longer implies that every non-empty subset $S \subset \mathbb{Z}^d$ is mixing. The first example of this nature was due to F. Ledrappier, who showed that there exists a $\mathbb{Z}^2$-action α by automorphisms of a compact, abelian group X which is mixing (with respect to Haar measure), but which has a non-mixing set of cardinality 3 ([56]).

EXAMPLES 27.1. (1) Let $f = 1 + u_1 + u_2 \in \mathfrak{R}_2^{(2)}$, and let $\alpha = \alpha^{\mathfrak{R}_2^{(2)}/(f)}$ be the shift-action of $\mathbb{Z}^d$ and λ_X the normalized Haar measure on $X = X^{\mathfrak{R}_2^{(2)}/(f)} \subset \mathbb{F}_2^{\mathbb{Z}^2}$ (cf. (6.19)–(6.20), and Examples 5.3 (5), 6.18 (5), 8.5 (1), and 22.5 (2)). The principal ideal $(f) \subset \mathfrak{R}_2^{(2)}$ contains the polynomial $f^{2^n} = 1 + u_1^{2^n} + u_2^{2^n}$ for every $n \geq 1$. According to (6.19),

$$X = \{x = (x_{\mathbf{m}}) \in \mathbb{F}_2^{\mathbb{Z}^2} : x_{(m_1,m_2)} + x_{(m_1+1,m_2)} + x_{(m_1,m_2+1)} = 0 \ (\mathrm{mod} \ 2)$$
$$\text{for all } \mathbf{m} = (m_1, m_2) \in \mathbb{Z}^2\}$$

$$= \left\{ x = (x_{\mathbf{m}}) \in \mathbb{F}_2^{\mathbb{Z}^2} : \sum_{\mathbf{n} \in \mathbb{Z}^2} c_g(\mathbf{n}) x_{\mathbf{m}+\mathbf{n}} = 0 \ (\mathrm{mod} \ 2) \right.$$
$$\left. \text{for every } g \in (f) \text{ and } \mathbf{m} \in \mathbb{Z}^2 \right\},$$

so that, in particular, $x_{(0,0)} + x_{(2^n,0)} + x_{(0,2^n)} = 0 \ (\mathrm{mod} \ 2)$ for every $n \geq 1$. If $F = \mathcal{S}(f) = \{(0,0),(1,0),(0,1)\}$, and if $B_{\mathbf{n}} = \{x \in X : x_0 = 0\}$ for every $\mathbf{n} \in F$, then $\lambda_X(B_{\mathbf{n}}) = \frac{1}{2}$ for all $\mathbf{n} \in F$, but

$$\lambda_X \left(\bigcap_{\mathbf{n} \in F} \alpha_{-k\mathbf{n}}(B_{\mathbf{n}}) \right) = \lambda_X(\{x \in X : x_{(0,0)} = x_{(0,2^k)} = x_{(2^k,0)} = 0\})$$
$$= \lambda_X(\{x \in X : x_{(0,0)} = x_{(0,2^k)} = 0\}) = \frac{1}{4}$$
$$\neq \frac{1}{8} = \prod_{\mathbf{n} \in F} \lambda_X(B_{\mathbf{n}})$$

for every $k \geq 1$. This shows that $F = \mathcal{S}(f)$ is non-mixing for α, although α is mixing by Theorem 6.5 (2).

(2) More generally, let p be a rational prime, $d \geq 2$, and let $\{0\} \neq \mathfrak{a} \subset \mathfrak{R}_d^{(p)}$ be an ideal. Exactly the same argument as in Example (1) shows that, for every non-zero element $f \in \mathfrak{a}$, the support $\mathcal{S}(f) \subset \mathbb{Z}^d$, as well as every finite set F' with $\mathcal{S}(f) \subset F' \subset \mathbb{Z}^d$, is non-mixing for $\alpha = \alpha^{\mathfrak{R}_d^{(p)}/\mathfrak{a}}$ (cf. Remark 6.19 (4)). $\boxdot$

The order of mixing, as well as the collection of all mixing (or non-mixing) sets, are measurable conjugacy invariants of $\mathbb{Z}^d$-actions, so that the explicit determination of these objects for a given $\mathbb{Z}^d$-action is of dynamical importance. For a general, measure preserving $\mathbb{Z}^d$-action T on a probability space $(Y, \mathfrak{T}, \mu)$ this is a problem of very considerable difficulty unless, of course, T has completely positive entropy (Theorem 20.14); however, for a $\mathbb{Z}^d$-action by automorphisms of compact, abelian groups, these questions become slightly more tractable, and we shall explore them in this and the following section. Example 27.1 shows that certain mixing $\mathbb{Z}^d$-action by automorphisms of compact, abelian groups can be mixing without being mixing of all orders, and that the breakdown of higher order mixing can occur in a particularly striking way. Although the $\mathbb{Z}^d$-actions in Example 27.1 all live on zero-dimensional groups, there is—a priori—no reason to suspect any link between connectedness (or lack thereof) and mixing properties. Nevertheless, this is exactly what turns out to be the case. In this section we prove that any mixing $\mathbb{Z}^d$-action α by automorphisms of a compact, connected, abelian group X is mixing of all orders, irrespective of any assumptions concerning entropy, and that $\mathbb{Z}^d$-actions by automorphisms of compact, zero-dimensional, abelian groups are mixing of all orders if and only if they have completely positive entropy. If such an action does not have completely positive entropy, then it has non-mixing sets, and the study of these sets in Section 27 will not only lead to further measurable conjugacy invariants for these actions, but also give insight into the nature of their measurable (self-)conjugacies (cf. Section 31).

In order to make the mixing behaviour of a $\mathbb{Z}^d$-action α by automorphisms of a compact, abelian group X more amenable to analysis we express the equations (20.9) and (27.1) in algebraic terms. If α is not r-mixing for some $r \geq 2$ then there exist sets $\mathbf{B}_1, \ldots, \mathbf{B}_r$ in $\mathfrak{B}_X$ and a sequence $(\mathbf{n}^{(k)} = (\mathbf{n}_1^{(k)}, \ldots, \mathbf{n}_r^{(k)})$, $k \geq 1)$ in $(\mathbb{Z}^d)^r$ such that $\mathbf{n}_1^{(k)} = \mathbf{0}$ for every $k \geq 1$, $\lim_{k\to\infty} \mathbf{n}_j^{(k)} - \mathbf{n}_i^{(k)} = \infty$ for $1 \leq i < j \leq r$, and

$$\lim_{k\to\infty} \mu\big(T_{-\mathbf{n}_1^{(k)}}(B_1) \cap \cdots \cap T_{-\mathbf{n}_r^{(k)}}(B_r)\big) \neq \mu(B_1) \cdot \ldots \cdot \mu(B_r). \qquad (27.2)$$

By expanding each of the indicator function 1_{B_j} in (27.2) as a Fourier series we conclude that there exist characters $\chi_1, \ldots, \chi_r$ in $\hat{X}$, not all of them trivial, such that

$$\lim_{k\to\infty} \int \big(\chi_1 \cdot \alpha_{\mathbf{n}_1^{(k)}}\big) \cdot \ldots \cdot \big(\chi_r \cdot \alpha_{\mathbf{n}_r^{(k)}}\big)\, d\lambda_X$$
$$\neq \prod_{j=1}^{r} \left(\int (\chi_j \cdot \alpha_{\mathbf{n}_j})\, d\lambda_X\right) = 0. \qquad (27.3)$$

In terms of the R_d-module $\mathfrak{M} = \hat{X}$ arising via Lemma 5.1, (27.3) amounts to saying that α is not r-mixing if and only if there exists a non-zero element $(a_1, \ldots, a_r) \in \mathfrak{M}^r$ and a sequence $(\mathbf{n}^{(k)} = (\mathbf{n}_1^{(k)}, \ldots, \mathbf{n}_r^{(k)})$, $k \geq 1)$ in $(\mathbb{Z}^d)^r$ such that $\mathbf{n}_1^{(k)} = \mathbf{0}$ for every $k \geq 1$, $\lim_{k\to\infty} \mathbf{n}_j^{(k)} - \mathbf{n}_i^{(k)} = \infty$ for $1 \leq i < j \leq r$, and

$$u^{\mathbf{n}_1^{(k)}} \cdot a_1 + \cdots + u^{\mathbf{n}_r^{(k)}} \cdot a_r = 0 \qquad (27.4)$$

for every $k \geq 1$. Similarly we see that a non-empty, finite set $F \subset \mathbb{Z}^d$ is non-mixing for α if and only if there exists a non-zero element $(a_{\mathbf{n}}, \mathbf{n} \in F) \in \mathfrak{M}^F$ and an increasing sequence of positive integers $(k_m, m \geq 1)$ such that

$$\sum_{\mathbf{n}\in F} u^{k_m \mathbf{n}} \cdot a_{\mathbf{n}} = 0 \qquad (27.5)$$

for every $m \geq 1$.

In analyzing the higher order mixing behaviour of a $\mathbb{Z}^d$-action α by automorphisms of a compact, abelian group X we follow our usual strategy of reducing the dynamical properties of α to conditions on the $\mathbb{Z}^d$-actions $\alpha^{\mathfrak{R}_d/\mathfrak{p}}$, where $\mathfrak{p}$ varies over the set of prime ideals associated with the $\mathfrak{R}_d$-module $\mathfrak{M} = \hat{X}$ defined by Lemma 5.1. In Proposition 6.6 we saw that α is mixing if and only if every prime ideal $\mathfrak{p}$ associated with $\mathfrak{M}$ is mixing, and that the latter condition is equivalent to saying that none of the prime ideals associated with $\mathfrak{M}$ contains a generalized cyclotomic polynomial. The higher order mixing properties of α can again be described in terms of the associated prime ideals of $\mathfrak{M}$.

THEOREM 27.2. *Let α be a $\mathbb{Z}^d$-action by automorphisms of a compact, abelian group X, and let $\mathfrak{M} = \hat{X}$ be the $\mathfrak{R}_d$-module arising from α via* Lemma 5.1.

> (1) *The following conditions are equivalent for every integer $r \geq 2$.*
>> (a) *α is r-mixing;*
>> (b) *$\alpha^{\mathfrak{R}_d/\mathfrak{p}}$ is r-mixing for every prime ideal $\mathfrak{p}$ associated with $\mathfrak{M}$.*
>
> (2) *The following conditions are equivalent for every finite subset $F \subset \mathbb{Z}^d$.*
>> (a) *F is mixing for α;*
>> (b) *For every prime ideal $\mathfrak{p}$ associated with $\mathfrak{M}$, F is mixing for $\alpha^{\mathfrak{R}_d/\mathfrak{p}}$.*

PROOF. Suppose that α is r-mixing. If $\mathfrak{p} \subset \mathfrak{R}_d$ is a prime ideal associated with $\mathfrak{M}$, then there exists an element $a \in \mathfrak{M}$ such that $\mathfrak{p} = \{f \in \mathfrak{R}_d : f \cdot a = 0\}$, and we set $\mathfrak{Y} = \mathfrak{R}_d \cdot a \subset \mathfrak{M}$. Then $\mathfrak{Y} \cong \mathfrak{R}_d/\mathfrak{p}$ and $Y = \hat{\mathfrak{Y}} = X/\mathfrak{Y}^\perp$, where $\mathfrak{Y}^\perp = \{x \in X : \langle x, a \rangle = 1 \text{ for all } a \in \mathfrak{Y}\}$ is the annihilator of $\mathfrak{Y}$. Since $\mathfrak{Y}$ is invariant under the $\mathbb{Z}^d$-action $\hat{\alpha} \colon \mathbf{n} \mapsto \hat{\alpha}_\mathbf{n}$ dual to α, $\mathfrak{Y}^\perp$ is a closed, α-invariant subgroup of X, and the $\mathbb{Z}^d$-action α^Y induced by α on Y is a factor of α and hence r-mixing. Since the $\mathfrak{R}_d$-module arising from α^Y is equal to $\hat{Y} = \mathfrak{Y} \cong \mathfrak{R}_d/\mathfrak{p}$ we conclude that $\alpha^{\mathfrak{R}_d/\mathfrak{p}}$ must be r-mixing.

Conversely, if α is not r-mixing, then (27.4) shows that there exists a non-zero element $(a_1, \ldots, a_r) \in \mathfrak{M}^r$ and a sequence $(\mathbf{n}^{(k)} = (\mathbf{n}_1^{(k)}, \ldots, \mathbf{n}_r^{(k)}), k \geq 1)$ in $(\mathbb{Z}^d)^r$ such that $\mathbf{n}_1^{(k)} = \mathbf{0}$ for every $k \geq 1$, $\lim_{k \to \infty} \mathbf{n}_j^{(k)} - \mathbf{n}_i^{(k)} = \infty$ for $1 \leq i < j \leq r$, and $u^{\mathbf{n}_1^{(k)}} \cdot a_1 + \cdots + u^{\mathbf{n}_r^{(k)}} \cdot a_r = 0$ for every $k \geq 1$. There exists a Noetherian submodule $\mathfrak{N} \subset \mathfrak{M}$ such that $\{a_1, \ldots, a_r\} \subset \mathfrak{N}$, and (27.4) implies that the $\mathbb{Z}^d$-action $\alpha^\mathfrak{N}$, which is a quotient of α, is not r-mixing.

Since $\mathfrak{N}$ is Noetherian, the set of prime ideals associated with $\mathfrak{N}$ is finite and equal to $\{\mathfrak{p}_1, \ldots, \mathfrak{p}_m\}$, say, and we choose a corresponding reduced primary decomposition $\mathfrak{W}_1, \ldots, \mathfrak{W}_m$ of $\mathfrak{N}$ (cf. (6.5)). The map $a \mapsto (a + \mathfrak{W}_1, \ldots, a + \mathfrak{W}_m)$ from $\mathfrak{N}$ into $\mathfrak{K} = \bigoplus_{i=1}^m \mathfrak{N}/\mathfrak{W}_i$ is injective, and the dual homomorphism from $\bar{X} = \hat{\mathfrak{K}}$ to $\hat{\mathfrak{N}} = X^\mathfrak{N}$ is surjective. Hence $\alpha^\mathfrak{N}$ is a factor of $\alpha^\mathfrak{K}$, so that $\alpha^\mathfrak{K}$ cannot be r-mixing. By applying (27.4) to the $\mathfrak{R}_d$-module $\mathfrak{K}$ we see that there exists a $j \in \{1, \ldots, m\}$ such that $\alpha^{\mathfrak{N}/\mathfrak{W}_j}$ is not r-mixing.

Put $\mathfrak{V} = \mathfrak{N}/\mathfrak{W}_j$, $\mathfrak{p} = \mathfrak{p}_j$, and use Proposition 6.1 to find integers $1 \leq t \leq s$ and submodules $\mathfrak{V} = \mathfrak{N}_s \supset \cdots \supset \mathfrak{N}_0 = \{0\}$ such that, for every $k = 1, \ldots, s$, $\mathfrak{N}_k/\mathfrak{N}_{k-1} \cong \mathfrak{R}_d/\mathfrak{q}_k$ for some prime ideal $\mathfrak{p} \subset \mathfrak{q}_k \subset \mathfrak{R}_d$, $\mathfrak{q}_k = \mathfrak{p}$ for $k = 1, \ldots, t$, and $\mathfrak{q}_k \supsetneq \mathfrak{p}$ for $i = t+1, \ldots, s$. We choose Laurent polynomials $g_k \in \mathfrak{q}_k \smallsetminus \mathfrak{p}$, $k = t+1, \ldots, s$, and set $g = g_{t+1} \cdots g_s$. Since $\alpha^\mathfrak{V}$ is not r-mixing, (27.4) implies the existence of a non-zero element $(a_1, \ldots, a_r) \in \mathfrak{V}^r$ and a sequence $(\mathbf{n}^{(k)} = (\mathbf{n}_1^{(k)}, \ldots, \mathbf{n}_r^{(k)}), k \geq 1)$ in $(\mathbb{Z}^d)^r$ such that $\mathbf{n}_1^{(k)} = \mathbf{0}$ for every $k \geq 1$, $\lim_{k \to \infty} \mathbf{n}_j^{(k)} - \mathbf{n}_i^{(k)} = \infty$ for $1 \leq i < j \leq r$, and $u^{\mathbf{n}_1^{(k)}} \cdot a_1 + \cdots + u^{\mathbf{n}_r^{(k)}} \cdot a_r = 0$ for every $k \geq 1$. Put $b_i = g \cdot a_i$, and note that $0 \neq (b_1, \ldots, b_r) \in (\mathfrak{N}_t)^r$, since $g \cdot a \neq 0$ for every non-zero element $a \in \mathfrak{V}$. There exists a unique integer $l \in \{1, \ldots, t\}$ such that $(b_1, \ldots, b_r) \in (\mathfrak{N}_l)^r \smallsetminus (\mathfrak{N}_{l-1})^r$, and by setting $b_i' = b_i + \mathfrak{N}_{l-1} \in \mathfrak{N}_l / \mathfrak{N}_{l-1} \cong \mathfrak{R}_d / \mathfrak{p}$ we obtain that $0 \neq (b_1', \ldots, b_r') \in (\mathfrak{N}_l / \mathfrak{N}_{l-1})^r \cong (\mathfrak{R}_d / \mathfrak{p})^r$ and $u^{\mathbf{n}_1^{(k)}} \cdot b_1' + \cdots + u^{\mathbf{n}_r^{(k)}} \cdot b_r' = 0$ for every $k \geq 1$, so that $\alpha^{\mathfrak{R}_d / \mathfrak{p}}$ is not r-mixing by (27.4). Since the prime ideal $\mathfrak{p}$ is associated with the submodule $\mathfrak{N} \subset \mathfrak{M}$, $\mathfrak{p}$ is also associated with $\mathfrak{M}$, and (1) is proved. The proof of (2) is identical, except that we use (27.5) instead of (27.4). $\square$

Theorem 27.2 shows that a $\mathbb{Z}^d$-action α by automorphisms of a compact, abelian group X is mixing of order $r \geq 2$ if and only if the $\mathbb{Z}^d$-actions $\alpha^{\mathfrak{R}_d / \mathfrak{p}}$ are r-mixing for all prime ideals $\mathfrak{p} \subset \mathfrak{R}_d$ associated with the $\mathfrak{R}_d$-module $\mathfrak{M} = \hat{X}$ defined by α. In order to apply this result we shall characterize those prime ideals $\mathfrak{p} \subset \mathfrak{R}_d$ for which $\alpha^{\mathfrak{R}_d / \mathfrak{p}}$ is r-mixing for every $r \geq 2$. For every prime ideal $\mathfrak{p} \subset \mathfrak{R}_d$ we define the characteristic $p(\mathfrak{p})$ of $\mathfrak{R}_d / \mathfrak{p}$ as in (6.2).

THEOREM 27.3. Let $d \geq 1$, and let $\mathfrak{p} \subset \mathfrak{R}_d$ be a mixing prime ideal (cf. Definition 6.16 and Theorem 6.5 (2)).

(1) If $p(\mathfrak{p}) > 0$ then $\alpha^{\mathfrak{R}_d / \mathfrak{p}}$ is r-mixing for every $r \geq 2$ if and only if $\mathfrak{p} = (p(\mathfrak{p})) = p(\mathfrak{p}) \mathfrak{R}_d$;
(2) If $p(\mathfrak{p}) = 0$ then $\alpha^{\mathfrak{R}_d / \mathfrak{p}}$ is r-mixing for every $r \geq 2$.

We postpone the proof of Theorem 27.3 for the moment and look instead at some of its consequences. If α is a $\mathbb{Z}^d$-action by automorphisms of a compact, abelian group X with completely positive entropy, then it is mixing of all orders by Theorem 20.14. If the group X is zero-dimensional, the reverse implication is also true.

COROLLARY 27.4. Let α be a $\mathbb{Z}^d$-action by automorphisms of a compact, abelian, zero-dimensional group X. The following conditions are equivalent.

(1) α has completely positive entropy;
(2) α is r-mixing for every $r \geq 2$.

PROOF. Since X is zero-dimensional, every prime ideal $\mathfrak{p}$ associated with the $\mathfrak{R}_d$-module $\mathfrak{M} = \hat{X}$ arising from α via Lemma 5.1 contains a non-zero constant, so that $p(\mathfrak{p}) > 0$. According to Corollary 18.5 and Theorem 20.8, α has completely positive entropy if and only if $\mathfrak{p} = p(\mathfrak{p}) \mathfrak{R}_d$ for every prime ideal $\mathfrak{p}$ associated with $\mathfrak{M}$, and the equivalence of (1) and (2) follows from Theorem 27.2 and Theorem 27.3 (1). $\square$

EXAMPLE 27.5. Let $d = 2$, and let $\mathfrak{I} = (4, 2 + 2u_1 + 2u_2) = 4\mathfrak{R}_2 + (2 + 2u_1 + 2u_2)\mathfrak{R}_2 \subset \mathfrak{R}_2$. The $\mathfrak{R}_2$-module $\mathfrak{M} = \mathfrak{R}_2 / \mathfrak{I}$ has two associated primes: $\mathfrak{p}_1 = (2) = 2\mathfrak{R}_2$, and $\mathfrak{p}_2 = (2, 1 + u_1 + u_2) = 2\mathfrak{R}_2 + (1 + u_1 + u_2)\mathfrak{R}_2$. Since $\mathfrak{R}_2 / \mathfrak{p}_1 \cong \mathfrak{R}_2^{(2)}$, $\alpha^{\mathfrak{R}_2 / \mathfrak{p}_1}$ has no non-mixing sets, and the collection of non-mixing sets of the $\mathbb{Z}^2$-action $\alpha = \alpha^{\mathfrak{M}}$ coincides with that of $\alpha^{\mathfrak{R}_2 / \mathfrak{p}_2} \cong \alpha^{\mathfrak{R}_2^{(2)} / (1 + u_1 + u_2)}$ (cf. Theorem 27.2). $\boxdot$

The next corollary shows that the higher order mixing behaviour of $\mathbb{Z}^d$-actions by automorphisms of compact, connected, abelian groups is quite different from the zero-dimensional case, and requires no assumptions concerning entropy.

COROLLARY 27.6. *Let $d \geq 1$, and let α be a mixing $\mathbb{Z}^d$-action on a compact, connected, abelian group X. Then α is r-mixing for every $r \geq 2$.*

PROOF. The group X is connected if and only if the dual group $\hat{X}$ is torsion-free, i.e. if and only if $na \neq 0$ whenever $0 \neq a \in \hat{X}$ and $0 \neq n \in \mathbb{Z}$. We write $\mathfrak{M} = \hat{X}$ for the $\mathfrak{R}_d$-module defined by α via Lemma 5.1, note that the connectedness of X implies that $p(\mathfrak{p}) = 0$ for every prime ideal $\mathfrak{p} \subset \mathfrak{R}_d$ associated with $\mathfrak{M}$, and apply the Theorems 27.2 and 27.3 (2). $\square$

COROLLARY 27.7. *Let $A_1, \dots, A_d$ be commuting automorphism of the n-torus $\mathbb{T}^n = \mathbb{R}^n / \mathbb{Z}^n$ with the property that the $\mathbb{Z}^d$-action α: $(m_1, \dots, m_d) \mapsto \alpha_{(m_1, \dots, m_d)} = A_1^{m_1} \cdots A_d^{m_d}$ is mixing. Then α is r-mixing for every $r \geq 2$.*

PROOF OF THEOREM 27.3 (1). Suppose that $p = p(\mathfrak{p}) > 0$, and that $\alpha^{\mathfrak{R}_d/\mathfrak{p}}$ is r-mixing for every $r \geq 2$. If $\mathfrak{p} \neq (p)$, then the prime ideal $\mathfrak{q} = \{f_{/p} : f \in \mathfrak{q}\} \subset \mathfrak{R}_d^{(p)}$ is non-zero, where $f_{/p} \in \mathfrak{R}_d^{(p)}$ is obtained by reducing the coefficients of a Laurent polynomial $f \in \mathfrak{R}_d$ modulo p (cf. (6.2)). Since $\mathfrak{R}_d/\mathfrak{p} \cong \mathfrak{R}_d^{(p)}/\mathfrak{q}$, we may identify the $\mathbb{Z}^d$-actions $\alpha^{\mathfrak{R}_d/\mathfrak{p}}$ and $\alpha^{\mathfrak{R}_d^{(p)}/\mathfrak{q}}$ (cf. Remark 6.19 (4)). We choose a non-zero element $g \in \mathfrak{q}$ and see as in Example 27.1 (1)–(2) that the support $\mathcal{S}(g)$ is a non-mixing set. In particular, $\alpha^{\mathfrak{R}_d/\mathfrak{p}}$ is not mixing of order $|\mathcal{S}(g)|$.

Conversely, if $p = p(\mathfrak{p})$ and $\mathfrak{p} = (p)$, then $\alpha^{\mathfrak{R}_d/\mathfrak{p}}$ is (conjugate to) the (Bernoulli) shift-action of $\mathbb{Z}^d$ on $\mathbb{F}_p^{\mathbb{Z}^d}$, and therefore mixing of all orders (cf. (6.19)). $\square$

The proof of Theorem 27.3 (2) depends on the following result concerning additive relations in fields of characteristic zero.

PROPOSITION 27.8 ([81]). *Let $\mathbb{F}$ be a field of characteristic zero, $G \subset \mathbb{F}^\times = \mathbb{F} \smallsetminus \{0\}$ a finitely generated, multiplicative subgroup, $n \geq 1$, and $a_1, \dots, a_n$ non-zero elements of $\mathbb{F}$. Then the equation*

$$a_1 x_1 + \cdots + a_n x_n = 1 \tag{27.6}$$

has only finitely many distinct solutions $(x_1, \dots, x_n)$ in G^n for which no proper subsum $a_{i_1} x_{i_1} + \cdots + a_{i_k} x_{i_k}$ vanishes.

PROOF OF THEOREM 27.3 (2). Suppose that $\alpha^{\mathfrak{R}_d/\mathfrak{p}}$ is not s-mixing for some $s > 2$, and that s is the smallest integer with this property. According to (27.4) there exists a non-zero element $(a_1, \dots, a_s) \in (\mathfrak{R}_d/\mathfrak{p})^s$ and a sequence $(\mathbf{n}^{(k)} = (\mathbf{n}_1^{(k)}, \dots, \mathbf{n}_s^{(k)}), k \geq 1)$ in $(\mathbb{Z}^d)^s$ such that $\mathbf{n}_1^{(k)} = \mathbf{0}$ for every $k \geq 1$, $\lim_{k \to \infty} \mathbf{n}_j^{(k)} - \mathbf{n}_i^{(k)} = \infty$ whenever $1 \leq i < j \leq s$, and $u^{\mathbf{n}_1^{(k)}} \cdot a_1 + \cdots + u^{\mathbf{n}_s^{(k)}} \cdot a_s = 0$

for every $k \geq 1$. For simplicity we assume that $\mathbf{n}_i^{(k)} \neq \mathbf{n}_i^{(l)}$ whenever $1 \leq k < l$ and $i \in \{2, \ldots, s\}$. The minimality of s is easily seen to imply that $a_i \neq 0$ for $i = 1, \ldots, s$. We write $\mathbb{F}$ for the field of fractions of the integral domain $\mathfrak{R}_d/\mathfrak{p}$, regard $\mathfrak{R}_d/\mathfrak{p}$ as a subring of $\mathbb{F}$, denote by G the multiplicative subgroup of $\mathbb{F}^{\times}$ generated by $\{u^{\mathbf{n}} + \mathfrak{p} : \mathbf{n} \in \mathbb{Z}^d\}$, and observe that G is a free abelian group on d generators, since $\alpha^{\mathfrak{R}_d/\mathfrak{p}}$ is mixing (Lemma 6.6 (2)). It follows that the equation

$$-\frac{a_2}{a_1}x_2 - \cdots - \frac{a_s}{a_1}x_s = 1$$

has infinitely many distinct solutions $(x_2, \ldots, x_s)$ in $G^{s-1} \subset (\mathbb{F}^{\times})^{s-1}$, and Corollary 27.8 implies that all but finitely many of these solutions have vanishing subsums. In particular there exists a subset $\{i_1, \ldots, i_m\} \subsetneq \{2, \ldots, s\}$ such that

$$u^{\mathbf{n}_{i_1}^{(k)}} \cdot a_{i_1} + \cdots + u^{\mathbf{n}_{i_m}^{(k)}} \cdot a_{i_m} = 0$$

for infinitely many $k \geq 1$, so that $\alpha^{\mathfrak{R}_d/\mathfrak{p}}$ is not mixing of order $m < s$. This contradiction to our choice of s implies that $\alpha^{\mathfrak{R}_d/\mathfrak{p}}$ is mixing of every order. $\square$

CONCLUDING REMARKS 27.9. (1) Theorem 27.2 is taken from [47] and [98], and Theorem 27.3 from [98]. Proposition 27.8 can also be found (with a different proof) in [21]; the reference [81] was pointed out to me by Ward. In [98] Proposition 27.8 is derived in a somewhat disguised form from the main result in [92] which may yield further information about the rate of multiple mixing of $\mathbb{Z}^d$-actions by automorphisms of finite-dimensional tori or solenoids. In order to describe Schlickewei's result in [92] we assume that $\mathbb{K}$ is an algebraic number field of degree D over $\mathbb{Q}$ and denote by $P(\mathbb{K})$ the set of places and $P_{\infty}(\mathbb{K})$ the set of infinite (or archimedean) places of $\mathbb{K}$ (cf. Section 7). For every $v \in P(\mathbb{K})$, $|\cdot|_v$ denotes the associated valuation, normalized so that $|a|_v = \mathrm{mod}_{\mathbb{K}_v}(a)$ for every $a \in \mathbb{K}$ (cf. Section 7). Let S, $P_{\infty}(\mathbb{K}) \subset S \subset P(\mathbb{K})$ be a finite set with cardinality $|S|$. An element $a \in \mathbb{K}$ is called an S-unit if $|a|_v = 1$ for every $v \in P(\mathbb{K}) \smallsetminus S$. Then the following is true: *For every $n \geq 1$, and for all non-zero elements $a_1, \ldots, a_n$ in $\mathbb{K}$, the equation*

$$a_1 x_1 + \cdots + a_n x_n = 1$$

has not more than

$$(4|S|D!)^{2^{36nD!}|S|^6} \tag{27.7}$$

solutions $(x_1, \ldots, x_n)$ in S-units such that no proper subsum $a_{i_1}x_{i_1} + \cdots + a_{i_k}x_{i_k}$ vanishes. A glance at the proof of Theorem 27.3 reveals that Corollary 27.7 also follows from the theorem of Schlickewei just quoted; however, Corollary 27.7 does not require the full strength of of the estimate (27.7), but only the finiteness assertion of Proposition 27.8. What further dynamical information about mixing $\mathbb{Z}^d$-actions by automorphisms of compact, connected, finite dimensional, abelian groups can be gained from (27.7)?

(2) The non-existence of non-mixing sets for mixing $\mathbb{Z}^d$-actions by auto-morphisms of compact, connected, abelian groups follows from a result in [66] (cf. [94]), which can be regarded as a precursor of Proposition 27.8: *Let $\mathbb{K}$ be an algebraic number field of degree D over $\mathbb{Q}$, and let $a_1, \ldots, a_n$ and $x_1, \ldots, x_n$ be non-zero elements of $\mathbb{K}$. If there exist infinitely many integers $0 < k_1 < k_2 < \ldots$ such that*

$$a_1 x_1^{k_i} + \cdots + a_n x_n^{k_i} = 0 \tag{27.8}$$

for every $i \geq 1$, then there exist integers $1 \leq k < l \leq n$ and $m > 0$ such that

$$x_k^m = x_l^m. \tag{27.9}$$

If p is a rational prime such that $|a_i|_v = 1$ for every $i \in \{1, \ldots, n\}$ and every valuation v of $\mathbb{K}$ which lies above p, then Mahler's proof gives an effective bound on m in terms of p and D. By using a straightforward argument described in [93] one can extend the implication (27.8)⇒(27.9) to an arbitrary field $\mathbb{K}$ of characteristic 0. If $\mathfrak{p} \subset \mathfrak{R}_d$ is a mixing prime ideal with $p(\mathfrak{p}) = 0$, and if $\mathbb{K}$ equal to the quotient field of $\mathfrak{R}_d/\mathfrak{p}$, then (27.5) and (27.8)–(27.9) together imply that every non-empty, finite set $F \subset \mathbb{Z}^d$ is mixing.

28. Masser's theorem and non-mixing sets

In Section 27 we saw that every mixing $\mathbb{Z}^d$-action by automorphisms of a compact, connected, abelian group X is mixing of all orders, so that every non-empty, finite set $F \subset \mathbb{Z}^d$ is mixing for α (cf. (27.1)). If X is not connected we denote by X° the connected component of the identity in X and write α^{X/X° for the $\mathbb{Z}^d$-action on the zero-dimensional group X/X° induced by α. Let $\mathfrak{M} = \hat{X}$ be the $\mathfrak{R}_d$-module arising from α, and observe that

$$\mathfrak{N} = \widehat{X/X^\circ} = \{a \in \mathfrak{M} : ma = 0 \text{ for some non-zero } m \in \mathbb{Z}\}.$$

Since every prime ideal $\mathfrak{p} \subset \mathfrak{R}_d$ with $p(\mathfrak{p}) > 0$ associated with $\mathfrak{M}$ is also associated with the submodule $\mathfrak{N} \subset \mathfrak{M}$ we see from the Theorems 27.2–27.3 that $\alpha = \alpha^{\mathfrak{M}}$ and $\alpha^{X/X^\circ} = \alpha^{\mathfrak{N}}$ have the same mixing sets. This reduces the general problem of determining the mixing sets for a $\mathbb{Z}^d$-action α on a compact, abelian group X to the special case where X is zero-dimensional.

Assume therefore from now on that α is a $\mathbb{Z}^d$-action by automorphisms of a compact, abelian, zero-dimensional group X, and let $\mathfrak{M} = \hat{X}$ be the $\mathfrak{R}_d$-module arising from α via Lemma 5.1. In order to determine whether a non-empty, finite set $F \subset \mathbb{Z}^d$ is mixing for α one has to go through the following steps.

(1) Find all prime ideals associated with the $\mathfrak{R}_d$-module $\mathfrak{M}$;
(2) For every prime ideal $\mathfrak{p} \subset \mathfrak{R}_d$ associated with $\mathfrak{M}$, check whether F is mixing for $\alpha^{\mathfrak{R}_d/\mathfrak{p}}$.

The first step is not really feasible unless $\mathfrak{M}$ is Noetherian or, equivalently, unless α is expansive (otherwise $\mathfrak{M}$ may have infinitely many distinct associated prime ideals). In order to understand what is involved in the second step we note that every prime ideal $\mathfrak{p} \subset \mathfrak{R}_d$ associated with $\mathfrak{M}$ satisfies that $p(\mathfrak{p}) > 0$. Assume therefore that $\mathfrak{p} \subset \mathfrak{R}_d$ is a prime ideal with $p = p(\mathfrak{p}) > 0$. If $\mathfrak{p} = (p)$, then we know from Theorem 27.3 (1) that $\alpha^{\mathfrak{R}_d/\mathfrak{p}}$ is mixing of all orders, and that every non-empty, finite set $F \subset \mathbb{Z}^d$ is mixing for $\alpha^{\mathfrak{R}_d/\mathfrak{p}}$. If $\mathfrak{p} \neq (p)$, the prime ideal

$$\mathfrak{q} = \{f_{/p} : f \in \mathfrak{p}\} \subset \mathfrak{R}_d^{(p)} \tag{28.1}$$

(cf. (6.2)) is non-zero, $\mathfrak{R}_d/\mathfrak{p} \cong \mathfrak{R}_d^{(p)}/\mathfrak{q}$, and we identify the $\mathbb{Z}^d$-actions $\alpha^{\mathfrak{R}_d/\mathfrak{p}}$ and $\alpha^{\mathfrak{R}_d^{(p)}/\mathfrak{q}}$ (cf. Remark 6.19). According to (27.5), a set $F = \{\mathbf{n}_1, \ldots, \mathbf{n}_r\} \subset \mathbb{Z}^d$ is mixing if and only if, for every $(f_1, \ldots, f_r) \in (\mathfrak{R}_d^{(p)})^r$ with $f_i \notin \mathfrak{q}$ for some $i \in \{1, \ldots, r\}$,

$$u^{k\mathbf{n}_1} f_1 + \cdots + u^{k\mathbf{n}_r} f_r \notin \mathfrak{q} \tag{28.2}$$

for all sufficiently large $k \geq 0$.

In order to gain insight into the meaning of (28.2) we denote by $\mathbb{F}_{p^k}$ the field with p^k elements and write $\overline{\mathbb{F}}_p \supset \mathbb{F}_p$ for the algebraic closure of the prime field $\mathbb{F}_p$. We shall need the following theorem by Masser ([70], [47]).

THEOREM 28.1. *Let $\mathbb{F}$ be a field of characteristic $p \neq 0$ with algebraic closure $\overline{\mathbb{F}} \supset \mathbb{F}$, $r \geq 1$, and let $S = \{x_1, \ldots, x_r\} \subset \mathbb{F}^\times$. The following conditions are equivalent.*

(1) *There exists a non-zero element $(\xi_1, \ldots, \xi_r) \in \mathbb{F}^r$ such that*

$$\xi_1 x_1^k + \cdots + \xi_r x_r^k = 0$$

for infinitely many $k \geq 0$;

(2) *There exist positive integers a, b, and an element $(y_1, \ldots, y_r) \in (\overline{\mathbb{F}}^\times)^r$, such that $x_i = y_i^a$ for $i = 1, \ldots, r$, and $\{y_1^b, \ldots, y_r^b\}$ is linearly dependent over $\overline{\mathbb{F}}_p$;*

(3) *If $\mathbb{E} = \mathbb{F} \cap \overline{\mathbb{F}}_p$, then there exist positive integers a, b, and elements $(\omega_1, \ldots, \omega_r) \in (\mathbb{E}^\times)^r$, $(z_1, \ldots, z_r) \in (\overline{\mathbb{F}}^\times)^r$, such that $x_i = \omega_i z_i^a$ for $i = 1, \ldots, r$, and $\{z_1^b, \ldots, z_r^b\}$ is linearly dependent over $\mathbb{E}$.*

In order to apply Theorem 28.1 to (28.2) we fix $d \geq 1$ and denote by

$$\mathcal{R}_d^{(p)} = \overline{\mathbb{F}}_p[u_1^{\pm \frac{1}{k}}, \ldots, u_d^{\pm \frac{1}{k}}, k \geq 1, p \nmid k] \supset \mathfrak{R}_d^{(p)} \tag{28.3}$$

the ring of Laurent polynomials in the variables $\{u_1^{\pm \frac{1}{k}}, \ldots, u_d^{\pm \frac{1}{k}}, k \geq 1, p \nmid k\}$ with coefficients in $\overline{\mathbb{F}}_p$, where $p \nmid k$ indicates that p does not divide k. Every

$f \in \mathcal{R}_d^{(p)}$ can be written as

$$f = \sum_{\mathbf{r}=(r_1,\ldots,r_d)\in(\mathbb{Q}^{(p)})^d} c_f(\mathbf{r})u^{\mathbf{r}}, \tag{28.4}$$

where $\mathbb{Q}^{(p)} = \{\frac{j}{k} : j \in \mathbb{Z}, k \geq 1, p \nmid k\}$, and where $u^{\mathbf{r}} = u_1^{r_1} \cdot \ldots \cdot u_d^{r_d}$ for every $\mathbf{r} \in (\mathbb{Q}^{(p)})^d$. We write

$$\mathcal{S}(f) = \{\mathbf{r} \in (\mathbb{Q}^{(p)})^d : c_f(\mathbf{r}) \neq 0\} \tag{28.5}$$

for the *support* of f, $\mathcal{C}(f) \subset \mathbb{R}^d$ for the convex hull of $\mathcal{S}(f)$, and denote by $\mathcal{E}(f) \subset (\mathbb{Q}^{(p)})^d$ the set of extreme points of $\mathcal{C}(f)$. If $f \neq 0$ there exists a unique smallest positive rational number $m^*(f)$ such that $m^*(f)\mathcal{S}(u^{-\mathbf{r}}f) \subset \mathbb{Z}^d$ for some (and hence for all) $\mathbf{r} \in \mathcal{S}(f)$, and we set $\mathcal{S}^*(f) = m^*(f)\mathcal{S}(u^{-\mathbf{r}}f)$ for some $\mathbf{r} \in \mathcal{S}(f)$ and write $\mathcal{C}^*(h)$ for the convex hull of $\mathcal{S}^*(f)$ in $\mathbb{R}^d$. Both $\mathcal{S}^*(f)$ and $\mathcal{C}^*(f)$ are determined only up to translation, but this will not cause any problems. Finally, if

$$\begin{aligned}\overline{\mathfrak{R}}_d^{(p,k)} &= \overline{\mathbb{F}}_p[u_1^{\pm\frac{1}{k}},\ldots,u_d^{\pm\frac{1}{k}}], \; k \geq 1, \; p \nmid k, \\ \overline{\mathfrak{R}}_d^{(p,1)} &= \overline{\mathfrak{R}}_d^{(p)},\end{aligned} \tag{28.6}$$

then

$$\mathcal{R}_d^{(p)} = \bigcup_{k \geq 1} \overline{\mathfrak{R}}_d^{(p,k)},$$

and there exists, for every non-zero $f \in \mathcal{R}_d^{(p)}$, a unique smallest integer $\mathfrak{r}(f)$ such that

$$u^{-\mathbf{r}}f \in \overline{\mathfrak{R}}_d^{(p,\mathfrak{r}(f))} \tag{28.7}$$

for some (and hence for all) $\mathbf{r} \in \mathcal{S}(f)$. In particular, if $c_f(\mathbf{0}) = 1$, then $f \in \overline{\mathfrak{R}}_d^{(p,\mathfrak{r}(f))}$.

LEMMA 28.2. *Let* $\mathfrak{q} \subset \mathfrak{R}_d^{(p)}$ *be a non-zero prime ideal. Then there exists a prime ideal* $\overline{\mathfrak{q}} \subset \mathcal{R}_d^{(p)}$ *such that* $\mathfrak{q} = \mathfrak{R}_d^{(p)} \cap \overline{\mathfrak{q}}$.

PROOF. Choose an increasing sequence $(k_m, m \geq 1)$ in $\mathbb{N}$ such that $k_1 = 1$, $p \nmid k_m$ and k_m divides k_{m+1} for all $m \geq 1$, and that every $k \in \mathbb{N}$ with $p \nmid k$ divides k_m for some $m \geq 1$. If $m < m'$, and if $\mathfrak{p} \subset \overline{\mathfrak{R}}_d^{(p,k_{m'})}$ is a prime ideal, then $\mathfrak{p} \cap \mathfrak{R}_d^{(p,k_m)}$ is a prime ideal in $\mathfrak{R}_d^{(p,k_m)}$.

We choose inductively a sequence of prime ideals $\mathfrak{p}_m \subset \overline{\mathfrak{R}}_d^{(p,k_m)}$ such that $\mathfrak{p}_{m'} \cap \overline{\mathfrak{R}}_d^{(p,k_m)} = \mathfrak{p}_m$ whenever $1 \leq m < m'$, and $\mathfrak{p}_1 \cap \mathfrak{R}_d^{(p)} = \mathfrak{q}$. Indeed, if $m = 1$, we write $\mathfrak{q}^{(1)} \subset \overline{\mathfrak{R}}_d^{(p)}$ for the ideal generated by $\mathfrak{q}$, and choose a minimal set $\{\mathfrak{a}_1,\ldots,\mathfrak{a}_s\}$ of prime ideals in $\overline{\mathfrak{R}}_d^{(p)}$ with $\bigcap_j \mathfrak{a}_j = \sqrt{\mathfrak{q}^{(1)}} = \{h \in \overline{\mathfrak{R}}_d^{(p)} : h^l \in$

$\mathfrak{q}^{(1)}$ for sufficiently large l}. Then $\bigcap_{j=1}^{s}(\mathfrak{a}_j \cap \mathfrak{R}_d^{(p)}) = \sqrt{\mathfrak{q}} = \mathfrak{q}$. By Proposition 1.1.11 in [5] there exists a $j_0 \in \{1,\ldots,s\}$ with $\mathfrak{a}_j \cap \mathfrak{R}_d^{(p)} = \mathfrak{q}$, and we set $\mathfrak{p}_1 = \mathfrak{a}_{j_0}$.

Suppose that $M \geq 1$, and that we have found prime ideals $\mathfrak{p}_m \subset \overline{\mathfrak{R}}_d^{(p,k_m)}$, $1 \leq m \leq M$, with the required properties. Let $\mathfrak{q}^{(M+1)} \subset \overline{\mathfrak{R}}_d^{(p,k_{M+1})}$ be the ideal generated by $\mathfrak{p}_m$, and repeat the argument in the preceding paragraph to find a prime ideal $\mathfrak{p}_{M+1} \subset \overline{\mathfrak{R}}_d^{(p,k_{M+1})}$ with $\mathfrak{p}_{M+1} \cap \overline{\mathfrak{R}}_d^{(p,k_M)} = \mathfrak{p}_M$. This completes the induction step and yields the promised sequence $(\mathfrak{p}_m, m \geq 1)$. The ideal $\overline{\mathfrak{q}} = \bigcup_{m \geq 1} \mathfrak{p}_m \subset \mathcal{R}_d^{(p)}$ is obviously prime, and satisfies that $\overline{\mathfrak{q}} \cap \mathfrak{R}_d^{(p)} = \mathfrak{p}_1 \cap \mathfrak{R}_d^{(p)} = \mathfrak{q}$. □

An element $h \in \mathfrak{R}_d^{(p)}$ is a *generalized polynomial in a single variable* if there exist $\mathbf{m}, \mathbf{n} \in \mathbb{Z}^d$ and a polynomial $h' \in \mathbb{F}_p[v]$ such that $\mathbf{n} \neq \mathbf{0}$ and $h(u_1,\ldots,u_d) = u^{\mathbf{m}} h'(u^{\mathbf{n}})$.

LEMMA 28.3. *Let $f \in \mathfrak{R}_d^{(p)}$ be an irreducible Laurent polynomial which is not a generalized polynomial in a single variable, and let $\mathfrak{q} = (f) = f\mathfrak{R}_d^{(p)}$. Then the prime ideal $\overline{\mathfrak{q}} \subset \mathcal{R}_d^{(p)}$ with $\overline{\mathfrak{q}} \cap \mathfrak{R}_d^{(p)} = \mathfrak{q}$ is principal (cf. Lemma 28.2).*

PROOF. Let $k \geq 1$ with $p \nmid k$, and let h be an irreducible factor of f in $\overline{\mathfrak{R}}_d^{(p,k)}$ with $c_h(\mathbf{0}) = 1$. Put

$$\Omega_k = \{\omega \in \overline{\mathbb{F}}_p : \omega^k = 1\}, \tag{28.8}$$

and define ring-automorphisms σ and θ_ω, $\omega \in \Omega_k^d$, of $\overline{\mathfrak{R}}_d^{(p,k)}$ by

$$\tau(g) = \sum_{\mathbf{n} \in \mathbb{Z}^d} c_g(\tfrac{\mathbf{n}}{k})^p u^{\frac{\mathbf{n}}{k}},$$

$$\theta_\omega(g) = \sum_{\mathbf{n} \in \mathbb{Z}^d} \omega^{\mathbf{n}} c_g(\tfrac{\mathbf{n}}{k}) u^{\frac{\mathbf{n}}{k}}, \tag{28.9}$$

for every $g = \sum_{\mathbf{n} \in \mathbb{Z}^d} c_g(\tfrac{\mathbf{n}}{k}) u^{\frac{\mathbf{n}}{k}} \in \overline{\mathfrak{R}}_d^{(p,k)}$ and $\omega = (\omega_1,\ldots,\omega_d) \in \Omega_k^d$, where $\omega^{\mathbf{n}} = \omega_1^{n_1} \cdot \ldots \cdot \omega_d^{n_d}$ for all $\mathbf{n} = (n_1,\ldots,n_d) \in \mathbb{Z}^d$. Then $g \in \overline{\mathfrak{R}}_d^{(p)} = \overline{\mathfrak{R}}_d^{(p,1)}$ if and only if $\theta_\omega(g) = g$ for every $\omega \in \Omega_k^d$, and $g \in \mathfrak{R}_d^{(p)}$ if and only if, in addition, $\tau(g) = g$.

Since $f \in \mathfrak{R}_d^{(p)}$ and h divides f, we know that $\tau^m \cdot \theta_\omega(h)$ divides f for every $\omega \in \Omega_k^d$ and $m \geq 0$. Put

$$\Phi_k(h) = \{\theta_\omega(h) : \omega \in \Omega_k^d\},$$

$$\overline{\Phi}_k(h) = \{\tau^m \cdot \theta_\omega(h) : \omega \in \Omega_k^d, m \geq 0\}, \tag{28.10}$$

denote by $h_1 = h, h_2, \ldots, h_L$ the distinct elements in $\overline{\Phi}_k(h)$, and put $h' = h_1 \cdot \ldots \cdot h_L$. Then $\tau^m \cdot \theta_\omega(h') = h'$ for all $m \geq 0$ and $\omega \in \Omega_b^d$, i.e. $h' \in \mathfrak{R}_d^{(p)}$. Since every $\phi \in \overline{\Phi}_k(h)$ has constant term 1, no two distinct elements of $\overline{\Phi}_k(h)$ can divide each other, so that h' divides f, and the irreducibility of f in $\mathfrak{R}_d^{(p)}$ implies that $h' = u^{\mathbf{m}} f$ for some $\mathbf{m} \in \mathbb{Z}^d$. For simplicity we assume that $\mathbf{m} = \mathbf{0}$, $h' = f$, and $\mathcal{C}(f) = |\overline{\Phi}_k(h^{(k)})|\mathcal{C}(h^{(k)})$.

For every $k \geq 1$ with $p \nmid k$ we have found an irreducible Laurent polynomial $h = h^{(k)} \subset \overline{\mathfrak{R}}_d^{(p,k)}$ such that f is the product of the (distinct) elements in $\overline{\Phi}_k(h^{(k)})$. Since all elements of $\overline{\Phi}_k(h^{(k)})$ have identical supports we obtain that

$$
\begin{aligned}
\mathcal{C}(f) &= |\overline{\Phi}_k(h^{(k)})|\mathcal{C}(h^{(k)}), \\
\mathfrak{o}(h^{(k)})N(f)\mathcal{C}(h^{(k)}) &= \mathcal{C}(f),
\end{aligned}
\tag{28.11}
$$

where $N(f)$ is the largest positive integer with

$$
\tfrac{1}{N(f)}\mathcal{E}(f) \subset \mathbb{Z}^d.
\tag{28.12}
$$

From (28.11) we obtain that

$$
|\overline{\Phi}_k(h^{(k)})| \leq \mathfrak{o}(h^{(k)})N(f).
\tag{28.13}
$$

We write $[F] \subset (\mathbb{Q}^{(p)})^d$ for the subgroup generated by a finite subset $F \subset (\mathbb{Q}^{(p)})^d$. Since f is not a generalized polynomial in a single variable, $[\mathcal{E}(f)] \cong \mathbb{Z}^{d'}$ for some $d' \in \{2, \ldots, d\}$. Since there are at least $\left(\mathbb{Z}^d + \tfrac{1}{\mathfrak{r}(h^{(k)})}[\mathcal{E}(f)]\right) / \mathbb{Z}^d$ distinct elements in $\Phi_k(h^{(k)})$, (28.13) implies that

$$
\left|\left(\mathbb{Z}^d + \tfrac{1}{\mathfrak{r}(h^{(k)})}[\mathcal{E}(f)]\right) / \mathbb{Z}^d\right| \leq \mathfrak{r}(h^{(k)})N(f),
\tag{28.14}
$$

where $N(f)$ is defined in (28.12). From (28.14) we obtain an upper bound

$$
K(f) = \max_{k \geq 1, \, p \nmid k} \mathfrak{o}(h^{(k)})
\tag{28.15}
$$

on the possible values of $\mathfrak{o}(h^{(k)})$, which is explicitly computable in terms of $\mathcal{C}(f)$. Furthermore, if $\mathbb{F}_{p^r}$ is the smallest field containing the coefficients of $h_f = h^{(K(f))}$, then f is divisible by $h_f \cdot \ldots \cdot \tau^r(h_f)$, and (28.13) shows that $r \leq K(f)N(f)$ and

$$
h_f \in \mathbb{F}_{p^{K(f)N(f)}}[u_1^{\pm\frac{1}{K(f)}}, \ldots, u_d^{\pm\frac{1}{K(f)}}]
\tag{28.16}
$$

By setting $\overline{\mathfrak{q}} = h_f \mathcal{R}_d^{(p)}$ we have proved the lemma. $\square$

REMARKS 28.4. (1) The assertion of Lemma 28.3 is wrong for generalized polynomials in a single variable. Indeed, $f = 1 - u_1 \in \mathfrak{R}_d^{(p)}$, and if $k \geq 1$ is not divisible by p, then $1 - u_1$ is divisible by $1 - \omega u_1^{\frac{1}{k}}$ for all $\omega \in \Omega_k$, and f has no irreducible divisors in $\mathcal{R}_d^{(p)}$.

(2) The bounds $K(f), N(f)$ in the proof of Lemma 28.3 can be computed quite easily (cf. Examples 28.5), so that the proof of Lemma 28.3 gives an explicit procedure for calculating the principal ideal $\bar{\mathfrak{q}}$.

EXAMPLES 28.5. (1) Let $f = 1 + u_1 + u_2 \in \mathfrak{R}_2^{(2)}$ (cf. Example 27.1). Then $\mathcal{S}(f) = \mathcal{E}(f) = \{(0,0), (1,0), (0,1)\}$, $N(f) = 1$, $[\mathcal{E}(f)] = \mathbb{Z}^2$, and any integer $M = \mathfrak{r}(h^{(k)}) \geq 1$ satisfying (28.14) has the property that $\left|(\mathbb{Z}^2 + \frac{1}{M}\mathbb{Z}^2) \,/\, \mathbb{Z}^2\right| \leq M$. This shows that $K(f) = N(f) = K(f)N(f) = 1$, and that f is irreducible in $\mathcal{R}_2^{(2)}$.

(2) Let $f = 1 + u_1 + u_2 + u_2^2 \in \mathfrak{R}_2^{(2)}$. Then $\mathcal{E}(f) = \{(0,0), (1,0), (0,2)\}$, $N(f) = 1$, $[\mathcal{E}(f)] = \{(m_1, 2m_2) : (m_1, m_2) \in \mathbb{Z}^2\}$, and condition (28.14) takes the form $\left|(\mathbb{Z}^2 + \frac{1}{M}[\mathcal{E}(f)]) \,/\, \mathbb{Z}^2\right| \leq M$. This shows that $M \leq 2$. Hence $K(f) = N(f) = K(f)N(f) = 1$, since $K(f)$ cannot be divisible by 2. As f is irreducible in $\mathfrak{R}_2^{(2)}$, f is also irreducible in $\mathcal{R}_2^{(2)}$.

(3) Let $f = 1 + u_1 + u_2^2 \in \mathfrak{R}_2^{(2)}$. As in Example (2) we see that $N(f) = 1$, $[\mathcal{E}(f)] = \{(m_1, 2m_2) : (m_1, m_2) \in \mathbb{Z}^2\}$, $K(f) = 1$, and that f is irreducible in $\mathcal{R}_2^{(2)}$.

(4) Let $f = 1 + u_1^2 + u_2 + u_1 u_2 + u_2^2 \in \mathfrak{R}_2^{(2)}$. Then

$$\mathcal{S}(f) = \begin{smallmatrix}\bullet\\\bullet\bullet\\\circ\ \circ\end{smallmatrix}\ ,\quad \mathcal{E}(f) = \begin{smallmatrix}\circ\ \circ\\ \\\circ\ \bullet\end{smallmatrix}$$

(cf. Example 25.2), $N(f) = 2$, $[\mathcal{E}(f)] = 2\mathbb{Z}^2$, and (28.14) implies that $K(f) = 1$.

The polynomial f is irreducible in $\mathfrak{R}_2^{(2)}$, but not in $\mathbb{F}_{2^{K(f)N(f)}}[u_1^{\pm 1}, u_2^{\pm 1}]$: if we write $\mathbb{F}_4$ as $\{0, 1, a, a^2\}$, then $f = (1 + u_1 + au_2)(1 + u_1 + a^2 u_2)$, and the polynomials $1 + u_1 + au_2, 1 + u_1 + a^2 u_2$ are the irreducible factors of f in $\mathcal{R}_2^{(2)}$.

(5) If $f = 1 + u_1 + u_1^2 + u_1^3 + u_2 + u_1 u_2 + u_1^2 u_2 + u_2^2 + u_1 u_2^2 + u_2^3 \in \mathfrak{R}_2^{(2)}$, then

$$\mathcal{S}(f) = \begin{smallmatrix}\bullet\\\bullet\ \bullet\\\bullet\ \bullet\ \bullet\\\bullet\ \bullet\ \bullet\ \bullet\end{smallmatrix}\ ,\quad \mathcal{E}(f) = \begin{smallmatrix}\bullet\\\circ\ \circ\\\circ\ \circ\ \circ\\\bullet\ \circ\ \circ\ \bullet\end{smallmatrix}\ ,$$

and $N(f) = K(f) = 3$. Although f is absolutely irreducible in $\mathfrak{R}_2^{(2)}$, it is not irreducible in $\overline{\mathfrak{R}}_2^{(2,3)}$: if $h = 1 + u_1^{\frac{1}{3}} + u_2^{\frac{1}{3}}$, then

$$f = \prod_{i=0}^{2}\prod_{j=0}^{2}(1 + a^i u_1^{\frac{1}{3}} + a^j u_2^{\frac{1}{3}}).$$

(6) Let $f = 1 + u_1^3 + u_2^3$. Then f is absolutely irreducible,

$$\mathcal{S}(f) = \mathcal{C}(f) = \begin{smallmatrix}\bullet\\\circ\ \circ\\\circ\ \circ\ \circ\\\bullet\ \circ\ \circ\ \bullet\end{smallmatrix}\ ,$$

and one can check as in the preceding examples that f is irreducible in $\mathcal{R}_2^{(2)}$. $\square$

In the notation of the Lemmas 28.2–28.3 we obtain the following corollary of Theorem 28.1.

COROLLARY 28.6. *Let* $\mathfrak{q} \subset \mathfrak{R}_d^{(p)}$ *be a non-zero prime ideal, and let* $F = \{\mathbf{n}_1, \ldots, \mathbf{n}_r\} \subset \mathbb{Z}^d$. *The following conditions are equivalent.*

(1) *There exist Laurent polynomials* $f_1, \ldots, f_r$ *in* $\mathfrak{R}_d^{(p)}$, *not all in* $\mathfrak{q}$, *such that*

$$f_1 u^{k\mathbf{n}_1} + \cdots + f_r u^{k\mathbf{n}_r} \in \mathfrak{q}$$

for infinitely many $k \geq 0$;

(2) *There exists Laurent polynomials* $g_1, \ldots, g_r$ *in* $\mathfrak{R}_d^{(p)}$, *not all in* $\mathfrak{q}$, *and an integer* $l \geq 1$ *such that*

$$g_1 u^{p^{lk}} + \cdots + g_r u^{p^{lk}} \in \mathfrak{q}$$

for every $k \geq 1$;

(3) *If* $\bar{\mathfrak{q}} \subset \mathcal{R}_d^{(p)}$ *is a prime ideal with* $\bar{\mathfrak{q}} \cap \mathfrak{R}_d^{(p)} = \mathfrak{q}$, *then there exists an element* $h \in \bar{\mathfrak{q}}$ *such that* $s\mathsf{S}(h) \subset F + \mathbf{m}$ *for some* $\mathbf{m} \in \mathbb{Z}^d$ *and some positive* $s \in \mathbb{Q}$.

PROOF. Suppose that (1) holds. Lemma 28.2 yields a prime ideal $\bar{\mathfrak{q}} \subset \mathcal{R}_d^{(p)}$ such that $\bar{\mathfrak{q}} \cap \mathfrak{R}_d^{(p)} = \mathfrak{q}$, and we denote by $\mathbb{K}$ the field of fractions of the integral domain $\mathcal{R}_d^{(p)}/\bar{\mathfrak{q}}$, write $\bar{\mathbb{K}} \supset \mathbb{K}$ for the algebraic closure of $\mathbb{K}$, set $v_j^s = u_j^s + \bar{\mathfrak{q}} \in \mathcal{R}_d^{(p)}/\bar{\mathfrak{q}} \subset \mathbb{K} \subset \bar{\mathbb{K}}$ and $x_j = v^{\mathbf{n}_j}$, $j = 1, \ldots, r$, for every $j = 1, \ldots, d$ and $s \in \mathbb{Q}^{(p)}$, and consider the $\bar{\mathbb{F}}_p$-linear ring-homomorphism $\rho' : \mathcal{R}_d^{(p)} \longmapsto \bar{\mathbb{K}}$ given by $\rho'(u^s) = v^s = v_1^{s_1} \ldots v_d^{s_d}$, $\mathbf{s} = (s_1, \ldots, s_d) \in (\mathbb{Q}^{(p)})^d$.

Our assumptions imply that there exist elements $\xi_1, \ldots, \xi_r$ in $\mathbb{K}$, not all equal to zero, such that

$$\xi_1 x_1^k + \cdots + \xi_r x_r^k = 0$$

for infinitely many $k \geq 1$, and the implication (1)$\Rightarrow$(2) in Theorem 28.1 allows us to find positive integers a, b in $\mathbb{Q}$ such that the set $\{x_1^{\frac{b}{a}}, \ldots, x_r^{\frac{b}{a}}\}$ is linearly dependent over $\bar{\mathbb{F}}_p$. One may obviously assume that $s = \frac{b}{a} \in \mathbb{Q}^{(p)}$.

We have proved that there exist elements $\omega_j \in \bar{\mathbb{F}}_p$, $j = 1, \ldots, r$, such that

$$h = \omega_1 u^{\frac{b}{a}\mathbf{n}_1} + \cdots + \omega_r u^{\frac{b}{a}\mathbf{n}_r} \in \bar{\mathfrak{q}} \cap \bar{\mathcal{R}}_d^{(p,a)},$$

which implies (3).

For the reverse implication (3)$\Rightarrow$(2)$\Rightarrow$(1) we assume for simplicity that $h \in \bar{\mathfrak{q}}$ satisfies that $s\mathsf{S}(h) \subset F$ for some positive rational s. If h is of the form $h = \omega_1 u^{\frac{1}{m}\mathbf{n}_1} + \cdots + \omega_r u^{\frac{1}{m}\mathbf{n}_r}$ with $m \geq 1$, $p \nmid m$, and with $\omega_1, \ldots, \omega_r$ in $\bar{\mathbb{F}}_p$, then the element $\omega_1 x_1^{\frac{1}{m}} + \cdots + \omega_r x_r^{\frac{1}{m}} \in \mathbb{K}$ is equal to zero, where x_i is the element in $\mathbb{K}$ corresponding to $u^{\mathbf{n}_i}$ (we are using the notation of the first paragraph in this proof). Then $\omega_1^{p^k} x_1^{p^k \frac{1}{m}} + \cdots + \omega_r^{p^k} x_r^{p^k \frac{1}{m}} = 0$ for every $k \geq 0$, and we can find an integer a and elements $\omega_j' \in \bar{\mathbb{F}}_p$ such that $p^k = a \pmod{m}$ and $\omega_j^{p^k} = \omega_j'$ for every $j = 1, \ldots, r$ and every k in an infinite arithmetic progression $K \subset \mathbb{N}$. Put $p^k = a + b_k m$ for all $k \in K$, and observe that

$$
\omega_1^{p^k} x_1^{p^k \frac{1}{m}} + \cdots + \omega_r^{p^k} x_r^{p^k \frac{1}{m}} = \omega_1' x_1^{a \frac{1}{m} + b_k} + \cdots + \omega_r' x_r^{a \frac{1}{m} + b_k}
$$
$$
= \omega_1' x_1^{\frac{a}{m}} x_1^{b_k} + \cdots + \omega_r' x_r^{\frac{a}{m}} x_r^{b_k} = 0
$$

for every $k \in K$. Hence $(\omega_1' x_1^{\frac{a}{m}}, \ldots, \omega_r' x_r^{\frac{a}{m}}) \in \mathbb{K}^r$ is orthogonal to $(x_1^{b_k}, \ldots, x_r^{b_k}) \in \mathbb{K}^r$ for every $k \in K$, and we conclude that there exists a non-zero element $(\xi_1, \ldots, \xi_r)$ in the subfield $\mathbb{K}' = \mathbb{F}_p(x_1, \ldots, x_d) \subset \mathbb{K}$ such that $\xi_1 x_1^{b_k} + \cdots + \xi_r x_r^{b_k} = 0$ for all $k \in K$.

Since $\mathbb{K}'$ is the field of fractions of the integral domain

$$
\mathbb{F}_p[u_1^{\pm \frac{1}{k}}, \ldots, u_d^{\pm \frac{1}{k}}]/\mathfrak{q}_k,
$$

where $\mathfrak{q}_k = \bar{\mathfrak{q}} \cap \mathbb{F}_p[u_1^{\pm \frac{1}{k}}, \ldots, u_d^{\pm \frac{1}{k}}]$, we can find $f_1, \ldots, f_r$ in $\mathbb{F}_p[u_1^{\pm 1}, \ldots, u_d^{\pm 1}]$ such that $f_i \notin \mathfrak{q}$ for some $i \in \{1, \ldots, r\}$, and $f_1 u^{b_k \mathbf{n}_1} + \cdots + f_r u^{b_k \mathbf{n}_r} \in \mathfrak{q}$ for every $k \in K$, so that $(3) \Rightarrow (2) \Rightarrow (1)$.

The final assertion is an immediate consequence of Lemma 28.3. $\square$

The following theorem summarizes the dynamical consequences of (27.5), Theorem 28.1, and Corollary 28.6. For notation we refer to the paragraph preceding Lemma 28.2.

THEOREM 28.7. *Let* $d \geq 2$, p *a rational prime,* $\mathfrak{q} \subset \mathfrak{R}_d^{(p)}$ *a non-zero prime ideal, and let* $\bar{\mathfrak{q}} \subset \mathcal{R}_d^{(p)} \supset \mathfrak{R}_d^{(p)}$ *be the prime ideal defined in* Lemma 28.2 *with* $\mathfrak{q} = \bar{\mathfrak{q}} \cap \mathfrak{R}_d^{(p)}$. *The following conditions are equivalent for every non-empty, finite set* $F \subset \mathbb{Z}^d$.

(1) F *is non-mixing for* $\alpha = \alpha^{\mathfrak{R}_d^{(p)}/\mathfrak{q}}$;
(2) $F \supset k S^*(h) + \mathbf{m}$ *for some* $h \in \bar{\mathfrak{q}}$, $k \geq 1$, *and* $\mathbf{m} \in \mathbb{Z}^d$.

In Example 27.1 we saw that, for every non-zero element $h \subset \mathfrak{q}$, the set $S(h)$ is non-mixing for $\alpha^{\mathfrak{R}_d^{(p)}/\mathfrak{q}}$. The main point of Theorem 28.7 is that, in addition to the supports of Laurent polynomials in $\mathfrak{q}$, other, more unexpected, non-mixing sets can arise, which do not contain the support of any non-zero Laurent polynomial in $\mathfrak{q}$. However, before discussing this phenomenon further we introduce two definitions.

DEFINITION 28.8. (1) A non-empty, finite set $F \subset \mathbb{Z}^d$ is an *extremal non-mixing set* of a measure preserving $\mathbb{Z}^d$-action T on a probability space $(Y, \mathfrak{T}, \mu)$ if the following conditions are satisfied.

(i) F is non-mixing for T;
(ii) Every subset $F' \subsetneq F$ is mixing for T;

(iii) If $\mathcal{C}(F) \subset \mathbb{R}^d$ is the convex hull of F, then every subset $F' \subset \mathcal{C}(F)$ with $\mathcal{C}(F') \neq \mathcal{C}(F)$ is mixing for T.

(2) A non-empty, finite set $F \subset \mathbb{Z}^d$ is a *minimal non-mixing set* of T if it satisfies the conditions (i) and (ii) in part (1) of this definition, and if

(iv) $\frac{1}{m}(F + \mathbf{m}) \not\subset \mathbb{Z}^d$ for every $m \geq 1$ and $\mathbf{m} \in \mathbb{Z}^d$.

(3) A non-empty, finite set $F \subset \mathbb{Z}^d$ is *incontractible* if $F - F = \{\mathbf{m} - \mathbf{m}' : \mathbf{m}, \mathbf{m}' \in F\}$ generates a primitive subgroup of $\mathbb{Z}^d$, and *contractible* otherwise.

Every extremal non-mixing set is also a minimal non-mixing set. Furthermore, if F is a non-mixing set of T, then there exists a minimal non-mixing set $F' \subset \mathbb{Z}^d$ such that $F \supset mF' + \mathbf{m}$ for some $m \geq 1$ and $\mathbf{m} \in \mathbb{Z}^d$. It may happen, however, that the set F' cannot be chosen to be extremal non-mixing (cf. Example 28.10 (7)). On the other hand, if F' is a minimal non-mixing set, then every finite set containing $aF' + \mathbf{m}$ for some $\mathbf{m} \in \mathbb{Z}^d$ and $a \geq 1$ is non-mixing. The need for introducing minimal non-mixing sets is due to the fact that the supersets of non-mixing sets are again non-mixing, and convey no further information about the $\mathbb{Z}^d$-action T. The reason for Definition 28.8 (1) will become clear from Proposition 28.9 and Examples 28.10.

PROPOSITION 28.9. *Let p be a rational prime, and let $f \in R_d^{(p)}$ be a non-zero, irreducible Laurent polynomial which is not a generalized polynomial in a single variable. Suppose that $\bar{\mathfrak{q}} = h_f \mathcal{R}_d^{(p)} \subset \mathcal{R}_d^{(p)}$ is a principal ideal with $\bar{\mathfrak{q}} \cap \mathfrak{R}_d^{(p)} = (f) = f\mathfrak{R}_d^{(p)}$ (Lemmas 28.2–28.3). Then $\mathbb{S}^*(h_f)$ is an extremal non-mixing set of the $\mathbb{Z}^d$-action $\alpha = \alpha^{\mathfrak{R}_d^{(p)}/(h_f)}$. If the set $\mathbb{S}^*(h_f)$ is incontractible (Definition 28.8 (3)), or if f is irreducible in $\mathcal{R}_d^{(p)}$, then $\mathbb{S}^*(h_f)$ is the only extremal non-mixing set of α. Otherwise there may be other extremal non-mixing sets, each of which is a translate of a set $F^{(k)} = \mathbb{S}^*(h^{(k)})$ for some irreducible factor $h^{(k)}$ of f in $\overline{\mathfrak{R}}_d^{(p,k)}$, where k is a divisor of $\mathfrak{o}(h_f)$.*

PROOF. Choose, for every $k \geq 1$ with $p \nmid k$, an irreducible divisor $h^{(k)}$ of f in $\overline{\mathfrak{R}}_d^{(p,k)}$ which is divisible by h_f in $\mathcal{R}_d^{(p)}$. Define $\mathfrak{r}(h_f)$ by (28.7), and assume without loss in generality that $c_{h_f}(\mathbf{0}) = 1$ and hence $h_f \in \overline{\mathfrak{R}}_d^{(p,\mathfrak{o}(h_f))}$. If $\bar{\Phi}_k(h_f), \Phi_k(h_f)$ are given by (28.10), then $|\Phi(h_f)| = 1$ for every k which does not divide $\mathfrak{r}(h_f)$, and

$$h_f^{(k)} = \prod_{\boldsymbol{\omega} = (\omega_1, \ldots, \omega_d) \in \Omega_a} h_f\left(\omega_1 u_1^{\frac{1}{\mathfrak{r}(h_f)}}, \ldots, \omega_d u_d^{\frac{1}{\mathfrak{r}(h_f)}}\right) \qquad (28.17)$$

whenever $\mathfrak{r}(h_f) = ak$ for some $a \geq 2$, $p \nmid a$. From the definition of $\mathfrak{r}(h_f)$ and Φ_a

it is clear that $|\Phi_a(h_f)| \geq a$, and that

$$
\begin{aligned}
\mathcal{C}^*(h_f^{(k)}) &\supsetneq \mathcal{C}^*(h_f) \text{ if } |\Phi_a(h_f)| > a, \\
\mathcal{C}^*(h_f^{(k)}) &= \mathcal{C}^*(h_f) \text{ if } |\Phi_a(h_f)| = a.
\end{aligned}
\tag{28.18}
$$

Suppose that $|\Phi_a(h_f)| = a$ for some divisor $a > 1$ of $\mathfrak{r}(h_f)$ with $\mathfrak{r}(h_f) = ak$. We assume without loss in generality that $\mathbf{0}$ is an extreme point of $\mathcal{S}(h_f)$. By looking at the various edges of the solid polytope $\mathcal{S}(h_f) \subset \mathbb{R}^d$ emanating from $\mathbf{0}$ we see that at most one of these edges can contain an element of $\mathcal{S}(h_f) \setminus \frac{a}{\mathfrak{r}(h_f)}\mathbb{Z}^d$ (otherwise $|\Phi_a(h_f)| > a$). Suppose that $\mathbf{v} \in \mathbb{Z}^d$ is a primitive element pointing in the direction of one of the edges of $\mathcal{C}(h_f)$ which start at $\mathbf{0}$ and contain only elements of $\mathcal{S}(h_f) \cap \frac{a}{\mathfrak{r}(h_f)}\mathbb{Z}\mathbf{v}$, and let b be the largest multiple of a such that $\mathcal{S}(h_f) \cap \frac{a}{\mathfrak{o}(h_f)}\mathbb{Z}\mathbf{v} \subset \frac{b}{\mathfrak{r}(h_f)}\mathbb{Z}\mathbf{v}$.

We regard the Laurent polynomial

$$
\psi = \sum_{\mathbf{r} \in \mathcal{S}(h_f) \cap \frac{a}{\mathfrak{r}(h_f)}\mathbb{Z}\mathbf{v}} c_{h_f}(\mathbf{r})u^{\mathbf{r}} = \sum_{\mathbf{r} \in \mathcal{S}(h_f) \cap \frac{b}{\mathfrak{r}(h_f)}\mathbb{Z}\mathbf{v}} c_{h_f}(\mathbf{r})u^{\mathbf{r}}
$$

as a polynomial ψ' in the variable $w = u^{\frac{b}{\mathfrak{r}(h_f)}\mathbf{v}}$. Suppose that the a-th power ψ'^a of ψ' can be written as a polynomial in the variable w^a. Then $\psi'^a(\omega w)^a = \psi'^a(\omega w) = \psi'^a(w)$ for every $\omega \in \Omega_a$, and by varying $\omega \in \Omega_a$ we conclude that $\psi'(\omega w) = \psi'(\omega)$ for all $\omega \in \Omega_a$. Hence ψ' can be written as a polynomial in the variable w^a, which is equivalent to saying that ψ is a polynomial in the variable $u^{\frac{ab}{\mathfrak{o}(h_f)}\mathbf{v}}$, contrary to our choice of b. It follows that ψ' cannot be written as a polynomial in w^a, and that

$$
\mathcal{S}(\psi^a) \not\subset \frac{ab}{\mathfrak{r}(h_f)}\mathbb{Z}\mathbf{v}.
$$

As

$$
\psi^a = \sum_{\mathbf{r} \in \mathcal{S}(h_f^{(k)}) \cap \frac{a}{\mathfrak{r}(h_f)}\mathbb{Z}\mathbf{v}} c_{h_f}(\mathbf{r})u^{\mathbf{r}},
$$

we conclude that

$$
\mathcal{S}^*(h_f^{(k)}) \cap \mathbb{Z}\mathbf{v} = \frac{\mathfrak{o}(h_f)}{a}\mathcal{S}(h_f^{(k)}) \cap \mathbb{Z}\mathbf{v} \not\subset \mathfrak{r}(h_f)\mathcal{S}(h_f) \cap \mathbb{Z}\mathbf{v} = \mathcal{S}^*(h_f) \cap \mathbb{Z}\mathbf{v},
$$

and hence that

$$
\mathcal{S}^*(h_f^{(k)}) \not\subset \mathcal{S}^*(h_f).
\tag{28.19}
$$

The set $F = \mathcal{S}(h_f)$ is non-mixing by Theorem 28.7. If F' is an extremal non-mixing set, then Theorem 28.7 allows us to find an element $h' = gh_f \in \bar{\mathfrak{q}} = h_f \mathcal{R}_d^{(p)}$ with $g \in \mathcal{R}_d^{(p)}$ and $\mathcal{S}^*(h') \subset F'$. If $\mathfrak{o}(g), \mathfrak{r}(h_f), \mathfrak{r}(h')$ are defined by (28.7), then we may assume without loss in generality that $g \in \overline{\mathfrak{R}}_d^{(p,\mathfrak{r}(g))}$, $h' \in \overline{\mathfrak{R}}_d^{(p,\mathfrak{r}(h'))}$, and the extremality of F' is easily seen to imply that $\mathfrak{o}(g) = \mathfrak{r}(h_f) = a\mathfrak{r}(h')$ for

some $a \geq 2$ with $p \nmid a$. Then $h'(\omega_1 u_1^{\frac{1}{\mathfrak{r}(h_f)}}, \ldots, \omega_d u_d^{\frac{1}{\mathfrak{o}(h_f)}}) = h'(u_1^{\frac{1}{\mathfrak{r}(h_f)}}, \ldots, u_d^{\frac{1}{\mathfrak{o}(h_f)}})$
for every $(\omega_1, \ldots, \omega_d) \in \Omega_a^d$ (cf. (28.8)). In particular h' must be divisible by
$h_f^{(k)}$ (cf. (28.17)), and Theorem 28.7 and the extremality of F' allow us to
assume without loss in generality that $h' = h_f^{(k)}$. From the second equation in
(28.18) we see that $\mathcal{C}(F') = \mathcal{C}^*(h') = \mathcal{C}^*(h_f^{(k)}) = \mathcal{C}^*(h_f) = \mathcal{C}(F)$. If $F' \neq F$,
then (28.19) implies that $F' = \mathcal{S}^*(h_f^{(k)}) \not\subset \mathcal{S}^*(h_f) = F$, so that F is also an
extremal non-mixing set of α. We conclude that every extremal non-mixing set
F' of α must be a translate of $\mathcal{S}^*(h_f^{(k)})$ for some $k \geq 1$ which divides $\mathfrak{r}(h_f)$.

If f is irreducible in $\mathcal{R}_d^{(p)}$, then $h_f = f$, $\mathfrak{o}(h_f) = 1$, and $F = \mathcal{S}^*(f) = \mathcal{S}^*(h_f)$
is the only extremal non-mixing set of α. If $F = \mathcal{S}^*(h_f)$ is incontractible and
$c_{h_f}(\mathbf{0}) = 1$, then one checks easily that $|\Phi_a(h_f)| > a$ for every divisor a of
$\mathfrak{r}(h_f)$, and (28.18) and the argument in the preceding paragraph imply that F
is again the only extremal non-mixing set. $\square$

EXAMPLES 28.10. (1) Let $p = d = 2$. The set $F = \mathcal{S}^*(h_f) = \{(0,0), (1,0),$
$(0,1)\}$ is the unique extremal non-mixing for $\alpha = \alpha^{\mathcal{R}_2^{(2)}/(f)}$ for each of the
polynomials f in Example 28.5 (1) and (4)–(6), since F is incontractible.

(2) If $f = 1 + u_1 + u_2 + u_2^2 \in \mathcal{R}_2^{(2)}$ in Example 28.5 (2), then f is irreducible
in $\mathcal{R}_2^{(2)}$, and $F = \mathcal{S}(f) = \{(0,0), (1,0), (0,1), (0,2)\}$ is the unique extremal non-
mixing set of $\alpha = \alpha^{\mathcal{R}_2^{(2)}/(f)} = \sigma^{F_6}$ (cf. Example 25.2).

(3) If $f = 1 + u_1 + u_2^2 \in \mathcal{R}_2^{(2)}$ in Example 28.5 (3), then f is irreducible in
$\mathcal{R}_2^{(2)}$, and $F = \{(0,0), (1,0), (0,2)\}$ is extremal non-mixing for $\alpha = \alpha^{\mathcal{R}_2^{(2)}/(f)} =$
σ^{F_7} (cf. Example 25.2). Note, however, that F is contractible. Since F is mixing
for the $\mathbb{Z}^2$-action arising from $f = 1 + u_1 + u_2 + u_2^2 \in \mathcal{R}_2^{(2)}$ in Example (2), we
can finally distinguish between the $\mathbb{Z}^2$-actions σ^{F_6} and σ^{F_7} in Example 25.2,
since they do not have the same non-mixing sets.

(4) The polynomial $f = 1 + u_1 + u_1^3 + u_1^5 + u_1^6 + u_2 \in \mathcal{R}_2^{(2)}$ is irreducible,
and $K(f) = 3, N(f) = 1$ (cf. Lemma 28.3 and Remark 28.4). If $h_f(u_1^{\frac{1}{3}}, u_2^{\frac{1}{3}}) =$
$1 + u_1 + u_1^2 + u_2^{\frac{1}{3}} \in \overline{\mathcal{R}}_2^{(2,3)}$, then

$$f = \prod_{\omega \in \Omega_3} h_f(u_1^{\frac{1}{3}}, \omega u_2^{\frac{1}{3}}),$$

and Lemma 28.3 implies that h_f is irreducible in $\mathcal{R}_2^{(2)}$. Proposition 28.9 shows
that

$$F = \mathcal{S}^*(h_f) = {\;\bullet\;\atop\;\bullet\;}\circ\circ\bullet\circ\circ\bullet$$

is an extremal non-mixing set for $\alpha = \alpha^{\mathcal{R}_2^{(2)}/(f)}$ (cf. Example 25.2). However,

F is contractible, and

$$F^{(1)} = \mathcal{S}^*(h_f^{(1)}) = \mathcal{S}(f) = \vdots \bullet\, \bullet\, \circ\, \bullet\, \circ\, \bullet\, \bullet$$

is the second candidate for an extremal non-mixing set for α. As $F^{(1)} \supset F$, F is again the only extremal non-mixing set of α (cf. Proposition 28.9).

(5) Let $f = 1 + u_1 + u_1^2 + u_1^5 + u_1^6 + u_1^7 + u_1^9 + u_2 \in \mathfrak{R}_2^{(2)}$. Then $N(f) = 1, K(f) = 3$, and

$$f = \prod_{\omega \in \Omega_3} h_f(u_1^{\frac{1}{3}}, \omega u_2^{\frac{1}{3}}),$$

where

$$h_f(u_1^{\frac{1}{3}}, u_2^{\frac{1}{3}}) = 1 + u_1 + u_1^3 + u_2^{\frac{1}{3}} \in \overline{\mathfrak{R}}_2^{(2,3)}$$

is irreducible in $\mathcal{R}_d^{(p)}$. From Proposition 28.9 we know that

$$F = \mathcal{S}^*(h_f) = \vdots\, \bullet\, \circ\, \bullet\, \circ\, \circ\, \circ\, \circ\, \circ\, \bullet \quad \text{and} \quad F^{(1)} = \mathcal{S}^*(h_f^{(1)}) = \vdots\, \bullet\, \bullet\, \circ\, \circ\, \bullet\, \bullet\, \bullet\, \circ\, \bullet$$

are the two candidates for extremal non-mixing sets of $\alpha = \alpha^{\mathfrak{R}_2^{(2)}/(f)}$; from Proposition 28.9 we know that F must be extremal non-mixing, and by comparing $F^{(1)}$ with F we see that $F^{(1)}$ is a second extremal non-mixing set of α.

(6) Let $f = 1 + u_1 + u_1^2 + u_1^3 + u_1 u_2 + u_2^2 \in \mathfrak{R}_2^{(2)}$. Then $N(f) = 1, K(f) = 3$, and

$$f = \prod_{\omega \in \Omega_3} h_f(u_1^{\frac{1}{3}}, \omega u_2^{\frac{1}{3}}),$$

where $h_f = 1 + u_1 + u_2^{\frac{1}{3}} + u_2^{\frac{2}{3}} \in \overline{\mathfrak{R}}_2^{(2,3)}$ is irreducible in $\mathcal{R}_2^{(2)}$. As in Example (5) we obtain two different extremal non-mixing sets of $\alpha = \alpha^{\mathfrak{R}_2^{(2)}/(f)}$:

$$F = \mathcal{S}^*(h_f) = \vdots\, \begin{smallmatrix}\bullet\\\circ\\\circ\end{smallmatrix}\, \bullet, \quad \text{and} \quad F^{(1)} = \mathcal{S}(f) = \begin{smallmatrix}\circ\\\bullet\end{smallmatrix}\, \vdots\, \bullet\, \bullet.$$

(7) If $f \in \mathfrak{R}_d^{(p)}$ is a non-zero, irreducible Laurent polynomial which is not a generalized polynomial in a single variable, then the minimal non-mixing sets of $\alpha^{\mathfrak{R}_d^{(p)}/(f)}$ (Definition 28.8 (2)) appear to bear little resemblance with the supports of the Laurent polynomials $h_f^{(k)}$ appearing in Proposition 28.9. Consider the polynomial $f = 1 + u_1 + u_2 \in \mathfrak{R}_2^{(2)}$. From Example (1) we know that $F = \mathcal{S}(f) = \mathcal{S}(h_f) = \{(0,0), (1,0), (0,1)\}$ is the only extremal non-mixing set of $\alpha = \alpha^{\mathfrak{R}_2^{(2)}/(f)}$ in the sense of Definition 28.8 (1). However, since $g = (1 + u_1 + u_1)(1 + u_1) = 1 + u_1^2 + u_2 + u_1 u_2 \in (f)$, the set

$$F' = \mathcal{S}(g) = \begin{smallmatrix}\bullet&\bullet\\\circ&\bullet\end{smallmatrix}\, \bullet$$

is non-mixing by Example 27.1 (2), and one checks easily that every proper subset of F' is mixing. Hence F' satisfies Definition 28.8 (2), but is obviously not extremal non-mixing. $\boxdot$

EXAMPLE 28.11. In order to illustrate possible applications of Theorem 28.7 and Proposition 28.9 we indicate how to find all irreducible polynomials $f \subset \mathfrak{R}_2^{(2)}$ such that $F = \{(0,0),(1,0),(0,1)\} = \mathcal{S}(1+u_1+u_2) \subset \mathbb{Z}^2$ is the unique extremal non-mixing set for the $\mathbb{Z}^2$-action $\alpha = \alpha^{\mathfrak{R}_2^{(2)}/(f)}$. In order to find the subsets of $\mathbb{Z}^2$ arising as supports of such polynomials $f \in \mathfrak{R}_2^{(2)}$ we choose, for every $k > 1$, an element $a_k \in \mathbb{F}_{2^k}$ such that $\mathbb{F}_{2^k} = \{0, 1 = a_k^0, a_k, a_k^2, \ldots, a_k^{2^k-2}\}$, and consider all polynomials in $\mathbb{F}_{2^k}[u_1, u_2]$ of the form $h_k^{i,j} = 1 + a_k^i u_1 + a_k^j u_2$ with $i, j \geq 0$. By letting the Galois group act on $h_k^{i,j}$ we obtain irreducible polynomials $f_k^{i,j} = h_k^{i,j} \cdot h_k^{2i,2j} \ldots h_k^{2^l i, 2^l j} \in \mathfrak{R}_2^{(2)}$, where $l \geq 0$ is the smallest integer such that $h_k^{2^{l+1} i, 2^{l+1} j} = h_k^{i,j}$, and we note that the set F is extremal non-mixing for $\alpha^{\mathfrak{R}_2^{(2)}/(f_k^{i,j})}$ by Proposition 28.9. For the following list of all such irreducible polynomials $f \in \mathfrak{R}_2^{(2)}$ with degree ≤ 4 in each of the variables u_1, u_2 we set $\mathbb{F}_4 = \mathbb{F}_2[x]/(1+x+x^2)$, $\mathbb{F}_8 = \mathbb{F}_2[x]/(1+x+x^3)$, and $\mathbb{F}_{16} = \mathbb{F}_2[x]/(1+x+x^4)$, put $a_k = x$ for $k = 2, 3, 4$, and use the graphical representation of Example 25.2.

Let us begin with the polynomials of degree ≤ 2.

$$F = F_1^{0,0} = F_2^{0,0} = \ \vcenter{\hbox{$\bullet$}\hbox{$\bullet\ \bullet$}} \ ,$$

$$F_2^{1,0} = \ \vcenter{\hbox{$\bullet$}\hbox{$\circ\ \bullet$}\hbox{$\bullet\ \bullet$}} \ , \qquad F_2^{0,1} = \ \vcenter{\hbox{$\bullet$}\hbox{$\bullet\ \bullet$}\hbox{$\bullet\ \circ\ \bullet$}} \ , \qquad F_2^{1,1} = \ \vcenter{\hbox{$\bullet$}\hbox{$\bullet\ \circ$}\hbox{$\bullet\ \bullet$}} \ , \qquad F_2^{1,2} = \ \vcenter{\hbox{$\bullet$}\hbox{$\bullet\ \bullet$}\hbox{$\bullet\ \bullet$}} \ .$$

Note that $F_2^{0,1}$ appeared in the Examples 28.5 (4) and 28.10 (1). There are 18 irreducible polynomials $f \in \mathfrak{R}_2^{(2)}$ of degree 3 such that F is non-mixing for $\alpha^{\mathfrak{R}_2^{(2)}/(f)}$. The supports of these polynomials are as follows.

$$F_3^{1,0} = \ \vcenter{\hbox{$\bullet\ \circ$}\hbox{$\bullet\ \circ\ \bullet$}\hbox{$\bullet\ \circ\ \bullet\ \bullet$}} \ , \qquad F_3^{3,0} = \ \vcenter{\hbox{$\bullet$}\hbox{$\bullet\ \circ\ \circ$}\hbox{$\bullet\ \circ\ \bullet$}} \ , \qquad F_3^{0,1} = \ \vcenter{\hbox{$\bullet$}\hbox{$\bullet\ \bullet$}\hbox{$\circ\ \circ\ \circ$}\hbox{$\bullet\ \bullet\ \bullet$}} \ , \qquad F_3^{1,1} = \ \vcenter{\hbox{$\bullet$}\hbox{$\bullet\ \bullet$}\hbox{$\circ\ \circ\ \bullet$}\hbox{$\circ\ \bullet\ \bullet$}} \ ,$$

$$F_3^{2,1} = \ \vcenter{\hbox{$\bullet\ \circ$}\hbox{$\circ\ \bullet\ \bullet$}\hbox{$\bullet\ \bullet\ \bullet$}} \ , \qquad F_3^{3,1} = \ \vcenter{\hbox{$\bullet\ \circ$}\hbox{$\circ\ \circ\ \bullet$}\hbox{$\bullet\ \bullet\ \bullet$}} \ , \qquad F_3^{4,1} = \ \vcenter{\hbox{$\bullet$}\hbox{$\bullet\ \bullet\ \circ$}\hbox{$\circ\ \bullet\ \bullet$}} \ , \qquad F_3^{5,1} = \ \vcenter{\hbox{$\bullet\ \circ$}\hbox{$\circ\ \bullet\ \bullet$}\hbox{$\bullet\ \bullet\ \circ\ \bullet$}} \ ,$$

$$F_3^{6,1} = \ \vcenter{\hbox{$\bullet$}\hbox{$\bullet\ \bullet\ \circ$}\hbox{$\circ\ \bullet\ \circ\ \bullet$}} \ , \qquad F_3^{0,3} = \ \vcenter{\hbox{$\bullet$}\hbox{$\circ\ \circ$}\hbox{$\bullet\ \circ\ \bullet$}\hbox{$\bullet\ \bullet\ \bullet$}} \ , \qquad F_3^{1,3} = \ \vcenter{\hbox{$\bullet$}\hbox{$\circ\ \bullet$}\hbox{$\bullet\ \circ\ \circ$}\hbox{$\circ\ \bullet\ \bullet$}} \ , \qquad F_3^{2,3} = \ \vcenter{\hbox{$\bullet$}\hbox{$\circ\ \bullet$}\hbox{$\bullet\ \bullet\ \circ$}\hbox{$\circ\ \bullet\ \bullet$}} \ ,$$

$$F_3^{3,3} = \begin{smallmatrix}\bullet\bullet\\\circ\bullet\\\bullet\circ\bullet\\\bullet\bullet\bullet\end{smallmatrix} \;,\quad F_3^{4,3} = \begin{smallmatrix}\bullet\\\circ\circ\\\bullet\bullet\circ\\\bullet\bullet\bullet\end{smallmatrix} \;,\quad F_3^{5,3} = \begin{smallmatrix}\bullet\\\circ\circ\\\bullet\bullet\bullet\\\bullet\bullet\circ\end{smallmatrix} \;,\quad F_3^{6,3} = \begin{smallmatrix}\bullet\\\circ\bullet\\\bullet\bullet\circ\\\bullet\bullet\bullet\end{smallmatrix} \;,$$

$$F_3 = \begin{smallmatrix}\bullet\\\bullet\bullet\\\bullet\bullet\bullet\\\bullet\bullet\bullet\bullet\end{smallmatrix} \;,\quad F_1^{0,0} = \begin{smallmatrix}\bullet\\\circ\circ\\\circ\circ\circ\\\circ\circ\bullet\end{smallmatrix} \;.$$

The polynomials f_3 and $\phi_3(f_1^{0,0})$ in $\mathfrak{R}_2^{(2)}$ with $\mathcal{F}(f_3) = F_3$ and $\mathcal{F}(\phi_3(f_1^{0,0})) = 3F_1^{0,0}$ were discussed in the Examples 28.5 (5)–(6) and 28.10 (1).

Similarly one can find the supports of all 54 irreducible polynomials $f \in \mathfrak{R}_2^{(2)}$ of degree 4 such that $F = \{(0,0),(1,0),(0,1)\}$ is non-mixing for $\alpha^{\mathfrak{R}_2^{(2)}/(f)}$. We begin by listing the supports of those polynomials which are symmetric in u_1 and u_2. These supports are unaffected if the coordinates are interchanged.

$$F_4^{1,1} = \quad,\quad F_4^{4,1} = \quad,\quad F_4^{3,3} = \quad,$$

$$F_4^{12,3} = \quad,\quad F_4^{7,7} = \quad,\quad F_4^{13,7} = \quad.$$

The supports of the remaining polynomials are written in pairs of the form $F_4^{i,j} = \bar{F}_4^{i',j'}$, where $\bar{F}_4^{i',j'}$ is the set obtained by switching the two coordinates of $F_4^{i',j'}$.

$$F_4^{1,0} = \bar{F}_4^{0,1} = \quad,\qquad F_4^{3,0} = \bar{F}_4^{0,3} = \quad,$$

$$F_4^{7,0} = \bar{F}_4^{0,7} = \quad,\qquad F_4^{2,1} = \bar{F}_4^{8,1} = \quad,$$

$$F_4^{3,1} = \bar{F}_4^{1,3} = \quad,\qquad F_4^{5,1} = \bar{F}_4^{1,5} = \quad,$$

$$F_4^{6,1} = \bar{F}_4^{8,3} = \quad,\qquad F_4^{7,1} = \bar{F}_4^{1,7} = \quad,$$

$$F_4^{9,1} = \bar{F}_4^{2,3} = \quad,\qquad F_4^{10,1} = \bar{F}_4^{2,5} = \quad,$$

$$F_4^{11,1} = \bar{F}_4^{2,7} = \quad , \qquad F_4^{12,1} = \bar{F}_4^{4,3} = \quad ,$$

$$F_4^{13,1} = \bar{F}_4^{4,7} = \quad , \qquad F_4^{14,1} = \bar{F}_4^{8,7} = \quad ,$$

$$F_4^{5,3} = \bar{F}_4^{3,5} = \quad , \qquad F_4^{6,3} = \bar{F}_4^{9,3} = \quad ,$$

$$F_4^{7,3} = \bar{F}_4^{3,7} = \quad , \qquad F_4^{10,3} = \bar{F}_4^{6,5} = \quad ,$$

$$F_4^{11,3} = \bar{F}_4^{6,7} = \quad , \qquad F_4^{13,3} = \bar{F}_4^{12,7} = \quad ,$$

$$F_4^{14,3} = \bar{F}_4^{9,7} = \quad , \qquad F_4^{7,5} = \bar{F}_4^{5,7} = \quad ,$$

$$F_4^{11,5} = \bar{F}_4^{10,7} = \quad , \qquad F_4^{11,7} = \bar{F}_4^{14,7} = \quad . \quad \square$$

Concluding Remarks 28.12. (1) Most of the material in this section comes from [47]. Masser's (unpublished) Theorem 28.1 appeared in [70]; a proof following [70] very closely can be found in [47]. A more systematic treatment of the factorization of (Laurent) polynomials $f \in \mathfrak{R}_d^{(p)}$ in $\mathcal{R}_d^{(p)}$ can be found in [80].

(2) A solution to the problem of determining all minimal non-mixing sets for an expansive $\mathbb{Z}^d$-action α on a compact, zero-dimensional, abelian group X seems currently out of reach, even if one imposes very restrictive conditions on the module $\mathfrak{M} = \hat{X}$ (like assuming that $\mathfrak{M} = \mathfrak{R}_d^{(p)}/\mathfrak{p}$ for some principal prime ideal $\mathfrak{p} \subset \mathfrak{R}_d^{(p)}$). The unresolved difficulty in finding all minimal non-mixing sets for $\alpha^{\mathfrak{R}_d^{(p)}/\mathfrak{p}}$, where $\mathfrak{p} \subset \mathfrak{R}_d^{(p)}$ is a prime ideal, is a purely algebraic one: it amounts to finding all Laurent polynomials in $\mathfrak{p}$ with the minimal supports (with respect to inclusion). Example 28.10 (6) indicates that this problem is non-trivial. However, as Example 28.11 shows, Theorem 28.7 *can* be used to determine—at least in principle—all irreducible Laurent polynomials $f \in \mathfrak{R}_d^{(p)}$ such that a given (small) non-empty set $S \subset \mathbb{Z}^d$ is extremal non-mixing for $\alpha^{\mathfrak{R}_d^{(p)}/(f)}$. If f is a generalized polynomial in a single variable of the form

$f = u^{\mathbf{m}}g(u^{\mathbf{n}})$ with $g \in \mathbb{Z}[v^{\pm 1}]$ and $\mathbf{m}, \mathbf{n} \in \mathbb{Z}^d$, then we may assume without loss in generality that $\mathbf{n}$ is primitive, and a set $S \subset \mathbb{Z}^d$ is non-mixing if and only if it contains the support of a Laurent polynomial g which is divisible by $u^{\mathbf{n}} - \omega$ for some $\omega \in \overline{\mathbb{F}}_p$. In particular, the only extremal non-mixing subset of $\alpha^{\mathfrak{R}_d^{(p)}/(f)}$ (up to translation) is $\{\mathbf{0}, \mathbf{n}\}$. If f is not a generalized polynomial in a single variable, we first have to determine the sets $\Phi(f), \overline{\Phi(f)}$ in (28.10), which is a finite problem in view of the bounds $K(f), N(f)$ appearing in the proof of Lemma 28.3 and Remark 28.4 (2). All elements $h \in \overline{\Phi(f)}$ have identical supports, and $S^*(h)$ is an extremal non-mixing set of $\alpha^{\mathfrak{R}_d^{(p)}/(f)}$ by Proposition 28.9.

(3) If $\mathfrak{q} \subset \mathfrak{R}_d^{(p)}$ is a non-zero prime ideal, the problem of determining the precise order of mixing of the $\mathbb{Z}^d$-action $\alpha = \alpha^{\mathfrak{R}_d^{(p)}/\mathfrak{q}}$ is of considerable interest, both dynamically and algebraically. Its algebraic solution would require some form of Proposition 27.8 for fields with positive characteristic, bearing in mind the non-trivial solutions of equation (27.6) arising from Masser's Theorem 28.1. As an example, consider the $\mathbb{Z}^2$-action $\alpha = \alpha^{\mathfrak{R}_2^{(2)}/(1+u_1+u_2+u_2^2)}$ in Example 28.5 (2). It is clear that α is mixing (Theorem 6.5 (2)), and that $S(f)$ is the unique extremal non-mixing set of α (Proposition 28.9 and Example 28.10 (2)). One can prove that every non-mixing set F of α has cardinality ≥ 4. Is α 3-mixing?

CHAPTER IX

Rigidity

29. Almost minimal $\mathbb{Z}^d$-actions and invariant measures

An ergodic $\mathbb{Z}^d$-action α by automorphisms of a compact, abelian group X usually has many distinct non-atomic, invariant, ergodic probability measures. For example, if $d \geq 1$, and if α is the shift-action (2.1) of $\mathbb{Z}^d$ on $X = G^{\mathbb{Z}^2}$, where G is a nontrivial, compact, abelian group, then there exist uncountably many distinct, α-invariant, mixing probability measures on X: for every probability measure ν on G, $\mu = \nu^{\mathbb{Z}^d}$ is an α-invariant probability measure which is mixing of every order, and different measures ν, ν' lead to inequivalent probability measures $\mu = \nu^{\mathbb{Z}^d}$ and $\mu' = \nu'^{\mathbb{Z}^d}$. There exist, however, $\mathbb{Z}^d$-actions by automorphisms of compact, abelian groups for which Haar measure is the only non-atomic, invariant, mixing probability measure, or the only invariant and ergodic probability measure on X with respect to which any $\alpha_{\mathbf{n}}$ has positive entropy (cf. Example 29.6 (1) and [89]).

The study of invariant measures for $\mathbb{Z}^d$-actions by automorphisms of compact, abelian groups can be simplified to some extent by excluding the Haar measures of closed, proper, invariant subgroups, which may confuse the picture. We introduce the following definition.

DEFINITION 29.1. A $\mathbb{Z}^d$-action α by automorphisms of an infinite, compact, abelian group X is *almost minimal* if every closed, α-invariant subgroup $Y \subsetneq X$ is finite.

For the terminology in the following theorem we refer to (6.2), Definition 6.16, Proposition 8.2 and Remark 8.4 (2).

THEOREM 29.2. *Let α be an ergodic $\mathbb{Z}^d$-action by automorphisms of a compact, abelian group X. Then α is almost minimal if and only if there exist an ergodic prime ideal $\mathfrak{p} \subset \mathfrak{R}_d$ and a continuous, surjective group homomorphism $\phi \colon X^{\mathfrak{R}_d/\mathfrak{p}} \longmapsto X$ with the following properties.*

(1) *Either $p(\mathfrak{p}) = r(\mathfrak{p}) = 0$, or $p(\mathfrak{p}) > 0$ and $r(\mathfrak{p}) = 1$;*
(2) $\phi \cdot \alpha_{\mathbf{n}}^{\mathfrak{R}_d/\mathfrak{p}} = \alpha_{\mathbf{n}} \cdot \phi$ *for every* $\mathbf{n} \in \mathbb{Z}^d$;
(3) *The kernel of ϕ is finite.*

PROOF. If $\mathfrak{p} \subset \mathfrak{R}_d$ is an ergodic prime ideal, then we claim that $\alpha^{\mathfrak{R}_d/\mathfrak{p}}$ is almost minimal if and only if $\mathfrak{p}$ satisfies condition (1).

Indeed, if $p(\mathfrak{p}) = r(\mathfrak{p}) = 0$, then $\mathfrak{p} = \mathfrak{j}_c$ for some $c \in (\overline{\mathbb{Q}}^\times)^d$ (cf. Proposition 7.2), and every ideal $\mathfrak{a}$ with $\mathfrak{p} \subsetneqq \mathfrak{a} \subset \mathfrak{R}_d$ must contain an element $h \in \mathfrak{q}$ with $h(c) \neq 0$. From Lemma 6.7 it is clear that $\mathfrak{a} \cap \mathbb{Z} \neq \{0\}$, so that $\mathfrak{R}_d/\mathfrak{a}$ is finite. Hence $\alpha^{\mathfrak{R}_d/\mathfrak{p}}$ is almost minimal whenever $r(\mathfrak{p}) = 0$.

If $p(\mathfrak{p}) = 0$ and $r(\mathfrak{p}) > 0$, then the proof of Proposition 8.2 implies that we can find an element $h \in \mathfrak{R}_d \smallsetminus \mathfrak{p}$ such that $V_{\mathbb{C}}(\mathfrak{a}) \neq \varnothing$, where $\mathfrak{a} = \mathfrak{p} + h\mathfrak{R}_d$. Then $\mathfrak{R}_d/\mathfrak{a}$ is infinite, $\mathfrak{a} \supsetneqq \mathfrak{p}$, and $\alpha^{\mathfrak{R}_d/\mathfrak{p}}$ cannot be almost minimal.

Similarly we see that $\alpha^{\mathfrak{R}_d/\mathfrak{p}}$ cannot be almost minimal if $p(\mathfrak{p}) > 0$ and $r(\mathfrak{p}) > 1$. If $p(\mathfrak{p}) > 0$ and $r(\mathfrak{p}) = 1$, then there exists a primitive subgroup $\Gamma = \{k\mathbf{n} : k \in \mathbb{Z}\} \cong \mathbb{Z}$ in $\mathbb{Z}^d$ and a finite set $Q \subset \mathbb{Z}^d$ satisfying the conditions (1)–(3) in Proposition 8.2. Then $X^{\mathfrak{R}_d/\mathfrak{p}} \cong F^{\mathbb{Z}}$ for some finite group F, and the $\mathbb{Z}$-action $k \mapsto \alpha_{k\mathbf{n}}^{\mathfrak{R}_d/\mathfrak{p}}$ corresponds to the shift-action of $\mathbb{Z}$ on $F^{\mathbb{Z}}$ and satisfies the d.c.c.

For every $g \in \mathfrak{R}_d$ we define $\alpha_g^{\mathfrak{R}_d/\mathfrak{p}}$ by (6.14) and note that $\alpha_g^{\mathfrak{R}_d/\mathfrak{p}} \colon X^{\mathfrak{R}_d/\mathfrak{p}} \longmapsto X^{\mathfrak{R}_d/\mathfrak{p}}$ is the homomorphism dual to multiplication by g on $\mathfrak{R}_d/\mathfrak{p}$. If $g \in \mathfrak{R}_d \smallsetminus \mathfrak{p}$, then multiplication by g on $\mathfrak{R}_d/\mathfrak{p}$ is injective, and $\alpha_g^{\mathfrak{R}_d/\mathfrak{p}} \colon X^{\mathfrak{R}_d/\mathfrak{p}} \longmapsto X^{\mathfrak{R}_d/\mathfrak{p}}$ is surjective. We set $Y(g) = \ker(\alpha_g^{\mathfrak{R}_d/\mathfrak{p}})$ and apply the addition formula (14.1) to see that $h((\alpha_{\mathbf{n}}^{\mathfrak{R}_d/\mathfrak{p}})^{Y(g)}) + h(\alpha_{\mathbf{n}}^{\mathfrak{R}_d/\mathfrak{p}}) = h(\alpha_{\mathbf{n}}^{\mathfrak{R}_d/\mathfrak{p}}) < \infty$, where $(\alpha_{\mathbf{n}}^{\mathfrak{R}_d/\mathfrak{p}})^{Y(g)}$ denotes the restriction of $\alpha_{\mathbf{n}}^{\mathfrak{R}_d/\mathfrak{p}}$ to $Y(g)$. Hence $h((\alpha_{\mathbf{n}}^{\mathfrak{R}_d/\mathfrak{p}})^{Y(g)}) = 0$, and Theorem 19.2 shows that $Y(g)$ is finite.

If $\{0_X\} \neq Y \subsetneqq X^{\mathfrak{R}_d/\mathfrak{p}}$ is a closed, $\alpha^{\mathfrak{R}_d/\mathfrak{p}}$-invariant subgroup, then $Y^{\perp} = \mathfrak{a}/\mathfrak{p}$ for some ideal $\mathfrak{a} \subset \mathfrak{R}_d$, and we choose elements $g_1, \ldots, g_m$ in $\mathfrak{a} \smallsetminus \mathfrak{p}$ such that $\mathfrak{a} = \mathfrak{p} + \sum_{i=1}^{m} g_i \mathfrak{R}_d$. From (6.16) we see that

$$Y = \sum_{i=1}^{m} \ker(\alpha_{g_i}^{\mathfrak{R}_d/\mathfrak{p}}) \subset \ker(\alpha_g^{\mathfrak{R}_d/\mathfrak{p}}) = Y(g),$$

where $g = \prod_{i=1}^{m} g_i \in \mathfrak{R}_d \smallsetminus \mathfrak{p}$, and the argument in the preceding paragraph implies that Y is finite. Since Y was arbitrary we have proved that $\alpha^{\mathfrak{R}_d/\mathfrak{p}}$ is almost minimal, as claimed. This completes the proof of the assertion that $\alpha^{\mathfrak{R}_d/\mathfrak{p}}$ is almost minimal if and only if $\mathfrak{p}$ satisfies (1). Furthermore, if $\mathfrak{p} \subset \mathfrak{R}_d$ is an ergodic prime ideal satisfying (1), and if $Z \subset X^{\mathfrak{R}_d/\mathfrak{p}}$ is a proper, closed, $\alpha^{\mathfrak{R}_d/\mathfrak{p}}$-invariant—and hence finite—subgroup, then the $\mathbb{Z}^d$-action induced by $\alpha^{\mathfrak{R}_d/\mathfrak{p}}$ on $X^{\mathfrak{R}_d/\mathfrak{p}}/Z$ is obviously again almost minimal.

Now let α be an ergodic and almost minimal $\mathbb{Z}^d$-action by automorphisms of a compact, abelian group X, and let $\mathfrak{M} = \hat{X}$ be the $\mathfrak{R}_d$-module defined by α via Lemma 5.1. From Definition 3.1 it is clear that α satisfies the d.c.c., and duality implies that $\mathfrak{M}$ is Noetherian (Proposition 5.4) and that, for every submodule $\mathfrak{N} \subset \mathfrak{M}$, $|\mathfrak{M}/\mathfrak{N}| < \infty$ and $\alpha^{\mathfrak{N}}$ is ergodic and almost minimal. By considering submodules of the form $\mathfrak{N} = \mathfrak{R}_d \cdot a$ for some $a \in \mathfrak{M}$ we see in particular that $\alpha^{\mathfrak{R}_d/\mathfrak{p}}$ is ergodic and almost minimal for every prime ideal $\mathfrak{p}$ associated with $\mathfrak{M}$ (cf. Theorem 6.5 (1)), and the first part of this proof implies that $\mathfrak{p}$ satisfies (1).

Our next task is to prove that $\mathfrak{M}$ is $\mathfrak{p}$-primary for a single prime ideal $\mathfrak{p} \subset \mathfrak{R}_d$. Write $\{\mathfrak{p}_1, \ldots, \mathfrak{p}_m\}$ for the set of associated primes of $\mathfrak{M}$, choose a corresponding reduced primary decomposition $\{\mathfrak{W}_1, \ldots, \mathfrak{W}_m\}$ of $\mathfrak{M}$ (cf. (6.5)), and consider the injective homomorphism $\theta \colon a \mapsto (a+\mathfrak{W}_1, \ldots, a+\mathfrak{W}_m)$ from $\mathfrak{M}$ to $\mathfrak{K} = \bigoplus_{i=1}^m \mathfrak{M}/\mathfrak{W}_i$. The dual homomorphism $\hat\theta \colon V = X^{\mathfrak{K}} = \bigoplus_{i=1}^m \widehat{\mathfrak{M}/\mathfrak{W}_i} \longmapsto X$ is surjective, and its kernel $Z = \ker(\hat\theta) \subset V$ is of the form $\hat{Z} = \mathfrak{K}/\theta(\mathfrak{M}) = \sum_{i=1}^m (\mathfrak{M}/\mathfrak{W}_i)/\theta(\mathfrak{M})$, where each $\mathfrak{M}/\mathfrak{W}_i$ is viewed as a subgroup of $\mathfrak{K}$ in the obvious manner. The proof of Lemma 21.13 shows that, for every $i = 1, \ldots, m$, every prime ideal $\mathfrak{q}$ associated with the submodule $(\mathfrak{M}/\mathfrak{W}_i)/\theta(\mathfrak{M}) \subset \mathfrak{K}/\theta(\mathfrak{M})$ contains $\mathfrak{p}_i + \prod_{j \neq i} \mathfrak{p}_j \supsetneq \mathfrak{p}_i$, and the almost minimality of $\alpha^{\mathfrak{R}_d/\mathfrak{p}_i}$ shows that $\mathfrak{R}_d/\mathfrak{q}$ is finite. Since $(\mathfrak{M}/\mathfrak{W}_i)/\theta(\mathfrak{M})$ is Noetherian this implies that $(\mathfrak{M}/\mathfrak{W}_i)/\theta(\mathfrak{M})$ is finite (Proposition 6.1), and by varying $i \in \{1, \ldots, m\}$ we obtain that $\mathfrak{K}/\theta(\mathfrak{M}) = \hat{Z}$ is finite. Hence Z is finite, and $\alpha^{\mathfrak{K}}$ is therefore almost minimal. By considering the special form of $\mathfrak{K}$ we see that $m = 1$, as claimed.

We write $\mathfrak{p}$ for the unique prime ideal associated with $\mathfrak{M}$ and apply Proposition 6.1 to find a prime filtration $\{0\} = \mathfrak{N}_0 \subset \cdots \subset \mathfrak{N}_s = \mathfrak{M}$ and an integer $t \in \{1, \ldots, s\}$ such that, for every $i = 1, \ldots, s$, $\mathfrak{N}_i/\mathfrak{N}_{i-1} \cong \mathfrak{R}_d/\mathfrak{q}_i$ for some prime ideal $\mathfrak{q}_i \subset \mathfrak{R}_d$, and $\mathfrak{q}_i = \mathfrak{p}$ for $i = 1, \ldots, t$ and $\mathfrak{q}_i \supsetneq \mathfrak{p}$ for $i = t+1, \ldots, s$. Since α is almost minimal we know that $t = 1$ (otherwise $\mathfrak{M}$ has proper submodules of infinite index). If $s > t$ we choose elements $g_i \in \mathfrak{q}_i \setminus \mathfrak{p}$ for $i = 2, \ldots, s$ and set $g = g_2 \cdot \ldots \cdot g_s$. The equation (6.4) shows that multiplication by g on $\mathfrak{M}$ is injective, so that $g \cdot \mathfrak{M} \subset \mathfrak{N}_1 \cong \mathfrak{R}_d/\mathfrak{p}$ is isomorphic to $\mathfrak{M}$. Hence $X = \widehat{\mathfrak{M}}$ is isomorphic to the quotient of $X^{\mathfrak{R}_d/\mathfrak{p}}$ by a closed, $\alpha^{\mathfrak{R}_d/\mathfrak{p}}$-invariant, and necessarily finite, subgroup. $\square$

Theorem 29.2 shows that almost minimal $\mathbb{Z}^d$-actions live either on zero-dimensional groups, or on finite-dimensional tori or solenoids (cf. Corollary 7.4). We begin our investigation by considering the invariant probability measures of almost minimal $\mathbb{Z}^d$-actions on compact, abelian, zero-dimensional groups, and by proving that every 'sufficiently mixing' invariant measure on such a group must be equal to Haar measure. If α is an almost minimal $\mathbb{Z}^d$-action by automorphisms of a compact, zero-dimensional group X with normalized Haar measure λ_X we call a non-empty, finite set $E \subset \mathbb{Z}^d$ λ_X-non-mixing or μ-non-mixing if it is non-mixing for the $\mathbb{Z}^d$-actions α on $(X, \mathfrak{B}_X, \lambda_X)$ or $(X, \mathfrak{B}_X, \mu)$ (cf. (27.1)).

LEMMA 29.3. *Let α be a mixing, almost minimal $\mathbb{Z}^d$-action by automorphisms of a compact, zero-dimensional, abelian group X, μ a non-atomic, α-*

invariant probability measure on X, and $E \subset \mathbb{Z}^d$ a finite, non-empty set which is λ_X-non-mixing. Then E is μ-non-mixing.

PROOF. Theorem 29.2 allows us to assume without loss in generality that $\alpha = \alpha^{\mathfrak{R}_d/\mathfrak{p}}$ and $X = X^{\mathfrak{R}_d/\mathfrak{p}}$ for some prime ideal $\mathfrak{p} \subset \mathfrak{R}_d$ satisfying the second alternative in condition (2) of that theorem. We set $p = p(\mathfrak{p})$, define $\mathfrak{q} \subset \mathfrak{R}_d^{(p)}$ by (28.1), and apply Corollary 28.6 to find elements $f_{\mathbf{n}} \in \mathfrak{R}_d^{(p)}$, $\mathbf{n} \in E$, not all in $\mathfrak{q}$, such that $\sum_{\mathbf{n} \in E} u^{k\mathbf{n}} f_{\mathbf{n}} \in \mathfrak{q}$ for all k in an infinite subset $K \subset \mathbb{N}$.

If μ is non-atomic, and if $f_{\mathbf{n}} \in \mathfrak{R}_d^{(p)} \smallsetminus \mathfrak{q}$ for some $\mathbf{n} \in E$, then the character $\chi_{f_{\mathbf{n}}+\mathfrak{q}} = \langle f_{\mathbf{n}}+\mathfrak{q}, \cdot \rangle$ of X defined by $f_{\mathbf{n}}$ is not μ-a.e. equal to a constant. Otherwise $|\hat{\mu}(f_{\mathbf{n}} + \mathfrak{q})| = 1$, and the shift-invariance of μ implies that the set $\mathfrak{a} = \{f \in \mathfrak{R}_d^{(p)} : |\hat{\mu}(f + \mathfrak{q})| = 1\}$ is an ideal with $\mathfrak{q} \subsetneq \mathfrak{a} \subset \mathfrak{R}_d^{(p)}$, and that $Z = \mathfrak{a}^{\perp} \subset X$ is a proper, closed, shift-invariant subgroup. By assumption, Z is finite, so that $\hat{Z} = \mathfrak{R}_d^{(p)}/\mathfrak{a}$ is finite. We set $\bar{\mu}(B) = \mu(-B)$ for every $B \in \mathfrak{B}_Y$ and conclude that the convolution $|\mu|^2 = \mu * \bar{\mu}$, whose Fourier transform is given by $\widehat{|\mu|^2} = |\hat{\mu}|^2$, is a probability measure with finite support. Hence μ has finite support, which is absurd.

Since at least one of the characters $\chi_{f_{\mathbf{n}}+\mathfrak{q}}$, $\mathbf{n} \in E$, is non-trivial, but $1 = \chi_{\sum_{\mathbf{n} \in E} u^{k\mathbf{n}} f_{\mathbf{n}}+\mathfrak{q}} = \prod_{\mathbf{n} \in E} \chi_{f_{\mathbf{n}}+\mathfrak{q}} \cdot \alpha_{k\mathbf{n}}$ for every $k \in K$, E is μ-non-mixing. $\square$

If every μ-non-mixing set is also λ_X-non-mixing, then μ has to be equal to λ_X; in other words, λ_X is the unique 'most mixing' measure for α.

THEOREM 29.4. *Let α be a mixing, almost minimal $\mathbb{Z}^d$-action by automorphisms of a compact, zero-dimensional, abelian group X, and let μ be a non-atomic, α-invariant probability measure on X. If there exists a minimal λ_X-non-mixing set $E \subset \mathbb{Z}^d$ which is also minimal μ-non-mixing, then $\mu = \lambda_X$.*

PROOF. As in the proof of Lemma 29.3 we assume that $\alpha = \alpha^{\mathfrak{R}_d/\mathfrak{p}} = \alpha^{\mathfrak{R}_d^{(p)}/\mathfrak{q}}$ and $X = X^{\mathfrak{R}_d/\mathfrak{p}} = X^{\mathfrak{R}_d^{(p)}/\mathfrak{q}}$ for suitable prime ideals $\mathfrak{p} \subset \mathfrak{R}_d$ and $\mathfrak{q} \subset \mathfrak{R}_d^{(p)}$. Let $E \subset \mathbb{Z}^d$ be a minimal λ_X-non-mixing set which is also minimal μ-non-mixing, and choose—again as in the proof of Lemma 29.3—a map $f \colon E \longmapsto \mathfrak{R}_d^{(p)} \smallsetminus \mathfrak{q}$ and an infinite subset $K \subset \mathbb{N}$ such that $\sum_{\mathbf{n} \in E} u^{k\mathbf{n}} f(\mathbf{n}) \in \mathfrak{q}$ for every $k \in K$. Put $a(\mathbf{n}) = f(\mathbf{n}) + \mathfrak{q} \in \mathfrak{R}_d^{(p)}/\mathfrak{q}$, $\mathbf{n} \in E$, and note that $u^{k\mathbf{n}} \cdot a(\mathbf{n}) = -\sum_{\mathbf{m} \in E \smallsetminus \{\mathbf{n}\}} u^{k\mathbf{m}} a(\mathbf{m})$ for every $k \in K$ and $\mathbf{n} \in E$. For every $\mathbf{n} \in E$, the set $E \smallsetminus \{\mathbf{n}\}$ is μ-mixing, and

$$\hat{\mu}(a(\mathbf{n})) = \lim_{\substack{k \to \infty \\ k \in K}} \hat{\mu}(u^{k\mathbf{n}} \cdot a(\mathbf{n})) = \lim_{\substack{k \to \infty \\ k \in K}} \hat{\mu}\left(-\sum_{\mathbf{m} \in E \smallsetminus \{\mathbf{n}\}} u^{k\mathbf{m}} \cdot a(\mathbf{m}) \right)$$

$$= \prod_{\mathbf{m} \in E \smallsetminus \{\mathbf{n}\}} \overline{\hat{\mu}(a(\mathbf{m}))}.$$

By varying $\mathbf{n}$ in E we obtain that either $|\hat{\mu}(a(\mathbf{n}))| = 1$ for every $\mathbf{n} \in E$, or $\hat{\mu}(a(\mathbf{n})) = 0$ for every $\mathbf{n} \in E$.

In the first case the set $\mathfrak{N} = \{a \in \mathfrak{R}_d^{(p)}/\mathfrak{q} : |\hat{\mu}(a)| = 1\}$ is a non-zero submodule of $\mathfrak{M} = \mathfrak{R}_d^{(p)}/\mathfrak{q}$. Since α is almost minimal, the group $Z = \mathfrak{N}^{\perp} \subset X$ is finite, and the probability measure ν on X/Z induced by μ is invariant under the $\mathbb{Z}^d$-action on X/Z induced by α. Since $|\hat{\nu}(a)| = 1$ for every $a \in \widehat{X/Z} = \mathfrak{N}$, ν is concentrated in a single point of X/Z, and the finiteness of Z implies that μ is atomic, which is impossible.

This contradiction shows that $\hat{\mu}(a(\mathbf{n})) = 0$ for every $\mathbf{n} \in E$, and by replacing the map $a\colon \mathbf{n} \mapsto a(\mathbf{n})$ from E to $\mathfrak{M}$ with $a'\colon \mathbf{n} \mapsto a(\mathbf{n})' = h \cdot a(\mathbf{n})$ for an arbitrary, but temporarily fixed element $h \in \mathfrak{R}_d^{(p)}$ we see that $\hat{\mu}(h \cdot a(\mathbf{n})) = 0$ whenever $\mathbf{n} \in E$, $h \in \mathfrak{R}_d^{(p)}$, and $h \cdot a(\mathbf{n}) \neq 0$. Fix $\mathbf{n} \in E$ for the moment and note that $\mathfrak{N}' = \{h \cdot a(\mathbf{n}) : h \in \mathfrak{R}_d^{(p)}\}$ is either equal to $\mathfrak{M}$, or to $\mathfrak{b}'/\mathfrak{q}$ for some ideal $\mathfrak{b}'$ with $\mathfrak{q} \subsetneq \mathfrak{b}' \subsetneq \mathfrak{R}_d^{(p)}$. We set $Z' = \mathfrak{N}'^{\perp} \subset X$ and conclude as above that the measure ν' induced by μ on X/Z' is equal to the Haar measure on X/Z'. Since α is topologically conjugate to a skew-product action of $\mathbb{Z}^d$ on $X/Z' \times Z'$, the invariance and ergodicity of μ and λ_X under α implies that $\mu = \lambda_X$. $\square$

Under an additional assumption one can obtain a little more information from the proof of Theorem 29.4.

COROLLARY 29.5. *Let α be a mixing, almost minimal $\mathbb{Z}^d$-action by automorphisms of a compact, zero-dimensional, abelian group X, and let μ be a non-atomic, shift-invariant probability measure on X. If there exists a non-zero element $f \in \mathfrak{q}$ with support $\mathfrak{S}(f)$ such that $\mathfrak{S}(f) \smallsetminus \{\mathbf{n}\}$ is μ-mixing for some $\mathbf{n} \in \mathfrak{S}(f)$, then $\mu = \lambda_X$.*

PROOF. We use the setting of the proof of Theorem 29.4. Since

$$\sum_{\mathbf{n} \in \mathfrak{S}(f)} c_f(\mathbf{n}) u^{p^k \mathbf{n}} \in \mathfrak{q}$$

for every $k \geq 1$ we obtain that, for every $a \in \mathfrak{R}_d^{(p)}/\mathfrak{q}$,

$$\hat{\mu}(a) = \lim_{k \to \infty} \hat{\mu}(u^{p^k \mathbf{n}} \cdot a)$$

$$= \lim_{k \to \infty} \hat{\mu}\left(-\sum_{\mathbf{m} \in \mathfrak{S}(f) \smallsetminus \{\mathbf{n}\}} \frac{c_f(\mathbf{m})}{c_f(\mathbf{n})} u^{p^k} \cdot a\right) = \prod_{\mathbf{m} \in \mathfrak{S}(f) \smallsetminus \{\mathbf{n}\}} \overline{\hat{\mu}\left(\frac{c_f(\mathbf{m})}{c_f(\mathbf{n})} a\right)}.$$

As $\mathbb{F}_p$ is finite and μ is non-atomic, we conclude as in the proof of Theorem 29.4 that $\hat{\mu}(a) = 0$ whenever $0 \neq a \in \mathfrak{R}_d^{(p)}/\mathfrak{p}$, and that $\mu = \lambda_X$. $\square$

EXAMPLES 29.6. In the following examples we consider irreducible Laurent polynomials $f \in \mathfrak{R}_2^{(2)}$ and set $\mathfrak{q} = f\mathfrak{R}_2^{(2)}$, $X = X^{\mathfrak{R}_2^{(2)}/\mathfrak{q}}$, and $\alpha = \alpha^{\mathfrak{R}_2^{(2)}/\mathfrak{q}}$.

(1) Let $f = 1 + u_1 + u_2 + u_1^2 + u_1 u_2 + u_2^2$ (cf. Example 28.11). Then $E = \{(0,0),(1,0),(0,1)\}$ is a minimal λ_X-non-mixing set, and Theorem 29.4 implies that every α-invariant probability measure μ on X for which $\alpha_\mathbf{n}$ is μ-mixing for every $\mathbf{n} \in \{(1,0),(0,1),(1,-1)\}$, is equal to λ_X.

(2) Let $f = 1 + u_1 + u_2$ (cf. Example 28.10 (1)). Then $\mathcal{S}(f) = \{(0,0),(1,0),(0,1)\}$ is minimal λ_X-non-mixing, and by letting $\mathbf{n}$ vary in $\mathcal{S}(f)$ we see from Corollary 29.5 that every probability measure μ on X for which any of the transformations $\alpha_{(1,0)}$, $\alpha_{(0,1)}$, or $\alpha_{(-1,1)}$ is mixing, must be equal to λ_X.

(3) Let $f = 1 + u_1 + u_1^{-1} + u_2$. Then $\mathcal{S}(f) = \{(-1,0),(0,0),(1,0),(0,1)\}$ is minimal λ_X-non-mixing, and by setting $\mathbf{n} = (0,1)$ we see from Corollary 29.5 that every non-atomic, α-invariant probability measure μ on X for which $\alpha_{(1,0)}$ is three-mixing, is equal to λ_X. $\boxdot$

REMARK 29.7. Let α be an ergodic, almost minimal $\mathbb{Z}^d$-action by automorphisms of a compact, zero-dimensional, abelian group X. Although every sufficiently mixing, α-invariant probability measure on X has to be equal to λ_X by Theorem 29.4, there exist infinitely many shift-invariant, ergodic, non-atomic probability measures μ on X which are different from λ_X. A method for obtaining such measures is described in [46] and [96]: assume that $\alpha = \alpha^{\mathfrak{R}_d/\mathfrak{p}}$ and $X = X^{\mathfrak{R}_d/\mathfrak{p}}$ for some prime ideal $\mathfrak{p} \subset \mathfrak{R}_d$ with $p = p(\mathfrak{p}) > 1$ and $r(\mathfrak{p}) = 1$, and set $\mathfrak{M} = \mathfrak{R}_d/\mathfrak{p}$. For every $k \geq 1$ we consider the subring $\mathcal{R}^{(k)} = \mathbb{F}_p[u_1^{\pm p^k}, \ldots, u_d^{\pm p^k}] \subset \mathfrak{R}_d^{(p)}$ and write $\mathfrak{M}^{(k)}$ instead of $\mathfrak{M}$ in order to emphasize that $\mathfrak{M} = \mathfrak{M}^{(k)}$ is to be viewed as an $\mathcal{R}^{(k)}$-module. Then $\mathfrak{M}^{(k)}$ has non-trivial $\mathcal{R}^{(k)}$-submodules of infinite index. If $\mathfrak{N}^{(k)} \subset \mathfrak{M}^{(k)}$ is such a submodule, then $Y^{\mathfrak{N}^{(k)}} = (\mathfrak{N}^{(k)})^{\perp} \subset X$ is an infinite, closed subgroup of X which is invariant under $\{\alpha_{p^k\mathbf{m}} : \mathbf{m} \in \mathbb{Z}^d\}$. Although the Haar measure $\lambda^{\mathfrak{N}^{(k)}} = \lambda_{Y^{\mathfrak{N}^{(k)}}}$ is not α-invariant, its orbit average

$$\mu^{\mathfrak{N}^{(k)}} = \frac{1}{p^{dk}} \sum_{m_1=0}^{p^k-1} \cdots \sum_{m_d=0}^{p^k-1} \lambda^{\mathfrak{N}^{(k)}} \cdot \alpha_{(m_1,\ldots,m_d)}$$

is a non-atomic, shift-invariant, ergodic probability measure on X which is obviously non-mixing under every $\alpha_\mathbf{n}$. By choosing sequences $\mathfrak{N}^{(1)} \subset \mathfrak{N}^{(2)} \subset \cdots$ of submodules $\mathfrak{N}^{(k)} \subset \mathfrak{M}^{(k)}$ we obtain sequences $(\mu^{\mathfrak{N}^{(k)}}, k \geq 1)$ of such measures, and every limit point μ of such a sequence is again shift-invariant.

EXAMPLES 29.8. As in Example 29.6 (2) we set $\mathfrak{p} = (2, 1 + u_1 + u_2)$, regard $\alpha = \alpha^{\mathfrak{R}_2/\mathfrak{p}}$ as the shift-action of $\mathbb{Z}^2$ on the shift-invariant subgroup $X = X^{\mathfrak{R}_2/\mathfrak{p}} \subset \mathbb{F}_2^{\mathbb{Z}^2}$ (cf. (6.19)), and put $\mathfrak{M} = \mathfrak{R}_2^{(2)}/\mathfrak{p}$.

(1) $\mathfrak{M}$ is a module over the ring $\mathcal{R}^{(1)} = \mathbb{F}_2[u_1^{\pm 2}, u_2^{\pm 2}]$, $\mathfrak{N}^{(1)} = (1 +$

$u_1)\mathcal{R}^{(1)}/\mathfrak{p} \subset \mathfrak{M}$ is an $\mathcal{R}^{(1)}$-submodule with infinite index, and

$$Y^{\mathfrak{N}^{(1)}} = (\mathfrak{N}^{(1)})^{\perp} = \{x = (x_{\mathbf{n}}) \in X :$$
$$x_{(2m,2n)} = x_{(2m+1,2n)} \text{ for all } (m,n) \in \mathbb{Z}^2\}.$$

The measure

$$\mu^{\mathfrak{N}^{(1)}} = \frac{1}{4}(\lambda_{Y^{\mathfrak{N}^{(1)}}} + \lambda_{Y^{\mathfrak{N}^{(1)}}} \cdot \alpha_{(1,0)} + \lambda_{Y^{\mathfrak{N}^{(1)}}} \cdot \alpha_{(0,1)} + \lambda_{Y^{\mathfrak{N}^{(1)}}} \cdot \alpha_{(1,1)})$$

on X is non-atomic, shift-invariant and ergodic, but non-ergodic under $\alpha_{(1,0)}$.

(2) Consider the Morse sequence

$$M = 0110100110010110\ldots$$

obtained by letting $b_0 = 0, b_1 = \overline{b_0} = 1, b_2 = \overline{b_0 b_1} = 10, b_n = \overline{b_0 \cdots b_{n-1}} = b_{n-1}\overline{b_{n-1}}$, for $n > 1$. Then

$$M = b_0 b_1 b_2 \ldots .$$

For each $k \geq 1$, let $Y^{(k)} \subset X$ be the subgroup consisting of all elements $x = (x_{\mathbf{n}}) \subset X$ with

$$(x_{(m2^k,n2^k)}, \ldots, x_{(m2^k,n2^k+2^k-1)}) \in \{(0,\ldots,0),(1,\ldots,1), b_k, \overline{b_k}\}$$

for every $(m,n) \in \mathbb{Z}^2$. Then $Y^{(k+1)} \subset Y^{(k)}$, and we set

$$Z^{(k)} = \bigcup_{m=0}^{2^k-1}\bigcup_{n=0}^{2^k-1} \alpha_{(m,n)}(Y^{(k)}),$$

$$\mu^{(k)} = \frac{1}{2^{2k}} \sum_{m=0}^{2^k-1}\sum_{n=0}^{2^k-1} \lambda_{Y^{(k)}}\alpha_{(-m,-n)},$$

$$Z = \bigcap_{k \geq 1} Z^{(k)}.$$

Then Z contains two fixed points, but no other periodic points, since $Z^{(k)}$ contains no points of period less than 2^k except the two fixed points. Furthermore, if $Z' = \{(m,0) : m \in \mathbb{Z}\} \subset \mathbb{Z}^2$, then $\pi_{Z'}(Z) \subset \mathbb{F}_2^{\mathbb{Z}}$ contains the Morse minimal set. In particular, Z is uncountable, and every limit point μ of the sequence $(\mu^{(k)}, k \geq 1)$ is non-atomic and concentrated on Z.

For every $k \geq 1$ we write $\alpha^{Z^{(k)}}$ for the restriction of the $\mathbb{Z}^2$-action α to the closed, α-invariant subset $Z^{(k)} \subset X$ and denote by α^Z the restriction of α to Z. Then the topological entropy $h(\alpha_{(1,0)}^{(k)})$ of $\alpha_{(1,0)}^{(k)}$ is given by

$$h(\alpha_{(1,0)}^{(k)}) = \frac{1}{k}\log 4$$

for every $k \geq 1$, and

$$h(\alpha_{(1,0)}^Z) = h_\mu(\alpha_{(1,0)}) = 0$$

for every limit point μ of the sequence $(\mu^{(k)}, \, k \geq 1)$.

(3) Let μ be an α-invariant probability measure on X. For every $n \geq 1$ and $(i_0, \dots, i_{n-1}) \in \mathbb{F}_2^n$ we consider the cylinder sets

$$C(i_0, \dots, i_{n-1}) = \{x = (x_\mathbf{n}) \in X : x_{(k,0)} = i_k \text{ for } k = 0, \dots, n-1\},$$
$$D(i_0, \dots, i_{n-1}) = \{x = (x_\mathbf{n}) \in X : x_{(0,k)} = i_k \text{ for } k = 0, \dots, n-1\}$$

in X and observe that there exists, for every $(i_0, \dots, i_n) \in \mathbb{F}_2^n$, a unique $(j_0, \dots, j_{n-1}) \in \mathbb{F}_2^{n+1}$ with

$$C(i_0, \dots, i_{n-1}) = D(j_0, \dots, j_{n-1}). \tag{29.1}$$

If

$$\mathcal{P}_n = \{C(i_0, \dots, i_{n-1}) : (i_0, \dots, i_{n-1}) \in \mathbb{F}_2^{n+1}\},$$
$$\mathcal{Q}_n = \{D(i_0, \dots, i_{n-1}) : (i_0, \dots, i_{n-1}) \in \mathbb{F}_2^{n+1}\},$$

then (29.1) shows that $\mathcal{P}_n = \mathcal{Q}_n$ for every $n \geq 1$, so that

$$h_\mu(\alpha_{(1,0)}) = \lim_{n \to \infty} \frac{1}{n} H_\mu(\mathcal{P}_n) = \lim_{n \to \infty} \frac{1}{n} H_\mu(\mathcal{Q}_n) = h_\mu(\alpha_{(0,1)}). \tag{29.2}$$

By refining the argument in Remark 29.7 and Example (2) one can construct, for every $c \in [1,2]$, an ergodic, α-invariant probability measure μ on X with

$$h_\mu(\alpha_{(1,0)}) = h_\mu(\alpha_{(0,1)}) = \log c. \quad \boxdot$$

Theorem 29.4 establishes a connection between invariant measures and mixing sets for $\mathbb{Z}^d$-actions by automorphisms of compact, zero-dimensional groups. For connected groups the picture is once again quite different.

THEOREM 29.9 ([89]). Let $\mathfrak{p} = (2 - u_1, 3 - u_2) \subset \mathfrak{R}_2$, $\alpha = \alpha^{\mathfrak{R}_2/\mathfrak{p}}$ and $X = X^{\mathfrak{R}_2/\mathfrak{p}}$ (cf. Example 5.3 (4)). If μ is an α-invariant and ergodic probability measure on X such that $h_\mu(\alpha_\mathbf{n}) > 0$ for some $\mathbf{n} \in \mathbb{Z}^2$, then $\mu = \lambda_X$.

Let $d \geq 2$, and let α be an expansive, mixing and almost minimal $\mathbb{Z}^d$-action by automorphisms of a compact, connected, abelian group X. Under certain additional conditions it is shown in [40] that every α-invariant and mixing probability measure μ on X with $h_\mu(\alpha_\mathbf{n}) > 0$ for some $\mathbf{n} \in \mathbb{Z}^d$ is equal to λ_X; under even stronger assumptions on α the same conclusion can be obtained for every α-invariant and ergodic probability measure μ on X. All these results deal with special cases of the following problem motivated by the seminal paper [23].

PROBLEM 29.10. Let α be an almost minimal, expansive and mixing $\mathbb{Z}^d$-action by automorphisms of a compact, abelian group X, and let μ be a non-atomic, α-invariant and ergodic probability measure on X.

(1) If μ is ergodic under some $\alpha_\mathbf{n}$, $\mathbf{n} \in \mathbb{Z}^d$ (and, in particular, if μ is mixing), is $\mu = \lambda_X$?

(2) If X is connected and $h_\mu(\alpha_\mathbf{n}) > 0$ for some $\mathbf{n} \in \mathbb{Z}^d$, is $\mu = \lambda_X$? Is the assumption that μ has positive entropy under some $\alpha_\mathbf{n}$ necessary?

The paper [23] raises a second and related problem concerning the nature of all closed, invariant subsets of an almost minimal $\mathbb{Z}^d$-action α by automorphisms of a compact, abelian group. In [23] it is shown that in Example 5.3 (4) every closed, α-invariant subset $Y \subsetneq X$ is finite. Further developments concerning this problem can be found in [6], [7], [8], [9], [18] and [46]. In view of the somewhat diverse evidence emerging from these papers a general solution to this problem seems currently out of reach.

30. Cohomological rigidity

Let A be a Polish (=complete, separable and metric) abelian group, $d \geq 1$, and let T be a continuous $\mathbb{Z}^d$-action on a compact, metrizable space Y. A continuous map $c \colon \mathbb{Z}^d \times Y \longmapsto A$ is a (*continuous* 1-) *cocycle* for T if

$$c(\mathbf{m}, T_\mathbf{n}(y)) + c(\mathbf{n}, y) = c(\mathbf{m} + \mathbf{n}, y) \tag{30.1}$$

for every $\mathbf{m}, \mathbf{n} \in \mathbb{Z}^d$ and $y \in Y$. A cocycle $c \colon \mathbb{Z}^d \times Y \longmapsto A$ for T is a *homomorphism* if $c(\mathbf{n}, \cdot)$ is constant for every $\mathbf{n} \in \mathbb{Z}^d$, and a *coboundary* if there exists a Borel map $b \colon Y \longmapsto A$ with

$$c(\mathbf{n}, \cdot) = b \cdot T_\mathbf{n} - b \quad \mu\text{-a.e.} \tag{30.2}$$

for every $\mathbf{n} \in \mathbb{Z}^d$. Two cocycles $c_1, c_2 \colon \mathbb{Z}^d \times Y \longmapsto A$ are *cohomologous* if they differ by a coboundary, i.e. if there exists a Borel map $b \colon Y \longmapsto A$ such that

$$c_1(\mathbf{n}, \cdot) - c_2(\mathbf{n}, \cdot) = b \cdot T_\mathbf{n} - b\mu \quad \text{-a.e.} \tag{30.3}$$

for every $\mathbf{n} \in \mathbb{Z}^d$. Finally, a cocycle $c \colon \mathbb{Z}^d \times Y \longmapsto A$ is *trivial* if it is cohomologous to a homomorphism.

The functions b in (30.2) and (30.3) are called the *cobounding function* of c and the *transfer function* of (c_1, c_2), respectively. The set $Z_c^1(T, A)$ of all continuous cocycles $c \colon \mathbb{Z}^d \times Y \longmapsto A$ is a Polish group under point-wise addition and uniform convergence on compact subsets of $\mathbb{Z}^d \times Y$, and the sets $B^1(T, A)$ and $B_c^1(T, A)$ of coboundaries, and of coboundaries with continuous cobounding function, are subgroups of $Z_c^1(T, A)$. The quotient group $H_c^1(T, A) = Z_c^1(T, A)/B_c^1(T, A)$ is called the continuous *first cohomology group* of T with values in A. In general, the subgroup $B_c^1(T, A) \subset Z_c^1(T, A)$ is not closed, and the cohomology group $H_c^1(T, A)$ has therefore no nice topological structure. The following well known proposition is an example of this general phenomenon.

PROPOSITION 30.1. *Let α be a mixing $\mathbb{Z}^d$-action by automorphisms of a compact, abelian group X. If $A = \mathbb{R}$ or $A = \mathbb{S} \cong \mathbb{T}$, then $B_c^1(\alpha, A)$ is not closed in $Z_c^1(\alpha, A)$.*

If one restricts the class of cocycles under consideration, the picture may change considerably. Consider, for example, the $\mathbb{Z}$-action $T: n \mapsto \alpha^n$ defined by the powers of a single expansive automorphism α of $X = \mathbb{T}^n$ for some $n \geq 1$. The continuous cocycles $c: \mathbb{Z} \times X \longmapsto \mathbb{R}$ for T are in one-to-one correspondence with the continuous, real-valued functions $f = c(1, \cdot): X \longmapsto \mathbb{R}$. If $c \in B_c^1(T, \mathbb{R})$, then $\int c(1, \cdot)\, d\mu = 0$ for every α-invariant probability measure μ on X. Proposition 30.1 implies that the converse is not true: there exist cocycles $c \in Z_c^1(T, \mathbb{R}) \smallsetminus B_c^1(T, \mathbb{R})$ with $\int c(1, \cdot)\, d\mu = 0$ for every α-invariant probability measure on X. However, if c is Hölder-continuous (i.e. if $c(1, \cdot)$ is Hölder-continuous), then Livshitz' theorem ([65]) implies that $c \in B_c^1(T, \mathbb{R})$ if and only if $\int c(1, \cdot)\, d\mu = 0$ for every α-invariant probability measure μ on X or, equivalently, if and only if $c(n, x) = 0$ for every $n \geq 1$ and every $x \in X$ with $\alpha^n(x) = x$. Furthermore, if $c \in B_c^1(T, \mathbb{R})$, then the cobounding function of c is again Hölder-continuous. In particular the set of Hölder-coboundaries is a closed subgroup of the group of Hölder-cocycles in the usual topology on that space.

The effect of Hölder continuity is even more dramatic if one considers cocycles for $\mathbb{Z}^d$-actions by automorphisms of compact, abelian groups with $d > 1$ (cf. [38], [97]). As the groups carrying these actions are in general not finite-dimensional tori, the classical notion of Hölder continuity is not meaningful; there is, however, a natural Hölder structure associated with a given continuous and *expansive* $\mathbb{Z}^d$-action T on a compact, metrizable space X, which coincides with the usual one for an expansive $\mathbb{Z}^d$-action on a finite-dimensional manifold (cf. [38]).

Let $d \geq 1$, and let T be a continuous $\mathbb{Z}^d$-action on a compact, metric space (X, δ). We write $\|\cdot\|$ and $\langle \cdot, \cdot \rangle$ for the Euclidean norm and inner product on $\mathbb{R}^d \supset \mathbb{Z}^d$, and put $\mathbf{B}(r) = \{\mathbf{m} \in \mathbb{Z}^d : \|\mathbf{m}\| \leq r\}$ for every $r \geq 0$. Suppose that A is a Polish abelian group, $\gamma: A \times A \longmapsto \mathbb{R}^+$ a distinguished invariant metric on A, and $f: X \longmapsto A$ a continuous function. Put, for every $\varepsilon, r \geq 0$,

$$\omega_r^{\delta, \gamma}(f, T, \varepsilon) = \sup_{\{(x,x') \in X \times X : \delta(T_\mathbf{m}(x), T_\mathbf{m}(x')) < \varepsilon \text{ for every } \mathbf{m} \in \mathbf{B}(r)\}} \gamma(f(x), f(x')).$$

The function f has *T-summable variation* if there exists an $\varepsilon > 0$ such that

$$\omega^{\delta, \gamma}(f, T, \varepsilon) = \sum_{r=1}^{\infty} \omega_r^{\delta, \gamma}(f, T, \varepsilon) < \infty,$$

and f is T-*Hölder* if there exist constants $\varepsilon, \omega' > 0$ and $\omega \in (0,1)$ with

$$\omega_r^{\delta,\gamma}(f, T, \varepsilon) < \omega'\omega^r$$

for every $r > 0$. These notions obviously depend on γ, but are independent of the metric δ on X, and every T-Hölder function has T-summable variation. If the group A is discrete, and if $\gamma(a, a') = 1$ for $a \neq a'$ and 0 otherwise, then a function $f\colon X \longmapsto A$ is Hölder if and only if it is continuous.

DEFINITION 30.2. Let α be a $\mathbb{Z}^d$-action by automorphisms of a compact, abelian group X, and let A be a Polish abelian group. A cocycle $c\colon \mathbb{Z}^d \times X \longmapsto A$ is *algebraic* if $c(\mathbf{n}, \cdot)\colon X \longmapsto A$ is a continuous homomorphism for every $\mathbf{n} \in \mathbb{Z}^d$, and *affine* if $c = c' + c''$, where c' is algebraic and c'' a homomorphism. Furthermore, if A has a distinguished, invariant metric $\gamma\colon A \times A \longmapsto \mathbb{R}^+$, and if α is expansive, then a cocycle $c\colon \mathbb{Z}^d \times X \longmapsto A$ for α is *Hölder*, or has *summable variation*, if the maps $c(\mathbf{n}, \cdot)\colon X \longmapsto A$ have the respective property for every $\mathbf{n} \in \mathbb{Z}^d$.

THEOREM 30.3 ([38]). *Let $d > 1$, and let α be an expansive and mixing $\mathbb{Z}^d$-action by automorphisms of a compact, abelian group X. Every cocycle $c\colon \mathbb{Z}^d \times X \longmapsto \mathbb{R}$ with α-summable variation is continuously cohomologous to a homomorphism. If c is Hölder, then the transfer function is again Hölder.*

THEOREM 30.4 ([97]). *Let $d > 1$, and let α be an expansive and mixing $\mathbb{Z}^d$-action by automorphisms of a compact, abelian group X. Every cocycle $c\colon \mathbb{Z}^d \times X \longmapsto \mathbb{S}$ with α-summable variation is continuously cohomologous to an affine cocycle. If c is Hölder, then the transfer function is again Hölder.*

COROLLARY 30.5 ([97]). *Let $d > 1$, and let α be an expansive and mixing $\mathbb{Z}^d$-action by automorphisms of a compact, abelian group X. If A is a discrete abelian group and $c\colon \mathbb{Z}^d \times X \longmapsto A$ a continuous cocycle, then there exists a discrete, abelian group $A' \supset A$ such that A'/A is finite and c is cohomologous to an affine cocycle $c'\colon \mathbb{Z}^d \longmapsto A'$, with continuous transfer function.*

If α is an expansive and mixing $\mathbb{Z}^d$-action by automorphisms of a compact, abelian group X and A a Polish abelian group with a distinguished, invariant metric γ, then we write $Z_H^1(\alpha, A)$ for the group of all Hölder cocycles $c\colon \mathbb{Z}^d \times X \longmapsto A$, denote by $B_H^1(\alpha, A) \subset Z_H^1(\alpha, A)$ the subgroup of all coboundaries with Hölder cobounding function, and set $H_H^1(\alpha, A) = Z_H^1(\alpha, A)/B_H^1(\alpha, A) \cong Z_H^1(\alpha, A)/B_c^1(\alpha, A)$. As in [38] and [97] one checks easily that $Z_H^1(\alpha, A) \cap B^1(\alpha, A) = B_H^1(\alpha, A)$, so that $H_H^1(\alpha, A) \cong Z_H^1(\alpha, A)/B^1(\alpha, A)$.

According to Theorem 30.4, every element of $Z_H^1(\alpha, \mathbb{S})$ is cohomologous to an affine cocycle. If $Z_a^1(\alpha, \mathbb{S})$ and $\mathrm{Hom}(\mathbb{Z}^d, \mathbb{S}) \cong \mathbb{S}^d$ are the subgroups of $Z_H^1(\alpha, \mathbb{S})$ consisting of all algebraic cocycles and all homomorphisms, respectively, then $\mathrm{Hom}(\mathbb{Z}^d, \mathbb{S}) \cap B_H^1(\alpha, \mathbb{S}) = \{1\}$, and $Z_a^1(\alpha, \mathbb{S}) \cdot \mathrm{Hom}(\mathbb{Z}^d, \mathbb{S}) \cdot B_H^1(\alpha, \mathbb{S}) = Z_H^1(\alpha, \mathbb{S})$. This makes it desirable to find an explicit formula for the group $Z_a^1(\alpha, \mathbb{S})$.

THEOREM 30.6. *Let $d > 1$, α a mixing $\mathbb{Z}^d$-action by automorphisms of a compact, abelian group X, and let $\mathfrak{M} = \hat{X}$ be the $\mathfrak{R}_d$-module defined by α. An algebraic cocycle $c\colon \mathbb{Z}^d \longmapsto \mathfrak{M}$ is a coboundary if and only if c is an algebraic coboundary, i.e. if and only if there exists an element $b \in \mathfrak{M}$ such that*

$$c(\mathbf{n}) = (u^{\mathbf{n}} - 1) \cdot b$$

for every $\mathbf{n} \in \mathbb{Z}^d$. Furthermore, if $B_a^1(\alpha, \mathbb{S}) \subset Z_a^1(\alpha, \mathbb{S})$ is the subgroup of algebraic coboundaries, then

$$H_a^1(\alpha, \mathbb{S}) = Z_a^1(\alpha, \mathbb{S})/B^1(\alpha, \mathbb{S}) \cong Z_a^1(\alpha, \mathbb{S})/B_a^1(\alpha, \mathbb{S}),$$

and

$$H_a^1(\alpha, \mathbb{S}) \cong \mathfrak{M}^* = \left(\bigcap_{i=1}^d \prod_{1 \le j \le d,\, j \ne i} (u_j - 1) \cdot \mathfrak{M} \right) \Big/ \left(\prod_{i=1}^d (u_j - 1) \cdot \mathfrak{M} \right).$$

EXAMPLES 30.7. (1) Let $d = 2$, $p > 1$ a rational prime, and let $\mathfrak{M} = \mathfrak{R}_2^{(p)}/f\mathfrak{R}_2^{(p)}$, where $0 \ne f \in \mathfrak{R}_2^{(p)}$ (cf. (6.1)). The ideal $\mathfrak{m} = (u_1 - 1)\mathfrak{R}_2^{(p)} + (u_2 - 1)\mathfrak{R}_2^{(p)} \subset \mathfrak{R}_2^{(p)}$ is maximal, and consists of all elements $g \in \mathfrak{R}_2^{(p)}$ with $g(1,1) = 0$ (mod p). We claim that $H_a^1(\alpha, \mathbb{S}) \cong \mathfrak{M}^* \ne \{0\}$ if and only if $f \in \mathfrak{m}$.

Indeed, if $f \in \mathfrak{m}$, then there exist elements $g_1, g_2 \in \mathfrak{R}_2^{(p)}$ such that $f = (u_1 - 1)g_1 + (u_2 - 1)g_2$, and we set $a_1 = g_2 + \mathfrak{a} \in \mathfrak{M}$, $a_2 = -g_1 + \mathfrak{a} \in \mathfrak{M}$, and observe that $(u_2 - 1) \cdot a_1 = (u_1 - 1) \cdot a_2 = a$, say. From (11.5)–(11.6) it is clear that there exists an algebraic cocycle $c\colon \mathbb{Z}^d \longmapsto \mathfrak{M}$ with $c(\mathbf{e}^{(i)}) = a_i$ for $i = 1, 2$. Since $\alpha^{\mathfrak{M}}$ is mixing, multiplication by $u_i - 1$ is injective on $\mathfrak{M}$ for $i = 1, 2$, so that $a_1 \notin (u_1 - 1) \cdot \mathfrak{M}$, $a = a(c) = (u_2 - 1) \cdot a_1 \in ((u_1 - 1) \cdot \mathfrak{M} \cap (u_2 - 1) \cdot \mathfrak{M}) \smallsetminus (u_1 - 1)(u_2 - 1) \cdot \mathfrak{M}$, and c is not a coboundary by Theorem 30.6.

If $f \notin \mathfrak{m}$ we apply Hilbert's Nullstellensatz to find elements $g_1, g_2, g_3 \in \mathfrak{R}_2^{(p)}$ with $(u_1 - 1)g_1 + (u_2 - 1)g_2 + fg_3 = 1$. In particular, multiplication by $u_i - 1$ on $\mathfrak{M}/(1 - u_j) \cdot \mathfrak{M}$ is invertible for $i, j \in \{1, 2\}$ and $i \ne j$. If $c\colon \mathbb{Z}^d \longmapsto \mathfrak{M}$ is an algebraic cocycle, and if $a_j = c(\mathbf{e}^{(j)})$ and $a(c) = (u_i - 1) \cdot a_j$ for all $i, j \in \{1, \ldots, d\}$ with $i \ne j$, then

$$a(c) = (u_1 - 1)g_1 \cdot a(c) + (u_2 - 1)g_2 \cdot a(c) = (u_2 - 1) \cdot a_1 = (u_1 - 1) \cdot a_2,$$

so that $(u_1 - 1)g_1 \cdot a(c) \in (u_2 - 1) \cdot \mathfrak{M}$. As mentioned earlier, multiplication by $u_1 - 1$ is invertible on $\mathfrak{M}/(u_2 - 1) \cdot \mathfrak{M}$, and we conclude that $g_1 \cdot a(c) \in (u_2 - 1) \cdot \mathfrak{M}$. Similarly we see that $g_2 \cdot a(c) \in (u_1 - 1) \cdot \mathfrak{M}$ and hence that $a(c) \in (u_1 - 1)(u_2 - 1) \cdot \mathfrak{M}$. From Theorem 30.6 we conclude that every algebraic cocycle $c\colon \mathbb{Z}^d \longmapsto \mathfrak{M}$ is a coboundary.

(2) Let $d = 2$, $\mathfrak{a} = (u_1 - p, u_2 - q) = (u_1 - p)\mathfrak{R}_2 + (u_2 - q)\mathfrak{R}_2$, where p, q are positive integers, and let $\mathfrak{M} = \mathfrak{R}_2/\mathfrak{a}$. Then $\mathfrak{M} \cong \mathbb{Z}[\frac{1}{pq}] = \{k/(pq)^l : k \in \mathbb{Z}, l \ge 0\}$, and this isomorphism carries multiplication by u_1 and u_2 to multiplication by p and q, respectively. Then $(u_1 - 1) \cdot \mathfrak{M} \cap (u_2 - 1) \cdot \mathfrak{M} = \mathrm{lcm}(p - 1, q - 1)\mathfrak{M}$,

$(u_1 - 1)(u_2 - 1) \cdot \mathfrak{M} = (p - 1)(q - 1)\mathfrak{M}$, and $H_a^1(\alpha, \mathbb{S}) \cong \mathfrak{M}^* = \{0\}$ if and only if $\operatorname{lcm}(p - 1, q - 1) = (p - 1)(q - 1)$. $\boxdot$

Many questions concerning the cohomology of $\mathbb{Z}^d$-actions by automorphisms of compact, abelian groups are unresolved. For example, the $\mathbb{Z}^2$-action α in Theorem 29.9 has no non-trivial, continuous cocycles with values in any compact, abelian group A (Example 30.7 (1)), but there exist non-trivial, continuous cocycles for α with values in certain non-abelian groups ([95]). Can one describe all such cocycles? Can they again be characterized in algebraic terms? The paper [37] investigates certain higher cohomology of $\mathbb{Z}^d$-actions by commuting toral automorphisms. In what form do these results extend to arbitrary expansive and mixing $\mathbb{Z}^d$-actions by automorphisms of compact, abelian groups?

31. Isomorphism rigidity

This section is devoted to another striking manifestation of rigidity: certain $\mathbb{Z}^d$-actions by automorphisms of compact, abelian group have the property that every measurable self-conjugacy of this action is (up to modification on a null-set) a continuous group automorphism. Since the extent of this phenomenon is not yet understood we restrict ourselves to a particularly simple example.

THEOREM 31.1. *Let* $\mathfrak{p} = (2, 1 + u_1 + u_2) \subset \mathfrak{R}_2$, $\alpha = \alpha^{\mathfrak{R}_2/\mathfrak{p}}$, $X = X^{\mathfrak{R}_2/\mathfrak{p}}$, *and let* $\phi \colon X \longmapsto X$ *be a measure preserving Borel map with* $\alpha_{\mathbf{n}} \cdot \phi = \phi \cdot \alpha_{\mathbf{n}}$ *for every* $\mathbf{n} \in \mathbb{Z}^2$. *Then* ϕ *is* λ_X-a.e. equal to a continuous, surjective group homomorphism $\psi \colon X \longmapsto X$.

For the proof of Theorem 31.1 we require an elementary lemma.

LEMMA 31.2. *Let* T *be a measure-preserving and mixing transformation of a probability space* $(Y, \mathfrak{T}, \mu)$. *Then*

$$\lim_{n \to \infty} \int f(y, T^n y) \, d\mu(y) = \int\!\!\int f \, d(\mu \times \mu) \tag{31.1}$$

for every $f \in L^1(Y \times Y, \mathfrak{T} \otimes \mathfrak{T}, \mu \times \mu)$.

PROOF. If f is of the form $f(y_1, y_2) = 1_A(y_1)1_B(y_2)$ for some $A, B \in \mathfrak{T}$ then (31.1) is an obvious consequence of mixing, and the proof for a general function $f \in L^1(Y \times Y, \mathfrak{T} \otimes \mathfrak{T}, \mu \times \mu)$ follows from an approximation argument. $\Box$

PROOF OF THEOREM 31.1. Since $1 + u_1^{2^n} + u_2^{2^n} \in \mathfrak{p}$ for every $n \geq 0$, the definition of X in (6.19) implies that

$$x + \alpha_{(2^n, 0)}(x) = \alpha_{(0, 2^n)}(x) \tag{31.2}$$

for every $x \in X$ and $n \geq 0$. For every finite subset $F \subset \mathbb{Z}^2$ we set $A(F) = \{x = (x_{\mathbf{n}}) \in X : x_{\mathbf{n}} = 0$ for every $\mathbf{n} \in F\}$, $B(F) = \phi^{-1}(A(F))$, and conclude from (6.19) and (31.2) that

$$A(F) \cap \alpha_{(-2^n,0)}(A(F)) \subset \alpha_{(0,-2^n)}(A(F)),$$
$$B(F) \cap \alpha_{(-2^n,0)}(B(F)) \subset \alpha_{(0,-2^n)}(B(F))$$

for every sufficiently large $n \geq 0$. Hence

$$\int 1_{B(F)}(x)1_{B(F)}(\alpha_{(2^n,0)}(x))1_{B(F)}(\alpha_{(0,2^n)}(x))\,d\lambda_X(x)$$
$$= \int 1_{B(F)}(x)1_{B(F)}(\alpha_{(2^n,0)}(x))\,d\lambda_X(x) \to \lambda_X(B(F))^2 \tag{31.3}$$

as $n \to \infty$. According to (31.2), the first term in (31.3) is equal to

$$\int 1_{B(F)}(x)1_{B(F)}(\alpha_{(2^n,0)}(x))1_{B(F)}(x + \alpha_{(2^n,0)}(x))\,d\lambda_X(x),$$

and by letting $n \to \infty$ and applying Lemma 31.2 we conclude from (31.3) that

$$\lim_{n \to \infty} \int 1_{B(F)}(x)1_{B(F)}(\alpha_{(2^n,0)}(x))1_{B(F)}(\alpha_{(0,2^n)}(x))\,d\lambda_X(x)$$
$$= \iint 1_{B(F)}(x)1_{B(F)}(y)1_{B(F)}(x+y)\,d(\lambda_X \times \lambda_X)(x,y)$$
$$= \int 1_{B(F)}(x)\left(\int 1_{B(F)}(y)1_{B(F)}(x+y)\,d\lambda_X(y)\right)d\lambda_X(x) \tag{31.4}$$
$$= \lambda_X(B(F))^2$$
$$= \int 1_{B(F)}(x)\left(\int 1_{B(F)}(x+y)\,d\lambda_X(y)\right)d\lambda_X(x).$$

Since all the functions involved are indicator functions we conclude that

$$\lambda_X((B(F)+x)\triangle B(F)) = \begin{cases} 0 & \text{for } \lambda_X\text{-a.e. } x \in B(F) \\ 2\lambda_X B(F) & \text{for } \lambda_X\text{-a.e. } x \in X \smallsetminus B(F). \end{cases}$$

The convolution $1_{B(F)} * 1_{B(F)}$ is thus λ_X-a.e. equal to 1_B, and a standard argument involving Fourier transform shows that $B(F)$ differs by a null-set from a subgroup of X. As every subgroup of X with positive Haar measure is open and closed we can modify ϕ on a null-set, if necessary, and assume that $\phi^{-1}(A(F)) = B(F)$ is an open and closed subgroup of X for every finite set $F \subset \mathbb{Z}^2$ with $\lambda_X(A(F)) > 0$. We put $Y = \bigcap_{\{F \subset \mathbb{Z}^2 : \lambda_X(A(F)) > 0\}} B(F)$ and observe that ϕ induces a continuous group isomorphism $\psi' : X/Y \longmapsto X$ and hence a continuous, surjective group homomorphism $\psi : X \longmapsto X$ with $\psi(x) = \phi(x)$ for λ_X-a.e. $x \in X$. $\square$

COROLLARY 31.3. *If α is the $\mathbb{Z}^2$-action on the compact, abelian group X defined in Theorem 31.1, then every measure preserving automorphism ϕ of the measure space $(X, \mathfrak{B}_X, \lambda_X)$ with $\alpha_{\mathbf{n}} \cdot \phi = \phi \cdot \alpha_{\mathbf{n}}$ λ_X-a.e. is λ_X-a.e. equal to $\alpha_{\mathbf{n}}$ for some $\mathbf{n} \in \mathbb{Z}^2$.*

PROOF. From Theorem 31.1 we know that ϕ is λ_X-a.e equal to a continuous group automorphism ψ of X which commutes with $\alpha_{\mathbf{n}}$ for every $\mathbf{n} \in \mathbb{Z}^2$. We write the dual group $\hat{X} = \mathfrak{R}_2/\mathfrak{p}$ as $\mathfrak{R}_2^{(2)}/\mathfrak{q}$ with $\mathfrak{q} = (1 + u_1 + u_2)\mathfrak{R}_2^{(2)}$ (cf. Remark 6.19 (4)) and note that the dual automorphism $\hat{\psi} \colon \mathfrak{R}_2^{(2)}/\mathfrak{q} \longmapsto \mathfrak{R}_2^{(2)}/\mathfrak{q}$ satisfies that $\hat{\psi}(1 + \mathfrak{q}) = g + \mathfrak{q}$ for some $g \in \mathfrak{R}_2^{(2)}$ and hence that $\hat{\psi}$ is equal to multiplication by g on $\mathfrak{R}_2^{(2)}/\mathfrak{q}$. By replacing ϕ, ψ and g with $\alpha_{\mathbf{m}} \cdot \phi$, $\alpha_{\mathbf{m}} \cdot \psi$ and $u^{\mathbf{m}}g$ for some $\mathbf{m} \in \mathbb{Z}^2$, if necessary, we may assume without loss in generality that g is a polynomial in u_1, u_2, i.e. that g involves no negative powers in these variables. Furthermore, since ψ is an automorphism of X, $\hat{\psi}$ is an automorphism of $\mathfrak{R}_2^{(2)}/\mathfrak{q}$; in particular, multiplication by g on $\mathfrak{R}_2^{(2)}/\mathfrak{q}$ is surjective, which is the same as saying that

$$g\mathfrak{R}_2^{(2)} + \mathfrak{q} = \mathfrak{R}_2^{(2)}. \tag{31.5}$$

Since g is determined only up to addition of an element in $\mathfrak{q}$ we may also take it that g is a function of u_1 and does not depend on u_2 (just replace all the terms u_2^n, $n \geq 1$, by $(1 + u_1)^n$). If the polynomial $g = g(u_1)$ has a zero $\xi \in \overline{\mathbb{F}}_2 \smallsetminus \mathbb{F}_2$, then every element $h \in \mathfrak{q} + g\mathfrak{R}_2^{(2)}$ vanishes at the point $(\xi, 1 + \xi) \in (\overline{\mathbb{F}}_2^\times)^2$, in violation of (31.5). Hence $g(u_1) = u_1^k(1 - u_1)^l$ for some $k, l \geq 0$, and $\psi = \alpha_{(k,l)}$, as claimed. $\square$

REMARK 31.4. Even if the continuous, surjective group homomorphism $\psi \colon X \longmapsto X$ is not bijective, its dual $\hat{\psi} \colon \mathfrak{R}_2^{(2)}/\mathfrak{q} \longmapsto \mathfrak{R}_2^{(2)}/\mathfrak{q}$ is equal to multiplication by g for some $g \in \mathfrak{R}_2^{(2)} \smallsetminus \mathfrak{q}$, and $\psi = \alpha_g$ (cf. (6.14)—if g were in $\mathfrak{q}$, then $\hat{\psi}$ would not be injective and ψ would not be surjective). By applying Hilbert's Nullstellensatz to the module $\mathfrak{R}_2^{(2)}/\mathfrak{q}$ we see that $|\ker(\psi)| = |\ker(\alpha_g)| = 2^{|V(\mathfrak{p}) \cap V(g)|}$.

EXAMPLES 31.5. (1) Let $g = 1 + u_1^3 + u_2^3$. As $g + \mathfrak{p} = u_1 u_2 + \mathfrak{p}$, $\alpha_g = \alpha_{(1,1)}$ is an automorphism of X.

(2) Let $g = u_1^2 + u_2$. Then

$$V(\mathfrak{p}) \cap V(g) = \{(\omega, 1 + \omega) : \omega \in \overline{\mathbb{F}}_2 \text{ and } 1 + \omega + \omega^2 = 0\},$$

and $|\ker(\alpha_g)| = 4$. $\boxdot$

CONCLUDING REMARKS 31.6. (1) Theorem 31.1 is originally due to Kitchens (unpublished) and to [99] (with somewhat different proofs). It is not difficult to extend Theorem 31.1 to $\mathbb{Z}^2$-actions of the form $\alpha = \alpha^{\mathfrak{R}_2^{(p)}/\mathfrak{q}}$, where

$p > 0$ is a rational prime and $\mathfrak{q} \subset \mathfrak{R}_2^{(p)}$ is an ergodic—and hence principal—prime ideal generated by a Laurent polynomial $f \in \mathfrak{R}_2^{(p)}$ with extremal non-mixing support (cf. Proposition 25.5 and Definition 28.8 (1)). Further extensions are a little more complicated, but still possible, but the class of prime ideals $\mathfrak{p} \subset \mathfrak{R}_d$ for which the $\mathbb{Z}^d$-action $\alpha = \alpha^{\mathfrak{R}_d/\mathfrak{p}}$ satisfies (the analogue of) Theorem 31.1 is still a mystery. Does Theorem 31.1 hold for every expansive, mixing, almost minimal $\mathbb{Z}^d$-action α by automorphisms of a compact, abelian group X?

(2) One of the remarkable aspects of Corollary 31.3 is that each $\alpha_{\mathbf{n}}$, $\mathbf{n} \neq \mathbf{0}$ is an ergodic automorphism of X and hence Bernoulli (cf. Theorem 23.1 and 23.21), so that the set of automorphisms of $(X, \mathfrak{B}_X, \lambda_X)$ commuting with each individual $\alpha_{\mathbf{n}}$, $\mathbf{n} \in \mathbb{Z}^2$, is very large. We also note that there exist infinitely many finite-to-one, measure preserving maps $\phi \colon X \longmapsto X$ which commute with the $\mathbb{Z}^2$-action α, so that α does not have minimal self-joinings.

Bibliography

[1] J.F. Adams, *Lectures on Lie groups*, Benjamin, New York, 1969.

[2] L.V. Ahlfors, *Complex analysis*, 2nd edn., McGraw-Hill, New York, 1966.

[3] N. Aoki, *A simple proof of the Bernoullicity of ergodic automorphisms of compact abelian groups*, Israel J. Math. **38** (1981), 189–198.

[4] N. Aoki and H. Totoki, *Ergodic automorphisms of T^∞ are Bernoulli transformations*, Publ. Res. Inst. Math. Sci. **10** (1975), 535–544.

[5] M. Atiyah and I.G. MacDonald, *Introduction to Commutative Algebra*, Addison-Wesley, Reading, Mass., 1969.

[6] D. Berend, *Multi-invariant sets on tori*, Trans. Amer. Math. Soc. **280** (1983), 509–532.

[7] _____, *Multi-invariant sets on compact abelian groups*, Trans. Amer. Math. Soc. **286** (1984), 505–535.

[8] _____, *Minimal sets on tori*, Ergod. Th. & Dynam. Sys. **4** (1984), 499–507.

[9] _____, *Ergodic semigroups of epimorphisms*, Trans. Amer. Math. Soc. **289** (1985), 393–407.

[10] K.R. Berg, *Convolution of invariant measures, maximal entropy*, Math. Sys. Th. **3** (1969), 146–150.

[11] J. Bochnak, M. Coste and M.-F. Roy, *Géométrie algébrique réelle*, Springer Verlag, Berlin-Heidelberg-New York, 1987.

[12] D. Boyd, *Kronecker's theorem and Lehmer's problem for polynomials in several variables*, J. Number Theory **13** (1981), 116–121.

[13] _____, *Speculations concerning the range of Mahler's measure*, Can. Math. Bull. **24** (1981), 453–469.

[14] R. Burton and R. Pemantle, *Local characteristics, entropy and limit theorems for spanning trees and domino tilings via transfer-impedances*, Ann. Probability (to appear).

[15] R. Burton and J.E. Steif, *Nonuniqueness of measures of maximal entropy for subshifts of finite type*, Preprint (1993).

[16] J.W.S. Cassels, *Local Fields*, Cambridge University Press, Cambridge, 1986.

[17] J.P. Conze, *Entropie d'un groupe abélien de transformations*, Z. Wahrscheinlichkeitstheorie verw. Geb. **25** (1972), 11–30.

[18] S.G. Dani, *On badly approximable numbers, Schmidt games and bounded orbits of flows*, Number theory and dynamical systems (York 1987), London Math. Soc. Lecture Note Series, vol. 134, Cambridge University Press, Cambridge, 1989.

[19] H. Dehling, M. Denker and W. Philipp, *Versik processes and very weak Bernoulli processes with summable rates are independent*, Proc. Amer. Math. Soc. **91** (1984), 618–624.

[20] S. Eilenberg and N. Steenrod, *Foundations of algebraic topology*, Princeton University Press, Princeton, N.J., 1952.

[21] J.-H. Evertse and K. Györy, *On the numbers of solutions of weighted unit equations*, Compositio Math. **66** (1988), 329–354.

[22] F. Fagnani, *Some results on the structure of abelian group subshifts*, Preprint (1993).

[23] H. Furstenberg, *Disjointness in ergodic theory, minimal sets, and a problem in diophantine approximation*, Math. Sys. Th. **1** (1967), 1–49.

[24] A.O. Gelfond, *Transcendental and algebraic numbers*, Dover, New York, 1960.

[25] W. Geller and J. Propp, *The fundamental group of a $\mathbb{Z}^2$-shift*, in preparation.

[26] L. Goodwyn, *Topological entropy bounds measure theoretic entropy*, Proc. Amer. Math. Soc. **23** (1969), 679–688.

[27] P.R. Halmos, *On automorphisms of compact groups*, Bull. Amer. Math. Soc. **49** (1943), 619–624.

[28] E. Hewitt and K.A. Ross, *Abstract Harmonic Analysis I*, Springer Verlag, Berlin-Heidelberg-New York, 1963.

[29] ———, *Abstract Harmonic Analysis II*, Springer Verlag, Berlin-Heidelberg-New York, 1970.

[30] G. Hochschild, *The structure of Lie groups*, Holden-Day, San Francisco, 1965.

[31] L. Hurd, J. Kari and K. Culik, *The topological entropy of cellular automata is uncomputable*, Ergod. Th. & Dynam. Sys. **12** (1992), 255–265.

[32] K. Iwasawa, *On group rings of topological groups*, Proc. Imp. Acad. Japan Tokyo **20** (1944), 67–70.

[33] B. Kamiński, *The theory of invariant partitions for $\mathbb{Z}^d$-actions*, Bull. Acad. Pol.: Math. **29** (1981), 349–362.

[34] J.W. Kammeyer, *A complete classification of two-point extensions of a multidimensional Bernoulli shift*, J. Analyse Math. **54** (1990), 113–163.

[35] I. Kaplansky, *Groups with representations of bounded degree*, Can. J. Math. **1** (1949), 105–112.

[36] T. Kato, *Perturbation theory for linear operators*, Springer Verlag, Berlin-Heidelberg-New York, 1966.

[37] A. Katok and S. Katok, *Higher cohomology for abelian groups of toral automorphisms*, Preprint (1993).

[38] A. Katok and K. Schmidt, *The cohomology of expansive $\mathbb{Z}^d$-actions by automorphisms of compact abelian groups*, Pacific J. Math. (to appear).

[39] A. Katok and R.J. Spatzier, *Differential rigidity of hyperbolic abelian actions*, Preprint (1992).

[40] ———, *Invariant measures for higher rank hyperbolic abelian actions*, Preprint (1992).

[41] Y. Katznelson, *Ergodic automorphisms of T^n are Bernoulli shifts*, Israel J. Math. **10** (1971), 186–195.

[42] Y. Katznelson and B. Weiss, *Commuting measure preserving transformations*, Israel J. Math. **12** (1972), 161–173.

[43] B. Kitchens, *Expansive dynamics on zero-dimensional groups*, Ergod. Th. & Dynam. Sys. **7** (1987), 249–261.

[44] B. Kitchens and K. Schmidt, *Periodic points, decidability and Markov subgroups*, Dynamical Systems, Proceeding of the Special Year, Lecture Notes in Mathematics, vol. 1342, Springer Verlag, Berlin-Heidelberg-New York, 1988, pp. 440–454.

[45] _____, *Automorphisms of compact groups*, Ergod. Th. & Dynam. Sys. **9** (1989), 691–735.

[46] _____, *Markov subgroups of* $(\mathbb{Z}/2)^{\mathbb{Z}^2}$, Contemp. Math. **135** (1992), 265–283.

[47] _____, *Mixing sets and relative entropies for higher dimensional Markov shifts*, Ergod. Th. & Dynam. Sys. (to appear).

[48] W. Krieger, *On entropy and generators of measure-preserving transformations*, Trans. Amer. Math. Soc. **149** (1970), 453–464; *Erratum*, **168** (1972), 519.

[49] L. Kronecker, *Zwei Sätze über Gleichungen mit ganzzahligen Coefficienten*, J. reine angew. Math. **53** (1857), 173–175.

[50] P.-F. Lam, *On expansive transformation groups*, Trans. Amer. Math. Soc. **150** (1970), 131–138.

[51] S. Lang, *Algebra*, (2nd edition), Addison-Wesley, Reading, Mass., 1984.

[52] W.M. Lawton, *The structure of compact connected groups which admit an expansive automorphism*, Recent advances in Topological Dynamics, Lecture Notes in Mathematics, vol. 318, Springer Verlag, Berlin-Heidelberg-New York, 1973, pp. 182–196.

[53] _____, *A generalization of a theorem of Kronecker*, J. Sci. Fac. Chiang Mai Univ. **4** (1977), 15–23.

[54] _____, *A problem of Boyd concerning geometric means of polynomials*, J. Number Theory **16** (1983), 356–362.

[55] R.R. Laxton and W. Parry, *On the periodic points of certain automorphisms and a system of polynomial identities*, J. Algebra **6** (1967), 388–393.

[56] F. Ledrappier, *Un champ markovien peut être d'entropie nulle et mélangeant*, C. R. Acad. Sc. Paris Ser. A **287** (1978), 561–562.

[57] D.H. Lehmer, *Factorization of cyclotomic polynomials*, Ann. of Math. **34** (1933), 461–479.

[58] D. Lind, *Ergodic automorphisms of the infinite torus are Bernoulli*, Israel J. Math. **17** (1974), 162–168.

[59] _____, *Ergodic group automorphisms and specification*, Lecture Notes in Mathematics, vol. 729, Springer Verlag, Berlin-Heidelberg-New York, 1979, pp. 93–104.

[60] _____, *Dynamical properties of quasihyperbolic toral automorphisms*, Ergod. Th. & Dynam. Sys. **2** (1982), 49–68.

[61] D. Lind and K. Schmidt, *Bernoullicity of solenoidal automorphisms and global fields*, Israel J. Math. **87** (1994), 33–35.

[62] _____, *Periodic components of* $\mathbb{Z}^d$*-actions*, in preparation.

[63] D. Lind, K. Schmidt and T. Ward, *Mahler measure and entropy for commuting automorphisms of compact groups*, Invent. math. **101** (1990), 593–629.

[64] D. Lind and T. Ward, *Automorphisms of solenoids and p-adic entropy*, Ergod. Th. & Dynam. Sys. **8** (1988), 411–419.

[65] A. Livshitz, *Cohomology of dynamical systems*, Math. U.S.S.R. Izvestija **6** (1972), 1278–1301.

[66] K. Mahler, *Eine arithmetische Eigenschaft der taylor-koeffizienten rationaler Funktionen*, Proc. Acad. Sci. Amsterdam **38** (1935), 50–60.

[67] _____, *An application of Jensen's formula to polynomials*, Mathematika **7** (1960), 89–100.

[68] _____, *On some inequalities for polynomials in several variables*, J. London Math. Soc. **37** (1962), 341–344.

[69] B. Marcus, *A note on periodic points of toral automorphisms*, Mh. Math. **89** (1980), 121–129.

[70] D. Masser, *Two letters to D. Berend*, dated 12th and 19th September 1985.

[71] G. Miles and R.K. Thomas, *The breakdown of automorphisms of compact topological groups*, Studies in Probability and Ergodic Theory, Advances in Mathematics Supplementary Studies, vol. 2, Academic Press, New York, 1987, pp. 207–218.

[72] _____, *Generalized torus automorphisms are Bernoullian*, Studies in Probability and Ergodic Theory, Advances in Mathematics Supplementary Studies, vol. 2, Academic Press, New York, 1987, pp. 231–249.

[73] M. Misiurewicz, *A short proof of the variational principle for a $\mathbb{Z}_+^N$-action on a compact space*, Astérisque **40** (1975), 147–157.

[74] C.C. Moore and K. Schmidt, *Coboundaries and homomorphisms for nonsingular actions and a problem of H. Helson*, Proc. London Math. Soc. **40** (1980), 443–475.

[75] M.A. Naimark, *Normed rings*, Wolters-Noordhoff, Groningen, 1964.

[76] D.S. Ornstein and B. Weiss, *Entropy and isomorphism theorems for actions of amenable groups*, J. Analyse Math. **48** (1987), 1–141.

[77] W. Parry, *Entropy and generators in ergodic theory*, Benjamin, New York, 1969.

[78] K.R. Parthasarathy, *Probability measures on metric spaces*, Academic Press, New York-London, 1967.

[79] D.S. Passman, *The algebraic structure of group rings*, Wiley, New York, 1977.

[80] A.J. van der Poorten, *Factorisation in fractional powers*, Acta Arith. (to appear).

[81] A.J. van der Poorten and H.P. Schlickewei, *Additive relations in fields*, J. Austral. Math. Soc. Ser. A **51** (1991), 154–170.

[82] M. Ratner, *Rigidity of horocycle flows*, Ann. of Math. **115** (1982), 597–614.

[83] M. Reid, *Undergraduate algebraic geometry*, London Mathematical Society Student Texts, vol. 12, Cambridge University Press, Cambridge, 1988.

[84] V.A. Rokhlin, *Lectures on ergodic theory*, Russian Math. Surveys **22** (1967), 1–52.

[85] _____, *Metric properties of endomorphisms of compact commutative groups*, Amer. Math. Soc. Transl. **64** (1967), 244–252.

[86] A. Rothstein, *Versik processes: first steps*, Israel J. Math. **36** (1980), 205–224.

[87] W. Rudin, *Fourier analysis on groups*, Wiley-Interscience, New York-London, 1962.

[88] _____, *Complex Analysis*, McGraw-Hill, New York, 1966.

[89] D.J. Rudolph, *×2 and ×3 invariant measures and entropy*, Ergod. Th. & Dynam. Sys. **10** (1990), 395–406.

[90] _____, *Fundamentals of measurable dynamics*, Clarendon Press, Oxford, 1990.

[91] D.J. Rudolph and K. Schmidt, *Almost block independence and Bernoullicity of $\mathbb{Z}^d$-actions by automorphisms of compact abelian groups*, Preprint (1994).

[92] H.P. Schlickewei, *S-unit equations over number fields*, Invent. math. **102** (1990), 95–107.

[93] K. Schmidt, *Mixing automorphisms of compact groups and a theorem by Kurt Mahler*, Pacific J. Math. **137** (1989), 371–384.

[94] _____, *Automorphisms of compact abelian groups and affine varieties*, Proc. London Math. Soc. **61** (1990), 480–496.

[95] _____, *The cohomology of higher-dimensional shifts of finite type*, Pacific J. Math. (to appear).

[96] _____, *Invariant measures for certain expansive Z^2-actions*, Israel J. Math. (to appear).

[97] _____, *Cohomological rigidity of algebraic $\mathbb{Z}^d$-actions*, Preprint (1993).

[98] K. Schmidt and T. Ward, *Mixing automorphisms of compact groups and a theorem of Schlickewei*, Invent. math. **111** (1993), 69–76.

[99] M.A. Shereshevsky, *On the ergodic theory of cellular automata and two-dimensional Markov shifts generated by them*, Ph.D Thesis, University of Warwick, 1992.

[100] C.J. Smyth, *A Kronecker-type theorem for complex polynomials in several variables*, Can. Math. Bull. **24** (1981), 447–452; *Addenda and errata*, **25** (1982), 504.

[101] _____, *On measures of polynomials in several variables*, Bull. Australian Math. Soc. **23** (1981), 49–63.

[102] K.B. Stolarski, *Algebraic numbers and diophantine approximation*, Dekker, New York, 1974.

[103] R.K. Thomas, *Metric properties of transformations of G-spaces*, Trans. Amer. Math. Soc. **160** (1971), 103–117.

[104] _____, *The addition theorem for the entropy of transformations of G-spaces*, Trans. Amer. Math. Soc. **160** (1971), 119–130.

[105] P. Walters, *An introduction to ergodic theory*, Graduate Texts in Mathematics, vol. 79, Springer Verlag, Berlin-Heidelberg-New York, 1982.

[106] _____, *Almost block independence for the three dot $\mathbb{Z}^2$ dynamical system*, Israel J. Math. **76** (1991), 237–256.

[107] T. Ward, *Periodic points for expansive actions of $\mathbb{Z}^2$ on compact abelian groups*, Bull. London Math. Soc. **24** (1992), 317–324.

[108] _____, *The Bernoulli property for expansive $\mathbb{Z}^2$-actions on compact groups*, Israel J. Math. **79** (1992), 225–249.

[109] A. Weil, *Basic Number Theory*, Springer Verlag, Berlin-Heidelberg-New York, 1974.

[110] R.F. Williams, *Classification of subshifts of finite type*, Ann. of Math. **98** (1973), 120–153; *Errata*, **99** (1974), 380–381.

[111] S.A. Yuzvinskii, *Metric properties of endomorphisms of compact groups*, Amer. Math. Soc. Transl. Ser. 2 **66** (1986), 63–98.

Index

Symbols

306